T0138318

A MANUAL OF THE MAMMALIA

A MANUAL OF THE

Mammalia

AN HOMAGE TO LAWLOR'S

Handbook to the Orders and

Families of Living Mammals

DOUGLAS A. KELT AND

JAMES L. PATTON

Skull Photos by BILL STONE

Taxon Illustrations by FIONA A. REID

THE UNIVERSITY OF CHICAGO PRESS

Chicago and London

The University of Chicago Press, Chicago 60637
The University of Chicago Press, Ltd., London

Published 2020

Printed in the United States of America

28 27 26 25 24 23 2 3 4 5

ISBN-13: 978-0-226-53300-1 (cloth)
ISBN-13: 978-0-226-53314-8 (e-book)
DOI: https://doi.org/10.7208/chicago/9780226533148.001.0001

Library of Congress Cataloging-in-Publication Data

Names: Kelt, Douglas A. (Douglas Alan), 1959– author. | Patton, James L., author. | Stone, Bill (Photographer), photographer. | Reid, Fiona, 1955– illustrator. | Based on: Lawlor, Timothy E. Handbook to the orders and families of living mammals.
Title: A manual of the mammalia : an homage to Lawlor's "Handbook to the orders and families of living mammals" / Douglas A. Kelt and James L. Patton ; skull photos by Bill Stone ; taxon illustrations by Fiona A. Reid.
Description: Chicago ; London : The University of Chicago Press, 2020. | Includes bibliographical references and index.
Identifiers: LCCN 2018019130 | ISBN 9780226533001 (cloth : alk. paper) | ISBN 9780226533148 (e-book)
Subjects: LCSH: Mammals—Handbooks, manuals, etc. | Mammals—Identification. | Mammals—Classification.
Classification: LCC QL703 .K35 2019 | DDC 599.01/2—dc23
LC record available at https://lccn.loc.gov/2018019130

⊗ This paper meets the requirements of ANSI/NISO Z39.48-1992 (Permanence of Paper).

Contents

Introduction

ORGANIZATION OF THIS MANUAL

This manual introduces all orders and families of living (= Recent) mammals and provides uniquely diagnostic characters at each level, where available, as well as lists of recognition characters that can be used, in combination, to separate any single taxon from others. We highlight in boldface those characters that have proved to be especially useful to students in the courses we have taught or still teach. We prefer this type of character-based approach over one relying on dichotomous keys because it encourages students to learn and understand not only characters that identify a particular taxon, but also those that provide information about the its members' way of life. In this approach, the manual follows the format of T. E. Lawlor's 1979 *Handbook to the Orders and Families of Living Mammals* (Mad River Press). We used this excellent manual in our courses until revisions to mammalian taxonomy left it too outdated to be useful, and Tim Lawlor's untimely passing in 2011 meant that further revisions were not forthcoming. It is noteworthy that when Lawlor's *Handbook* was last published, taxonomists had named about 4,170 species of mammals in 20 orders (Honacki et al. 1982); the latest synthesis has raised this number to 6,495 (Burgin et al. 2018), and we recognize 28 orders here.

A basic understanding of anatomical terms and of the elements of the cranial and postcranial skeleton is essential for the identification of mammals at all hierarchical levels, from species to class. We thus begin this manual with a synopsis of mammalian anatomy, including lists and descriptions of individual bony elements or their parts and illustrations of various skulls and postcranial skeletal elements. We then follow with a brief review of external traits, such as facial vibrissae and foot pads, and the terms used to describe them. We end with a description of basic mammalian tooth anatomy, the terminology used for key occlusal surface features (cusps), and the various modifications of teeth for the diverse food habits exhibited by mammals.

The anatomy section is then followed by detailed accounts of living mammals of the world, organized sequentially by clade and hierarchical taxon. We begin by distinguishing between the definition and the diagnosis of the class Mammalia, then provide a synoptic classification above the family level that serves as the organizing principle for the taxon accounts that follow. The taxonomy we use is the traditional Linnaean categorical naming system supplemented with explicitly unranked clade-based names in common usage. The latter could be formalized within the Linnaean hierarchy, as was done, for example, by McKenna and Bell (1997), but we chose not to do so here. Our focus is on the 28 orders and 140 families of living mammals (29 orders if the whales—Cetacea—are treated apart from the Artiodactyla) and not the intricacies of mammalian classification per se. We provide phylogenetic hypotheses for most higher-level taxa (typically family and above), primarily to allow the student to trace character change within and among lineages, but also as an aid in grouping mammalian diversity into more recognizable, and manageable, units.

We organize the sequence of taxon accounts largely following the current understanding of phylogenetic relationships at the ordinal or subordinal levels, but arrange families within each higher category alphabetically in most cases. We illustrate key characteristics that diagnose many of these higher-level taxa, and we provide an overview of their members' salient biological attributes; illustrations (line drawings and/or digital images) of skulls, teeth, and other important characteristics; lists of diagnostic characters (where apparent) and/or other important characteristics of the hard and soft anatomy that can be used to identify members; a range map; and a synopsis of taxonomic diversity. Coverage of the biological details and characters of each group is not exhaustive; students are thus encouraged to consult the very large and recent literature (such as D. E. Wilson and R. A. Mittermeier's edited series *The Handbook of the Mammals of the World*) or web-based compendia (such as the mammal section of P. Myers's Animal Diversity Web, http://animaldiversity.org/accounts/Mammalia/).

We are very well aware that there is far more information provided in this manual than any individual can absorb in a single semester- or quarter-long course. Do not be overwhelmed; simply revel in the diversity that is the Mammalia.

ACKNOWLEDGMENTS

As the taxonomy in Lawlor's *Handbook* became increasingly outdated, JLP developed the initial version of this manual for students in IB 173/173L at UC Berkeley in the fall 2001 and spring 2003 semesters and for those in the mammalogy class taught by Matina Kalcounis-Rüppell at California State University, Sacramento, in fall 2002. Alan Shabel read and helped edit the entire manual in these early stages of its development, and Michael Nachman and students from his mammalogy classes at UC Berkeley provided substantial advice during our preparation of the final manuscript. JLP loaned a copy of the manual to DAK for his pedagogical efforts at UC Davis, and DAK incorporated additional materials and revised the manual to reflect the taxonomy and order presented in D. E. Wilson and D. R. Reeder's *Mammal Species of the World*, 3rd ed. (2005). DAK prodded JLP to publish the manual and make it available to a broader audience, and in the end we agreed to pursue this project in collaboration. As with all such efforts, it consumed far more time than either of us anticipated, but we hope that it will provide a useful manual for students just learning the remarkable diversity of Mammalia as well as a handbook for both professional and lay mammalogists. We also welcome input to further improve this product, which is a work in progress. If you find errors or material that could be presented more clearly, please contact either DAK at dakelt@ucdavis.edu or JLP at patton@berkeley.edu.

Fiona A. Reid drew all illustrations of exemplars for each family or subfamily; both her incredibly skilled hand and her artist's eye are widely recognized, and we feel privileged to have had her assistance on this project. Bill Stone photographed virtually all the skulls with similar skill and quality; Bill has retired from the University of California system but remains a passionate proponent of evolutionary biodiversity. Jake Esselstyn kindly provided the skull photograph of the nearly edentulous muroid *Paucidentomys vermidax*. Kris Helgen and Nicole Edmison of the United States National Museum loaned us skulls, as did Rob Voss, Neil Duncan, and Brian O'Toole of the American Museum of Natural History. Judith Eger and the Royal Ontario Museum kindly made available high-resolution photographs of the skull of *Mystacina*. Paula Jenkins of the Mammal Group and Isabel Martin of Image Resources, both at the Natural History Museum in London, facilitated our use of photographs of the rare rodent *Laonastes*. We are incredibly grateful to each.

Finally, we owe deep appreciation to staff at the University of Chicago Press. Christie Henry helped to shepherd our book through the initial stages of acceptance before she moved on to other pastures; we wish her the very best. Miranda Martin, Kelly Finefrock-Creed, and Mary Corrado have all helped to guide this effort from submission to actual product. Norma Sims Roche deserves particular credit for an outstanding job in editing our original text. The authors take full responsibility for any errors in this manual, but the Press deserves credit for helping to ferret out potentially confusing issues and generally making this a clearer and more useful manual.

Basics of Mammalian Anatomy

We assume that the user of this book has a fundamental understanding of directional terminology, but in recognition that basic anatomy courses are no longer required of many students learning mammalian biology, we outline a few key terms here.

When a dog stands on four feet, its head is at the **anterior** end, while its tail is at the **posterior** end. Its belly is **ventral**, while its back is **dorsal**. In the primate and human literature (in which the subject frequently is not "looking" anteriorly), **cranial** and **caudal** are functionally equivalent to anterior and posterior, respectively. The upper limbs are **proximal** relative to the lower limbs, which are **distal**. Body parts may be **medial** (closer to the core) or **mesial** (away from the core, = lateral); these terms are particularly useful for limbs, which are assumed to be in their relaxed position, such that the thumb (pollex) or "big toe" (hallux) is medial. Especially with dental terminology, **labial** refers to the side of a structure closer to the lips (*labia* = [Latin] "lip"), while **lingual** (*lingua* = [Latin] "tongue") refers to the side closer to the tongue; that side may also be termed **buccal**.

CRANIAL AND POSTCRANIAL ANATOMY

THE MAMMALIAN SKULL

The skull comprises the **cranium** and **mandible**. The cranium includes the **braincase** and the **rostrum**. Much of the following description comes from Lawlor (1979).

Dorsal aspect

Bones

nasal bones—paired bones forming the anteriormost roof of the nasal cavities.

premaxillary bones—paired bones forming the lower margin of the outer nasal openings and the anteriormost portion of the palate; composed of a **nasal process** or **branch** (elongate process extending dorsally along one side of the nasal cavity) and a **palatal process** or **branch** (meets the other premaxilla at the midline of the palate); one branch missing in some bat groups; fused with maxillae in some mammals; holds the upper incisor teeth.

maxillary bones—paired bones that bear all upper teeth except the incisors; constitute a large part of the sides of the rostrum and palate posterior to the premaxillae; often include a posterior process (**zygomatic process of the maxilla**) making up part of the zygomatic arch. In many species, there may be a notable gap (**diastema**) between teeth; for example, in rodents, lagomorphs, and many ungulates, canines are absent and cheek teeth are separated from incisors by a diastema.

frontal bones—paired bones located posterior to the nasals and dorsal to the maxillae; support antlers and horns in taxa with these structures; many mammals develop a **postorbital process** that projects laterally at the posterior border of the orbit; this may join with a similar process on the zygomatic arch to form a **postorbital bar or plate**; in other taxa, a fanlike **supraorbital process** extends laterally over the orbit.

parietal bones—paired bones located posterior to the frontals, forming the majority of the roof of the braincase; the suture that joins the frontal and parietal bones is the coronal suture.

interparietal bone—often indistinct or obliterated, may fuse with the supraoccipital bone; when visible, located at the posterior border of the parietals.

squamosal bones—paired bones located lateral and ventral to the parietals; ventral surface bears an articular surface, the **mandibular fossa** (which may also be called a glenoid fossa), where the jaw articulates; the posteroventral border of the mandibular fossa may wrap around the condyloid process of the dentary, forming a **postmandibular process** that provides for a more stable articulation (e.g., see account for family Mustelidae); include an anterior process (**zygomatic process of squamosal**) making up part of the zygomatic arch.

jugal bones—paired bones that form the central portion of the zygomatic arch; may possess a postorbital process that extends dorsally and may meet with the postorbital process of the frontal to form a postorbital bar or plate; occasionally in contact with the lacrimal or premaxilla; in marsupials, the jugal extends posterior to constitute part of the mandibular fossa. In human and primate anatomy these may be called the **zygomatic**, or **malar**, bones.

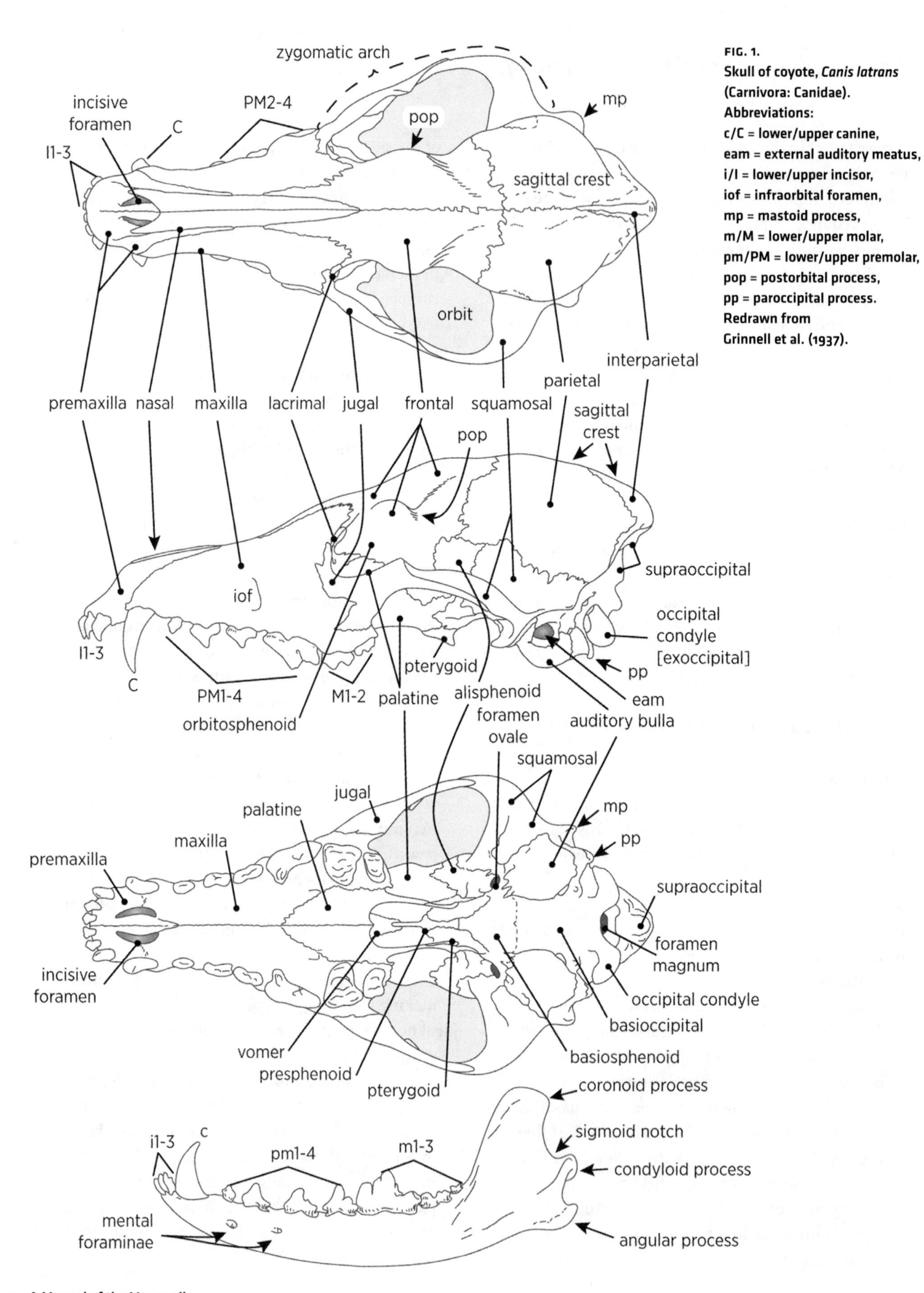

FIG. 1.
Skull of coyote, *Canis latrans* (Carnivora: Canidae). Abbreviations: c/C = lower/upper canine, eam = external auditory meatus, i/I = lower/upper incisor, iof = infraorbital foramen, mp = mastoid process, m/M = lower/upper molar, pm/PM = lower/upper premolar, pop = postorbital process, pp = paroccipital process. Redrawn from Grinnell et al. (1937).

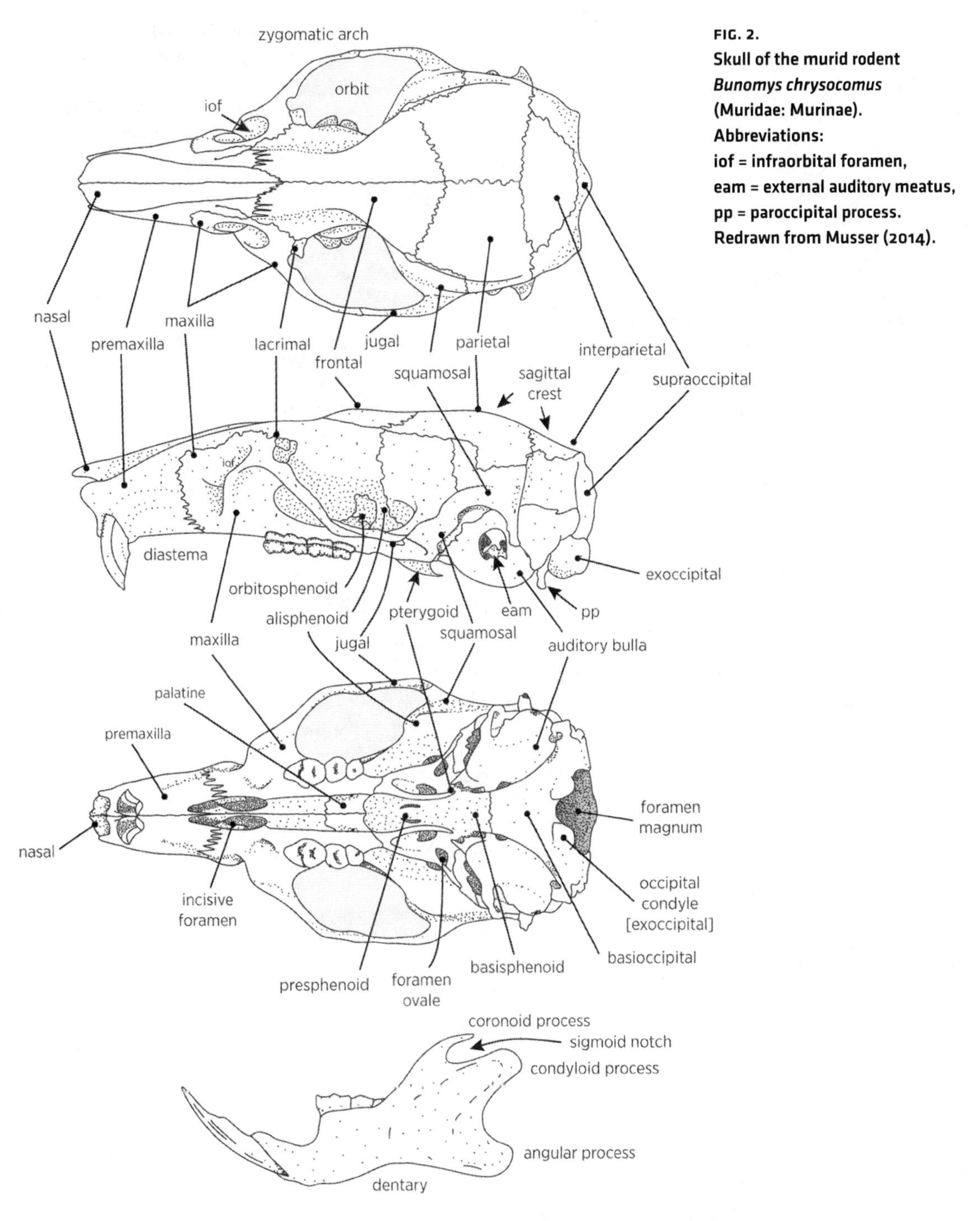

FIG. 2.
Skull of the murid rodent *Bunomys chrysocomus* (Muridae: Murinae). Abbreviations: iof = infraorbital foramen, eam = external auditory meatus, pp = paroccipital process. Redrawn from Musser (2014).

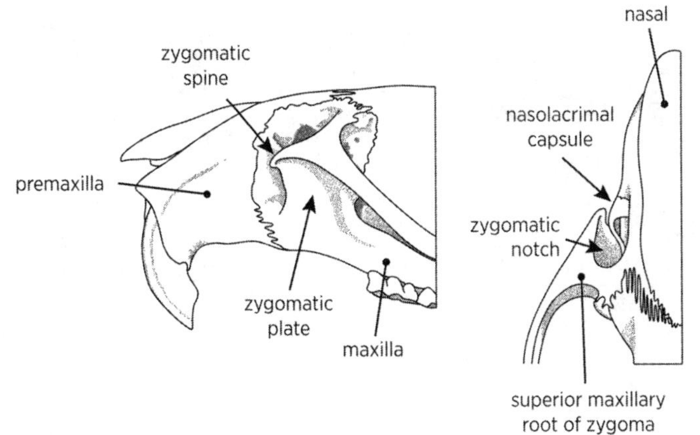

FIG. 3. Lateral views of the rostral and zygomatic regions of the cricetid rodent *Pseudoryzomys simplex* (Cricetidae: Sigmodontinae). Redrawn from Weksler (2006).

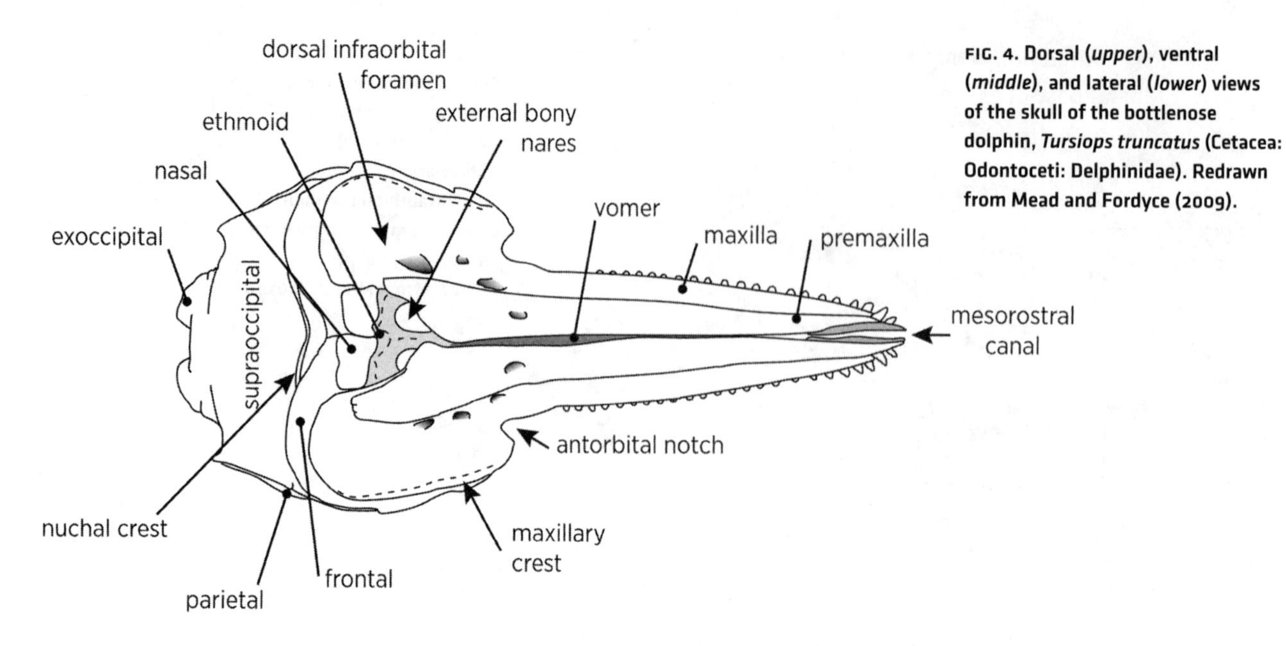

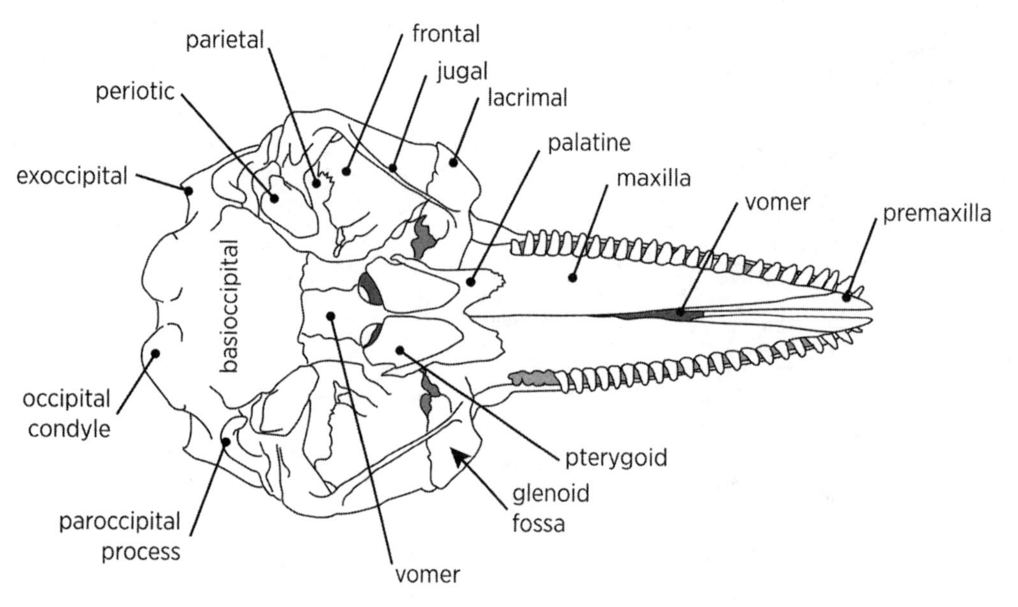

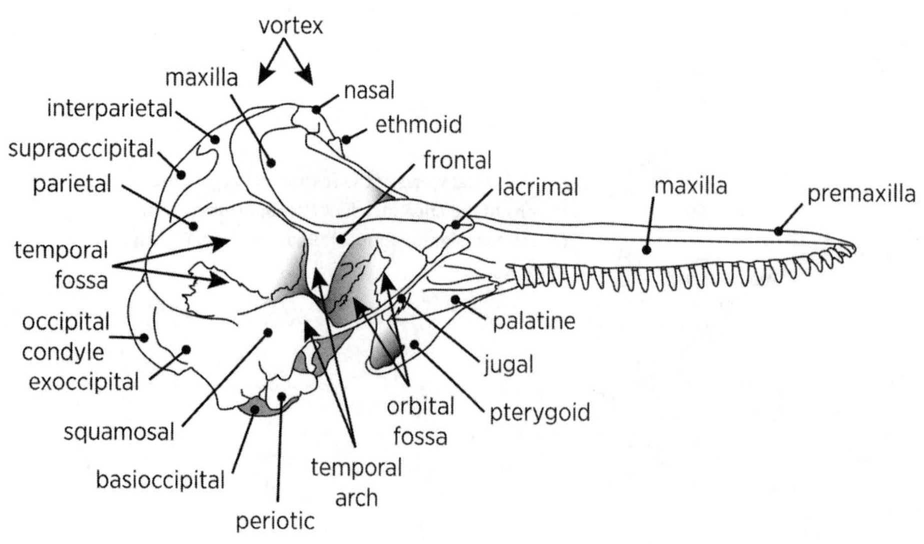

FIG. 4. Dorsal (*upper*), ventral (*middle*), and lateral (*lower*) views of the skull of the bottlenose dolphin, *Tursiops truncatus* (Cetacea: Odontoceti: Delphinidae). Redrawn from Mead and Fordyce (2009).

lacrimal bones—paired bones located on or adjacent to the anterodorsal base of the zygomatic arch; possess a **lacrimal foramen**, through which the tear duct passes.

turbinal bones—visible in anterior view only, these bones occur within the nasal passage; consist of three sets of fragile bones inserting on the nasal bones (nasoturbinals), ethmoid bones (ethmoturbinal), and maxillary bones (maxilloturbinal); these bones are important to thermoregulation, water balance, and olfaction.

Structures

zygomatic arch—the conspicuous arch of bone extending laterally on either side of the cranium; forms the lateral and ventral borders of the orbit (anterior) and temporal fossa (posterior); may be composed of maxillary, jugal, squamosal, and lacrimal bones; serves for the origin for the masseter jaw musculature; may include a dorsally oriented postorbital process that defines the posteroventral border of the orbit. In many rodents, the anterior portion of the arch is angled upward and forms a broad **zygomatic plate**. In some rodents, there is an anterior extension of the zygomatic plate, termed the **zygomatic spine**.

orbits—socket-like depressions in which the eyes are housed; bordered ventrolaterally by the zygomatic arch (when present) and posteriorly by the temporal fossa.

postorbital bar—coalescence of the postorbital processes of the frontal bone and zygomatic arch that separates the orbit from the temporal opening in lateral view while the two openings remain confluent anteroposteriorly. Characteristically present in several different groups of mammals, such as Scandentia, strepsirrhine primates, Equidae among the Perissodactyla, and camelid and ruminant Artiodactyla.

postorbital plate—a bony "wall" separating the orbit from the temporal opening; composed of expanded frontal, alisphenoid, and orbitosphenoid bones that contact anterior elements of the zygomatic arch; characteristic of haplorrhine primates.

temporal opening—the lateral opening of the skull immediately posterior to the orbit (the posterior border of which is often defined structurally by the postorbital process[es] of the frontal and/or jugal/zygomatic arch) that is evolutionarily derived from the temporal fenestra of synapsids and serves as space for the temporal muscle where it inserts onto the coronoid process of the mandible. This space may confusingly be referred to as the temporal fossa, which we define separately.

temporal fossa—the large, shallow depression (the fossa temporalis) on the side of the cranial vault that is bounded dorsally by the temporal line (or temporal ridge, if present), caudally by the occipital (or lambdoidal) crest, ventrally by the squamosal root of the zygomatic arch, and anteriorly by the postorbital process of the frontal bone; this is the area from which the temporal muscle originates.

occipital (= **lambdoidal** or **nuchal**) **crest**—a ridge of bone across the posterodorsal margin of the cranium; usually a part of the supraoccipital bone; important place for attachment of neck muscles and nuchal ligament in large-headed animals (e.g., Perissodactyla, Artiodactyla).

sagittal crest—a ridge of bone located along the midline of the posterodorsal aspect of the skull; serves for additional (temporal) muscle attachment and is therefore most prominent in taxa requiring large temporal muscles (e.g., some Chiroptera, Carnivora).

Posterior aspect

Bones

occipital bone—constitutes the posterior wall of the braincase; formed by the fusion of the supraoccipital, basioccipital, and two exoccipitals.

supraoccipital bone—dorsal to the foramen magnum.

basioccipital bone—ventral to the foramen magnum; extends anteriorly on the ventral surface of the cranium between the auditory bullae.

exoccipital bones—lateral to the foramen magnum and bearing the occipital condyles.

mastoid bones—small and often inconspicuous bones adjacent to the paroccipital processes and at the posterior margins of the auditory bullae; a part of the otherwise concealed periotic bone; may protrude to form a mastoid process (in certain Carnivora) or become expanded to form part of the auditory bulla (in Heteromyidae and other rodents).

Structures

foramen magnum—the large opening in the occipital bone; conduit for the spinal cord and vertebral arteries.

occipital condyles—smooth raised surfaces on the exoccipital bones; serve as articulation points for the skull, setting into depressions in the first vertebra (the atlas).

paroccipital process—ventrally projecting process of the occipital bone; close to and just posterior to the auditory bulla; serves for insertion of certain chewing muscles and for muscles that open the jaw.

mastoid process—ventrally projecting process of the mastoid bone.

Ventral aspect

Bones

auditory bullae—swollen structures on either side of the basioccipital and posteroventral to the squamosal; formed by the tympanic (= ectotympanic) with the alisphenoid contributing in marsupials, and sometimes either, or both, the entotympanic and petrosal bones in placentals; serve to protect the middle ear and facilitate sound transmission; may be absent (e.g., monotremes) or hypertrophied (e.g., kangaroo rats and other rodents); in some taxa (e.g., some Primates) may form a ring or fuse into a tube extending laterally.

basisphenoid bone—anterior to the basioccipital and located in the ventral midline of the skull.

presphenoid bone—also in the ventral midline, located anterior to the basisphenoid.

orbitosphenoid bones—extend laterally from the presphenoid bone; may be fused and indistinguishable from the presphenoid.

alisphenoid bones—winglike bones in the walls of the temporal fossae posterior to the frontals and orbitosphenoids and anterior to the squamosals; in marsupials, the alisphenoids extend posteriorly to participate in the auditory bullae; the alisphenoid canal (which transmits part of cranial nerve V) may penetrate a bony shelf at the ventral base of the alisphenoid and is important in some mammalian groups (e.g., some Carnivora).

pterygoid bones—paired bones posterior to the internal opening of the nasal passages; often with an elongate process (the **hamular process**) extending posteriorly from the ventral surface of each pterygoid; between the pterygoids is the mesopterygoid fossa, while lateral to each pterygoid is a parapterygoid fossa.

palatine bones—paired bones forming the posterior portion of the palate, between the cheek teeth and posterior to the maxillae.

vomer bone—a single bone in the ventral midline of the skull, anterior to the alisphenoids and forming the posteroventral portion of the wall separating the two sides of the nasal passages; occasionally extends ventrally to constitute a minor part of the palate.

The lower jaw, or **mandible**, is composed of a single pair of bones in mammals, the **dentary**; these bones may be strongly fused anteriorly or they may be somewhat loosely connected.

Structures

coronoid process—dorsal process of the dentary, extends into the temporal fossa; serves for attachment of portions of the temporal muscle (which originates in the temporal fossa).

condyloid process—process located ventral to the coronoid process; bears the **mandibular condyle**, which articulates with the cranium in the mandibular fossa. The notch between the coronoid and the condyloid process is the **sigmoid notch.**

angular process—process of the dentary that forms the posteroventral angle of the jaw and is of variable size; may be enlarged for insertion of powerful masseter and other jaw musculature.

masseteric fossa—the lateral depression on either side of the dentary and ventral to the coronoid process; serves for insertion of much of the masseter muscle.

capsular process—a bony capsule that contains the root of the lower incisor; when visible, it is present on the labial side of the mandible, anterior to the condyloid process and ventral to the coronoid process.

POSTCRANIAL SKELETON

Major bones of the appendicular skeleton

Elements of the pectoral girdle and forelimb

scapula (including **spinous process**, **coracoid process**, and **acromion process**)—the **glenoid fossa** is the depression where the head of the humerus articulates (authors that use "glenoid fossa" for the mandibular articulation generally refer to the scapular structure as the "glenoid fossa of the scapula" or as the "glenoid cavity").

Table 1. Foramina of the skull.

Name	*Bone(s) penetrated*	*Traversing structures*
infraorbital	maxilla, ventromedial to orbit	trigeminal and infraorbital nerves and blood vessels
lacrimal	lacrimal	tear duct from orbit to pharynx
mental	lateral body of mandible (usually 2)	mental nerves and blood vessels
mandibular	medial surface of mandible	inferior alveolar nerve and vessels
optic	center of orbitosphenoid	optic nerve (cranial nerve II)
orbital fissure	orbitosphenoid-alisphenoid junction	oculomotor (III), trochlear (IV), abducens (VI), ophthalmic branch of trigeminal (V_1) nerves
incisive	premaxilla-maxilla, between canines	palatine vessels, trigeminal and nasopalatine nerves
palatine	maxilla-palatine junction	palatine vessels and trigeminal nerve
sphenopalatine	palatine, vertical wing	sphenopalatine nerves and vessels
rotundum	alisphenoid	maxillary branch of trigeminal (V_2)
ovale	alisphenoid	mandibular branch of trigeminal (V_3)
hypoglossal	occipital, lateral to end of condyle	hypoglossal nerve
stylomastoid	between mastoid and styloid processes	facial nerve
magnum	occipital, center	spinal cord
eustachian tube	tympanic bulla	from middle ear
olfactory	ethmoid, cribriform plate	olfactory nerves

clavicle—a structurally simple bone that, when present, extends from the acromion process of the scapula to the interclavicle or sternum; prominent in arboreal, fossorial, aerial, and generalized mammals; reduced in Canidae and Felidae; reduced or absent in cursorial mammals; absent in Artiodactyla, Perissodactyla, Cetacea.

coracoid, **procoracoid**, and **interclavicle** (of the shoulder girdle)—found only in monotremes.

humerus—the large, proximal bone of the forelimb. The **head** is the smooth projection that articulates with the glenoid fossa of the scapula; adjacent to the head are the **greater tuberosity** (= **trochiter** in bats), a lateral process, and the **lesser tuberosity** (= **trochin** in bats), located medially.

ulna—the dominant bone of the forelimb in most mammals (but not bats); includes the **trochlear notch**, which articulates with the humerus; proximal to the trochlear notch is the **olecranon process**.

radius—a small, usually slender bone that articulates proximally with the humerus and distally with the carpal bones of the wrist; smaller than the ulna in most mammals; the radius is the dominant forelimb bone of bats (a standard measure of size in bats is the forearm length—from the olecranon process to the tip of the radius).

manus (manal)—the forefoot.

carpals—small bones arranged into two rows—proximal and distal; the proximal row includes (medial to mesial) the scaphoid, lunate (the latter two may fuse as the scapholunar), triquetrum (= cuneiform, or ulnar carpal bone), and pisiform bones; the distal row includes the trapezium, trapezoid, central, capitate (= magnum), and hamate (= unciform); in some artiodactyls, the trapezoid and capitate are fused.

metacarpals (numbered 1–5)—may be greatly lengthened in cursorial species; may converge morphologically on metatarsals due to similar function (as a group called **metapodials**).

phalanges (numbered 1–5)—the first digit (thumb) is termed the **pollex**.

Elements of the pelvic girdle and hind limbs

ilium, ischium, pubis—these three bones together form the **innominate** (= **pelvis** or **os coxae**). Note the **acetabulum**, or socket for articulation with femur, and **obturator foramen**; the pubic bones fuse into the **pubic symphysis** in most mammals, forming the ventral boundary to the birth canal.

epipubic—paired bones in monotremes and marsupials, not present in eutherians; thought to serve as levers within hypaxial musculature, maintaining muscular tonus during locomotion and respiration (Reilly and White 2003).

femur—large proximal bone of the hind limb; the large **head** serves to articulate with the acetabulum of the innominate; the **greater trochanter** is a large projection lateral to the head; **lesser trochanter** is a smaller projection located on the posteromedial side below the head; **third trochanter** (e.g., in Perissodactyla; see order account) is on the shaft.

patella (= kneecap)—a sesamoid bone (e.g., formed in a tendon, not from a cartilaginous template).

tibia—the dominant bone of the lower hind limb, located medial or anterior to the fibula.

fibula—smaller bone of the lower hind limb, lateral to the tibia; free and well developed in arboreal and generalized walking mammals, reduced in cursorial mammals, where it becomes part of the ankle elements in Artiodactyla; may fuse with tibia to form a **tibiofibula** (e.g., in rodents and rabbits).

pes (pedal)—the hind foot.

tarsals—bones arranged in three rows (proximal, intermediate, and distal); the two proximal bones are the **calcaneus** and **astragalus** (= talus); the intermediate contains only the navicular bone; the distal row contains the cuboid, which fuses with the navicular (cubonavicular) in some artiodactyls, and three cuneiform (= wedge-shaped) bones—the medial (= first, internal, or entocuneiform), middle (= second, intermediate, or mesocuneiform), and lateral (= third, external, or ectocuneiform).

metatarsals (numbered 1–5)—may be greatly lengthened, and reduced in number, in cursorial species; may converge morphologically on metacarpals due to similar function (as a group called **metapodials**).

phalanges (numbered 1–5)—the first digit (big toe) is termed the **hallux**.

Major elements of the axial skeleton

sternum—consists of a series of bony segments (**sternebrae**); anteriormost is the **manubrium**, posteriormost is the **xiphoid process**.

ribs—elongate, curved bones extending laterally and ventrally from the vertebrae to the sternum; generally associated only with thoracic vertebrates (monotremes and sloths [Bradypodidae, Choloepodidae] have cervical ribs); articulate with ribs via two processes, the posterodorsal **tuberculum** and the anteroventral head, or **capitulum**; posteriormost ribs may not meet the sternum and are called **floating ribs**.

vertebral elements:

- **cervical vertebrae** (including **atlas** and **axis**)—typically 7 in most mammals, but may be fewer; often compressed or fused in fossorial, saltatorial, and aquatic mammals. Characterized by transverse foramina, small transverse processes, and lack of articulation facets for ribs. **Atlas** (first vertebra) lacks a **centrum**; **axis** (second vertebra) has an anterior projection, the **dens** (= **odontoid process**).
- **thoracic vertebrae**—usually 12–15 in number; characterized by rib articulation facets, no transverse foramina, presence of **transverse processes** where head of rib articulates, and generally large and posteriorly sloping **dorsal** (**neural**) **spines** (= **spinous processes**).
- **lumbar vertebrae**—usually 6–7 elements, but 20 in odontocete cetaceans; characterized by neural spines (= spinous processes) that typically slope anteriorly (these may be especially large in saltators, serving to support a long and massive tail); especially large transverse processes that slope anteroventrally.
- **sacral vertebrae**—3–5 vertebrae typically fused into a single element to form the **sacrum**, which provides rigid support for the pelvic girdle and hind limbs. Number may be as high as 10 in Xenarthra, but they are few in number and not differentiated from lumbar vertebrate in mammals with reduced hind limbs (such as Cetacea, Sirenia).
- **caudal vertebrae**—vary widely in number, depending on presence and length of tail; typically rather simple, lack a neural arch, and, except for first few, lack transverse processes and a neural spine; may include ventral chevron bones.
- **zygapophyses**—articulation facets between each pair of adjacent vertebrae, located on both anterior and posterior ends.

baculum (= os penis, penis bone): a sesamoid bone located in the penis of some mammals (female analog, when present, termed the baubelum); quite variable in some taxa; found only in Primates [mostly prosimians], Rodentia [most], Insectivora sensu lato [most], Carnivora [nearly all], Chiroptera [nearly all], and the North American pika (Lagomorpha:

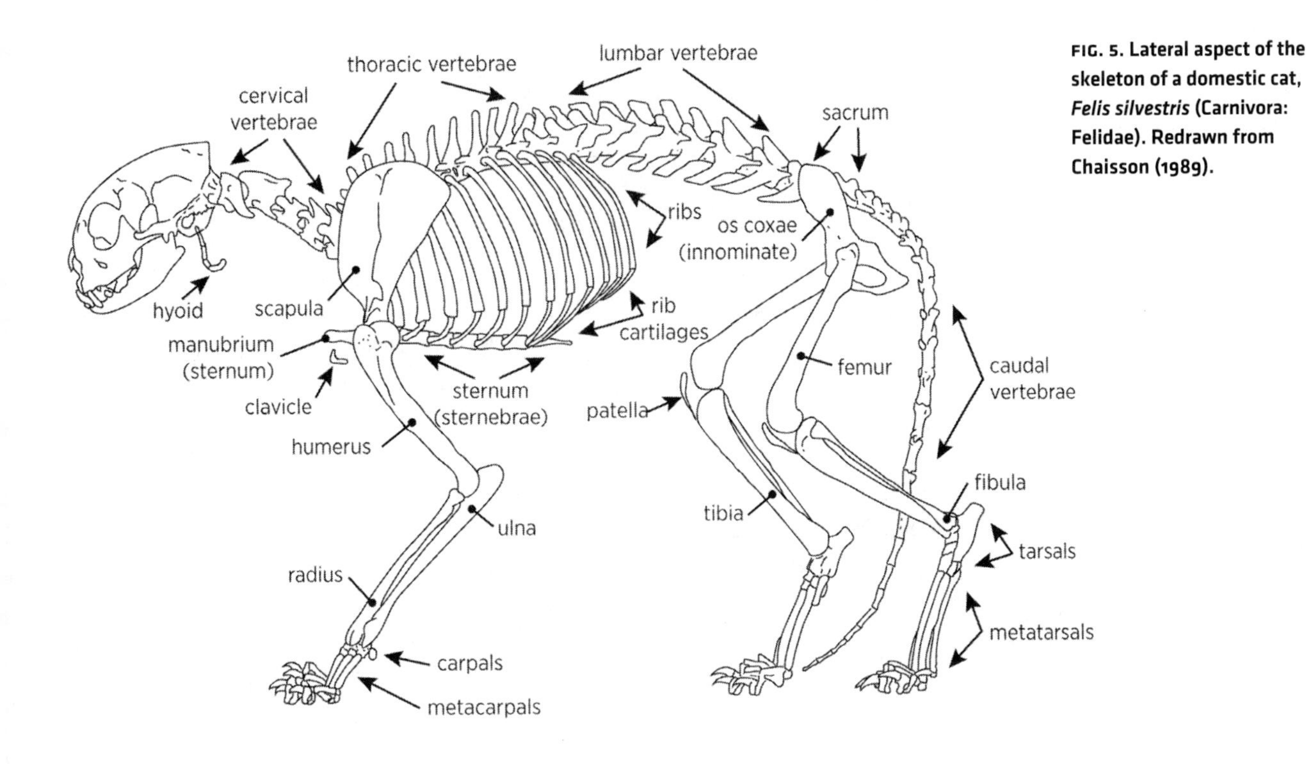

FIG. 5. Lateral aspect of the skeleton of a domestic cat, *Felis silvestris* (Carnivora: Felidae). Redrawn from Chaisson (1989).

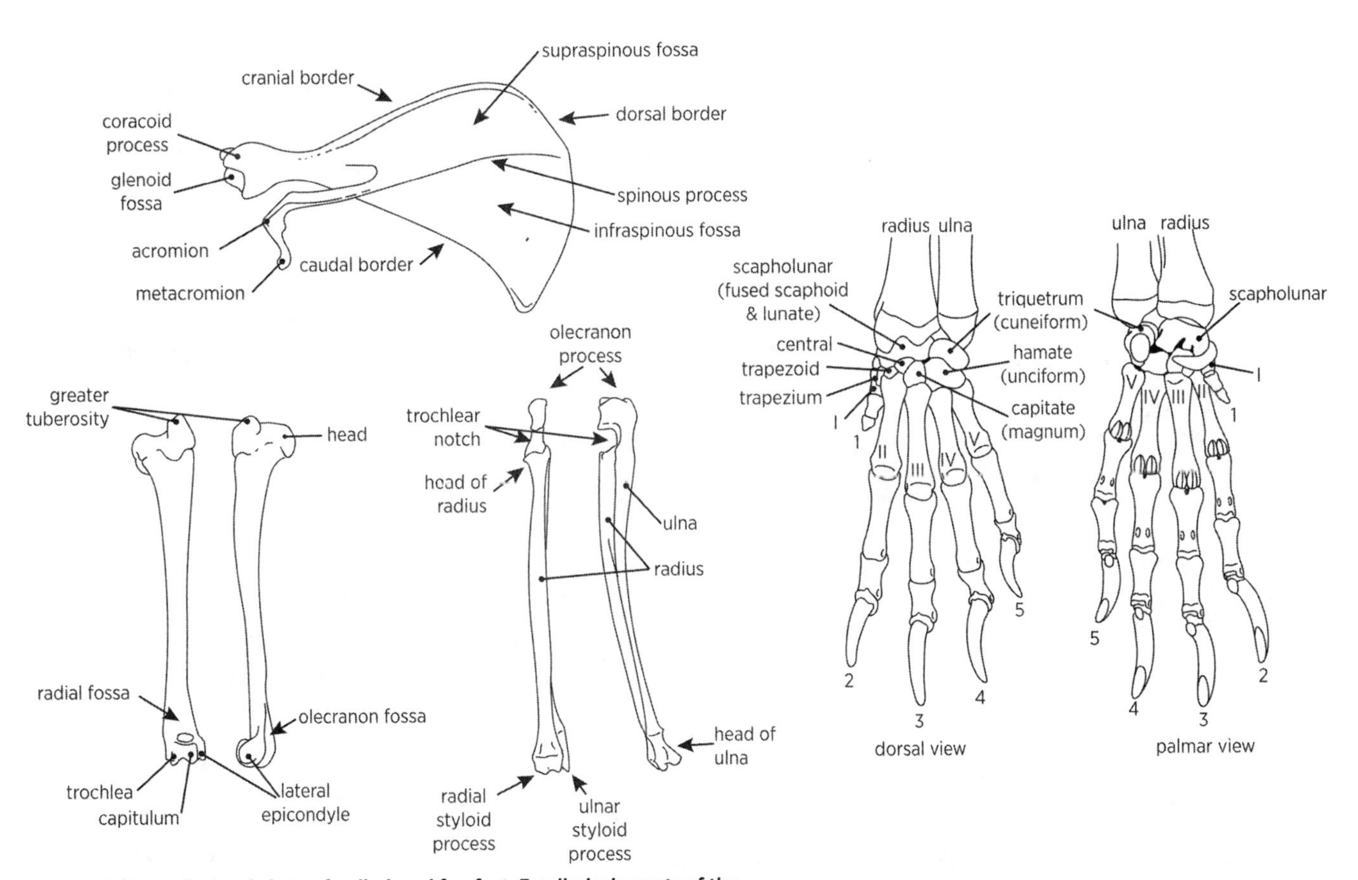

FIG. 6. Appendicular skeleton: forelimb and forefeet. Forelimb elements of the generalized mammal: scapula (*upper left*), humerus (*lower left*; anterior and lateral views), radius-ulna (*lower middle*; anterior and lateral views), and manus (*right*) of the woodchuck, *Marmota monax*. I–V = metacarpals, 1–5 = digits. Redrawn from McLaughlin and Chaisson (1990) and Bezuidenhout and Evans (2005).

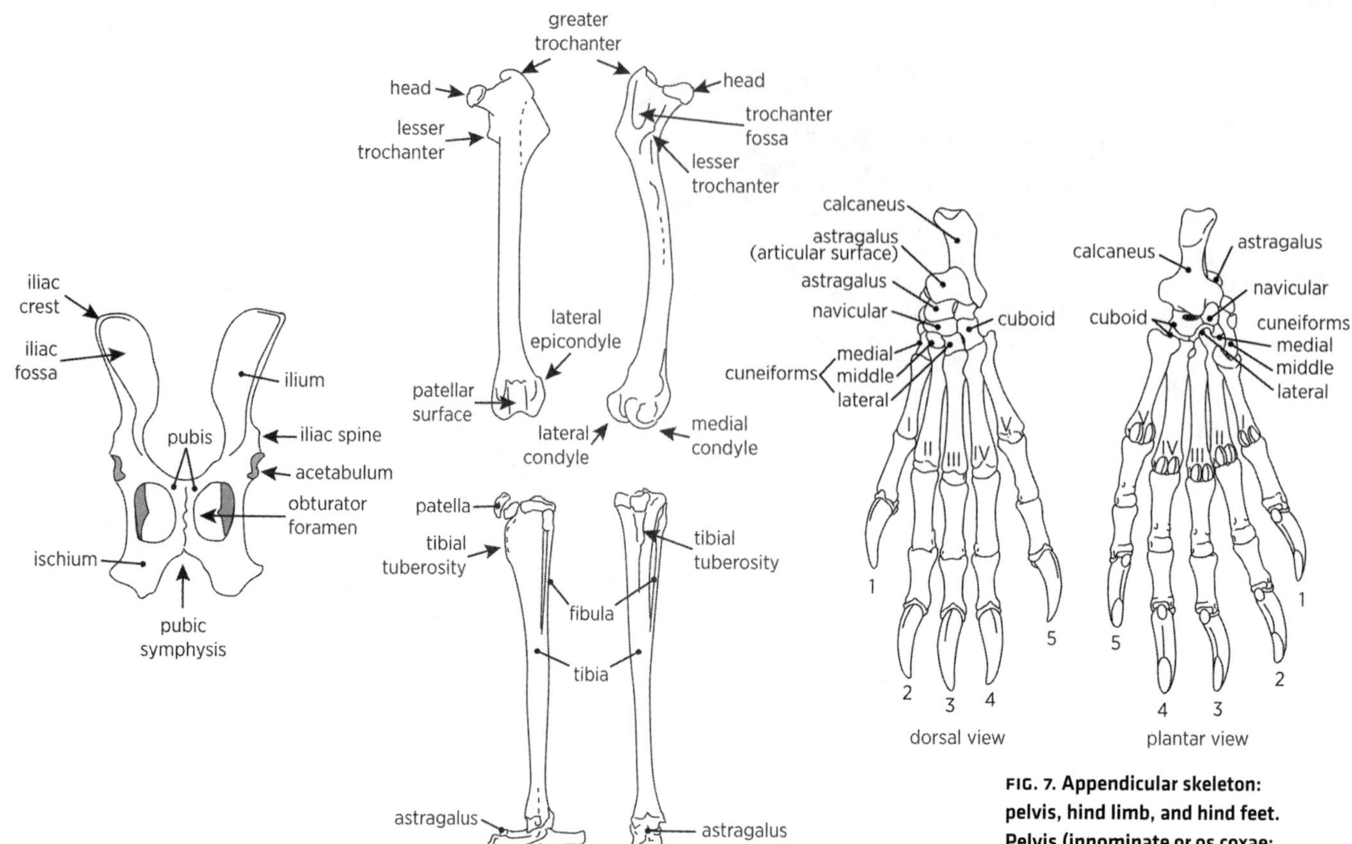

FIG. 7. Appendicular skeleton: pelvis, hind limb, and hind feet. Pelvis (innominate or os coxae; *left*) and hind limb elements of the generalized mammal: femur (*upper middle*; anterior and lateral views), tibiofibula (*lower middle*; anterior and lateral views), and pes (*right*) of the woodchuck, *Marmota monax*. I–V = metatarsals, 1–5 = digits. Redrawn from McLaughlin and Chaisson (1990) and Bezuidenhout and Evans (2005).

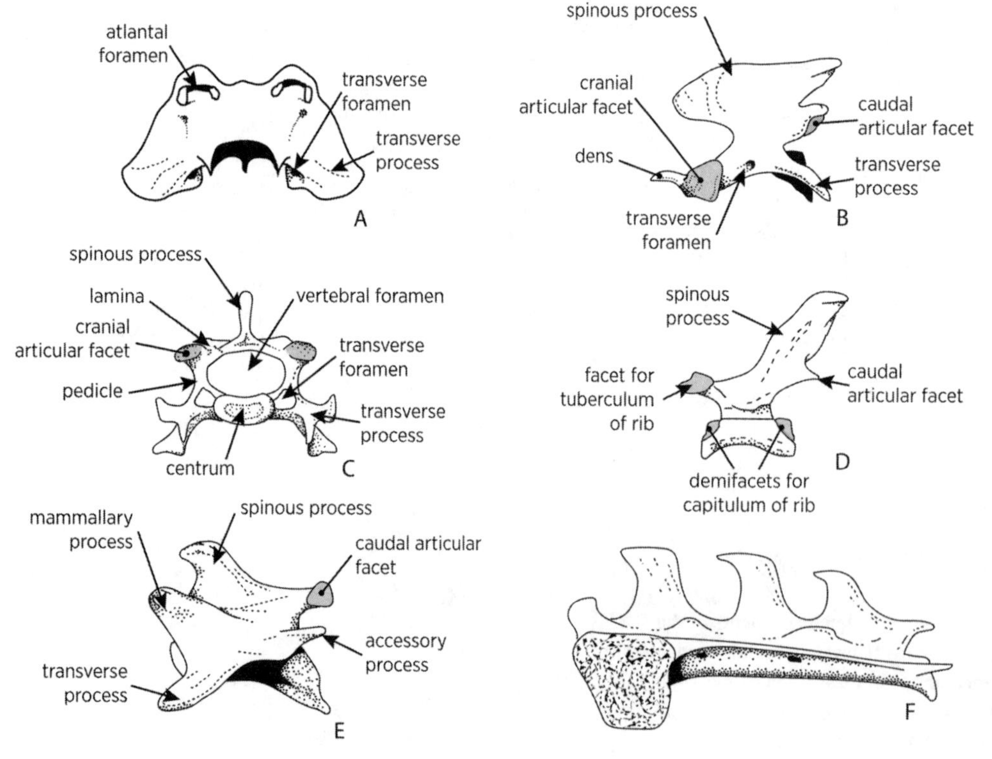

FIG. 8. Axial skeleton: vertebral elements. *A*, dorsal view of the atlas (first cervical vertebra); *B*, lateral view of the axis (second cervical vertebra); *C*, frontal view of a cervical vertebra; *D*, lateral view of a thoracic vertebra; *E*, lateral view of a lumbar vertebra; *F*, lateral view of the sacrum. Redrawn from Radke and Chaisson (1998).

Ochotonidae; Weimann et al. 2014). With the discovery of a baculum in the pika, the traditional mnemonic PRICC, used by generations of students to remember this group, requires revision, e.g., to PRICCL. Schultz et al. (2016) reviewed the distribution of bacula among living mammals.

THE INTEGUMENT

Mammalian skin includes two layers. **Dermis** tissue comprises a thick layer of connective tissue including blood vessels and sensory receptors. Leather is dermis. Above the dermis is **epidermis**, composed of squamous epithelial cells, only the basal layer of which is living and reproducing. Toward the surface of the skin, layers of epithelial cells are keratinized (strengthened by addition of a fibrous protein, keratin) and continually flake off as they are replaced by growth from below. Modified epithelial cells form a variety of important structures, including epidermal scales, hair, horn, and claws.

Epidermal scales come in a variety of forms. Pangolins (Pholidota: Manidae) are covered with imbricated epidermal scales attached at their base to thick skin for protection (body hair is greatly reduced or absent). Armadillos (Xenarthra: Cingulata: Dasypodidae) are covered with bony **scutes** (formed in the dermal layer, and in turn covered with epidermal scales, or plates) forming an armor typically composed of five shields: cephalic, scapular, dorsal, pelvic, and caudal. As with pangolins, armadillo scales provide protection, and body hair is generally reduced. Whereas pangolins have flexible armor, however, armadillos are more constrained by the rigidity of their shields; these shields overlap, and between them are bands of flexible skin. Many mammals have small epidermal scales on their feet and tails; beaver tails (Rodentia: Castoridae) are particularly clear examples.

MAMMALIAN HAIR

Hair is unique to the Mammalia, and all mammals have some hair at some time in their lives. The total body covering of hair (= **pelage**) of the typical mammal is composed of different morphologically recognizable types of hairs, usually termed **vibrissae**, **guard hairs**, and **underfur**. Most people are familiar with the whiskers of cats, dogs, and many other mammals; these hairs are vibrissae. Guard hairs are the relatively long hairs that give the pelt its color and texture. Underfur is found beneath the protective guard hairs and serves as insulation. Each of these types can change over the body of an individual and ontogenetically during life. Moreover, in many species, there is gradation between hair types, so division into types is often quite subjective.

Structure of hair—typical hairs are made of three layers of keratin material: a central core, or **medulla**; a layer of **cortex** surrounding the medulla; and an outermost layer called the **cuticle**. All three layers are composed of dead cells and all show their cellular nature under a microscope. The shape, arrangement, and relative sizes of all three layers are useful in hair identification, as is the pattern of pigment deposition in the medulla. The cuticle typically presents a scaled texture that may vary greatly, even on the same individual hair. The form of the scale margin, the distance between scale margins, and the scale pattern itself may be diagnostic of genera or species of mammals.

Hairs with **definitive** growth are shed and replaced once a given length has been reached. **Angora** hairs grow continuously and may or may not be shed.

Vibrissae are long, stiff hairs that serve as important tactile receptors in mammals. They have well-innervated bases and an abundant blood supply. Facial vibrissae (fig. 9) fall into several distinct categories, but vibrissae may also occur elsewhere on the body, most notably the limbs or ankles. On the forelimb, Brown and Yalden (1973, following Jones 1923) defines these as anconeal (near the elbow), medial antebrachial (on the forelimb proper), and ulnar carpal (just above the wrist); on the hind limb may be calcaneal (inside of ankle) vibrissae.

Guard hairs may be variably modified. Most common are **awns**, which make up the outer layer of fur for most mammals; awns exhibit definitive growth and are expanded distally and weakened basally so that they are lost relatively readily. Many species exhibit **bristles**, firm hairs with angora growth that form the manes of lions and other taxa. In several groups (New and Old World porcupines, hedgehogs, echidnas, various groups of "spiny rats"), they are stiff and enlarged as **spines** (or **quills**); in New World porcupines, the cuticular scales of guard hairs are elongate and serve as proximally directed barbs that help to lodge the spine in the skin of a potential predator.

In some groups (e.g., "spiny rats" of the Neotropical Echimyidae), authors (Moojen 1948; Patton et al. 2015) routinely refer to aristiform and setiform hairs. The former are "varyingly stiffened spines" (Patton et al.

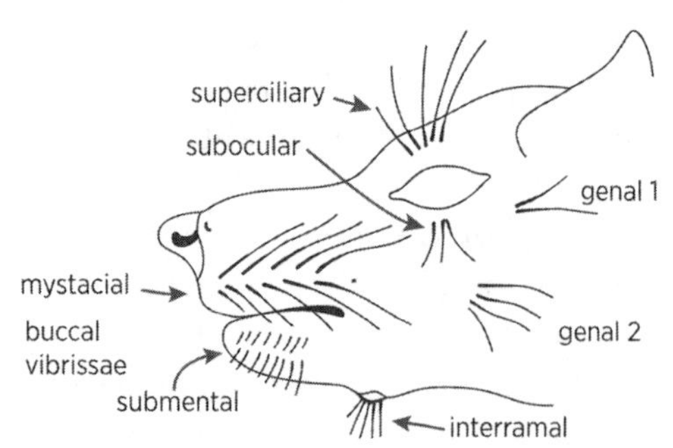

FIG. 9. Facial vibrissae of a generalized mammal head. Redrawn from Brown (1971).

2000) whereas the latter are softer hairs, which Moojen simply referred to as "over hairs."

Underfur comes in two general forms. **Wool** is long, soft, and often curly underfur with angora growth. **Fur** is relatively short and has definitive growth.

Hairs wear over time, and the need for hairs may change seasonally. Consequently, mammals replace some or all of their hair in a process called **molting**. Some molting occurs continuously in many species, with small numbers of hairs being lost and replaced daily. Three other types of molting merit consideration. Young mammals often undergo a **post-juvenal molt** in which they replace their juvenile pelage with adult pelage. Such molts occur in a predictable wavelike pattern that differs among species. In temperate and higher latitudes, most mammals undergo rapid **annual molts**, which allow them to replace hairs that are worn through use. Finally, some species undergo **seasonal molts**, which allow for seasonally variable pelage characters; some species incorporate seasonal camouflage into such molts, with white winter coats and brown summer coats.

Mammal tails may be scaly and only sparsely haired, or they may be covered with fine, velvety hairs. Many mammals have a **tuft** of elongate hairs extending from the tip of the tail, which may serve in communication or as a counter-balance in richochetal locomotion (e.g., *Dipodomys*, family Heteromyidae). A terminal extension of more sparsely distributed and shorted hairs is often referred to as a **pencil** (i.e., the tail is **penicillate**, as in *Perognathus*, family Heteromyidae). Tails may also possess a **crest**, a linear array of progressively elongate hairs extending from the dorsal surface toward the tip to merge with the terminal tuft (as in most *Chaetodipus*, family Heteromyidae). The condition in which longer hairs circumscribe the entire terminal portion of the tail is referred to as a **brush**.

FOOT POSTURE AND FOOT PADS

Mammals exhibit substantial variation in foot posture, which reflects their modes of locomotion. Many mammals are **plantigrade**, which means that they place the entire foot—digits to ankle or wrist—on the ground (e.g., humans, bears). In contrast, many mammals walk only on the digits, and have elevated the carpals/tarsals off the ground (e.g., dogs, cats); this foot posture is **digitigrade**. Ungulate mammals (most Artiodactyla and Perissodactyla) bear their weight on the terminal phalange and specifically on the hooves; this foot posture is **unguligrade**. Finally, the term

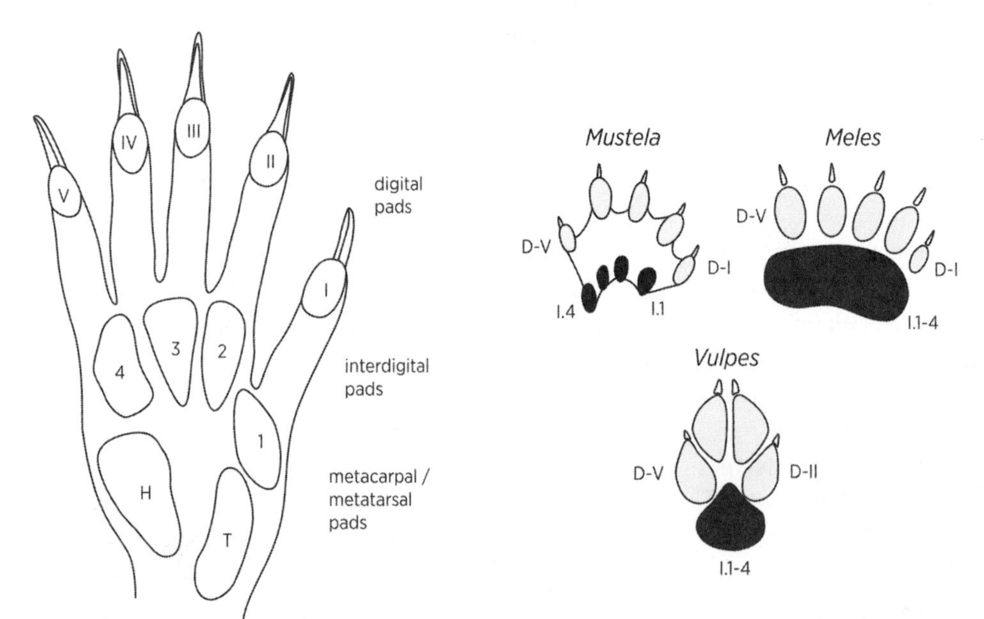

FIG. 10. ***Left*: Generalized right pes/manus to illustrate terminology associated with foot pads. T and H refer to the thenar and hypothenar pad, respectively. *Right*: Right hind footprints of a plantigrade (*Mustela*) and two digitigrade mammals (*Meles* and *Vulpes*) to illustrate relationship between footprints and stance. Digital pads are denoted by D-I, etc., and interdigital pads by I.1, etc. Both digitigrade species illustrate fusion of interdigital pads. Redrawn from Brown and Yalden (1973).**

subunguligrade is often used for mammals that walk on their toe tips [digitigrade] plus a digital pad (camels, tapirs, rhinoceroses, elephants, hyraxes).

Foot (or plantar) pads are areas of cornified epidermis that provide traction as well as protection from punctures, scratches, and so forth. In plantigrade feet they generally occur in three sets (digital, interdigital, and metacarpal/metatarsal; fig. 10), and the animal's weight typically is borne on all of these pads. Note that the thenar pad is on the medial side of the foot and the hypothenar is lateral.

In digitigrade mammals, the weight is borne on the digital and interdigital pads, while the thenar and hypothenar pads are reduced or absent. Interdigital pads are often fused in digitigrade species.

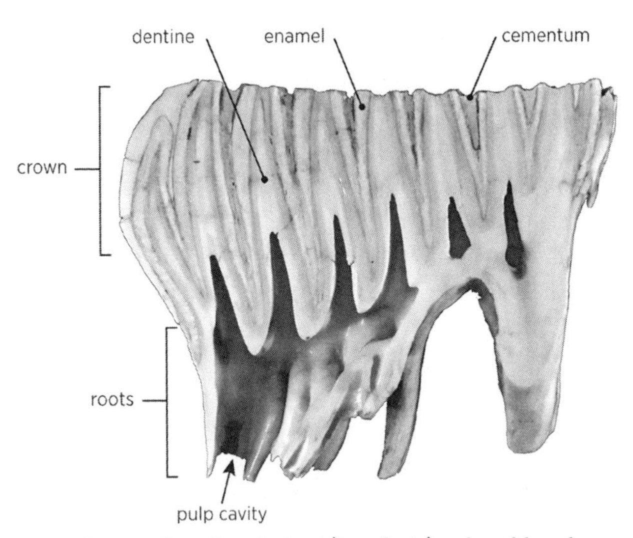

FIG. 11. Cross section of an elephant (*Loxodonta*) molar, with each of the major parts labeled. The infolded enamel filled with coronal cementum is especially evident (see description of the lophodont occlusal pattern below).

TEETH

The **dental arcade**, or **arch**, refers to the crescentic arrangement of teeth on each side of the upper and lower jaw. The mammalian dentition is uniquely characterized as **thecodont** (anchored to the upper and lower jaws by placement in deep sockets), **heterodont** (containing a diversity of tooth types, each of which serves a generally different function; note, however, that a few mammals have **homodont** dentitions, where all teeth are of similar shape [e.g., armadillos, otariid pinnipeds, and toothed whales]), and **diphyodont** (having two generations, where an initial deciduous tooth is lost when the adult tooth erupts). In general, incisors, canines, and premolars are diphyodont, although exceptions where deciduous teeth are retained rather than replaced in the adult animal are found throughout the Mammalia (e.g., the deciduous fourth premolar is retained in some rodents, as in all Echimyidae); whereas molars are **monophyodont** (lack deciduous precursors).

Teeth are composed of an outer layer of **enamel** covering the tooth crown, made up of hydroxyapatite prisms that constitute one of the hardest organic substances known, and an inner layer of softer **dentin** (or **dentine**). The dentin surrounds the **pulp cavity**, the center of the tooth made up of living connective tissue (fig. 11). A layer of **cementum**, secreted by specialized cells within the root and similar to alveolar bone, covers the outer surface of the roots. Cementum also fills the open spaces between the deeply infolded enamel of **lophodont** teeth specialized for grazing (in which case it is called **coronal cementum**).

The number and types of teeth may be useful in

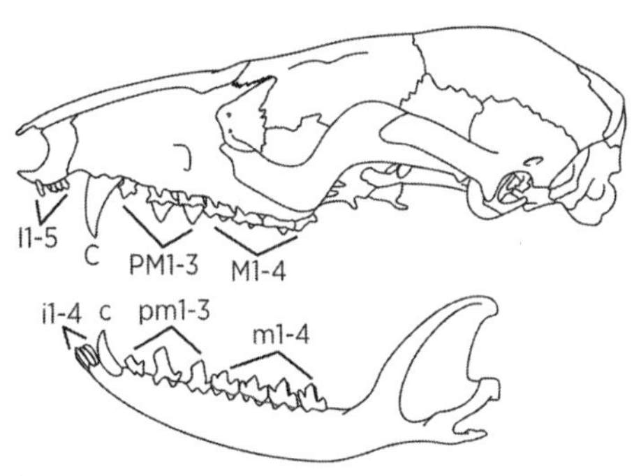

FIG. 12. Lateral view of skull and mandible of a short-tailed opossum, *Monodelphis* (Metatheria: Didelphidae), with teeth identified. Modified from Rossi et al. (2010).

characterizing mammalian orders and families, so mammalogists have developed a shorthand way of describing them.

DENTAL FORMULAE

Because the number of each tooth type (incisor, canine, premolar, molar) is identical on the right and left sides of the mouth, we count the number of each tooth type on one side of the upper and lower jaws and record those numbers.

Using a *Monodelphis* skull (fig. 12) as an example, we can see that the upper jaw includes five incisors, one canine, three premolars, and four molars. The lower jaw holds four incisors (one barely visible here), one canine, three premolars, and four molars. Given these counts, the total number of teeth is 50.

Letters and numbers may be used to designate each

specific tooth: upper and lower teeth are commonly distinguished either by case (e.g., M1 = first upper molar, whereas m2 = second lower molar) or by superscripts and subscripts (e.g., M^1 or m^1 = first upper molar; M_2 or m_2 = second lower molar). Note also that deciduous teeth are designated with a "d," as in dPM3 (or dPM^3 or dpm^3) for the third deciduous upper premolar. The entire dental formula for *Monodelphis* may be written in several different ways, of which the following are the most common:

5/4 1/1 3/3 4/4 = 50

5I 1C 3PM 4M/4i 1c 3pm 4m = 50

$$\frac{5}{4}\ \frac{1}{1}\ \frac{3}{3}\ \frac{4}{4} = 50$$

We use the third of these alternatives throughout the accounts in this book.

Living Metatheria and Eutheria differ in their respective ancestral dental formulae, namely, 5/4 1/1 3/3 4/4 = 50 for marsupials (as illustrated for *Monodelphis* above) versus 3/3 1/1 4/4 3/3 = 44 for eutherians. However, this does not reflect actual homologies in all cases. For the evolution of these differences from the common therian ancestor, see O'Leary et al. (2013) and Williamson et al. (2014). Note that teeth of each type in living mammals are usually numbered sequentially from front to back and begin with the number 1 (e.g., M1 . . . M4 for the ancestral metatherian and M1 . . . M3 for the ancestral eutherian).

TOOTH MORPHOLOGY

Dental terminology may confuse students at first: some terms refer to the pattern of cusps, whereas others define the structure or height of cusps. Most cusps and other structures have specific names, some of which are fairly self-evident (e.g., the anterocone is the anteriormost cusp) while others require knowledge of a specific lexicon. At its extreme, the terminology used to describe teeth is rich and extensive, and this diversity of terms reflects the importance of teeth in both paleontological and neontologic studies of mammalian diversity. Evans and Sanson (2003; see also commentary by Weil 2003) and Ungar (2010) provide excellent overviews of the functional morphology of different types of mammalian teeth. In this section, we outline terms and structures that are fundamental in describing mammalian teeth as well as in relating that structure to function.

Early therian mammals developed a molar tooth type referred to as the **tribosphenic molar** (fig. 13), from which most contemporary molariform teeth are thought to be derived (although the tribosphenic molar evidently evolved more than once; see Luo et al. 2001). Molar terminology is complicated at best, but some simple patterns are present. The upper tribosphenic molar has three principal cusps arranged in a triangle, termed the **trigon**; these cusps include the **protocone**, located on the lingual side of the tooth, which is often connected to the **paracone** (anterior and labial) and **metacone** (posterior and labial) by crests, or **cristae**. A general term for any tooth with three primary cusps, such as the tribosphenic, is **tritubercular**. Secondary cusps are designated with the suffix **–ule** (e.g., **conule, metaconule**). Many teeth have a shelflike ridge of enamel (the **cingulum**) that borders one or more margins of the tooth. The **stylar shelf** is a ledge located along the labial side of the tooth, an expanded portion of the cingulum, which may support additional cusps (**styles**, e.g., **parastyle**, **mesostyle**, **metastyle**; see below); some authors consider similar shelves on the lingual side to be stylar shelves as well (e.g., labial vs. lingual stylar shelves).

The lower tribosphenic molar is more complicated, as it includes an anterior raised trio of cusps (**trigonid**; the suffix **–id** is used for all structural elements of the lower tooth) and a posterior "heel" (**talonid**) against which the upper tooth can rest. The trigonid is arranged opposite that of the upper tooth: the **protoconid** is labial, the **paraconid** is anterior and lingual, and the **metaconid** is posterior and lingual. The talonid extends posteriorly from the crest connecting the protoconid and metaconid; it typically bears two to three principal cusps, the **entoconid** (lingual edge), **hypoconid** (labial edge), and sometimes, a **hypoconulid** (posterior edge). Structurally, the trigonid is elevated substantially above the talonid, and the **talonid basin**, created by these elevated peripheral cusps, receives the protocone of the opposing upper molar during mastication; together, this combination of structural features produces both shearing action (as the trigon and trigonid slice by each other) and a mortar-and-pestle-like crushing action (as the trigon settles into the talonid basin) that is capable of processing a wide variety of foods. Lower teeth may have a **cingulid** as well.

Many mammals developed more complex dental topography, and one key evolutionary feature (Hunter and Jernvall 1995) was the addition of a fourth cusp, the

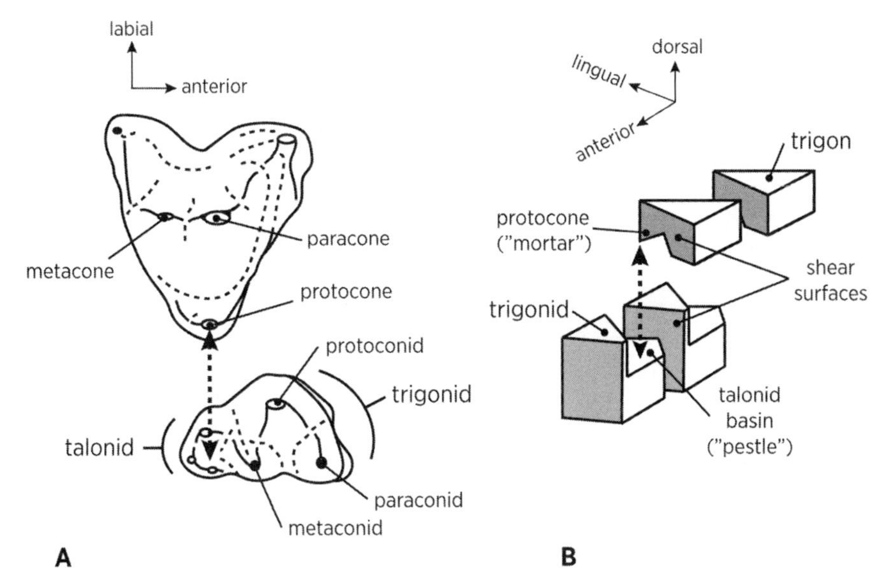

FIG. 13. Tribosphenic molars. *A*, depiction of the upper and lower early **tribosphenic molar** pair of the Cretaceous boreosphenidan genus *Kielantherium*, with the major cusps and structures labeled; these same names apply to the wide range of derived molariform teeth of living mammals. *B*, schematic of the upper-lower occlusion of adjacent tribosphenic molar pairs, showing the dual functions of slicing, achieved by the paired vertical shearing surfaces formed by connections between the principal cusps of the trigon and trigonid, and crushing, achieved by the protocone of the upper tooth and the talonid basin of the opposing lower tooth, during mastication.

hypocone (and associated **hypoconid**), which lies at the posterolingual side of the tooth (e.g., see description of quadritubercular occlusal pattern below).

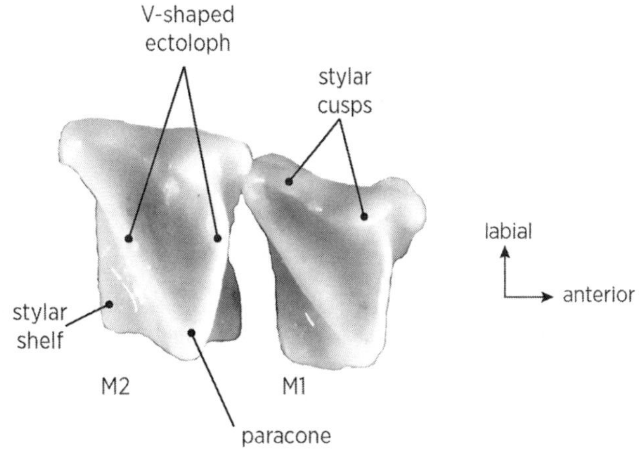

FIG. 14. Zalambdodont right upper first and second molars of the greater hedgehog tenrec, *Setifer setosus* (Afrotheria: Tenrecidae), with the V-shaped ectoloph, major cusps, and structures identified.

TYPES OF MOLAR OCCLUSAL PATTERNS

1. **zalambdodont** pattern (fig. 14; basic pattern of the marsupial mole [Notoryctidae], Tenrecoidea insectivores [Tenrecidae and Chrysochloridae], and Lipotyphla insectivores [Solenodontidae]). These teeth have a distinct chevron- or V-shaped crest (**ectoloph**). The apex of the lingual ridge is the paracone (or possibly a fused paracone and metacone; DO NOT confuse this cusp with the protocone!). A ridge with stylar cusps borders the lateral edge of the tooth. The protocone is degenerate (e.g., Solenodontidae) or nonexistent (e.g., Tenrecidae); if present, it occurs on a low-lying stylar shelf on the labial side of the tooth.
2. **dilambdodont** pattern (fig. 15; basic pattern of most Lipotyphla insectivores [e.g., Soricidae, Talpidae], many bats, dermopterans, and tupaiids). These teeth are characterized by a W-shaped ectoloph, the base of which includes the anterior paracone and posterior metacone. Three stylar cusps include (anterior to posterior) the parastyle, mesostyle, and metastyle. The protocone forms a lingual chevron when the hypocone is not developed or is only weakly developed.
3. **quadritubercular** (figs. 16, 17), a pattern with four primary cusps; often called **euthemorphic**, and if squarish in shape may be called **quadrate**; most

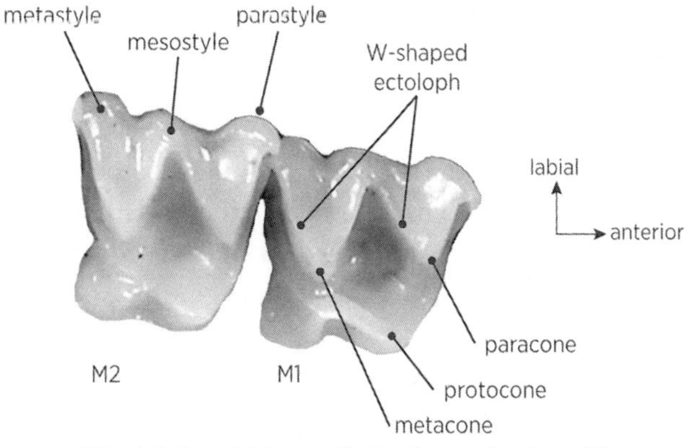

FIG. 15. Dilambdodont right upper first and second molars of the greater bonneted bat, *Eumops perotis* (Chiroptera: Molossidae), with the W-shaped ectoloph, labial styles, and major cusps labeled.

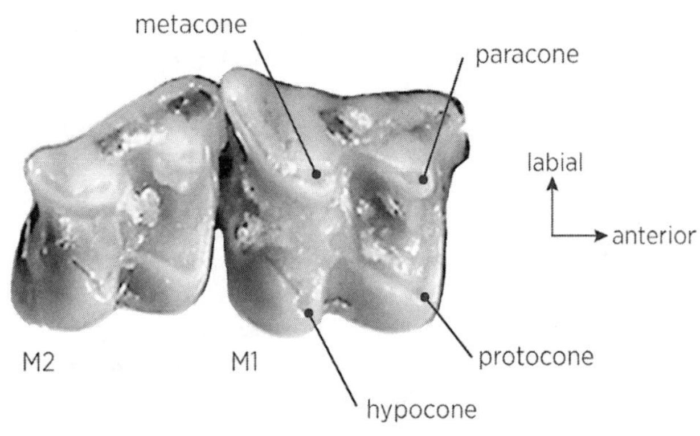

FIG. 16. Quadrate right upper first and second molars of Brandt's hedgehog, *Paraechinus hypomelas* (Lipotyphla: Erinaceidae), with principal cusps identified. Note the presence of the posterolingual hypocone, the addition of which gives the crown its quadrate shape.

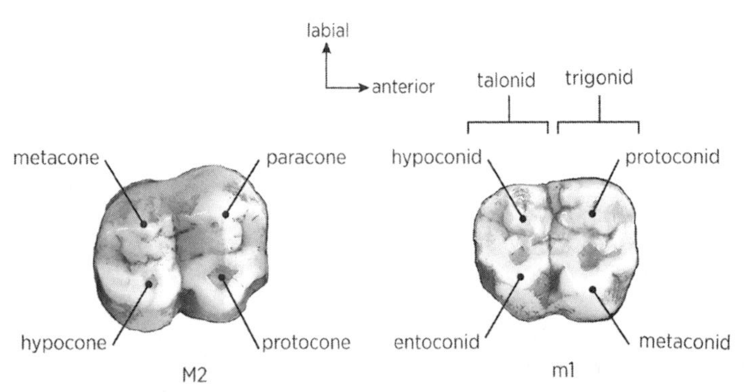

FIG. 17. Quadrate upper second molar (M2; *left*) and lower first molar (m1; *right*) of the hamadryas baboon, *Papio hamadryas* (Primates: Cercopithecidae), with the four major cusps of both teeth labeled. Note the size equivalence of the trigonid and talonid cusps.

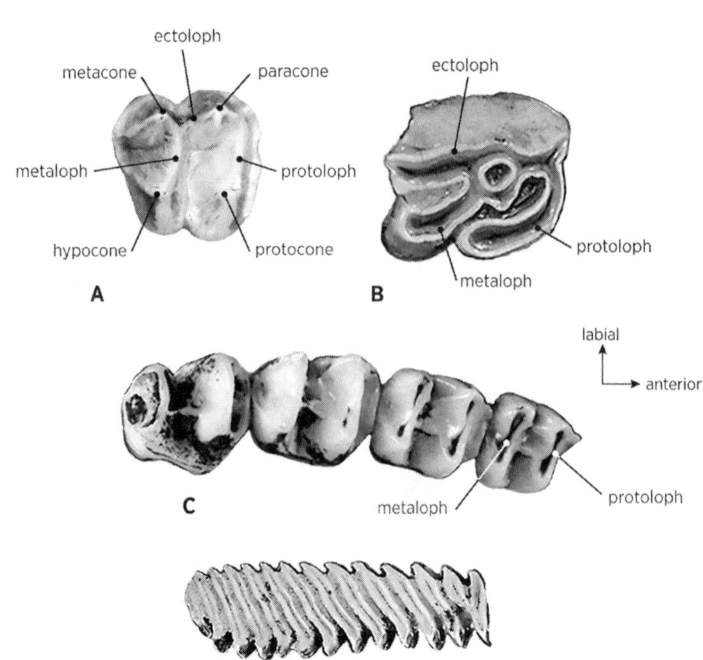

FIG. 18. Lophodont molars. *A–B*, the π-shaped lophodont occlusal pattern of Perissodactyla. *A*, upper right PM4 of a lowland tapir, *Tapirus terrestris* (Perissodactyla: Tapiridae); *B*, M1 of a white rhinoceros, *Ceratotherium simus* (Perissodactyla: Rhinocerotidae). *C*, the transverse bilophodont pattern of the upper right molar series (M1–M4) of the dusky pademelon, *Thylogale brunii* (Diprotodontia: Macropodiformes: Macropodidae). *D*, multiple transverse lophs of the loxodont upper right third molar (M3) of the capybara, *Hydrochoerus hydrochaeris* (Rodentia: Caviomorpha: Caviidae).

living mammals have some modification of this pattern. These teeth are formed by the addition of a fourth cusp (the hypocone) on the posterolingual corner of the upper tooth, which results in a more or less square surface.

Quadritubercular lower molars may also become squared (quadrate) with the evolutionary loss of the paraconid of the trigonid and expansion of the talonid, which becomes elevated to the level of the trigonid, and where the entoconid and hypoconid (and hypoconulid if present) are equal in size to the protoconid and metaconid of the trigonid.

4. **lophodont** pattern (fig. 18; basic pattern of most herbivores). The cusps are joined into low ridges, called **lophs**; these may be oriented anteroposteriorly (e.g., joining the paracone and metacone into an ectoloph), transversely (e.g., paracone and protocone join to form the **protoloph** [sometimes called the **anteroloph**] and metacone and hypocone into the **metaloph** [or **posteroloph**]), or a combination of the two directions (termed **π-shaped**, as in Perissodactyla). **Transverse bilophodonty** (two lophs) characterizes several marsupials (e.g., wombats, kangaroos, and wallabies), lagomorphs, many rodents, and some primates. In some mammals, a series of transverse lophs gives the appearance of a washboard; such **loxodonty** is found in modern elephants (hence the generic name of the African elephant, *Loxodonta*) and some grazing rodents. Loph orientation is indicative of the major axis of jaw movement during mastication (e.g., the jaw slides fore and aft in mammals with transversely oriented lophs or from side to side in those with anteroposteriorly directed lophs). In specialized grazers, loph surfaces wear quickly to expose the underlying softer dentine, thus forming a series of harder enamel ridges that function to shear tough plant materials during mastication.
5. **selenodont** pattern (fig. 19; characteristic of a few marsupials [e.g., Phalangeridae], hyraxes

[Procaviidae], and especially Tylopoda [Camelidae] and Ruminantia [Tragulidae, Giraffidae, Antilocapridae, Moschidae, Cervidae, and Bovidae] within Artiodactyla). These teeth exhibit an anteroposterior expansion of the four individual cusps that characterize the quadritubercular molar into crescent-shaped ridges, called **selenes**. Early wear on each selene exposes the underlying softer dentine, which produces anteroposterior ridges of hard enamel that act like rasps on a file to grind abrasive foods. Because jaw action during mastication is generally perpendicular to the axis of these enamel ridges, most mammals with selenodont teeth chew with side-to-side motion.

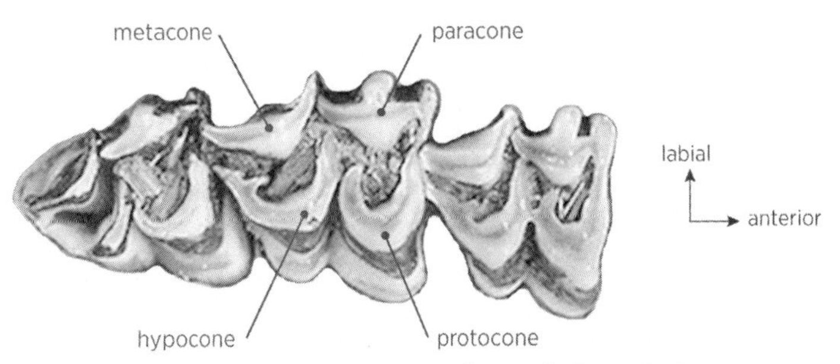

FIG. 19. Selenodont upper right molar series (M1–M3) of the Siberian musk deer, *Moschus moschiferus* (Artiodactyla: Ruminantia: Moschidae), with the four principal cusps labeled.

SPECIALIZED MOLAR CUSPS

1. **bunodont** teeth (e.g., fig. 20) have low, rather rounded cusps that serve to crush food items largely by vertical action of the jaws, and thus characterize generalized omnivores, such as primates, several carnivorans (especially raccoons and bears), and a few artiodactyls (such as pigs, peccaries, and hippopotamuses). This type of cusp is often, but not always, found on quadritubercular molars.
2. **secodont** teeth (fig. 21) are bladelike and designed for cutting or slicing. One specific type of secodont teeth is **carnassial** teeth, characteristic of many Carnivora, especially Canidae, Mustelidae, Felidae, Hyaenidae, Viverridae, and Herpestidae, as well as some Procyonidae. In all living Carnivora, the **carnassial pair** is the fourth upper premolar (PM4) and first lower molar (m1). These teeth represent extreme modification for shearing by anteroposterior expansion of the paracone and metacone, reduction of the protocone, and addition of the parastyle in some genera (e.g., Felidae, Hyaenidae) to lengthen the slicing blade of PM4, as well as expansion of the paraconid and protoconid with reduction to loss of the metaconid in the opposing lower molar. In the highly specialized Felidae, the talonid is completely lost. Note that in shortening the rostrum for greater biting force, many felids have evolutionarily lost PM1, so that in extant species the third upper PM is homologous to PM4.

Another general term for bladelike teeth is **plagiaulacoid** (fig. 22), but its use is usually restricted to the grooved bladelike teeth found among some living marsupials (e.g., order Diprotodontia).

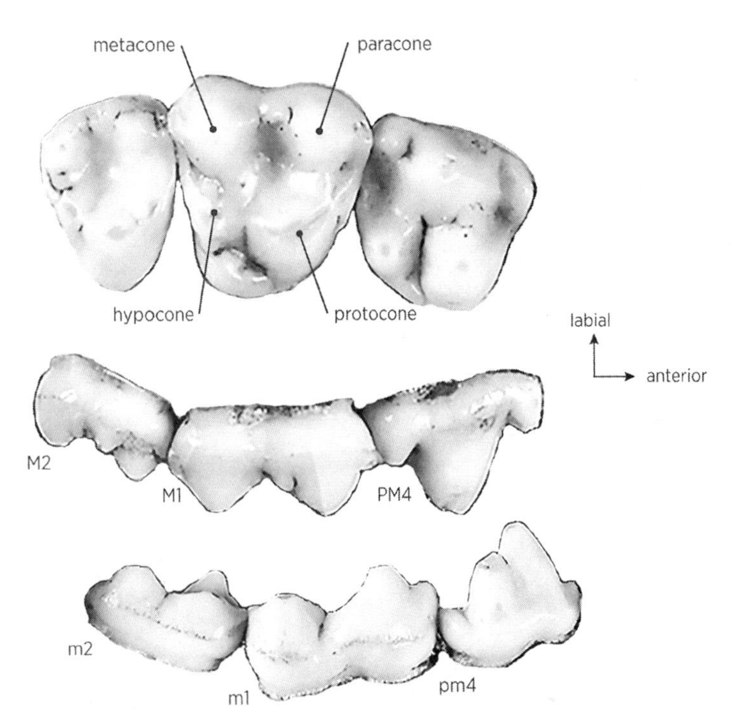

FIG. 20. Occlusal (*top*) and lateral (*middle*) views of the bunodont upper right PM4–M2 and lateral view (*bottom*) of the lower right pm4–m2 of the raccoon, *Procyon lotor* (Carnivora: Procyonidae).

CROWN HEIGHT AND ROOT DEVELOPMENT

The specialized occlusal surfaces of mammalian cheek teeth (molars and premolars), which are generally designed for the hardness and abrasiveness of the types of foods eaten, have coevolved with the height of the crown itself as well as with the size and length of the roots that anchor each tooth in bony alveoli in the upper and lower jaws. Low-crowned, or **brachydont**, teeth (figs. 23, 24) have well-developed and typically elongate roots, with each root often occupying a separate alveolus. Teeth with zalambdodont, dilambdodont, and quadritubercular occlusal surfaces are well anchored in the jaw and are designed to withstand the forces generated by the vertical jaw

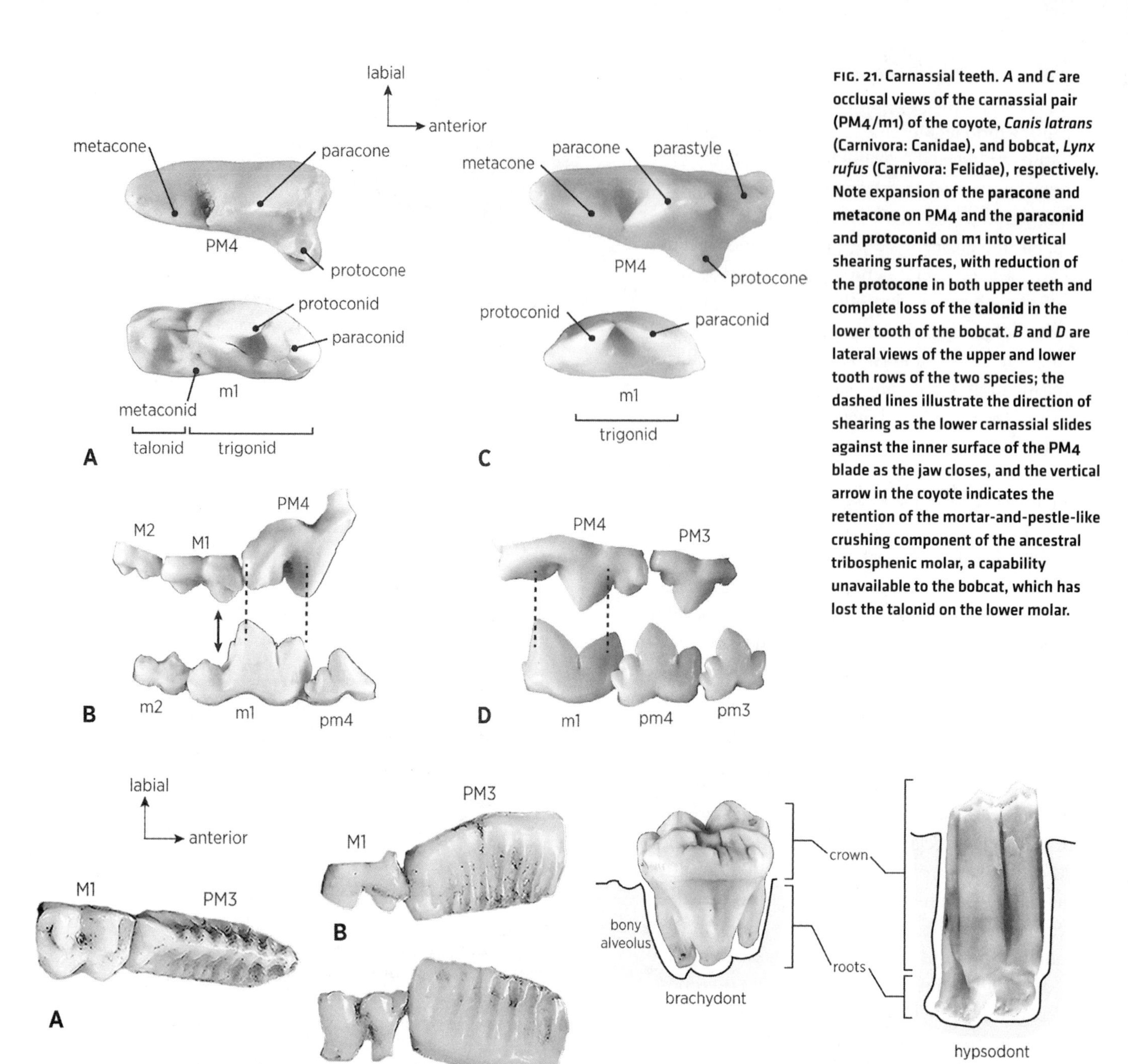

Fig. 21. Carnassial teeth. *A* and *C* are occlusal views of the carnassial pair (PM4/m1) of the coyote, ***Canis latrans*** (Carnivora: Canidae), and bobcat, ***Lynx rufus*** (Carnivora: Felidae), respectively. Note expansion of the **paracone** and **metacone** on PM4 and the **paraconid** and **protoconid** on m1 into vertical shearing surfaces, with reduction of the **protocone** in both upper teeth and complete loss of the **talonid** in the lower tooth of the bobcat. *B* and *D* are lateral views of the upper and lower tooth rows of the two species; the dashed lines illustrate the direction of shearing as the lower carnassial slides against the inner surface of the PM4 blade as the jaw closes, and the vertical arrow in the coyote indicates the retention of the mortar-and-pestle-like crushing component of the ancestral tribosphenic molar, a capability unavailable to the bobcat, which has lost the talonid on the lower molar.

Fig. 22. The third upper and lower **plagiaulacoid** premolars of the rufous rat-kangaroo, *Aepyprymnus rufescens* (Macropodiformes: Potoroidae). *A*, occlusal view of PM3; *B*, lateral view of PM3; *C*, lateral view of pm3.

Fig. 23. Contrast between labial views of a **brachydont** molar (*left*) of a bear (Carnivora: Ursidae), with its short crown situated entirely above the bony alveolus and its elongate and massive roots sunk in bony sockets, and a **hypsodont** molar (*right*) of a sheep (Artiodactyla: Bovidae), in which the elongate crown extends deeply into the alveolus and the roots, when they form, are very short.

action of crushing or the specialized slicing of food by carnassial teeth. In contrast, herbivorous mammals, with lophodont or selenodont cheek teeth designed for grinding plant materials, typically have high-crowned, or **hypsodont**, teeth. The latter come in two categories. In **euhypsodont** teeth (figs. 24, 25), a substantial portion of the crown extends down into the alveolus, and the tooth continues to grow as the crown surface itself is worn by feeding on abrasive plant materials. These teeth may never form roots, or do so only late in life. They contrast with **coronal hypsodont** teeth, in which the crown is high, but extends largely above the alveolus, and has well-developed roots. Note, however, that many authors refer to *any* high-crowned tooth as

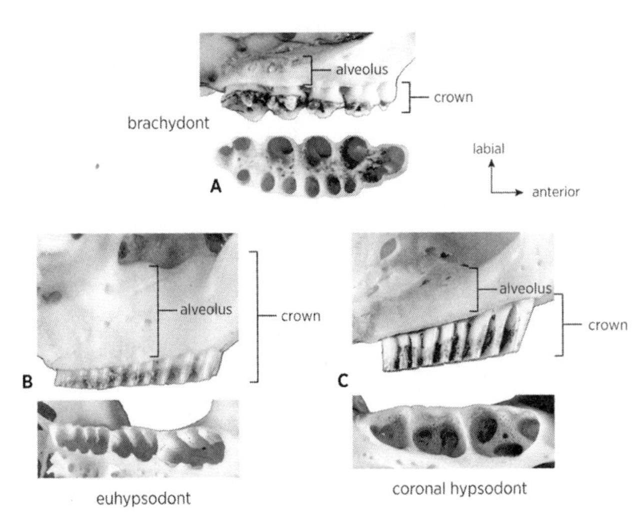

FIG. 24. Relationship between crown height and both the depth of the bony alveolus and evidence of roots in three species of rodents. *A*, a brachydont squirrel (Sciuromorpha: Sciuridae: *Sciurus*), with short crowns positioned above the bone and a shallow maxillary alveolus with multiple individual root sockets. *B*, a euhypsodont arvicoline (Myomorpha: Cricetidae: Arvicolinae: *Microtus*), with an exceedingly deep maxilla to accommodate the ever-growing cheek teeth and lacking root sockets in the alveoli. *C*, a coronal hypsodont cricetid (Myomorpha: Cricetidae: Neotominae: *Neotoma*), with the crown largely above the alveolus, a relatively shallow maxilla, and sockets for individual roots, which are especially evident in the alveolus of M1.

hypsodont, including rodent incisors, elephant tusks, carnivore carnassials, and so on. Von Koenigswald (2011) provides an excellent review of the diverse terminology applied to high-crowned and ever-growing teeth.

Perhaps unsurprisingly, the elongate crowns of hypsodont teeth are reflected by the very deep maxillae and mandibular rami that bear the cheek teeth, in comparison to the distinctly more shallow elements housing brachydont teeth.

For specialized grazers (e.g., selenodont artiodactyls and lophodont horses, elephants, and many rodents), the inter-selene or inter-loph areas of the tooth fold deeply into the crown (see fig. 11). This folding, in combination with rapid early wear to expose ridges of enamel on the occlusal surface and continual growth of the tooth, ensures that these ridges remain exposed on the occlusal surface essentially throughout life. The entire array of cheek teeth forms a continuous grinding platform, with minimal (or no) differentiation of premolars and molars.

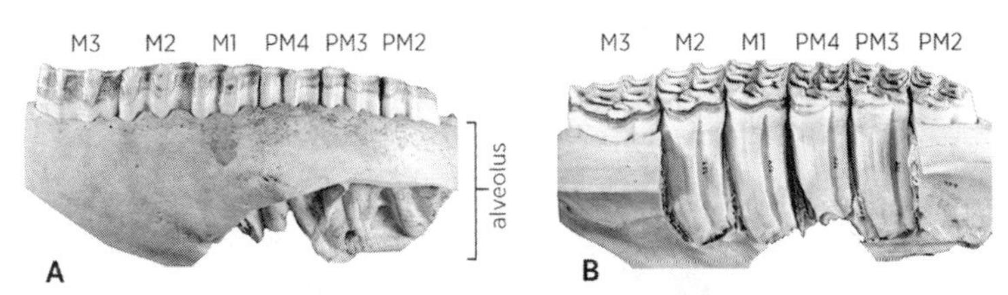

FIG. 25. Labial (*A*) and lingual (*B*) cutaway views of the upper lophodont premolar-molar series of a horse, *Equus* (Perissodactyla: Equidae). The deep alveoli accommodating the extremely elongate crowns of each tooth are readily apparent, as is the lack of, or only shallow, roots in *B*.

INCISOR PROCUMBENCY

Most dental terminology focuses on the cheek teeth. One important exception is the orientation of the incisor teeth, which varies as a function of use and is especially important in the descriptions of many rodent species. Specifically, **procumbency** refers to the position of the cutting edge of the upper incisor relative to the vertical plane of the tooth (fig. 26). In **proodont** incisors, the cutting edge extends anterior to the vertical plane, whereas **orthodont** incisors present the cutting edge perpendicular to the plane. In **opisthodont** incisors, the cutting edge lies posterior to the plane.

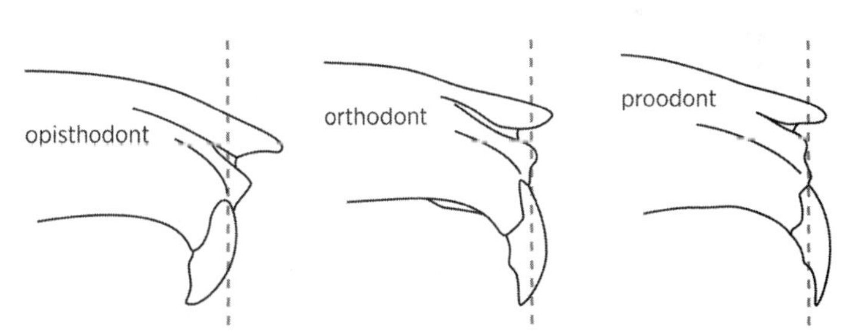

FIG. 26. Illustrations of opisthodont, orthodont, and proodont upper incisors of rodents. Redrawn from Hershkovitz (1962).

CLASSIFICATION OF LIVING MAMMALS

Substantive changes in the classification of mammals at higher categorical levels have occurred over the past half century due to two primary factors:

1. The philosophy of classification has shifted from one in which categorical rank may be based on both hypothesized phylogenetic relationships and recognition of major differences in lifestyle, to one that emphasizes phylogeny solely and thus reflects the splitting events in a phylogeny. The classification of G. G. Simpson (1945) is the best example of the former style of classification, whereas that of McKenna and Bell (1997) reflects the latter; both adhere to the classical Linnaean categorical rank system.
2. Hypotheses of relationships among major groups of mammals have also changed substantially, again for two basic reasons: First, cladistic principles are now universally used to distinguish between shared-primitive (= symplesiomorphic) and shared-derived (= synapomorphic) morphological traits, and the establishment of phyletic propinquity is based solely on the latter. Second, the rapid acquisition of molecular data, primarily DNA sequences from both mitochondrial and nuclear genes, has allowed for more comprehensive assessments of relationships. As will be evident in the accounts that follow, molecular phylogenies have often proved to be radically different from "traditional" views, and thus many hypotheses of phylogenetic relationships among groups of mammals remain hotly debated. As molecular data accumulate, in both numbers of genes and taxa sampled, the classification of mammals is likely to change even more in the coming decade. Thus, do not be surprised if you see new articles about mammalian relationships that differ from the hypotheses that we depict here.

The classification used in earlier mammalogical textbooks followed primarily the Linnaean tradition employed by Simpson (1945), although the most recent texts (Feldhamer et al. 2015; Vaughan et al. 2015) update groupings according to advances in our knowledge of phylogenetic relationships. For the most part, this manual follows the taxonomy of Wilson and Reeder (2005), with modifications based on more recent molecular and morphological phylogenetic studies. The classification we use herein includes the traditional Linnaean categorical ranks (class, order, suborder, etc.), but also employs names that are commonly used to denote clades without designation of a specific Linnaean hierarchical rank (Afrotheria, Boreoeutheria, Laurasiatheria, etc.). We also include alternative names for a hierarchical rank that are commonly found in the literature (e.g., infraclass Marsupialia for Metatheria).

We emphasize that both the data used to estimate relationships and the philosophy of classification are very active areas of research, and that fundamental changes in our perceptions of the major groups of mammals, how these groups are related, and thus both the evolutionary transformations of major mammalian characters and the resulting classifications, will continue to change in the near future. Gatesy et al. (2017), for example, illuminate the areas of constancy across the mammalian tree of life and highlight those nodes that remain in conflict due to different character sets, different methods of analysis, or insufficient data.

We use the following classification for the hierarchical organization of living mammals in the accounts that follow:

- Class Mammalia
  - Subclass Prototheria
    - Order Monotremata (platypus, echidnas)
  - Subclass Theria
    - Infraclass Metatheria (= Marsupialia)
      - Order Didelphimorphia
      - Order Paucituberculata
      - Order Microbiotheria
      - Order Notoryctemorphia
      - Order Dasyuromorphia
      - Order Peramelemorphia (= Peramelina)
      - Order Diprotodontia
        - Suborder Vombatiformes
        - Suborder Phalangeriformes
          - Superfamily Phalangeroidea
          - Superfamily Petauroidea
        - Suborder Macropodiformes
    - Infraclass Eutheria or Placentalia
      - Clade Atlantogenata
        - Clade Xenarthra
          - Order Cingulata
          - Order Pilosa
            - Suborder Folivora
            - Suborder Vermilingua
        - Clade Afrotheria
          - Clade Afroinsectiphilia
            - Order Tubulidentata
            - Clade Afroinsectivora
              - Order Macroscelidea
              - Order Tenrecoidea
                - Suborder Chrysochloridea
                - Suborder Tenrecomorpha
          - Clade Paenungulata (= Subungulata)
            - Order Hyracoidea
            - Clade Tethytheria
              - Order Proboscidea
              - Order Sirenia
      - Clade Boreoeutheria
        - Clade Euarchontoglires
          - Clade Glires
            - Order Lagomorpha
            - Order Rodentia
              - Suborder Sciuromorpha
              - Suborder Castorimorpha
              - Suborder Myomorpha (= Myodonta)
                - Superfamily Dipodoidea
                - Superfamily Muroidea
              - Suborder Anomaluromorpha
              - Suborder Hystricomorpha (= Ctenohystrica)

Infraorder Ctenodactylomorphi
Infraorder Hystricognathi
Clade Caviomorpha
Superfamily Erethizontoidea
Superfamily Chinchilloidea
Superfamily Cavioidea
Superfamily Octodontoidea
Clade Euarchonta
Order Primates
Suborder Strepsirrhini
Infraorder Lemuriformes
Superfamily Cheirogaleoidea
Superfamily Lemuroidea
Infraorder Chiromyiformes
Infraorder Lorisiformes
Suborder Haplorrhini
Infraorder Tarsiiformes
Infraorder Simiiformes (= Anthropoidea)
Parvorder Platyrrhini
Parvorder Catarrhini
Superfamily Cercopithecoidea
Superfamily Hominoidea
Clade Sundatheria
Order Dermoptera
Order Scandentia
Clade Laurasiatheria
Clade Lipotyphla (= Eulipotyphla)
Order Erinaceomorpha
Order Soricomorpha
Clade Scrotifera
Order Chiroptera
Clade Yinpterochiroptera (= Pteropodiformes)
Superfamily Rhinolophoidea
Clade Yangochiroptera (= Vespertilioniformes)
Superfamily Emballonuroidea
Superfamily Noctilionoidea
Superfamily Vespertilionoidea
Clade Ferae
Order Pholidota
Order Carnivora
Suborder Feliformia
Suborder Caniformia
Superfamily Canoidea
Superfamily Arctoidea
Clade Musteloidea
Clade Pinnipedia
Clade Euungulata
Order Perissodactyla
Superorder Cetartiodactyla

Order Artiodactyla
- Suborder Suina
- Suborder Whippomorpha
  - Infraorder Ancodonta
- Suborder Tylopoda
- Suborder Ruminantia
  - Infraorder Tragulina
  - Infraorder Pecora
    - Superfamily Cervoidea
    - Superfamily Giraffoidea
    - Superfamily Bovoidea

Clade Cetacea (= Cete)
- Subclade Mysticeti
- Subclade Odontoceti

Class MAMMALIA

Following the seminal paper by Rowe (1988), the class Mammalia is now **defined** as the crown group that contains the most recent common ancestor of living mammals (monotremes, marsupials, and eutherians) as well as all descendants that are now extinct (such as the eutriconodonts, symmetrodonts, and multituberculates; fig. 27). Mammalia is nested within the clade Mammaliaformes, which arose and diversified during the Triassic, and which is a subset of Cynodontia ("mammal-like reptiles"), which in turn is a subset of Synapsida, one of the principal branches of terrestrial vertebrates that arose in the late Paleozoic.

With this as a definition, the class Mammalia can be **diagnosed** by the following set of anatomical characters. Many of these characters involve features of the jaw and ear bones; these elements are identified, and their evolutionary shift from primitive synapsids to mammals described, in figure 28.

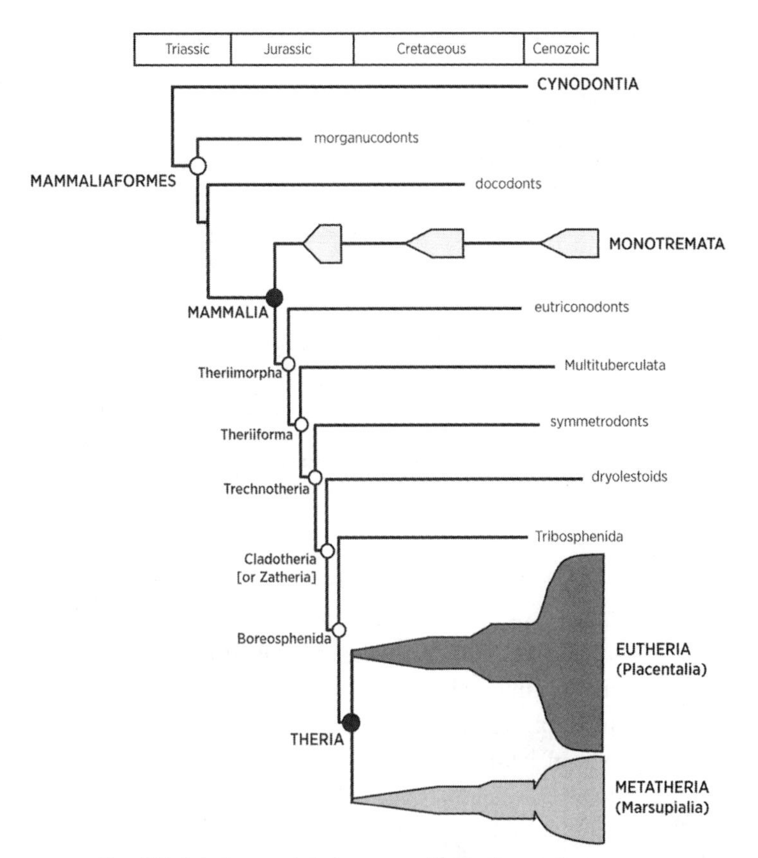

FIG. 27. Simplified cladogram and chronogram illustrating major lineages of mammals and their immediate ancestors from the basal Cynodontia. Each successive node on the tree unites groups by their exclusive shared-derived characters (= synapomorphies). Relative diversity of the three modern mammalian groups from their origin in the early Cenozoic is illustrated (see Kielan-Jaworowska et al. 2004; Luo 2007, 2011; Luo et al. 2011; Bi et al. 2014; Brusatte and Luo 2016; redrawn from Williamson et al. 2014). Note the deep temporal origin (Middle Jurassic) of all three modern mammalian lineages (crown group Mammalia) and the rapid diversification of both Eutheria and Metatheria at the beginning of the Cenozoic.

SHARED-DERIVED CHARACTERS:

- accessory jaw bones shifted away from the cranio-mandibular joint (CMJ) in adults to become associated with the cranium alone (these bones include the middle ear bones [malleus = reptilian articular; incus = reptilian quadrate; stapes = reptilian columellar auris] as well as other bones formerly associated with the jaw [mammalian tympanic or ectotympanic = reptilian angular])
- stapes very small relative to skull size
- atlas intercentrum and neural arches fused to form single, ring-shaped osseous structure
- epiphyses on the long bones and girdles
- heart completely divided into four chambers with a thick, compact myocardium (muscular wall)
- heart with atrioventricular node ("pacemaker") and Purkinje fibers
- single aortic trunk
- pulmonary artery with three semilunar valves
- erythrocytes lack nuclei at maturity
- endothermy
- central nervous system covered by three meninges
- cerebellum folded
- nerve filaments with three polypeptides
- brain with divided optic lobes
- strong representation of the facial nerve field in the motor cortex of the brain
- superficial musculature expanded onto the face and differentiated into muscle groups associated with the eye, ear, and snout

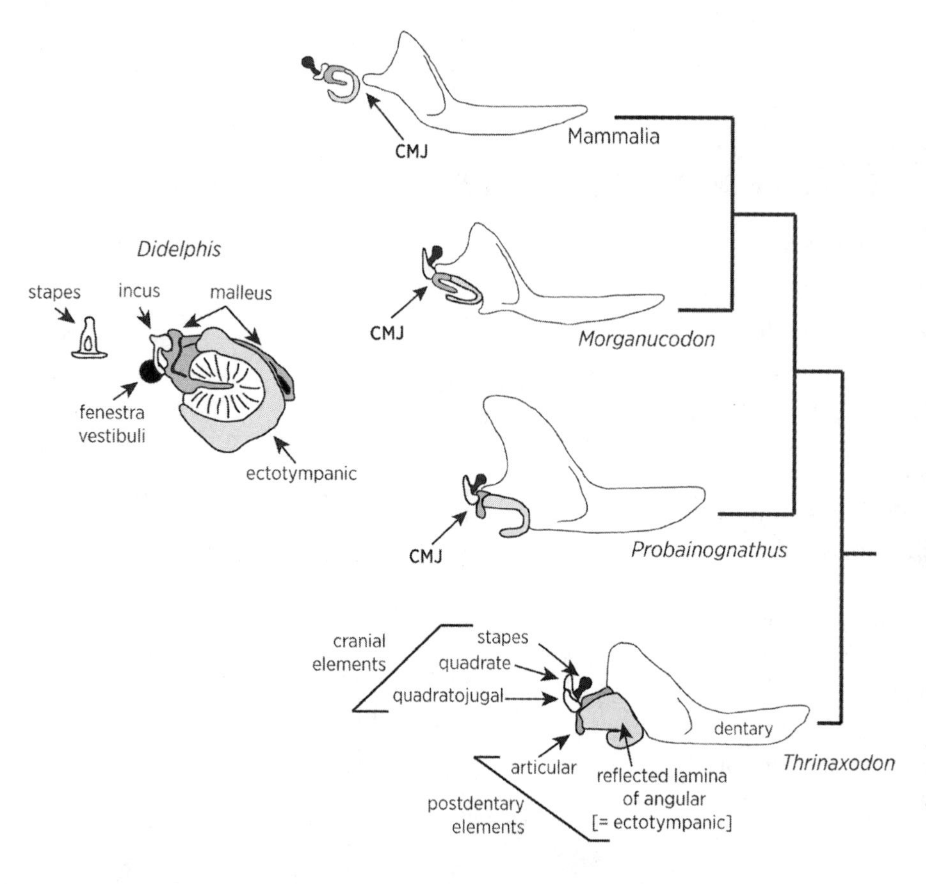

FIG. 28. Major steps in the evolution of the mammalian jaw joint and ear region. Homologous bones are shaded similarly from one taxon to the next. View is of lateral side. Dentary bones are shown diagrammatically, without teeth. Postdentary bones of the nonmammalian cynodont (e.g., *Thrinaxodon*) became modified as middle ear bones of Mammalia (reflected lamina of the angular bone = ectotympanic; articular = malleus). Cranial bones (quadrate = incus, quadratojugal, and stapes) are shown without the rest of the cranium for simplicity. The tympanic membrane is shown only in the inset of *Didelphis* (*left*). The fenestra vestibuli is the opening into the inner ear in which the footplate of the stapes is normally fitted. CMJ = cranio-mandibular joint between skull and lower jaw; note shift in CMJ from the articular-quadrate joint of the nonmammalian cynodont (*Thrinaxodon*) to the condyle of the dentary in Mammalia. For a thorough review of the evolution of mammal ears, see Luo (2011). Redrawn from Rowe (1996).

- muscular diaphragm separates pleural (lung) and abdominal cavities, with consequent development of diaphragmatic breathing
- complex lung structure, division of the lungs into lobes, bronchioles, alveoli
- epiglottis
- skin with erector muscles and dermal papillae
- hair
- sebaceous glands
- sweat glands
- mammary glands

Subclass PROTOTHERIA

Order MONOTREMATA

The order Monotremata consists of five living species in three genera, including the platypus (*Ornithorhynchus*) in the family Ornithorhynchidae and the echidnas (*Tachyglossus* and *Zaglossus*) in the family Tachyglossidae. All Monotremata are highly specialized mammals that retain many primitive mammalian or even amniote characteristics (oviparity; shell-covered eggs; mammary glands without a nipple; epipubic bones; coracoid, procoracoid [= epicoracoid], and interclavicle in the shoulder girdle; splayed limb posture; and a cloaca [the basis for the name Monotremata, which means "single hole"]). Monotremes are endothermic, but have low metabolic rates compared with eutherian mammals; all either hibernate or are periodically inactive in winter; all are long-lived, with captive echidnas living more than 50 years.

DIAGNOSTIC CHARACTERS:

- **pectoral girdle with large procoracoids, coracoids, and interclavicle** (these bones are absent from all other mammals; see fig. 29)
- **females lay eggs** (oviparous)

RECOGNITION CHARACTERS

1. limbs modified for digging or swimming
2. ankle in males with horny spur
3. no vibrissae
4. epipubic bones present
5. skull birdlike in shape, sutures usually obliterated by fusion of bones in adults
6. **no auditory bulla**
7. premaxillae separated for at least part of their length
8. jugal reduced or absent
9. **no lacrimal**
10. palate extending far posteriorly
11. **no teeth in adults** (adult platypus has horny pads only)
12. cloaca present (absent in other mammals, with few exceptions)
13. penis within cloaca, used only for passage of sperm
14. mammae without pendulous teats (nipples)

Family ORNITHORHYNCHIDAE (platypus)

"An unbeliever in everything beyond his own reason, might exclaim: 'Surely two distinct Creators must have been [at] work.'" [C. Darwin, January 19, 1836; in Keynes 2001]

Medium-sized (body length 300–450 mm; mass up to 2 kg); body specialized for aquatic and semifossorial life, streamlined, relatively slender; tail flattened dorsoventrally, about one-third length of body; external pinnae lacking; feet broad, webbed, digits 5/5, each with strong claws for digging; web on manus extends well beyond claw tips, but not to base of claws on pes; spur at base of hind foot in both sexes when young, disappears in adult females; spur grooved for passage

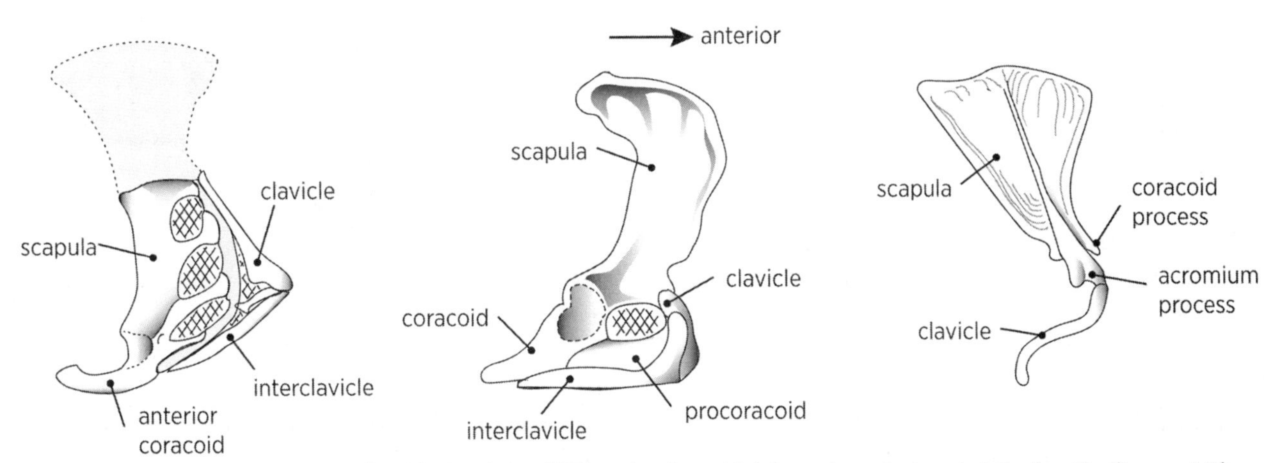

FIG. 29. The monotreme shoulder girdle (*Ornithorhynchus*; *middle*) retains the multiple bony elements characteristic of reptiles (*Iguana*; *left*); the eutherian shoulder girdle (*Erinaceus*; *right*) is simplified, with only the scapula and clavicle present (the other elements, such as the coracoid process, are fused to the scapula). Redrawn from Vaughan et al. (2015).

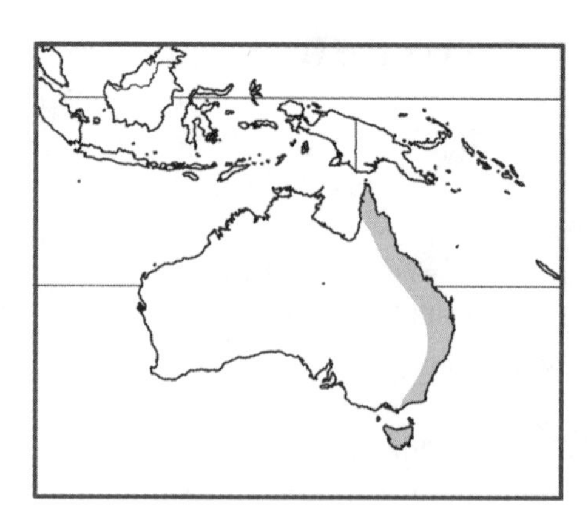

FIG. 31. Geographic range of family Ornithorhynchidae: eastern Australia, Tasmania.

FIG. 30. Ornithorhynchidae; platypus, *Ornithorhynchus anatinus*.

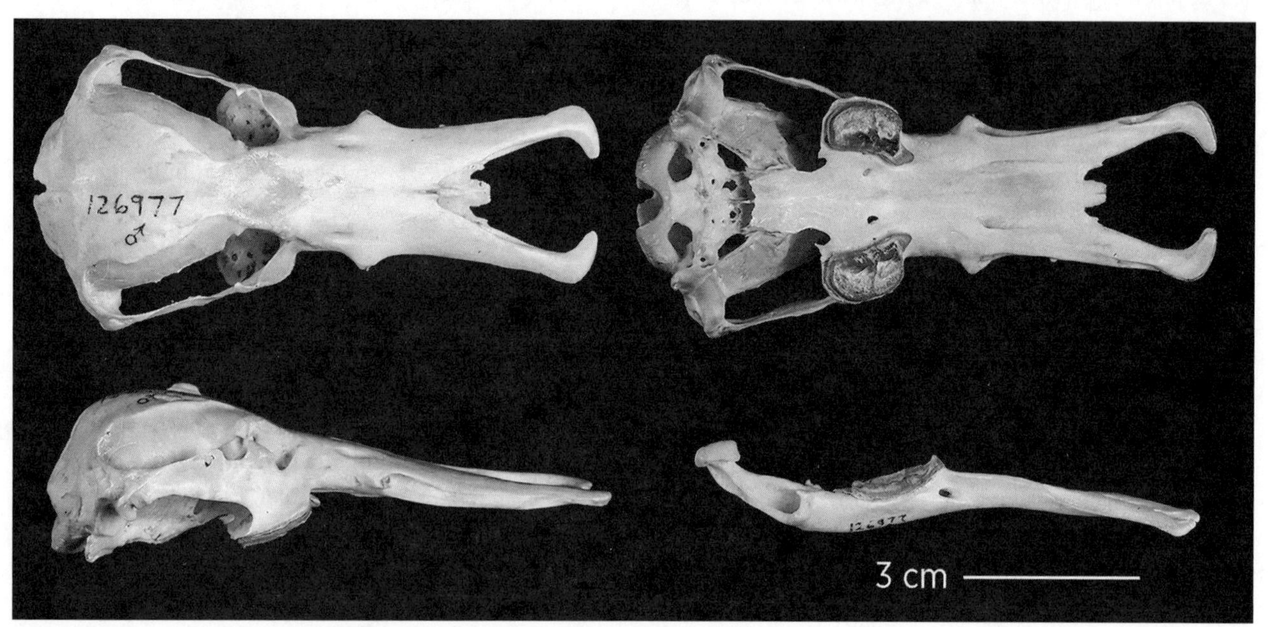

FIG. 32. Ornithorhynchidae; skull and mandible of the platypus, *Ornithorhynchus anatinus*.

of poisonous glandular secretion; pelage soft, dense, with woolly underfur. Forage mainly for aquatic invertebrates on stream bottom; unique "bill" houses electromechanical sensory apparatus allowing prey detection in murky waters, with edible items apparently located by touch; prey stored in internal cheek pouches while submerged. Nest in self-dug burrows in stream or pond banks. Females lack a pouch; incubate eggs with body. Live in freshwater, from clear mountain streams to muddy, sluggish coastal streams, ponds, and lakes.

RECOGNITION CHARACTERS:

1. **pelage of soft hairs, no spines**
2. **feet webbed, with moderately large claws**
3. no distinct preening digit on hind foot
4. **snout broad, "duck-billed"**
5. **no pinna**
6. **tail well developed, flattened**
7. **tongue flattened**
8. cranium elongate, relatively broad, not narrow anteriorly
9. **premaxillae expanded laterally and separate anteriorly**
10. lower jaw moderate in size

DENTAL FORMULA: No teeth (embryonic teeth present; only horny plates present on jaws of adults).

TAXONOMIC DIVERSITY: The family is monotypic, with a single species, *Ornithorhynchus anatinus* (platypus).

Family TACHYGLOSSIDAE (echidnas)

Medium-sized (body length 350–775 mm; mass 2.5–10 kg); body robust, muscular; limbs subequal, powerful, adapted for digging; tail short; feet broad, with well-developed, curved, spatulate claws (claws on all five digits in *Tachyglossus*, restricted to middle three digits in *Zaglossus*); eyes small; pelage of coarse hair and hollow, thin-walled, sturdy, barbless spines on dorsal and lateral surfaces reaching up to 60 mm (more numerous and well developed in *Tachyglossus*); snout elongate and narrow, about half to two-thirds length of head. Females develop incubation patch in breeding season. Generally solitary, primarily crepuscular or nocturnal; terrestrial, semifossorial, powerful and rapid diggers; insectivorous. Habitat includes open forests, scrublands, semidesert sand plains, and tropical montane forests.

RECOGNITION CHARACTERS:

1. **body covered with thickened, stiff spines**
2. **feet not webbed, modified for digging, claws large**
3. 2nd digit of hind foot elongate, modified for preening
4. **snout slender, long**
5. **pinna well developed**
6. **tail very small**
7. **tongue thin, elongate (wormlike)**
8. cranium elongate, slender anteriorly
9. **premaxillae separated except at anterior ends**
10. lower jaw slender, rodlike

DENTAL FORMULA: No teeth

TAXONOMIC DIVERSITY: Four species in two genera, *Tachyglossus* (short-beaked echidna; eastern New Guinea and Australia) and *Zaglossus* (long-beaked echidnas; interior Papua New Guinea and Irian Jaya).

FIG. 33. Tachyglossidae; short-beaked echidna, *Tachyglossus aculeatus*.

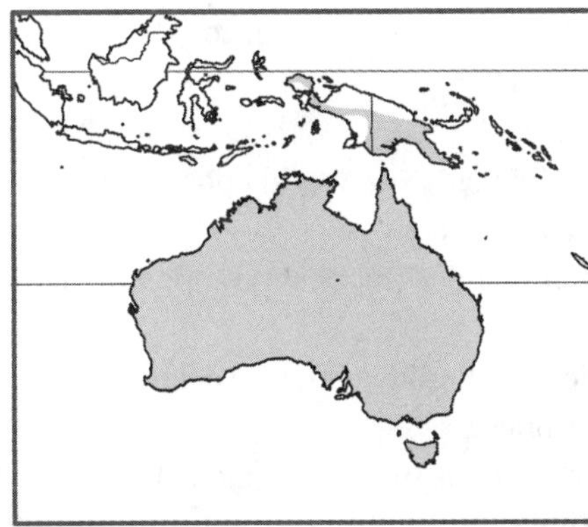

FIG. 34. Geographic range of family Tachyglossidae: Australasia (Australia, including Tasmania, and New Guinea).

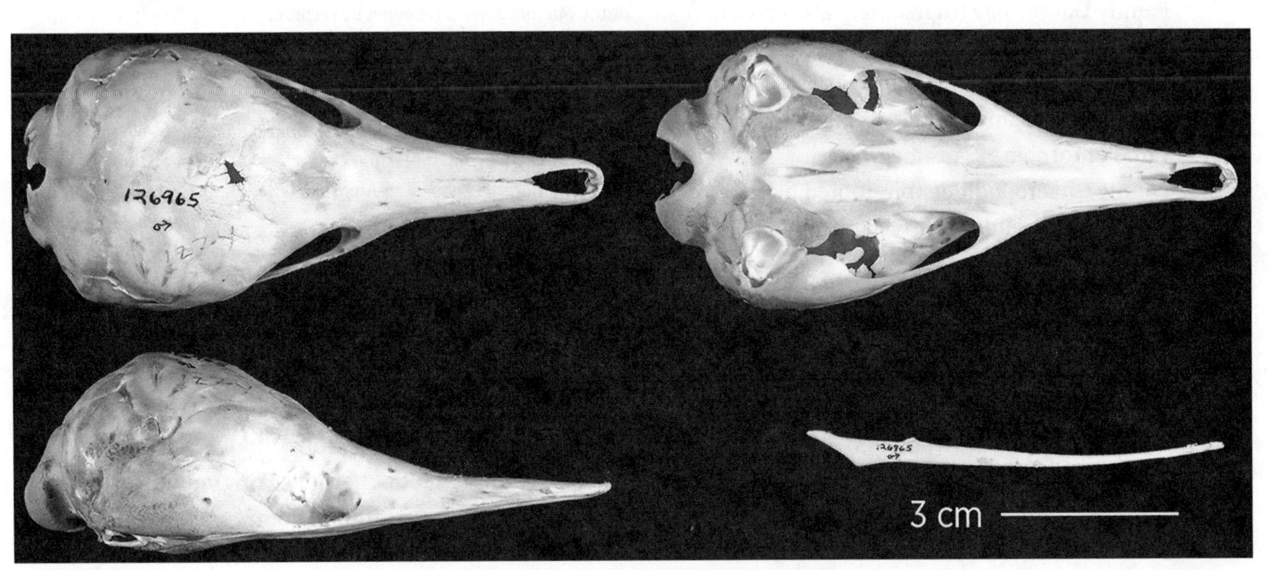

FIG. 35. Tachyglossidae; skull and mandible of the short-beaked echidna, *Tachyglossus aculeatus*.

Subclass THERIA

Infraclass METATHERIA (= Marsupialia)

Older classifications (e.g., Simpson 1945) placed marsupials in the single order Marsupialia, reflecting a view that all marsupials were simple variants on a common theme defined by their reproductive anatomy and life history. More recent classifications recognize seven orders (e.g., Wilson and Reeder 2005; Jackson and Groves 2015; Wilson and Mittermeier 2015) to reflect extensive diversity in lifestyles and major adaptations nearly equivalent to those seen among terrestrial eutherian orders. Some recent classifications organize metatherian orders into two superordinal groups: cohort Ameridelphia, comprising the New World Didelphimorphia + Paucituberculata, and cohort Australidelphia, comprising the New World Microbiotheria and the four Australasian orders (see fig. 36 and Marshall et al. 1990; Case et al. 2005).

Order Didelphimorphia
- Family Didelphidae (New World opossums)

Order Paucituberculata
- Family Caenolestidae (rat or shrew opossums)

Order Microbiotheria
- Family Microbiotheriidae (monito del monte)

Order Notoryctemorphia
- Family Notoryctidae (marsupial moles)

Order Dasyuromorphia
- Family Dasyuridae (marsupial carnivores and mice)
- Family Myrmecobiidae (numbat)
- †Family Thylacinidae (marsupial wolf)

Order Peramelemorphia (= Peramelina)
- Family Chaeropodidae (pig-footed bandicoots)
- Family Peramelidae (bandicoots and echymiperas)
- Family Thylacomyidae (bilbies)

Order Diprotodontia
- Suborder Vombatiformes
  - Family Phascolarctidae (koala)
  - Family Vombatidae (wombats)
- Suborder Phalangeriformes
  - Superfamily Phalangeroidea
    - Family Burramyidae (pygmy possums)
    - Family Phalangeridae (brush-tailed possums, cuscuses)
  - Superfamily Petauroidea
    - Family Acrobatidae (feathertail glider, feather-tailed possum)
    - Family Petauridae (sugar gliders, Leadbeater's possum, striped possums)
    - Family Pseudocheiridae (ring-tailed possums)
    - Family Tarsipedidae (honey possum)
- Suborder Macropodiformes
  - Family Hypsiprymnodontidae (musky rat-kangaroo)

A
Ameridelphia
Australidelphia
Didelphimorphia
Paucituberculata
Peramelemorphia
Diprotodontia
Notoryctemorphia
Dasyuromorphia
Microbiotheria

B
Australidelphia
Paucituberculata
Didelphimorphia
Microbiotheria
Diprotodontia
Peramelemorphia
Dasyuromorphia
Notoryctemorphia

C
Australidelphia
Didelphimorphia
Paucituberculata
Peramelemorphia
Microbiotheria
Notoryctemorphia
Dasyuromorphia
Diprotodontia

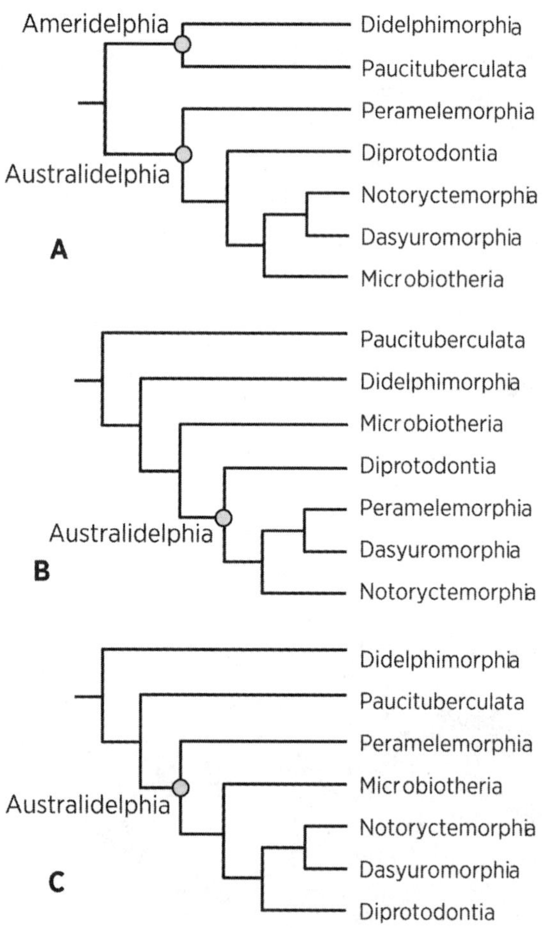

FIG. 36. Three hypotheses of relationships among metatherian orders. *A*, tree based on morphological characters (e.g., Marshall et al. 1990; Case et al. 2005). This tree posits an Australasian origin of the Microbiotheria followed by recolonization of South America, thus dividing living marsupials into two cohorts. *B*, tree based on DNA and protein sequences (Meredith et al. 2011). Relationships among New World marsupials are uncertain, but Australasian orders are monophyletic with Diprotodontia basal. *C*, tree based on morphological and molecular characters (Asher et al. 2004). Peramelemorphia is basal to Australasian radiation that includes New World Microbiotheria.

Family Macropodidae (kangaroos, wallabies, and relatives)
Family Potoroidae (bettongs, potoroos, rat-kangaroo)

The common name "opossum" is used only in reference to New World taxa; the common name "possum" refers only to Australasian taxa.

PHYLOGENETIC RELATIONSHIPS AMONG METATHERIAN ORDERS

While seven metatherian orders are now widely accepted, some interordinal relationships remain unresolved. Three slightly different phylogenetic hypotheses, which reflect morphological, molecular, and combined data, are shown in figure 36.

DIAGNOSTIC CHARACTERS OF METATHERIA (SEE O'LEARY ET AL. 2013; WILLIAMSON ET AL. 2014):

1. SOFT ANATOMY
   - **pouch (marsupium) usually present**, absent in some forms
   - vaginae paired
   - cloaca generally absent; short if present
   - penis external, forked, transmits both urine and sperm
   - testes descend through body wall to scrotum
   - cerebrum relatively small
2. DENTITION
   - postcanine teeth primitively comprise three premolars and four molars; first two premolars monophyodont but last premolar diphyodont; first molariform cheek tooth presumed to be evolutionarily derived from a premolar
   - lateral divergence of lower canines
3. SKULL
   - **braincase small**
   - **jugal forming part of mandibular fossa**
   - alisphenoid large, forming anterior part of the auditory bulla
   - **angular process of lower jaw inflected**
   - palatine bones with large vacuities (fenestrated)
   - presence of a posterior masseteric shelf on dentary
   - ventral exposure of presphenoid
4. POSTCRANIAL
   - **epipubic bones present** (also found in monotremes)

RECOGNITION CHARACTERS:

1. POLYPROTODONTY VERSUS DIPROTODONTY (FIG. 37; TABLE 2)

Multiple lower incisors subequal in size (= polyprotodont) versus medial incisor greatly enlarged (= diprotodont). Diprotodont taxa may have multiple small posterior incisors (e.g., Caenolestidae,

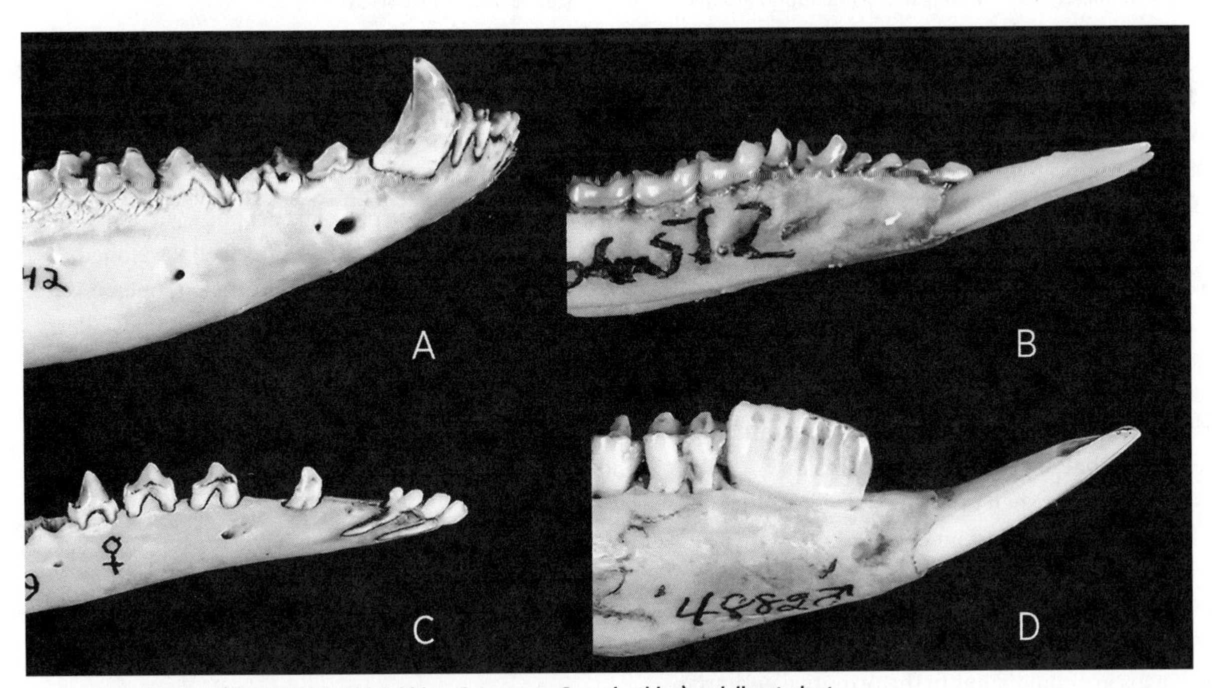

FIG. 37. Polyprotodont (*A*, *Didelphis*, Didelphidae; *C*, *Lestoros*, Caenolestidae) and diprotodont (*B*, *Isoodon*, Peramelidae; *D*, *Aepyprymnus*, Potoroidae) lower incisors in marsupials.

Phalangeridae) or only the enlarged medial pair (e.g., Vombatidae, Potoroidae, Macropodidae).

2. POLYDACTYLY VERSUS SYNDACTYLY (FIG. 38; TABLE 2)

All digits of the hind foot are free, and approximately equal in size (polydactyly), or digits 2 and 3 are distinctly smaller and encased in a common sheath of skin (syndactyly).

Most marsupials are either polydactylous and polyprotodont or syndactylous and diprotodont; two orders—one in South America and one in Australasia—are exceptions to this generality (table 2).

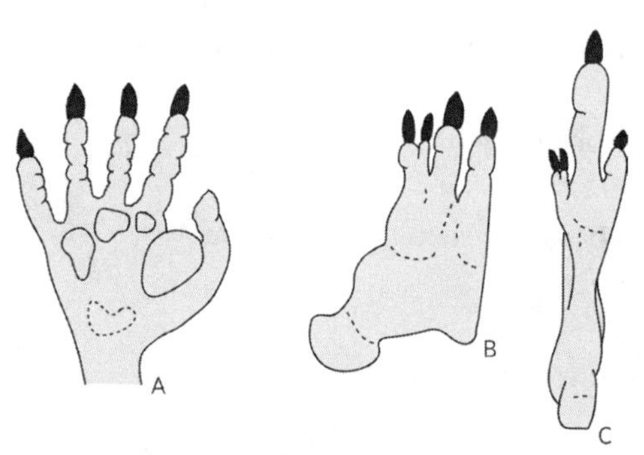

FIG. 38. Polydactylous (*A, Marmosops*, Didelphidae) and syndactylous (*B, Phascolarctos*, Phascolarctidae; *C, Aepyprymnus*, Potoroidae) hind feet in marsupials.

Table 2. Distribution of polyprotodonty/diprotodonty and polydactyly/syndactyly character states among living marsupial groups.

	Polyprotodont	*Diprotodont*
Polydactylous	Didelphimorphia (NW) Microbiotheria (NW) Dasyuromorphia (Aust.) Notoryctemorphia (Aust.)	Paucituberculata (NW)
Syndactylous	Peramelemorphia (Aust.)	most Diprotodontia (Aust.)

Note: NW = New World; Aust. = Australasian.

3. TRIBOSPHENIC MOLAR

The primitive tribosphenic upper molar of metatherians differs from that of primitive eutherians in possessing a stylar shelf on the labial side of the tooth on which several stylar cusps are present. The primitive tribosphenic lower molar of metatherians has the ectoconid and hypoconulid closely appressed and distant from the hypoconid on the talonid; these three cusps are equidistant in the primitive eutherian tribosphenic molar (see Fig. 39).

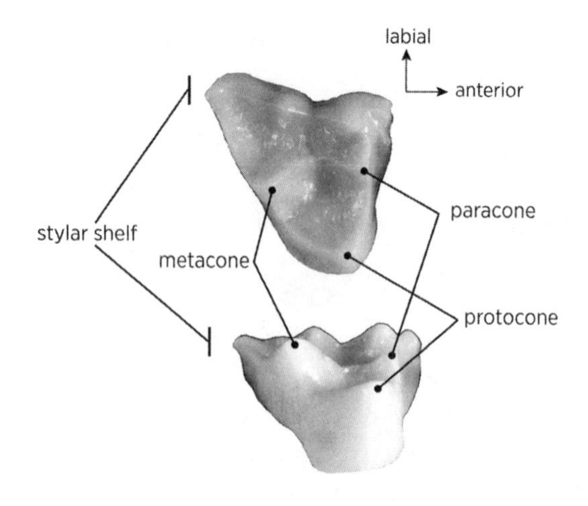

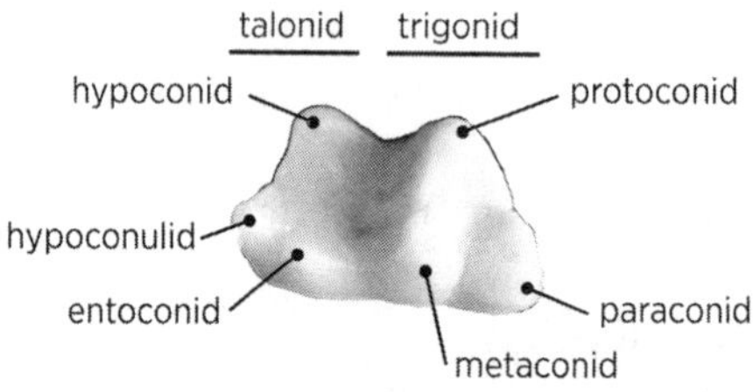

FIG. 39. The basic tribosphenic molar of a marsupial. *Top, middle*: Occlusal (*top*) and lingual (*middle*) views of the right M3; note stylar shelf on labial side of tooth, typically with five distinct cusps (termed stylar cusps A [anterior] through E [posterior]). *Bottom*: Left m3; entoconid and hypoconulid on talonid are closely appressed and widely separated from hypoconid.

Order DIDELPHIMORPHIA

Family DIDELPHIDAE (New World opossums)

Small to medium-sized marsupials (10 g to >2 kg); arboreal to terrestrial, scansorial and semiaquatic; **tail usually longer than head and body** (except in *Lestodelphys, Monodelphis*, and some *Thylamys*), **usually naked**, prehensile in some (*Caluromys, Didelphis, Marmosa*); foot posture plantigrade; polydactylous, digits 5/5, subequal in length; hallux well developed, opposable, without claw; **pouch, if present, opens anteriorly.** Usually solitary and nocturnal, crepuscular, or diurnal; generalized body plan considered ancestral for metatherians. Habitats include pampas grasslands, high-elevation Andean shrublands, and both temperate and tropical forests.

DIAGNOSTIC CHARACTERS:

- polyprotodonty with **5/4 incisors**
- polydactyly
- small, uninflated auditory bullae

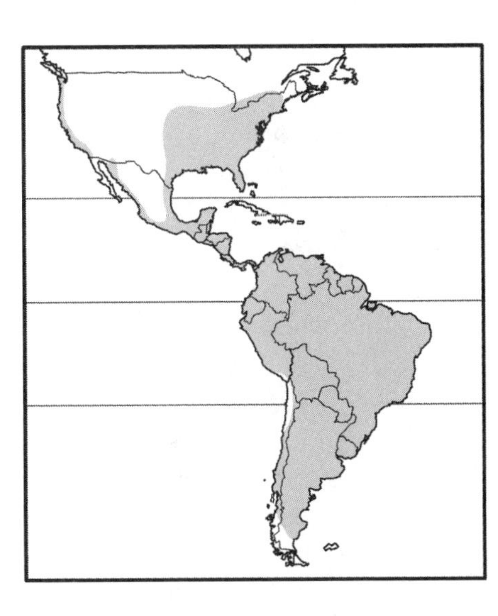

FIG. 41. Geographic range of family Didelphidae: New World: most are Neotropical (Mexico south to southern South America); one species (*Didelphis virginiana*) extends into temperate North America.

FIG. 40. Didelphidae: Didelphinae; water opossum, *Chironectes minimus* (*top*); Robinson's mouse opossum, *Marmosa robinsoni* (*left*); northern three-striped opossum, *Monodelphis americana* (*right*); and Virginia opossum, *Didelphis virginiana* (*bottom*).

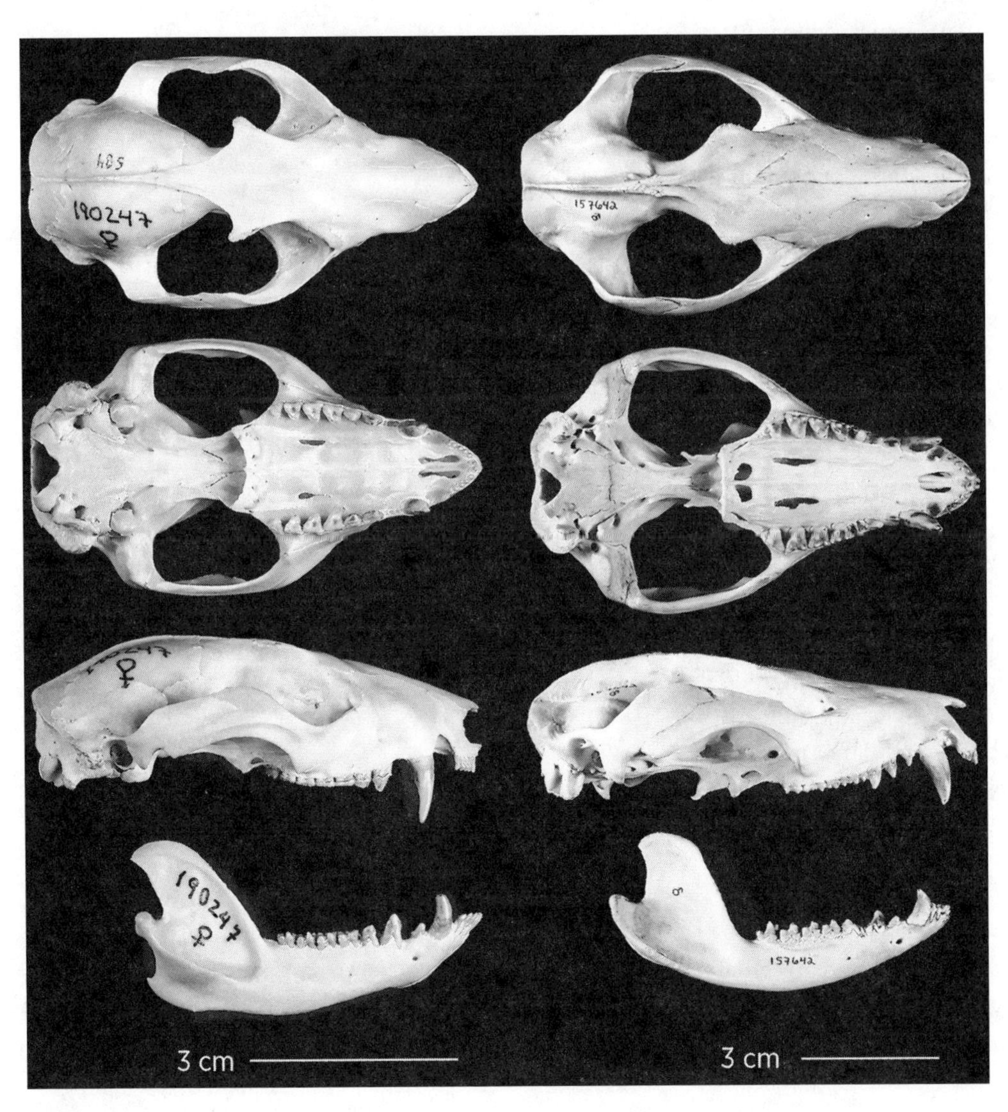

FIG. 42. Didelphidae; skull and mandible of the woolly opossum, *Caluromys lanatus* (Caluromyinae; *left*), and the common opossum, *Didelphis marsupialis* (Didelphinae; *right*).

RECOGNITION CHARACTERS:

1. cranium relatively long and slender
2. **sagittal crest often well developed**
3. zygomatic arch relatively slender
4. paroccipital process small
5. **lower incisors subequal in size, not procumbent**
6. canines well developed
7. molars tritubercular (tribosphenic)

DENTAL FORMULA: $\frac{5}{4}\ \frac{1}{1}\ \frac{3}{3}\ \frac{4}{4} = 50$

TAXONOMIC DIVERSITY: 18 genera in four subfamilies (Voss and Jansa 2009; Voss et al. 2014):

Caluromyinae—*Caluromys* (woolly opossums) and *Caluromysiops* (black-shouldered opossum)

Didelphinae—14 genera including *Chacodelphys* (Chacoan pygmy opossum), *Chironectes* (water opossum), *Cryptonanus* (mouse opossums), *Didelphis* (common opossums), *Gracilinanus* (gracile mouse opossums), *Lestodelphys* (Patagonian opossum), *Lutreolina* (lutrine opossums), *Marmosa* (mouse opossums), *Marmosops* (mouse opossums), *Metachirus* (brown four-eyed opossum), *Monodelphis* (short-tailed opossums), *Philander* (gray and black four-eyed opossums), *Thylamys* (fat-tailed mouse opossums), and *Tlacautzin* (grayish mouse opossum)

Glironiinae—*Glironia* (bushy-tailed opossum)

Hyladelphinae—*Hyladelphys* (mouse opossum)

Order PAUCITUBERCULATA

Family CAENOLESTIDAE (rat or shrew opossums)

Small (head and body length 90–140 mm; mass 20–40 g), terrestrial, shrewlike, with pointed snout, small eyes, long vibrissae; tail long and thin or short and thick, sparsely haired; foot posture plantigrade, polydactylous; hallux present, weakly opposable, with claw; **females lack a pouch**. Crepuscular or nocturnal; live in underground burrows and surface runways; active hunters of insects, earthworms, and small vertebrates.

DIAGNOSTIC CHARACTERS:

- diprotodonty (**medial lower incisors greatly enlarged and strongly procumbent**)
- polydactyly (digits 5/5, subequal in length)

RECOGNITION CHARACTERS:

1. cranium elongate; rostrum long and narrow
2. no sagittal crest
3. zygomatic arch relatively slender
4. paroccipital process very small
5. canines may be well developed or reduced in size
6. upper molars quadritubercular, with moderately developed hypocone

DENTAL FORMULA: $\frac{4}{3\text{–}4}\ \frac{1}{1}\ \frac{3}{3}\ \frac{4}{4} = 46\text{–}48$

TAXONOMIC DIVERSITY: About six species in three genera (*Caenolestes*, *Lestoros*, and *Rhyncholestes*).

FIG. 43. Caenolestidae; Andean caenolestid, *Caenolestes condorensis*.

FIG. 44. Geographic range of family Caenolestidae: South America: Andean cloud forest from Venezuela to northern Peru (*Caenolestes*), southern Peru to Bolivia (*Lestoros*), and temperate rainforest of southern Chile (*Rhyncholestes*).

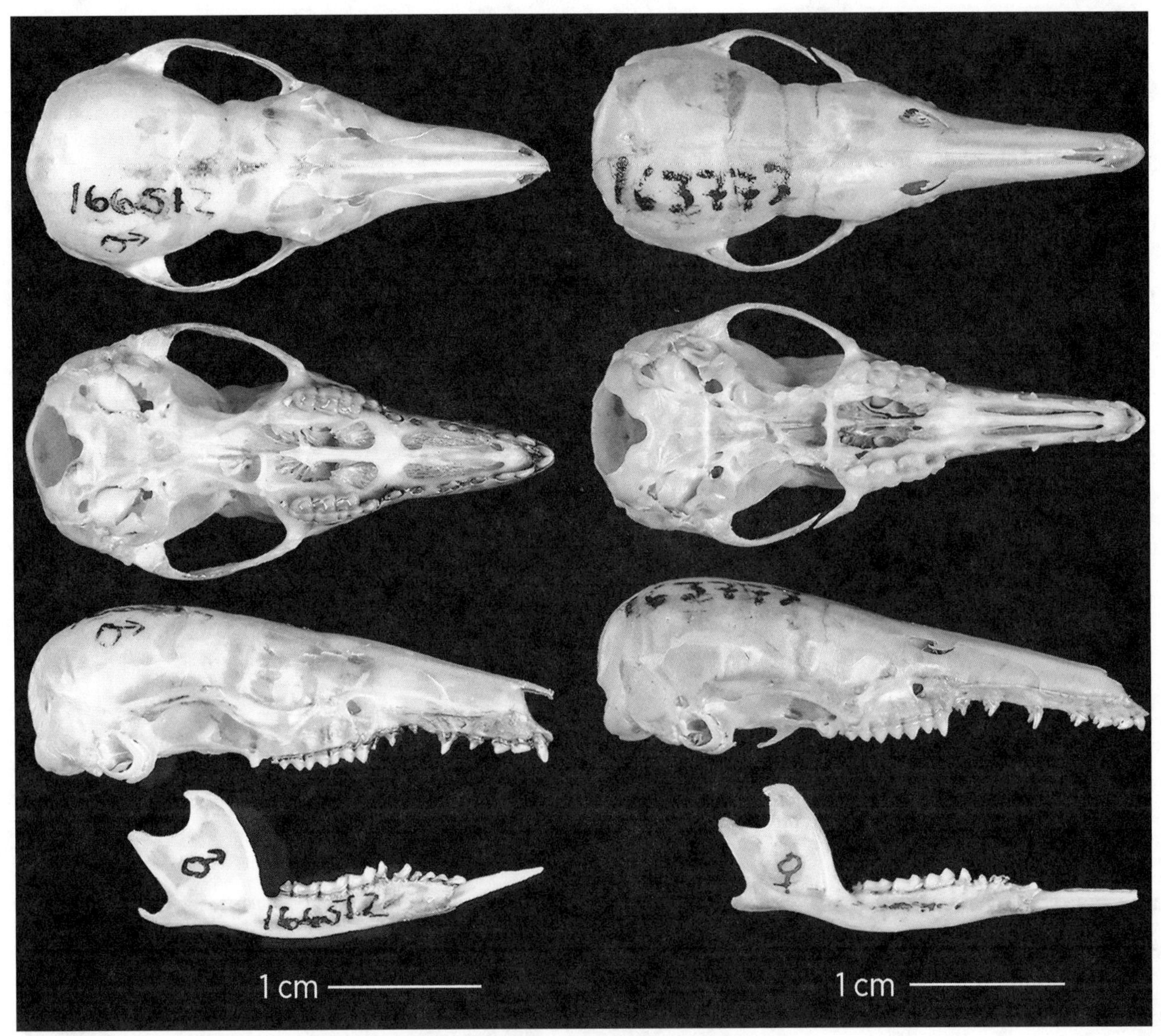

FIG. 45. Caenolestidae; skull and mandible of the Incan rat opossum, *Lestoros inca* (*left*), and of the Chilean shrew opossum, *Rhyncholestes raphanurus* (*right*).

Order MICROBIOTHERIA
Family MICROBIOTHERIIDAE
(monito del monte)

Small, mouselike (total length 195–250 mm; mass 16–50 g); ears round, well furred; tail subequal in length to head and body, furred to tip except for naked ventral strip near tip, and prehensile; tail stores fat in lean periods; polydactylous; pelage short and silky, fur dense; females with well-developed pouch, breed seasonally with one litter of two to four young; nocturnal, arboreal, terrestrial, and scansorial; omnivorous, but primarily insectivorous in most seasons, consumes vertebrates opportunistically, increasingly frugivorous in summer; an important disperser of mistletoe seeds.

DIAGNOSTIC CHARACTERS:

- polyprotodonty, with 5/4 incisors
- polydactyly
- greatly inflated auditory bullae with the inclusion of the entotympanic bone

RECOGNITION CHARACTERS:

1. skull without sagittal crest
2. bullae greatly inflated, together two-thirds as wide as braincase

3. incisors 5/4, as in Didelphidae, polyprotodont; lower incisors subequal in size, medial pair not procumbent
4. canines relatively small
5. molars tritubercular (tribosphenic)

DENTAL FORMULA: $\frac{5}{4}\ \frac{1}{1}\ \frac{3}{3}\ \frac{4}{4} = 50$

TAXONOMIC DIVERSITY: The family is monotypic, with a single species, *Dromiciops gliroides*.

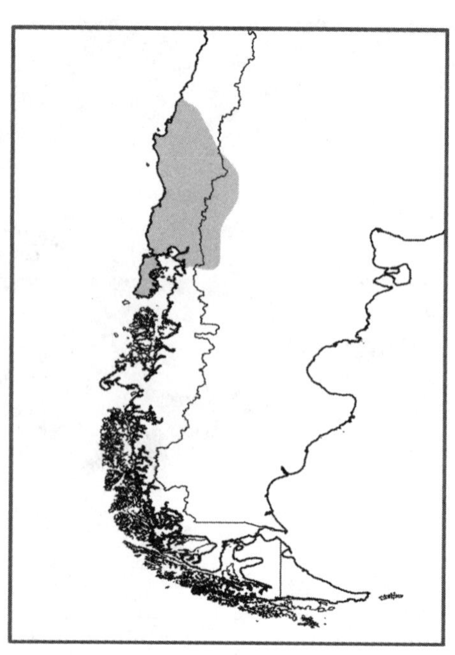

FIG. 47. Geographic range of family Microbiotheriidae: restricted to temperate rainforests of central Chile and adjacent parts of Argentina.

FIG. 46. Microbiotheriidae; monito del monte, *Dromiciops gliroides*.

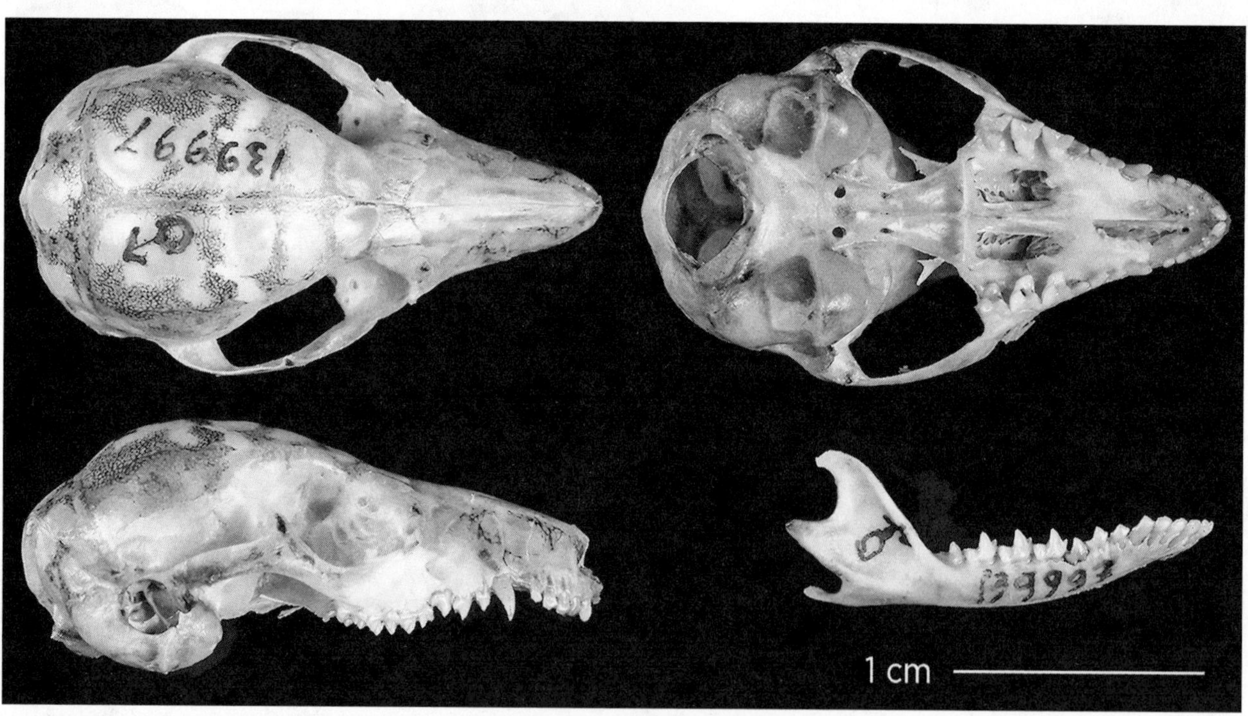

FIG. 48. Microbiotheriidae; skull and mandible of the monito del monte, *Dromiciops gliroides*.

Order NOTORYCTEMORPHIA

Family NOTORYCTIDAE (marsupial moles)

Small (body length 120–160 mm; mass 40–60 g), subterranean marsupial with cone-shaped head, **leathery shield over muzzle**, tubular body; **stubby tail, covered with leathery skin folded into concentric rings**; fur short, dense, and uniformly pale cream to white with an iridescent golden sheen; females with pouch, opens posteriorly; limbs short; digits 5/5, polydactylous and plantigrade; **3rd and 4th digits of forefeet greatly enlarged, bearing large triangular claws, 1st and 2nd opposed to 3rd and 4th**; hind feet flattened, hallux well developed; all digits bear claws except the 5th digit of the manus (bears a small nail) and the 1st and 5th digits of the pes (short pointed nail and small rudimentary nail, respectively); neck vertebrae fused for head rigidity during digging; insectivorous to carnivorous, feeding primarily on soil invertebrates such as beetle larvae but will take small lizards.

DIAGNOSTIC CHARACTERS:

- **eyes atrophied, not visible**
- no pinna

FIG. 49. Notoryctidae; marsupial mole, *Notoryctes typhlops*.

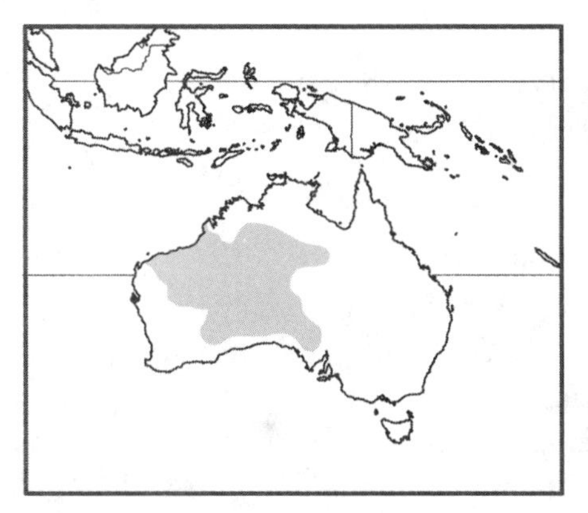

FIG. 50. Geographic range of family Notoryctidae: southern and western Australia; typically found in red sandy desert with sparse shrub cover.

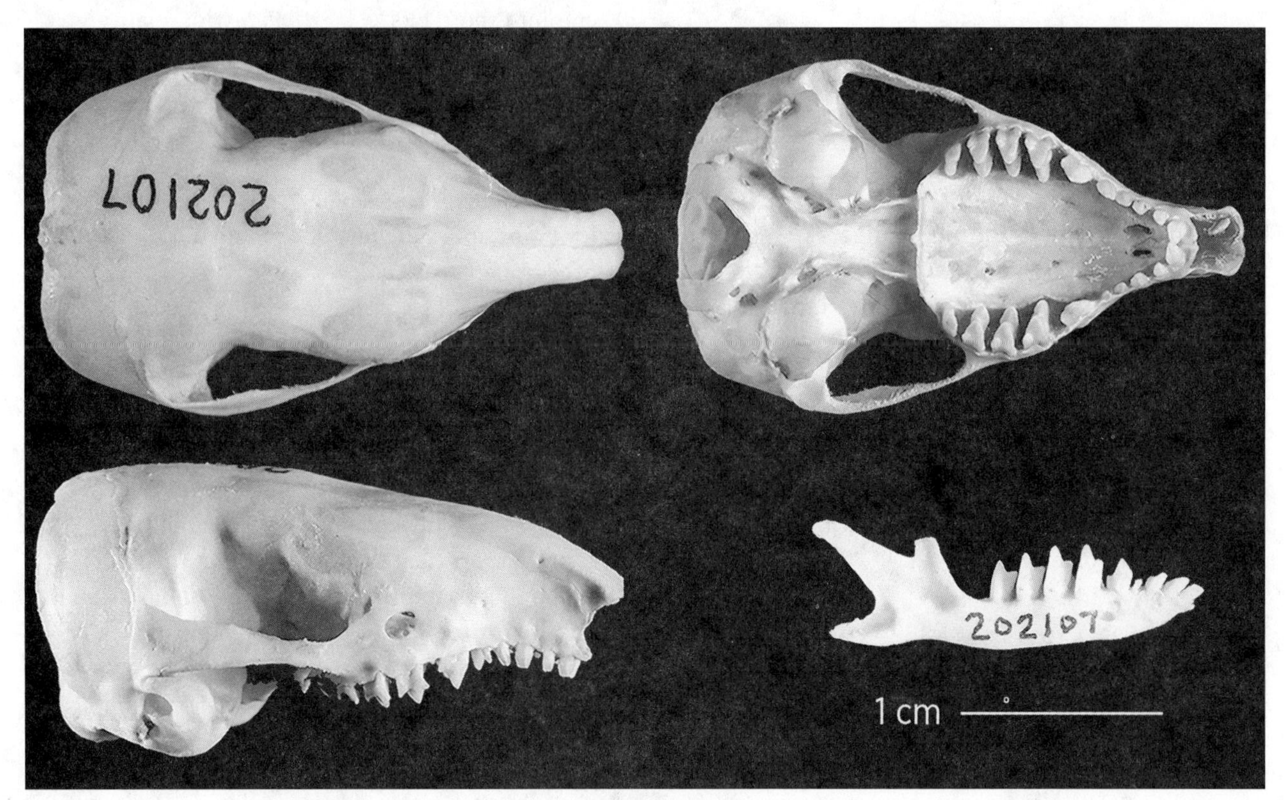

FIG. 51. Notoryctidae; skull and mandible of the marsupial mole *Notoryctes typhlops*.

RECOGNITION CHARACTERS:

1. **cranium elongate, conical**
2. no sagittal crest
3. zygomatic arch relatively robust, wide
4. paroccipital process very small
5. polyprotodont: lower incisors subequal in size, procumbent
6. canines reduced
7. molars simplified; occlusal surface zalambdodont (not tritubercular)

DENTAL FORMULA: $\frac{3\text{–}4}{3}\ \frac{1}{1}\ \frac{2}{2\text{–}3}\ \frac{4}{4} = 40\text{–}44$

TAXONOMIC DIVERSITY: The family is monotypic, with two species in the genus *Notoryctes*.

Order DASYUROMORPHIA

The Dasyuromorphia can be diagnosed by a combination of polydactyly and polyprotodonty (the only Australian marsupials to share these traits), with three or fewer premolars and four molars. It contains three families, which were combined into the Dasyuridae in older classifications.

DIAGNOSTIC CHARACTERS:

- **polyprotodonty with 4/3 incisors**
- **polydactyly**

Family DASYURIDAE (quolls, mulgaras, dasyures, dibbler, Tasmanian devil, phascogales, antechinuses, dunnarts, planigales, ningauis)

Highly diverse in body size (mouse-sized, 5–6 g, to medium-sized, up to 8 kg), but similar in shape, with moderately long body, long pointed head, long and usually well-furred tail, and short to medium-length legs; feet polydactylous and foot posture plantigrade; females of many species lack a pouch, with mammary area bordered by simple folds.

FIG. 52. Dasyuridae; fat-tailed dunnart, *Sminthopsis crassicaudata* (Sminthopsinae; *left*), and Eastern quoll, *Dasyurus viverrinus* (Dasyurinae; *right*).

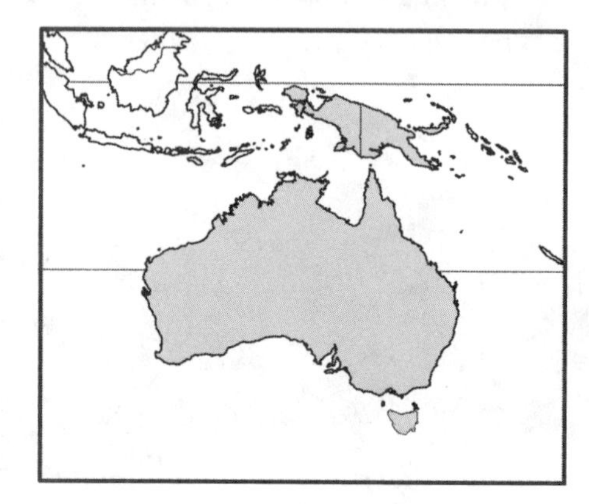

FIG. 53. Geographic range of family Dasyuridae: Australia, Tasmania, New Guinea, and many neighboring small islands; elevations from sea level to above 3,000 m (New Guinea), in all major terrestrial and some arboreal habitats.

RECOGNITION CHARACTERS

1. body ranges from very small (45–60 mm) to large (950–1,300 mm)
2. **pouch, if present, opening posteriorly**
3. **tail usually long and well haired**; never prehensile
4. foot posture plantigrade or digitigrade
5. polydactylous; digits 5/4 or 5/5, subequal in length
6. **hallux weakly opposable or absent, without claw**
7. cranium short, blocky (*Sarcophilus*), or relatively long and slender
8. sagittal crest usually absent (present in *Sarcophilus*)
9. zygomatic arch relatively slender (robust in *Sarcophilus*)
10. paroccipital process small
11. **lower incisors subequal in size and not procumbent**
12. canines well developed
13. molars tritubercular (tribosphenic)

DENTAL FORMULA: $\frac{4}{3}\ \frac{1}{1}\ \frac{2\text{–}3}{2\text{–}3}\ \frac{4}{4} = 42\text{–}46$

TAXONOMIC DIVERSITY: Includes about 21 genera and 70 species divided into four subfamilies (some authors treat these as tribes within two subfamilies: Dasyurinae = Dasyurini + Phascogalini; Sminthopsinae = Sminthopsini + Planigalinae). Key genera include the following:

Dasyurinae—*Dasycercus* (mulgaras), *Dasyurus* (quolls), *Phascolosorex* (marsupial shrews), *Pseudantechinus* (false antechinuses), and *Sarcophilus* (Tasmanian devil)

Phascogalinae—*Antechinus* (antechinuses) and *Phascogale* (phascogales)

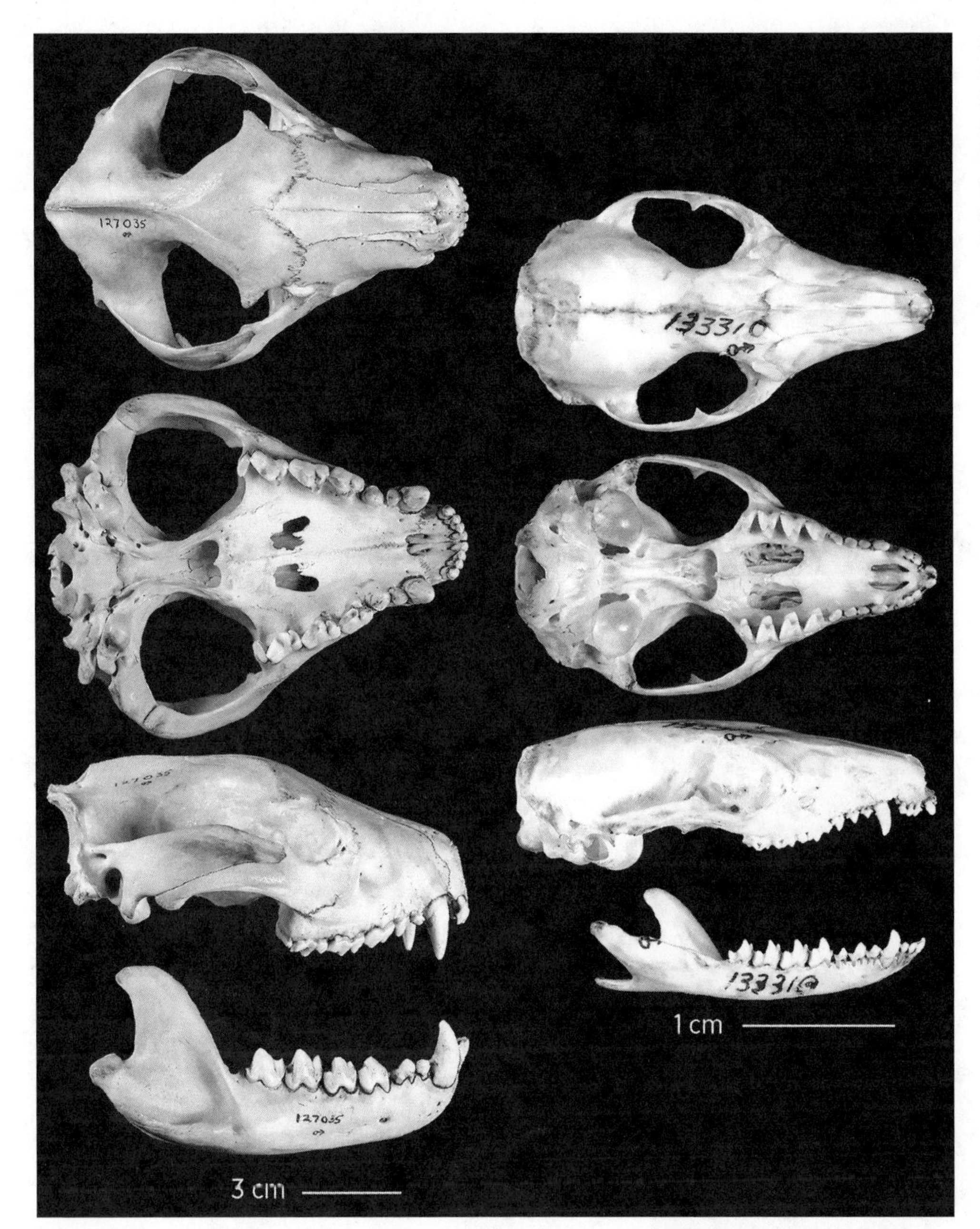

FIG. 54. Dasyuridae; skull and mandible of the Tasmanian devil, *Sarcophilus harrisii* (Dasyurinae; *left*), and the yellow-footed antechinus, *Antechinus flavipes* (Phascogalinae; *right*).

Planigalinae—*Planigale* (planigales)
Sminthopsinae—*Antechinomys* (kultarr), *Ningaui* (ningauis), and *Sminthopsis* (dunnarts)

Family MYRMECOBIIDAE (numbat or banded anteater)

Small to medium-sized (300–700 g) with pointed head, small ears, long, bushy tail, and distinctly banded across back and rump; foot posture digitigrade, feet polydactylous, digits 5/4 (**no hallux**), subequal in length and with large and strong claws; tongue markedly long and slender to extract ants and termites from galleries; **females lack a pouch**; dentition reduced in size and extra (supernumerary) cheek teeth may be present, giving counts of 7–8 above and 8–9 below; distinctive cranial characters include posterior elongation of hard palate, reduction in size and number of palatal vacuities, large postorbital processes of frontal bones, and unfused palatal branches of the premaxillae.

DIAGNOSTIC CHARACTERS:

- **pelage grayish or reddish brown, with alternating white and black stripes on lower back and rump**
- **five or six upper and lower molars; dentition totaling 50–54 teeth**

FIG. 55. Myrmecobiidae; numbat, *Myrmecobius fasciatus*.

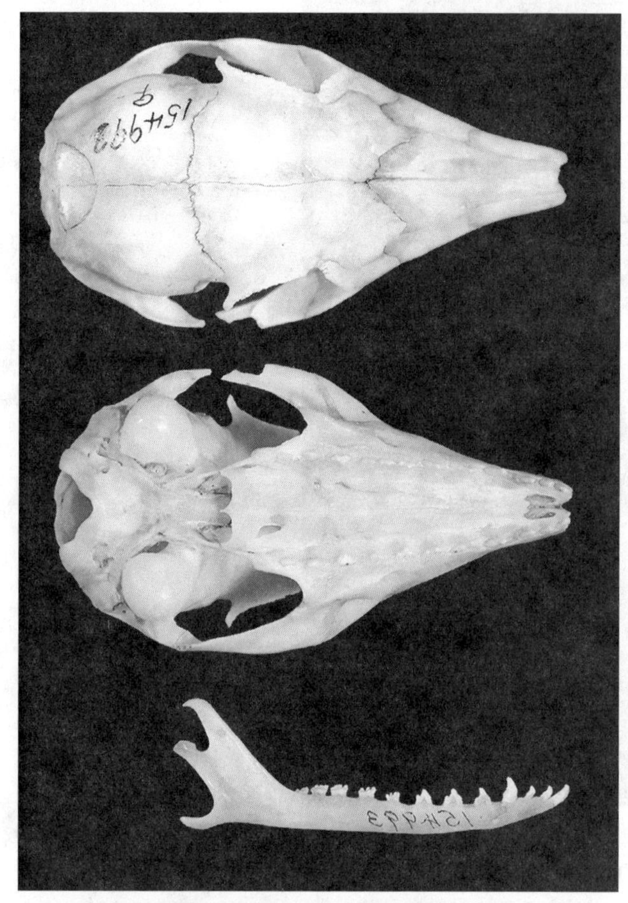

FIG. 56. Myrmecobiidae; skull and mandible of the numbat, *Myrmecobius fasciatus*. Note reduced dentition, small palatal fenestrae, and well-developed postorbital processes.

RECOGNITION CHARACTERS:

1. **cranium long, slender**
2. no sagittal crest
3. zygomatic arch slender
4. paroccipital process small
5. lower incisors subequal in size, not procumbent
6. canines relatively small
7. molars tritubercular

DENTAL FORMULA: $\frac{4}{3}\ \frac{1}{1}\ \frac{3}{3}\ \frac{5\text{–}6}{5\text{–}6} = 50\text{–}54$

RANGE: Extinct over about 99% of a historical range that encompassed most of southern and western Australia, now limited to a few sites in southwestern Australia and reintroduced into several reserves within the historical range.

TAXONOMIC DIVERSITY: The family is monotypic, with a single species, *Myrmecobius fasciatus* (numbat, or banded anteater).

†Family THYLACINIDAE (thylacine)

Large, doglike marsupial (mass about 35 kg); body slim and elongate, back and rump with transverse stripes, limbs long, foot posture digitigrade; polyprotodont and polydactylous, no hallux. Extinct as of about 1936 due to predator control efforts.

DENTAL FORMULA: $\frac{4}{3}\ \frac{1}{1}\ \frac{3}{3}\ \frac{4}{4} = 46$

Order PERAMELEMORPHIA (= Peramelina)

The combination of **polyprotodonty and syndactyly** uniquely diagnoses members of the Peramelemorphia. The two traditional families (Peramelidae and Peroryctidae) have recently been grouped into the single family Peramelidae; the genus *Macrotis* (the bilby) is now placed in its own family (Thylacomyidae), and the genus *Chaeropus* (the pig-footed bandicoot) is sometimes recognized as its own family (Chaeropodidae). For simplicity, and because all bandicoots share a suite of obvious morphological characters, we treat all of the Peramelemorphia together.

Families †CHAEROPODIDAE, PERAMELIDAE, and THYLACOMYIDAE (bandicoots and bilbies)

Small to medium-sized (body length 300–850 mm; mass <500 g–2 kg); head with elongate and pointed muzzle; tail short, sparsely haired or crested, not prehensile; hind limbs and hind feet elongate and adapted for quadrupedal saltatorial or running gait; **foot posture digitigrade; forefoot with three functional digits, foreclaws enlarged for digging; hind foot with four main digits, inner two syndactylous, outer two large and free; hallux vestigial and without claw, or absent**; females with well-developed pouch, opening posteriorly; embryos form a chorioallantoic placenta, a character shared among metatherians only with koalas and wombats, but in utero gestation is temporally brief. Habitats range from open plains to riverine and swamp edge, thick scrub, and temperate and tropical forests. The pig-footed bandicoot almost certainly is extinct, a consequence of predation by introduced red foxes (*Vulpes vulpes*) and feral cats (*Felis catus*).

DIAGNOSTIC CHARACTERS:

- **polyprotodonty with 4–5/3 incisors**
- **syndactyly of digits 2 and 3 of hind foot**
- **vestigial hallux**

FIG. 57. Peramelidae: Peramelinae; long-nosed bandicoot, *Perameles nasuta*.

RECOGNITION CHARACTERS:

1. **cranium very long, narrow**
2. no sagittal crest
3. **zygomatic arch very slender**
4. paroccipital process small
5. polyprotodont; lower incisors subequal in size but slightly procumbent
6. canines reduced
7. molars tritubercular to slightly trapezoidal or squared

DENTAL FORMULA: $\frac{4\text{–}5}{3}\ \frac{1}{1}\ \frac{3}{3}\ \frac{4}{4} = 46\text{–}48$

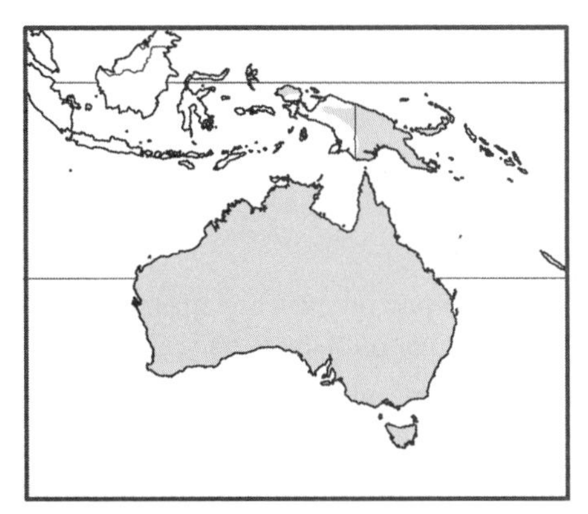

FIG. 58. Geographic range of family Peramelidae: Australia, Tasmania, New Guinea, Bismarck Archipelago, and adjacent islands, at elevations from sea level to above 4,000 m (in New Guinea), in hot, arid regions to cool, humid, montane rainforests.

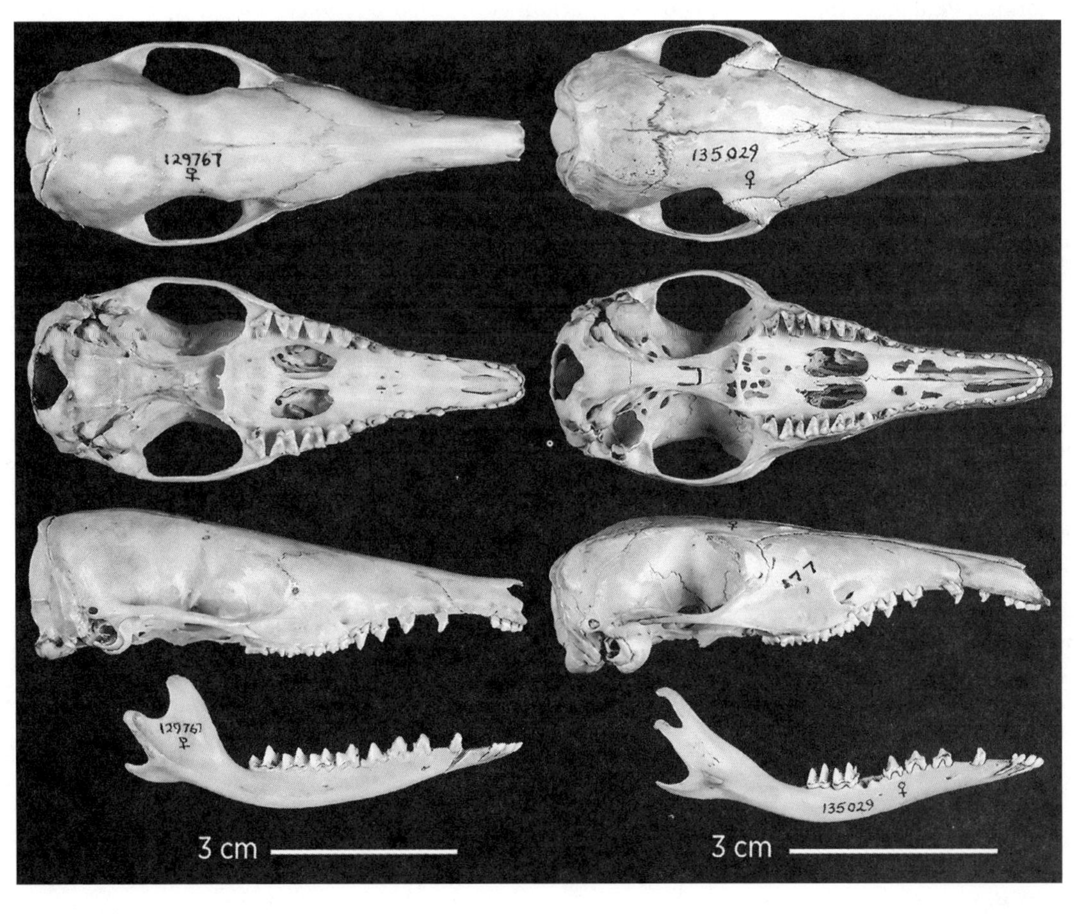

FIG. 59. Peramelidae; skull and mandible of the common echymipera, *Echymipera kalubu* (Echymiperinae; *left*), and the eastern barred bandicoot, *Perameles gunnii* (Peramelinae; *right*).

TAXONOMIC DIVERSITY: About 20 species in eight genera in three families, one with three subfamilies (Groves 2005a; Jackson and Groves 2015):

†Chaeropodidae: This family was monotypic, with a single species, *Chaeropus ecaudatus* (pig-footed bandicoot)

Peramelidae:

Echymiperinae—*Echymipera* (echymiperas), *Microperoryctes* (striped, mouse, and Papuan bandicoots), and *Rhynchomeles* (Seram bandicoot)

Peramelinae—*Isoodon* (bandicoots), *Perameles* (bandicoots)

Peroryctinae—*Peroryctes* (bandicoots)

Thylacomyidae: *Macrotis* (greater and lesser bilbies, the latter extinct)

Order DIPROTODONTIA

All families in the order Diprotodontia are characterized by **diprotodonty** and **most by syndactyly**. They also share a superficial thymus gland and a fasciculas aberrans (a bundle of nerve fibers connecting the two hemispheres of the brain; Archer 1984). Diprotodontids are commonly divided into three suborders: Vombatiformes, Phalangeriformes (with two superfamilies), and Macropodiformes (fig. 60).

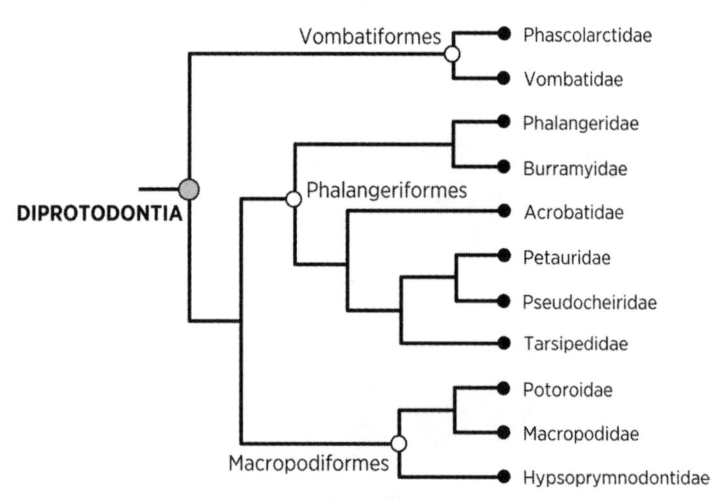

FIG. 60. Hypothesized phylogeny of families of diprotodont marsupials based on mitochondrial and nuclear DNA sequences (Meredith et al. 2009, 2011; note, however, that Gatesy et al. [2017] were unable to resolve relationships among the two superfamilies of Phalangeriformes and the Macropodiformes).

Suborder VOMBATIFORMES

The suborder Vombatiformes unites the koalas and the wombats. These species share a number of "soft" features, such as a nearly vestigial tail, a pouch opening posteriorly, cheek pouches, a unique gastric gland in the stomach, reduced paraflocculus in the temporal region of the brain, and hook-shaped spermatozoa (Archer 1984).

Family PHASCOLARCTIDAE (koala)

Relatively large (body length 600–850 mm; mass 4–15 kg), arboreal and herbivorous, easily recognized by **stout "teddy-bear-like" body** with **rudimentary (functionally absent) tail**; round and well-furred ears; large, spoon-shaped nose; densely woolly fur; foot posture plantigrade; digits 5/5, subequal in size; 2nd and 3rd digits of hind foot syndactylous; hallux well developed, opposable, without claw. Typical inhabitant of open eucalypt woodlands with eucalypt leaves constituting majority of diet; pouch present in females, opening posteriorly; stomach long and folded, cecum very large; females with well-developed pouch that opens posteriorly; placenta formed briefly during gestation.

RECOGNITION CHARACTERS:

1. cranium robust, broad
2. no sagittal crest
3. zygomatic arch robust, wide
4. paroccipital process large
5. **1st upper incisor much larger than other two**; diprotodont, medial lower incisor procumbent
6. upper canine small
7. **molars squarish, with crescentic ridges (pseudo-selenodont)**

DENTAL FORMULA: $\frac{3}{1}\ \frac{1}{0}\ \frac{1}{1}\ \frac{4}{4} = 30$

TAXONOMIC DIVERSITY: The family is monotypic, with a single species, *Phascolarctos cinereus* (koala).

FIG. 61. Phascolarctidae; koala, *Phascolarctos cinereus*.

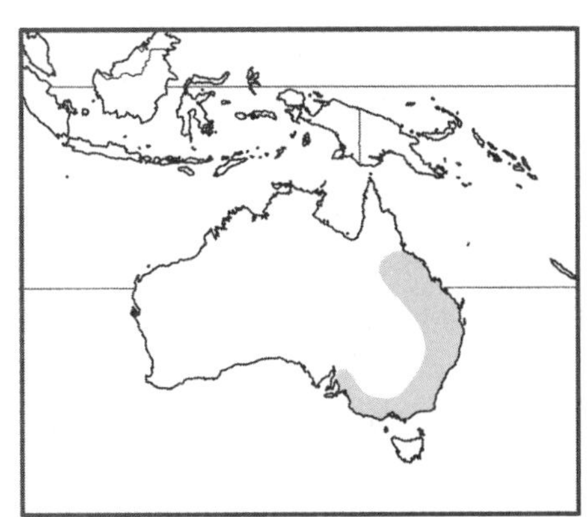

FIG. 62. Geographic range of family Phascolarctidae: dry sclerophyll forests and woodlands of eastern Australia.

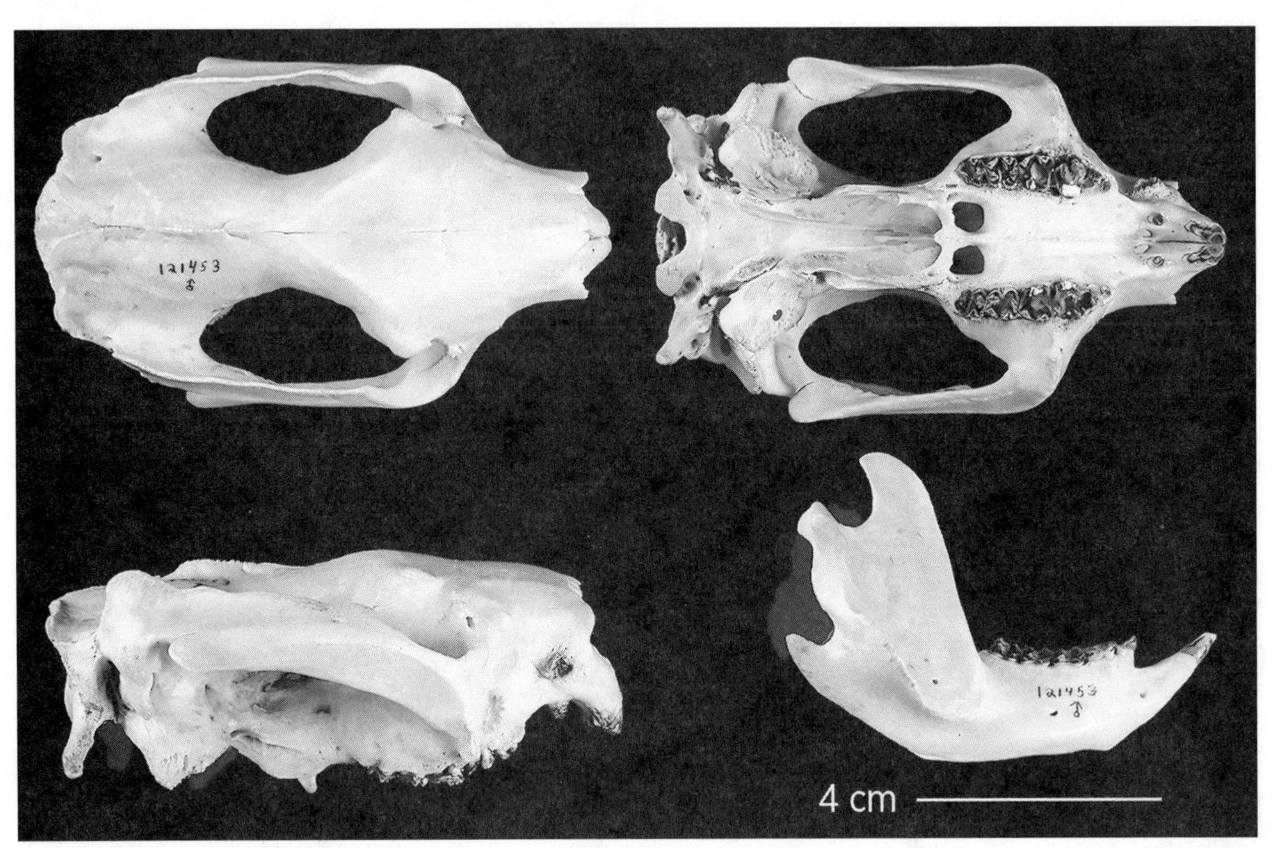

FIG. 63. Phascolarctidae; skull and mandible of the koala, *Phascolarctos cinereus*.

Family VOMBATIDAE (wombats)

Large (length up to 1 m or more; mass 20–35 kg), short legged, muscular, and quadrupedal, with a **short, stubby tail**; digs extensive burrow systems with rodent-like incisors and enlarged claws on powerful forefeet; foot posture plantigrade; digits 5/5, subequal in size, two innermost digits on hind feet syndactylous; hallux vestigial; crepuscular and nocturnal; herbivorous; females with pouch opening posteriorly, chorioallantoic placenta present, give birth to single young.

FIG. 64. Vombatidae; common wombat, *Vombatus ursinus*.

DIAGNOSTIC CHARACTER:

- **single pair of large, chisel-shaped incisors above and below**

RECOGNITION CHARACTERS:

1. **cranium robust, broad**
2. no sagittal crest
3. zygomatic arch relatively robust, wide
4. paroccipital process small
5. **incisors 1/1; diprotodont, lower incisor procumbent**
6. **canines absent**
7. molars transversely bilophodont, with lophs conical in shape

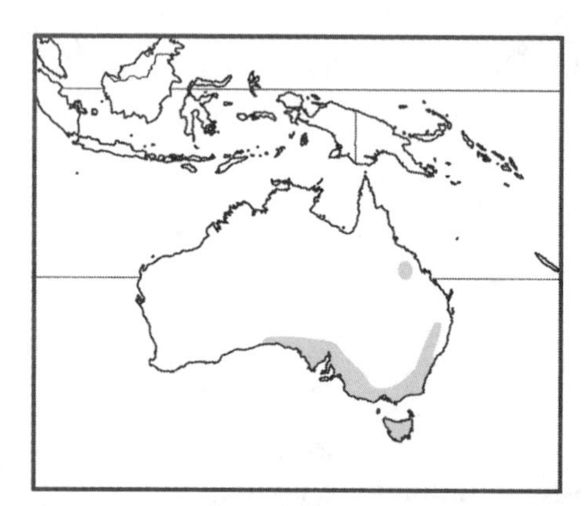

FIG. 65. Geographic range of family Vombatidae: southern and southeastern Australia in dry grasslands, savannas, and wet and dry sclerophyll forests.

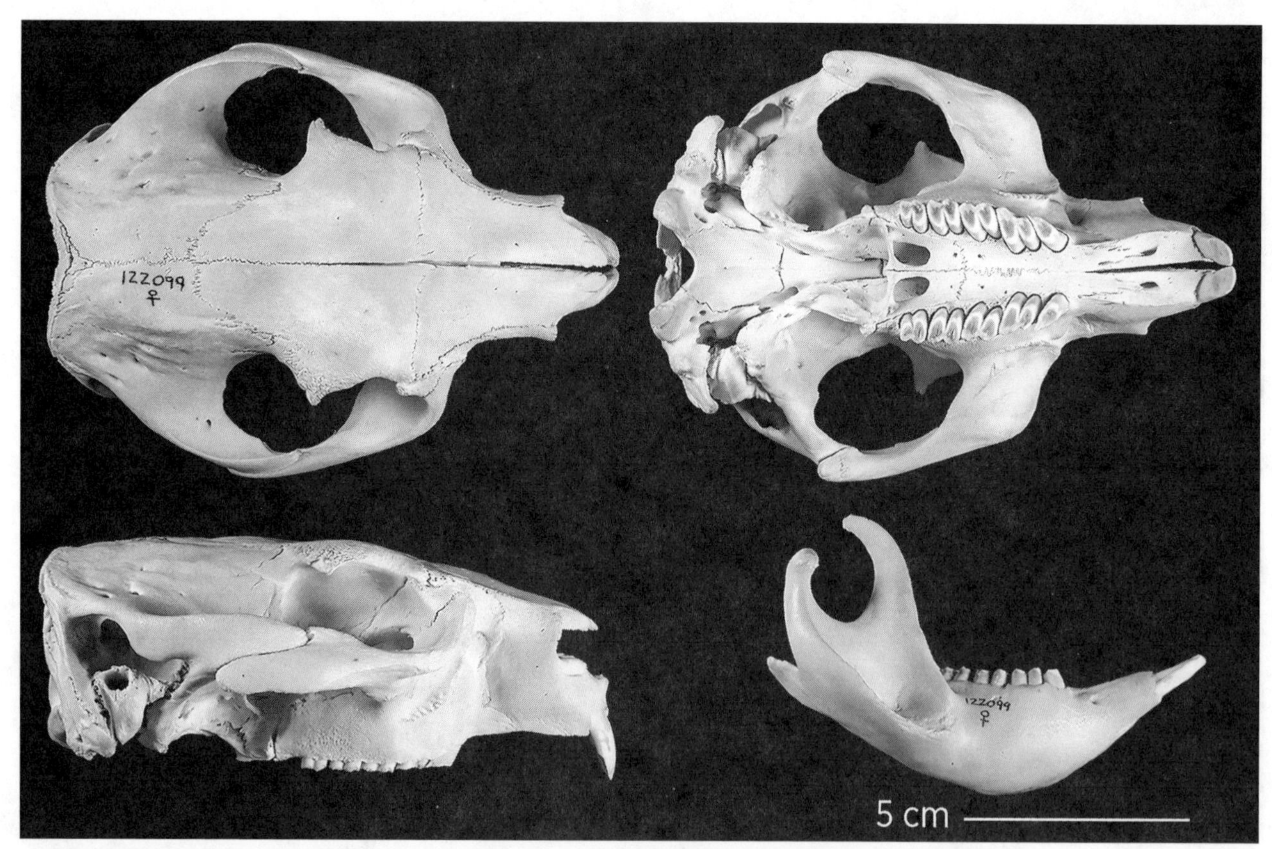

FIG. 66. Vombatidae; skull and mandible of the southern hairy-nosed wombat, *Lasiorhinus latifrons*.

DENTAL FORMULA: $\frac{1}{1}\ \frac{0}{0}\ \frac{1}{1}\ \frac{4}{4} = 24$

TAXONOMIC DIVERSITY: Two genera and three species: *Vombatus* (common wombat) and *Lasiorhinus* (hairy-nosed wombats).

FIG. 67. Burramyidae; eastern pygmy possum, *Cercartetus nanus*.

Suborder PHALANGERIFORMES

The Phalangeriformes includes all of the extant possums. The tooth formula of most members of the suborder is variable, but all are diprotodont with an enlarged, highly procumbent anterior lower incisor, and most possess a cheek tooth row with at least one well-developed premolar, modified as a plagiaulacoid slicing blade in some, and 3–4 molars. Teeth between the anterior incisor and the cheek teeth are generally reduced to small, undifferentiated pegs; on the lower jaw, it is debatable whether these pegs represent posterior incisors, a canine, or even anterior premolars. The Tarsipedidae possess the most simplified teeth, with all except the diprotodont lower first incisor reduced to a variable number of pegs.

Superfamily PHALANGEROIDEA

Family BURRAMYIDAE (pygmy possums)

Small (total length 50–120 mm; mass 10–50 g), with conical head, short muzzle, large eyes, and short, rounded ears; arboreal to scansorial, with long, slender, naked or well furred and prehensile tails, never with lateral fringes of hair; hind foot syndactylous; nocturnal; omnivorous (diet includes invertebrates, fruit, seeds, nectar, and pollen); females with anteriorly opening, well-developed pouch; embryonic diapause; only marsupial known to hibernate (although many small species employ torpor).

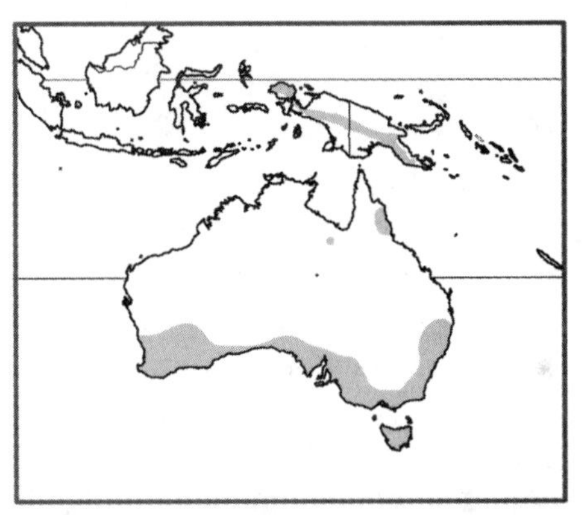

FIG. 68. Geographic range of family Burramyidae: southern, southeastern, and northeastern Australia, Tasmania, highlands of New Guinea, in both wet and dry sclerophyll forest with understory of shrubs or low trees, and in tropical rainforest.

RECOGNITION CHARACTERS:

1. posterior lower premolar enlarged, bladelike (plagiaulacoid) in *Burramys*, but not in *Cercartetus*
2. diprotodont, lower incisor procumbent

DENTAL FORMULA: $\frac{3}{1\text{–}3}\ \frac{1}{0}\ \frac{2\text{–}3}{2\text{–}3}\ \frac{3\text{–}4}{3\text{–}4} = 30\text{–}42$

Total count: *Cercartetus* = 10–11/8–9 = 36–40; *Burramys* = 10/9 = 38

2nd and 3rd lower incisors very small; homology not clear.

TAXONOMIC DIVERSITY: About five living species in two genera: *Burramys* (mountain pygmy possum) and *Cercartetus* (little pygmy possums).

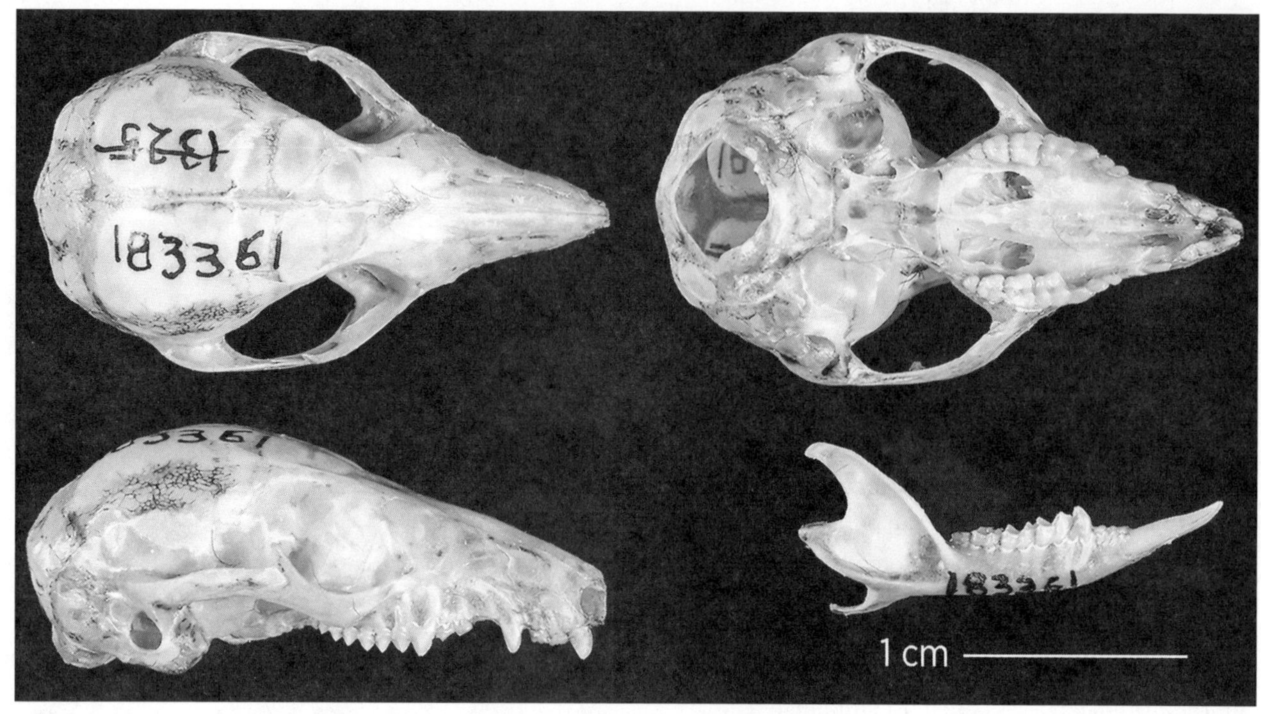

FIG. 69. Burramyidae; skull and mandible of the long-tailed pygmy possum, *Cercartetus caudatus*.

Family PHALANGERIDAE (phalanger, cuscuses, brush-tailed possums, scaly-tailed possum)

Medium-sized to large (total length 350–700 mm; mass up to 5 kg); body stocky and muscular; face short, eyes directed forward; rhinarium prominent; tail relatively long, well-haired to partly naked, and prehensile in most; all but *Trichosurus* arboreal, with 1st two digits of forefeet opposable to other three; foot posture plantigrade; digits 5/5, subequal in length; 2nd and 3rd digits of hind foot syndactylous; hallux well developed, opposable, without claw; pouch present in females, opening anteriorly; nocturnal or crepuscular; omnivorous. Habitat includes open plains, woodland to savanna forest, scrub, wet and dry sclerophyll forest, subalpine forest, rainforest, and oak-beech forests.

RECOGNITION CHARACTERS:

1. **cranium relatively broad**
2. sagittal crest often present
3. **zygomatic arch robust**
4. paroccipital process small to large
5. diprotodont; **1st lower incisor much larger than others, procumbent**
6. canines reduced in size
7. last upper and lower premolar usually enlarged, projecting above molar crowns (fig. 71, arrow)
8. molars bilobed (transversely lophodont), cusps rounded

DENTAL FORMULA: $\frac{3}{2}\ \frac{1}{0}\ \frac{2\text{–}3}{2\text{–}3}\ \frac{4}{4} = 36\text{–}40$

2nd upper premolar peg-like if present; 3rd upper premolar usually enlarged; 2nd lower incisor and 1st two lower premolars small and peg-like (see fig. 71)

Total count: *Trichosurus* = 10/7 = 34; *Phalanger* = 10–11/8–9 = 36–40

TAXONOMIC DIVERSITY: About 27 species in six genera allocated to three subfamilies:

Ailuropinae—*Ailurops* (bear cuscuses) and *Strigocuscus* (small cuscuses)

Phalangerinae—*Phalanger* (cuscuses) and *Spilocuscus* (cuscuses)

Trichosurinae—*Trichosurus* (brush-tailed possums), and *Wyulda* (scaly-tailed possum)

FIG. 70. Phalangeridae: Phalangerinae; common spotted cuscus, *Spilocuscus maculatus*.

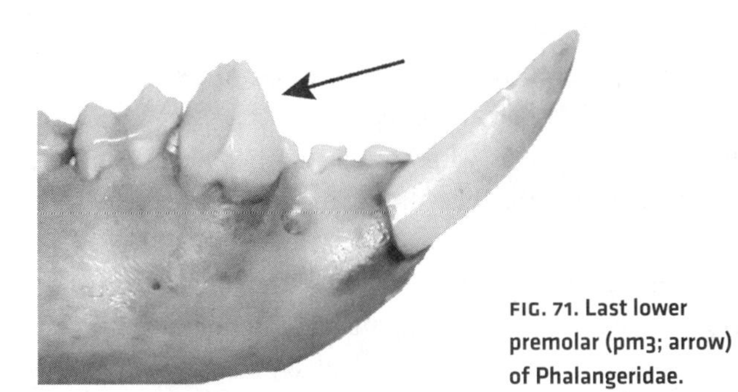

FIG. 71. Last lower premolar (pm3; arrow) of Phalangeridae.

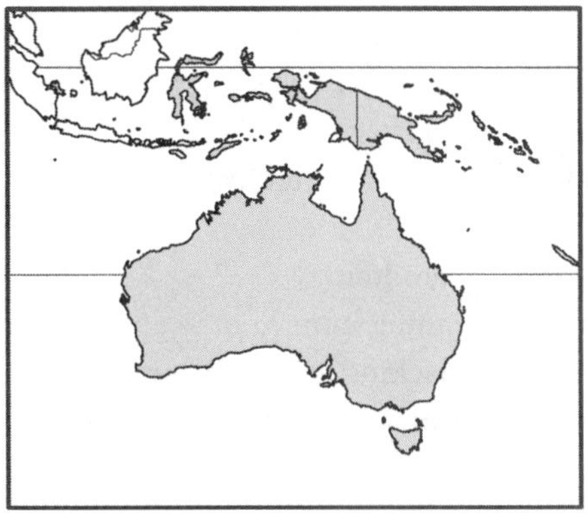

FIG. 72. Geographic range of family Phalangeridae: Australia, New Guinea, and adjacent islands (e.g., Solomon Islands), and Sulawesi and Timor in Indonesia.

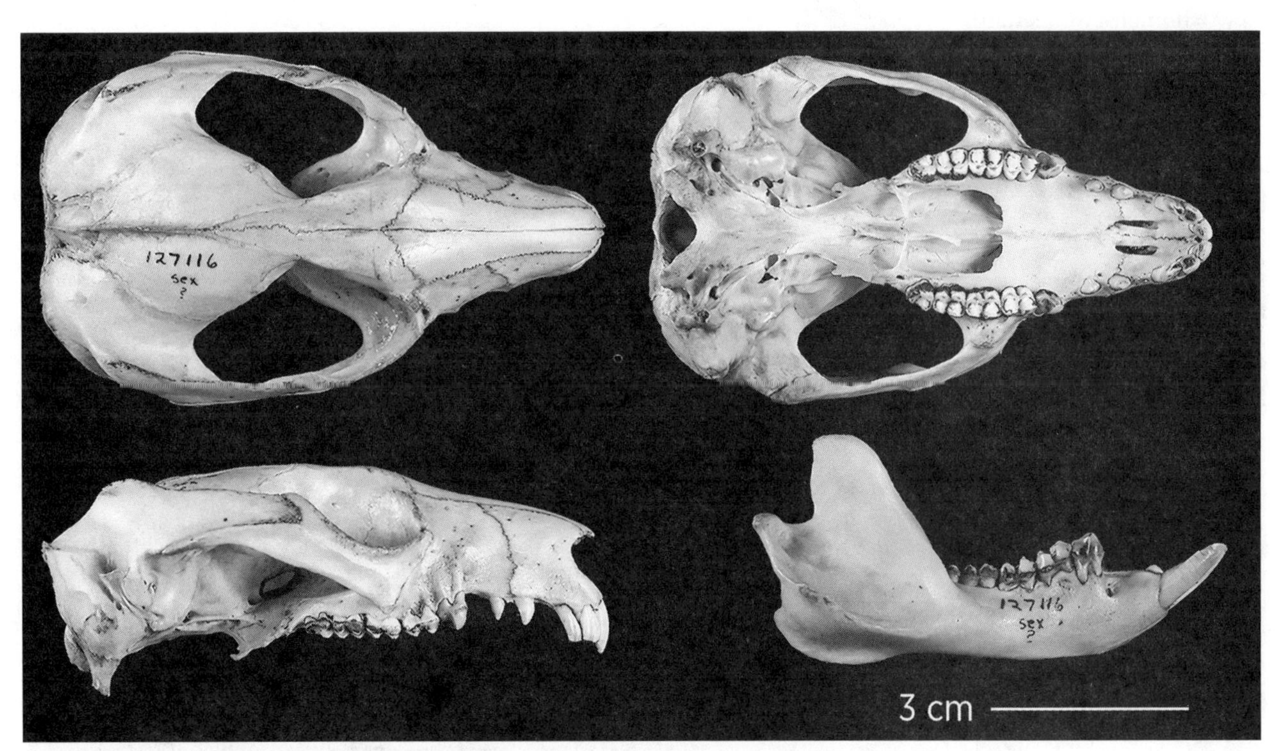

FIG. 73. Phalangeridae: Trichosurinae; skull and mandible of the brushtail possum, *Trichosurus vulpecula*.

Superfamily PETAUROIDEA

Family ACROBATIDAE (feather-tail gliders and feather-tailed possum)

Very small; arboreal, with partially prehensile tails; *Acrobates* (Australia) has gliding membrane stretched from wrist to ankle; *Distoechurus* (New Guinea) lacks membrane; largely nectar feeders with elongate, brush-tipped tongues for retrieving nectar and pollen from flowers. Habitat primarily wet and dry sclerophyll forests and woodlands; in New Guinea, primarily tropical rainforest and woodlands.

FIG. 74. Acrobatidae; feather-tail glider, *Acrobates pygmaeus*.

RECOGNITION CHARACTERS:

1. very small size (mass 10–50 g)
2. **tail feather-like** (hairs extend laterally from shaft as an aid in gliding)
3. syndactylous and diprotodont
4. **gliding membrane** running from wrist to ankle in *Acrobates*; membrane lacking in *Distoechurus*
5. **molars quadritubercular; only three above and below**

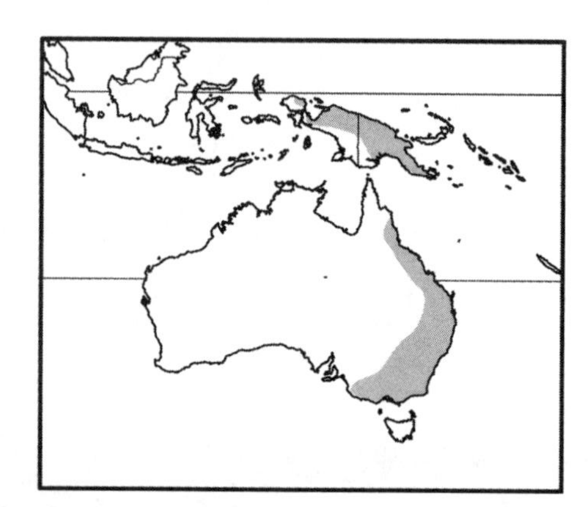

FIG. 75. Geographic range of family Acrobatidae: eastern Australia (*Acrobates*) and New Guinea (*Distoechurus*).

DENTAL FORMULA: Poorly known and variable (? = homology uncertain)

Acrobates: $\frac{3}{1}\ \frac{1}{1?}\ \frac{3}{3?}\ \frac{3}{3}$; total count: 10/8 = 36

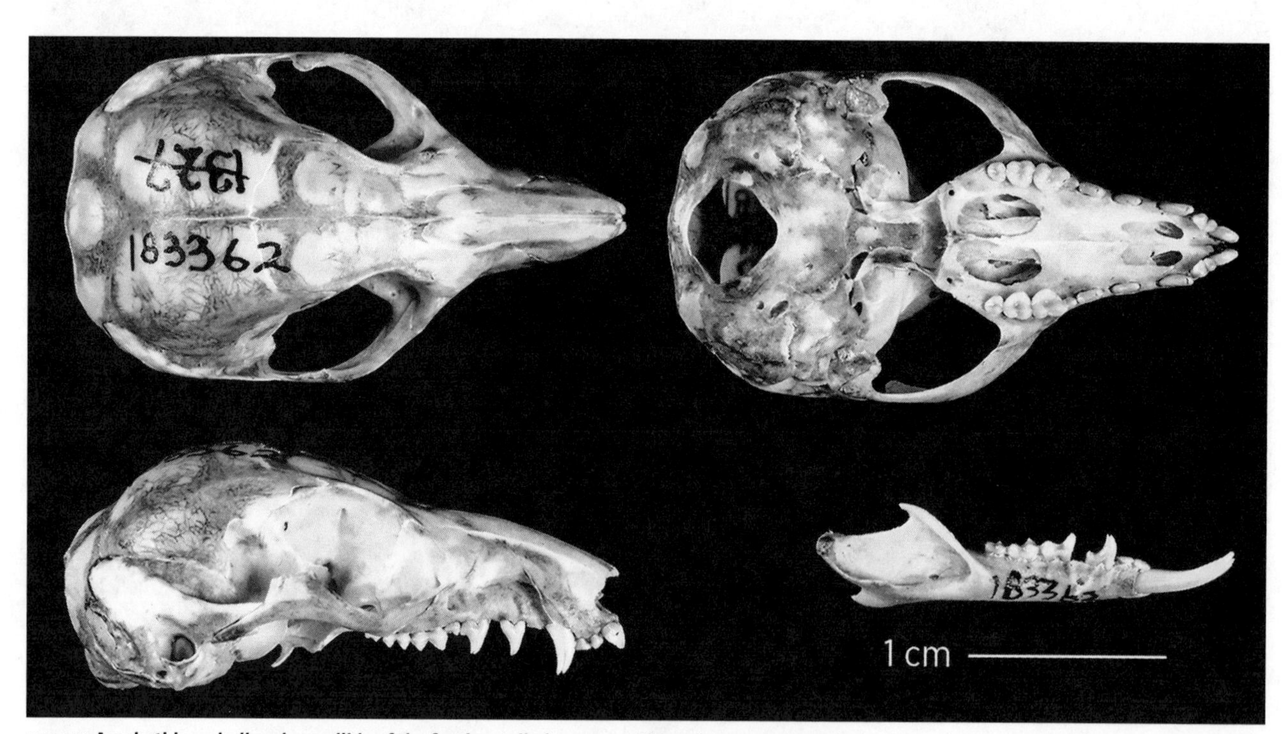

FIG. 76. Acrobatidae; skull and mandible of the feather-tailed possum, *Distoechurus pennatus*.

Distoechurus: $\frac{3}{1}\ \frac{1}{1?}\ \frac{2\text{–}3}{3?}\ \frac{3}{3}$; total count 9–10/8 = 34–36

TAXONOMIC DIVERSITY: Three species in two genera: *Acrobates* (narrow-toed and broad-toed feather-tail gliders) and *Distoechurus* (feather-tailed possum).

FIG. 77. Petauridae: Petaurinae; squirrel glider, *Petaurus norfolcensis*.

Family PETAURIDAE (sugar or lesser glider, striped possums, Leadbeater's possum)

Small to medium-sized (body length 150–270 mm; mass 95–430 g); prominent dorsal stripe that extends from the rump to the forehead; tail long (equal to or much longer than body length), well haired, and prehensile; *Petaurus* have well-developed gliding membrane, other genera either lack or have vestigial membrane; arboreal, with digits 1–2 of forefoot opposable to 3–5, and well-developed hallux on hind foot; syndactylous; females with well-developed pouch opening anteriorly; arboreal, feeding on insects and sap and gum of eucalypts and acacias, obtaining sap by scratching bark with incisors; *Dactylopsila* has elongate 4th digit on forefeet used to extract insects from cavities and bark.

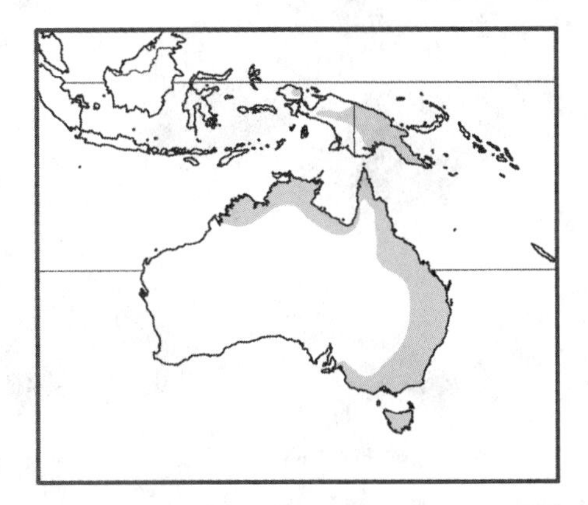

FIG. 78. Geographic range of family Petauridae: forested areas of northern, eastern, and southeastern Australia, Tasmania, and New Guinea.

RECOGNITION CHARACTERS:

1. **medial lower incisor greatly enlarged, nearly as long as rest of tooth row**
2. canine reduced
3. lower 3rd premolar (posterior peg tooth) reduced in size
4. upper molars successively smaller from front to back
5. tail usually well furred to tip, or ventral surface naked at tip only

DENTAL FORMULA: $\frac{3}{1\text{–}2}\ \frac{1}{0}\ \frac{3\ \text{(peg-like)}}{1\text{–}3\ \text{(peg-like)}}\ \frac{4}{4} = 34\text{–}40$

TAXONOMIC DIVERSITY: Three genera and about 11 species generally treated in two subfamilies:

Dactylopsilinae—*Dactylopsila* (striped possums) and *Gymnobelideus* (Leadbeater's possum)

Petaurinae—*Petaurus* (gliders)

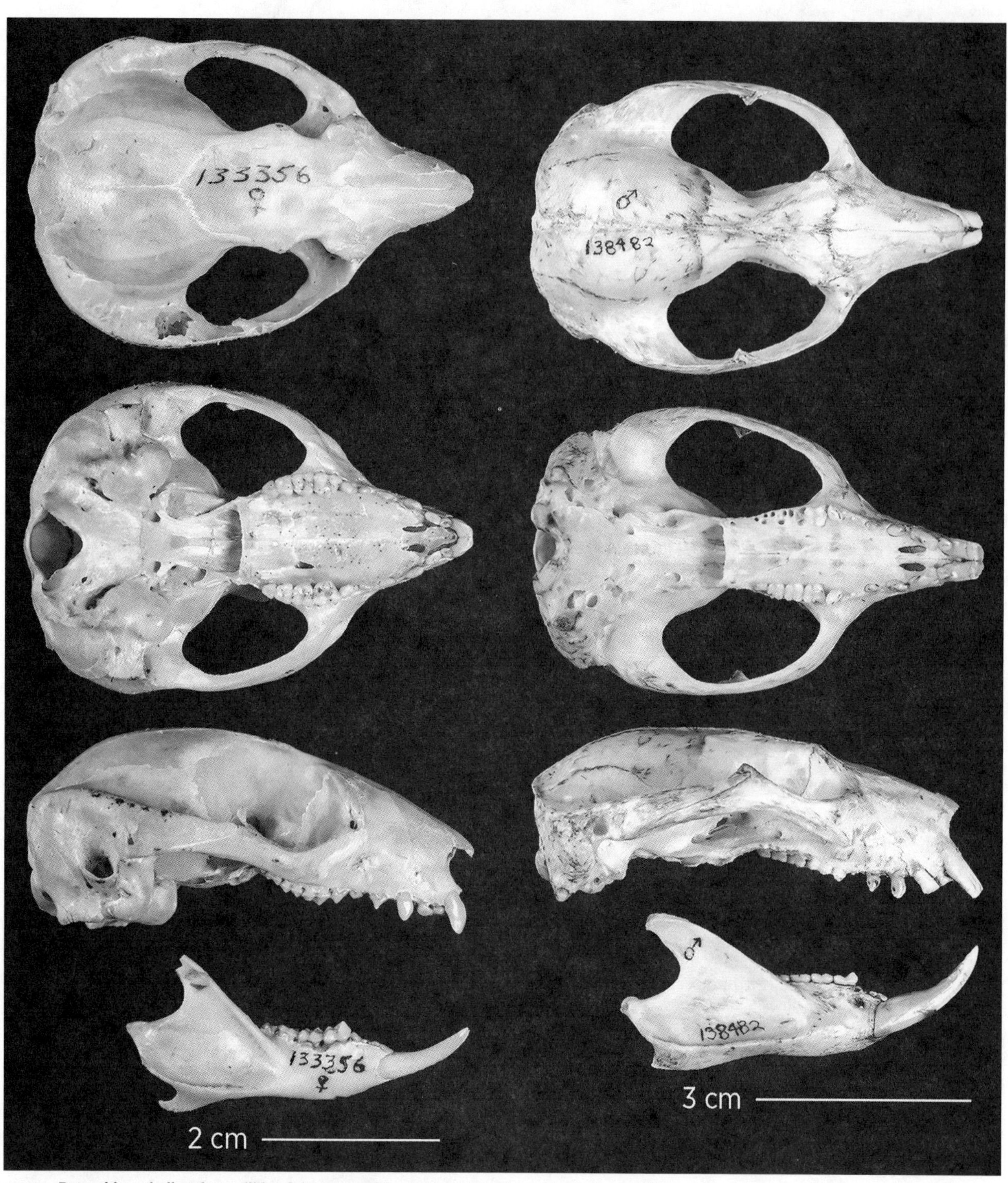

FIG. 79. Petauridae; skull and mandible of the sugar glider, *Petaurus breviceps* (Petaurinae; *left*), and the striped possum, *Dactylopsila trivirgata* (Dactylopsilinae; *right*).

Family PSEUDOCHEIRIDAE (ring-tailed possums, greater glider)

Medium-sized (mass 0.5–2 kg); most have well-developed prehensile tail; one genus (*Petauroides*) with gliding membrane that extends from ankle to elbow; digits 1–2 on forefeet opposable to 3–5; well-developed hallux on hind foot; pouch large, opening anteriorly; generally herbivorous, feeding on leaves; cecum large for microbial digestion; some species coprophagous; mostly arboreal and solitary. Habitat includes savanna forest, wet and dry sclerophyll forest, subalpine forest, rainforest, and oak and beech forests.

FIG. 80. Pseudocheiridae: Hemibelideinae; greater glider, *Petauroides volans*.

RECOGNITION CHARACTERS: Size and general characters similar to those of Phalangeridae; differences include the following:

1. **last upper and lower premolars not projecting above molar crowns** (fig. 81, arrow), **elongate, and multicuspidate**
2. canine small
3. 0–2 peg teeth on lower jaw between enlarged medial incisor and large premolar 3
4. molars with crescentic or triangular ridges

DENTAL FORMULA: $\frac{1\text{–}3}{1\text{–}2}\ \frac{1}{0}\ \frac{1\text{–}3}{1\text{–}3}\ \frac{4}{4} = 26\text{–}40$

2nd and 3rd upper incisors peg-like, similar to upper canine and to 1st upper premolars; 2nd lower incisor and 1st and 2nd lower premolars very small (see fig. 81)

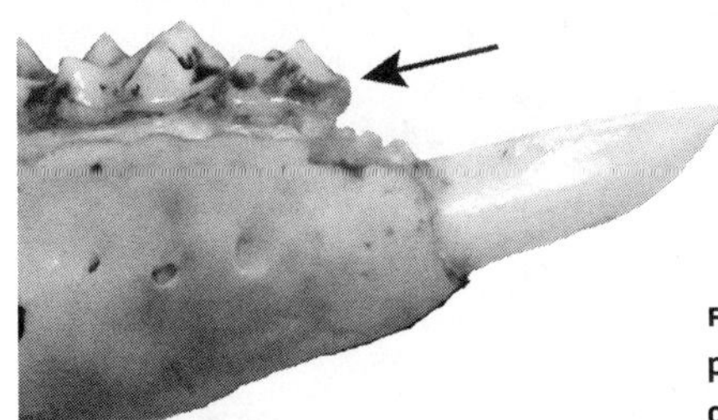

FIG. 81. Last lower premolar (pm3; arrow) of Pseudocheiridae.

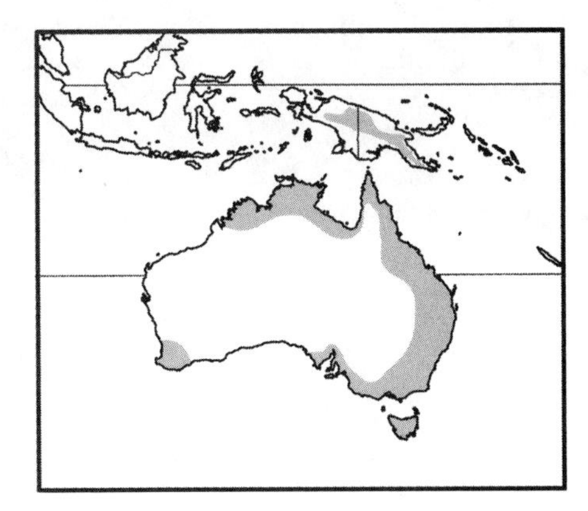

FIG. 82. Geographic range of family Pseudocheiridae: Australia, Tasmania, New Guinea.

TAXONOMIC DIVERSITY: Six genera and about 17 species in three subfamilies:

Hemibelideinae—*Hemibelideus* (lemuroid ring-tailed possum) and *Petauroides* (greater gliders)

Pseudocheirinae—*Pseudocheirus* (eastern and western ring-tailed possums) and *Pseudochirulus* (ring-tailed possums)

Pseudochiropsinae—*Petropseudes* (rock ring-tailed possum) and *Pseudochirops* (ring-tailed possums)

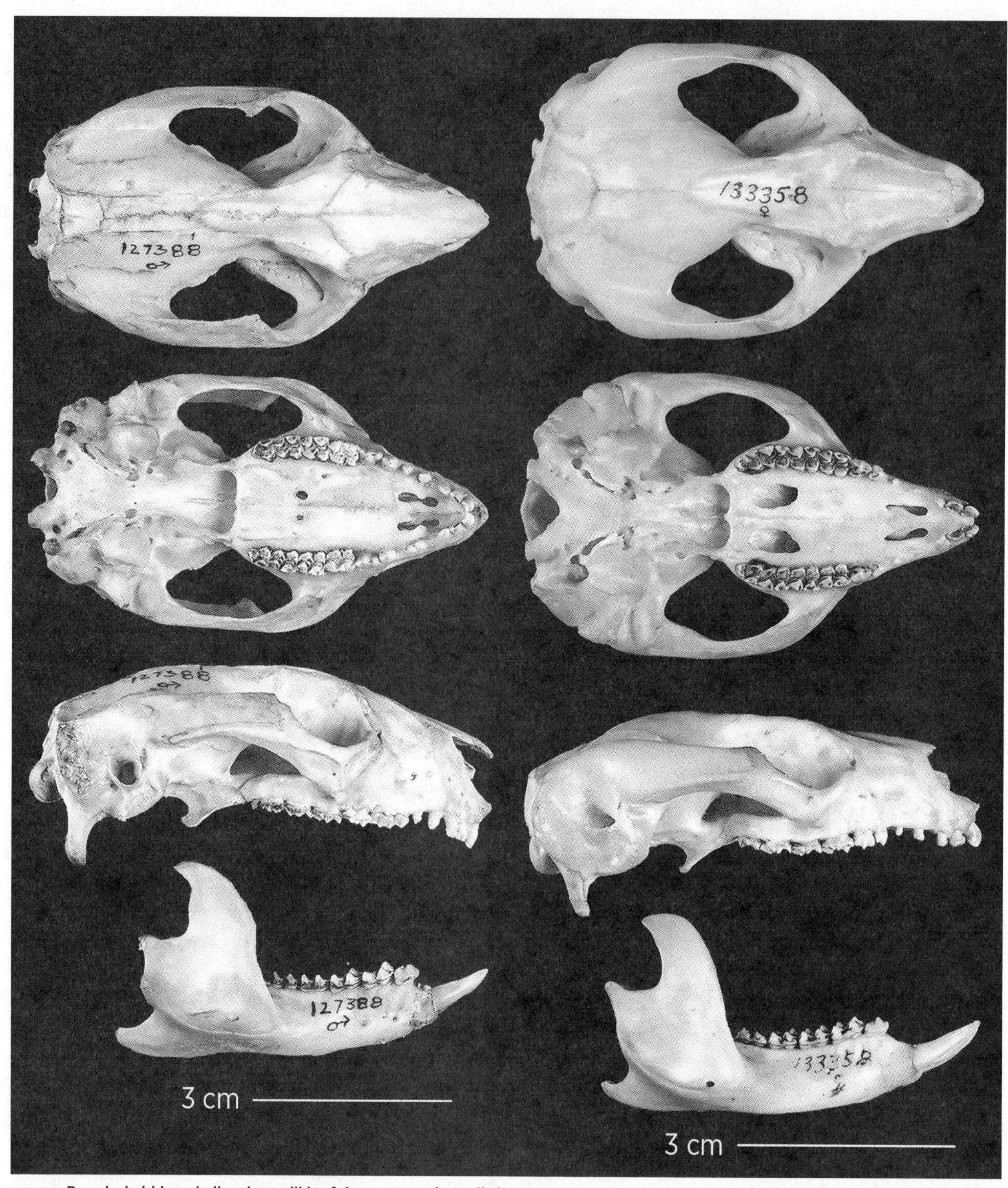

FIG. 83. Pseudocheiridae; skull and mandible of the common ring-tailed possum, *Pseudocheirus peregrinus* (Pseudocheirinae; *left*), and the greater glider, *Petauroides volans* (Hemibelideinae; *right*).

Family TARSIPEDIDAE (noolbenger or honey possum)

Tiny to small (body length 150–190 mm; mass 7–16 g); one of very few completely nectarivorous mammals, with elongate snout and long, protrusible tongue with brush tip used to gather pollen and nectar; elongate, sparsely haired tail and grasping feet that aid in climbing; digits 5/5, subequal in size, 2nd and 3rd digits of hind foot syndactylous; hallux well developed; pouch present in females, opening anteriorly. Nocturnal; become torpid when food is scarce or when cold; territorial; males with largest testes relative to body mass of any known mammal. Limited to flowering sand plain and scrub flora.

RECOGNITION CHARACTERS:

1. **cranium very long, narrow**
2. no sagittal crest
3. zygomatic arch very slender
4. paroccipital process minute or absent
5. **dentition degenerate—upper incisors small; single lower incisor, diprotodont, extremely procumbent**
6. canines reduced in size
7. **cheek teeth peg-like, vestigial**

DENTAL FORMULA: $\frac{2}{1}\ \frac{1}{0}\ \frac{1}{0}\ \frac{3}{3} = 22$

(all teeth small in size and reduction to a maximum of 10 teeth is not uncommon)

TAXONOMIC DIVERSITY: The family is monotypic, with a single species, *Tarsipes rostratus*.

FIG. 84. Tarsipedidae; honey possum or noolbenger, *Tarsipes rostratus*.

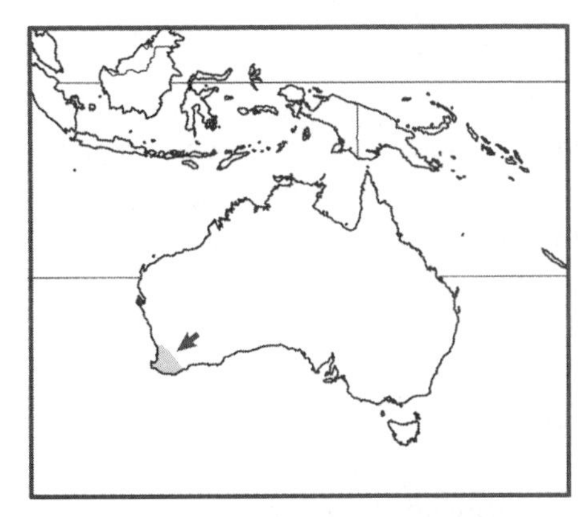

FIG. 85. Geographic range of family Tarsipedidae: extreme southwestern Australia.

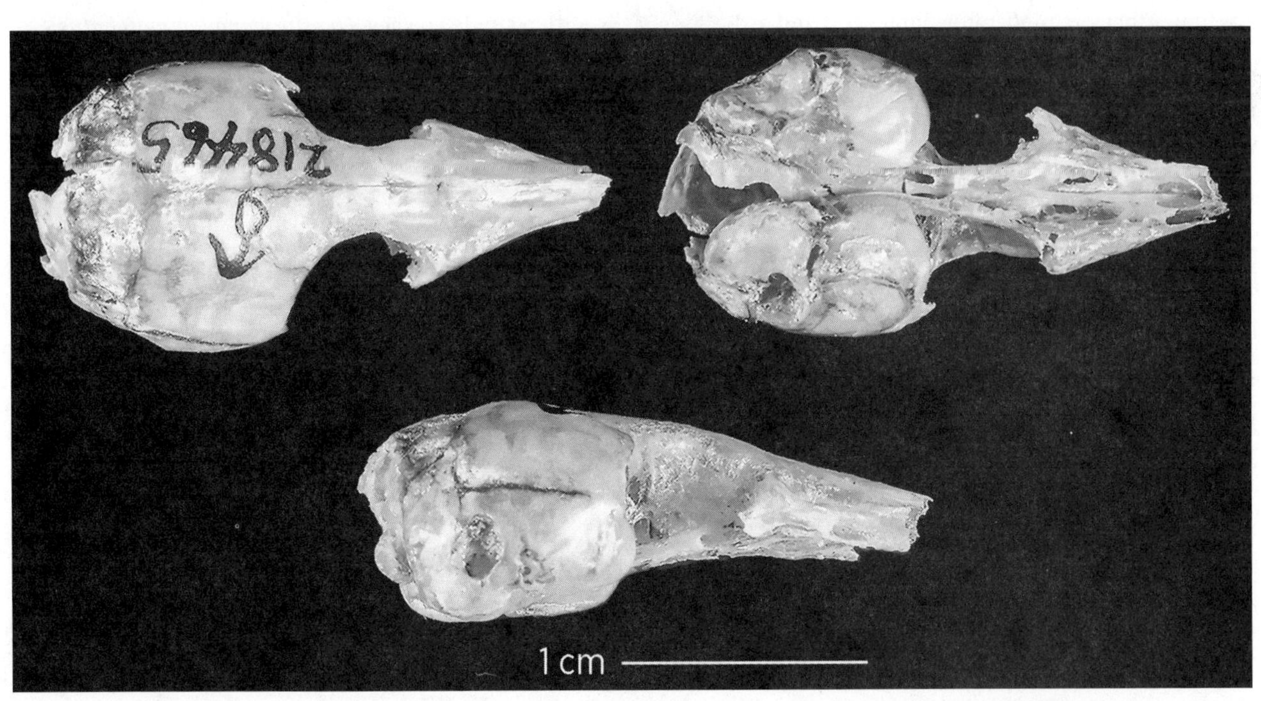

FIG. 86. Tarsipedidae; skull of the noolbenger, *Tarsipes rostratus*; note that the zygomatic arches have broken off this specimen.

Suborder MACROPODIFORMES

The three families in the suborder Macropodiformes share a common set of morphological attributes and were often grouped into a single family in older marsupial classifications. Because their shared traits collectively distinguish them from other diprotodonts, we list these traits first, then provide uniquely distinguishing features as well as general form and life history details for each family in the following accounts.

DIAGNOSTIC CHARACTERS:

- **hind limbs longer than forelimbs**
- **locomotion saltatorial**

RECOGNITION CHARACTERS:

1. body medium-sized to large (length 370–2,700 mm)
2. pouch present, opening anteriorly
3. tail long
4. **foot posture plantigrade**
5. **digits 5/4** (5/5 in *Hypsiprymnodon*); **two innermost digits of hind foot small and syndactylous** (2nd and 3rd in *Hypsiprymnodon*), **outer digits large**
6. **hallux usually vestigial or absent** (present in *Hypsiprymnodon*)
7. skull elongate
8. sagittal crest usually absent
9. zygomatic arch relatively slender
10. **paroccipital process very large**
11. deep masseteric fossa on lower jaw
12. upper incisors moderately large; 2nd and 3rd smaller and behind larger 1st
13. single pair of lower incisors, diprotodont, very procumbent
14. **upper canine, if present, relatively small**
15. large diastema separating incisors from cheek teeth
16. molars quadritubercular or bilophodont

Family HYPSIPRYMNODONTIDAE (musky rat-kangaroo)

Smallest of all macropods, with body length 150–270 mm, tail length 120–160 mm, and mass up to 680 g. Common name stems from musky odor emitted by both sexes. Dorsal color rich brown to rusty gray with pelage mostly short, velvety underfur; underside creamy tan. Tail naked, scaly, and prehensile; ears naked, thin, rounded, and dark in color. Both forefeet and hind feet have five toes with small and weak claws unequal in length. Hind limbs longer than forelimbs, as with other Macropodiformes, but proportionally shorter than in either Potoroidae or Macropodidae; locomotion by quadrupedal saltation. Females with four mammae in well-developed pouch. Both males and females reach sexual maturity at 1 year; litter size 2; postpartum pouch development lasts 270 days, at which time young have reached about half their adult weight. Diet diverse, including insects and worms, palm fruit, and tubers. Individuals generally solitary, but feeding aggregations of a few animals have been observed. Largely rainforest specialists, often present in dense vegetation bordering rivers and lakes. Clearing of forests for agricultural development has severely impacted their range.

RECOGNITION CHARACTERS:

1. small size (mass up to 680 g)
2. **tail naked and scaly, partially prehensile**
3. **digits 5/5**
4. movable, clawless hallux, but not opposable to other digits
5. 4th upper and lower premolar bladelike (plagiaulacoid); protruding above molar tooth rows
6. molars quadritubercular

DENTAL FORMULA: $\frac{3}{1}\ \frac{0\text{–}1}{0}\ \frac{2}{2}\ \frac{4}{4} = 32\text{–}34$

TAXONOMIC DIVERSITY: The family is monotypic, with a single species, *Hypsiprymnodon moschatus* (musky rat-kangaroo).

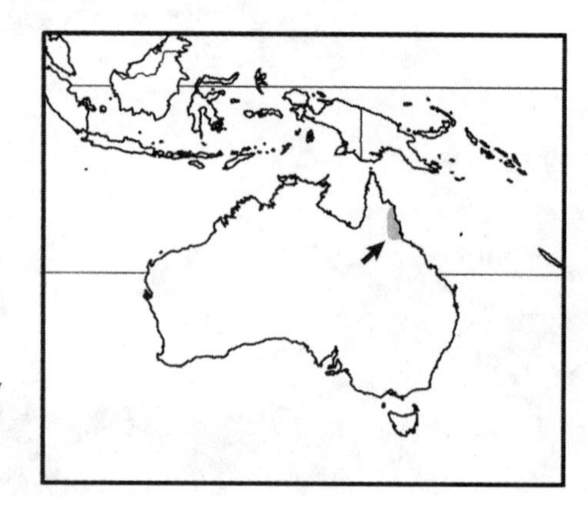

FIG. 87. Geographic range of family Hypsiprymnodontidae: tropical forests of northeastern Queensland, Australia (arrow).

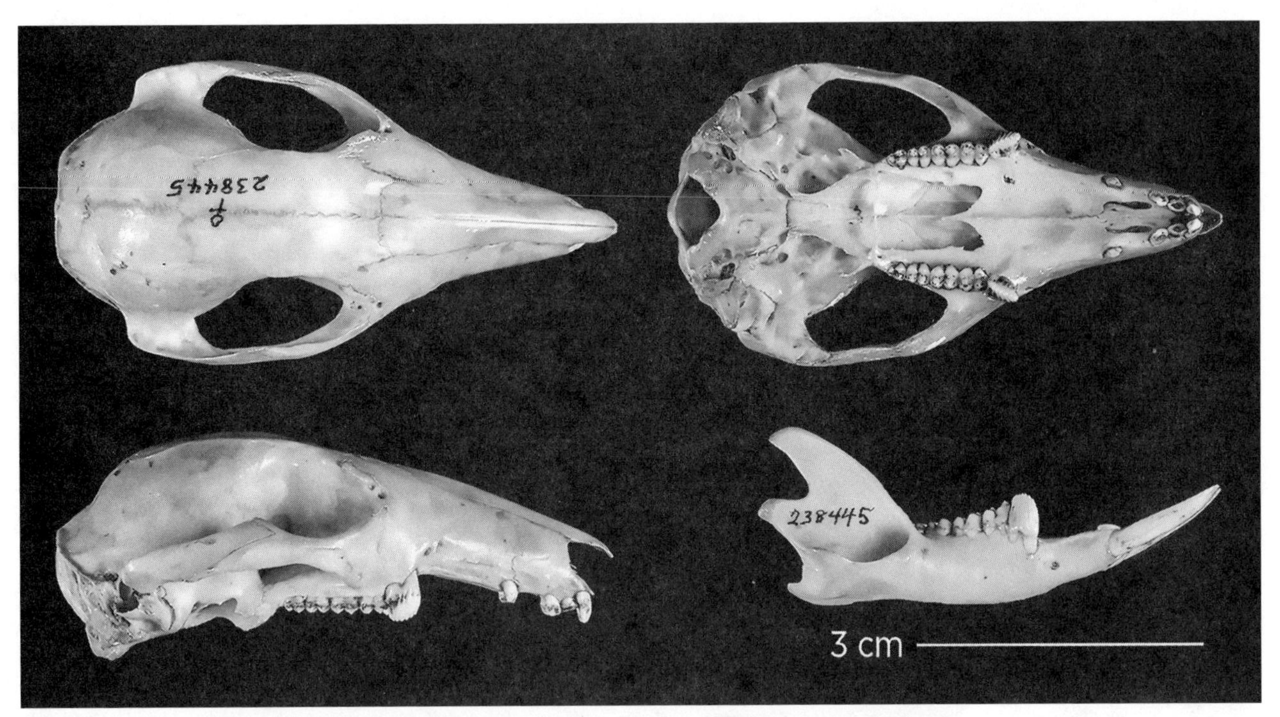

FIG. 88. Hypsiprymnodontidae; skull and mandible of the musky rat-kangaroo, *Hypsiprymnodon moschatus*.

Family MACROPODIDAE (kangaroos, pademelons, quokka, and wallabies)

Highly variable in size, but mostly medium-sized to large marsupials (mass 1–90 kg) with highly developed hind limbs and proportionally long and narrow hind feet (short and broad in the climber *Dendrolagus*) and notable for long, heavy, and strong tails used for balance both when in locomotion or when still; both characteristics of bipedal saltation (*Thylogale* often walk with quadrupedal gait). Hallux usually absent; 4th hind toe longest and strongest, transmitting main thrust of hopping (somewhat reduced in rock wallabies and tree kangaroos); 5th hind toe also notably large; syndactylous 2nd and 3rd hind toes have separate claws used in grooming. Pelage mostly gray to brown or reddish brown with lighter underside; tones vary markedly from pale to very dark in relation to habitat. Female reproductive cycle characterized by period of embryonic diapause when blastocyst suspends implantation; at times, females of most species may support young of three litters, one in the uterus, one residing full time attached to a nipple in the pouch, and a third living out of the pouch but returning to it to nurse or for security. All macropodids are grazers and browsers with a complex sacculated stomach for gastric fermentation by microorganisms; some species regurgitate food for additional mastication. Procumbent lower incisors do not occlude with upper incisors but press into tough pad on roof of mouth just posterior to upper incisors, as in bovid and cervid Artiodactyla. Most are nocturnal, a few are either diurnal or crepuscular. Species may be solitary, associating only for mating; mostly solitary but congregate at rich food patches; or gregarious, forming groups of conspecifics with coordinated behavior. Mating system poorly known for most, but some species are strongly polygynous, with a single, dominant male responsible for a large proportion of offspring.

RECOGNITION CHARACTERS:

1. medium-sized to large (mass up 90 kg)
2. **tail haired; non-prehensile**
3. bipedal saltator
4. **digits 5/4** (hallux usually missing)
5. upper incisors subequal; arcade rounded with 2nd and 3rd lateral to 1st
6. upper canine lacking or small
7. premolars less bladelike; occlusal surface in line with molar tooth rows
8. molars typically transversely bilophodont
9. molars move forward with successive eruption to eventually push out premolars

FIG. 89. Macropodidae: Macropodinae; red kangaroo, *Macropus rufus*.

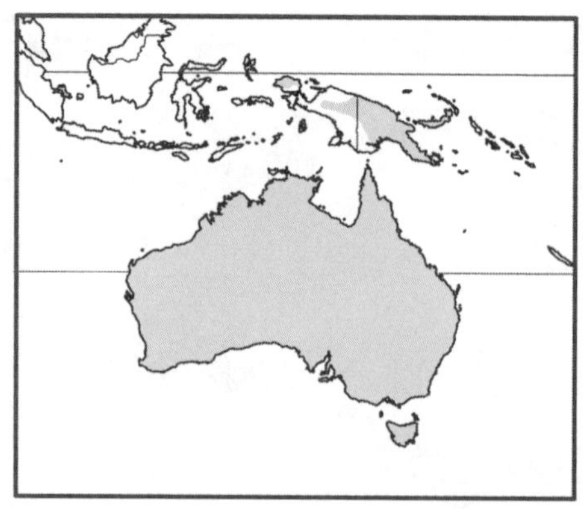

FIG. 90.
Geographic range of family Macropodidae: throughout Australia and New Guinea and nearby islands in sandy to stony deserts, grasslands, dry woodlands, and subtropical forests and rainforests.

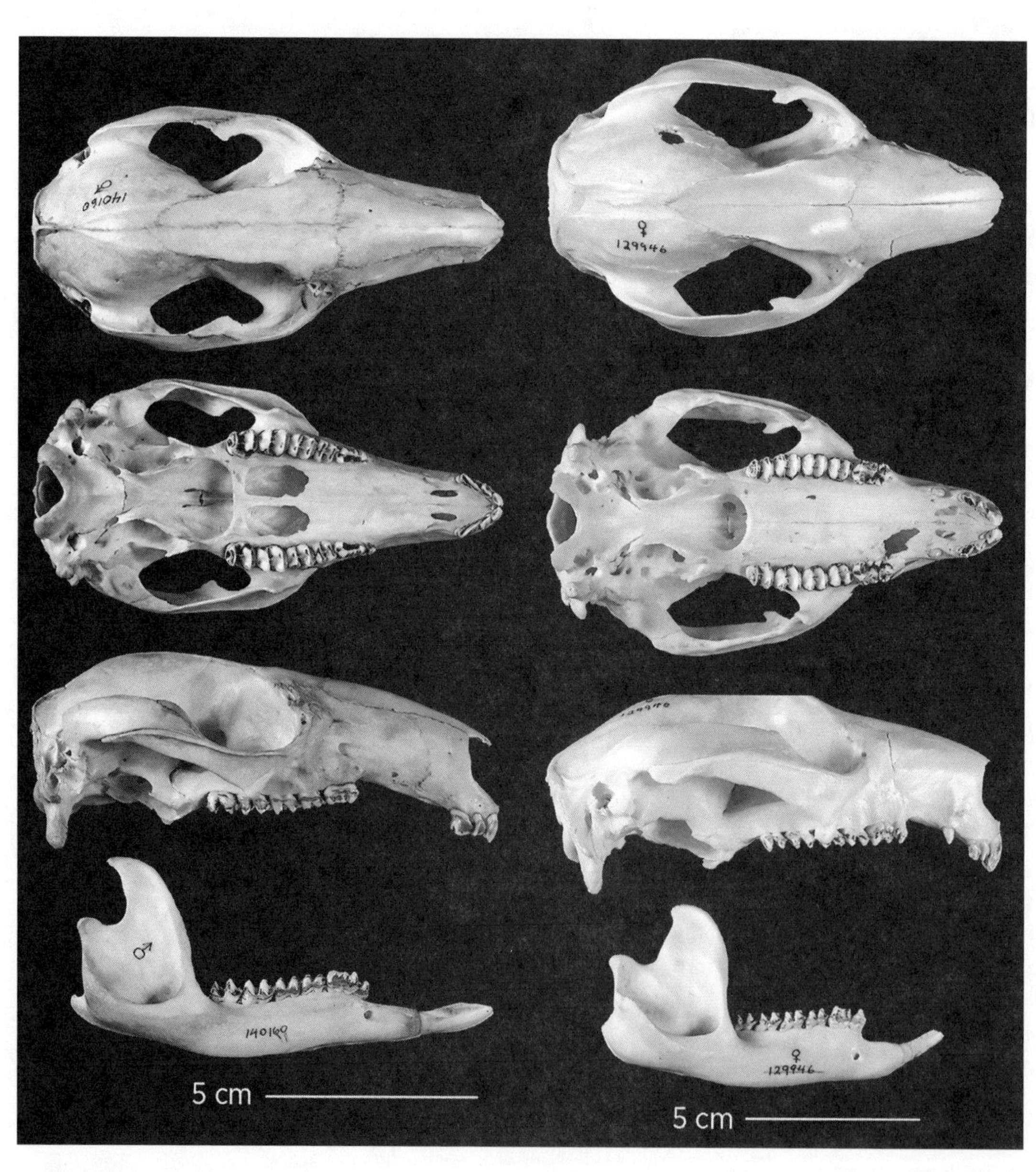

FIG. 91. Macropodidae: Macropodinae; skull and mandible of the dusky pademelon, *Thylogale brunii* (*left*), and Goodfellow's tree kangaroo, *Dendrolagus goodfellowi* (*right*).

DENTAL FORMULA: $\frac{3}{1}\ \frac{0\text{–}1}{0}\ \frac{2}{2}\ \frac{4}{4} = 32\text{–}34$
(young have two bladelike deciduous upper premolars, both of which are replaced by a single bladelike permanent premolar)

TAXONOMIC DIVERSITY: Second most speciose family of marsupials (after Didelphidae), with about 60 species in 13 genera divided into two subfamilies:
Lagostrophinae—*Lagostrophus* (banded hare-wallaby)
Macropodinae—*Dendrolagus* (tree kangaroos), *Dorcopsis* (forest wallabies), *Dorcopsulus* (forest wallabies), *Lagorchestes* (hare-wallabies), *Macropus* (grey kangaroos), *Notamacropus* (wallabies), *Onychogalea* (nail-tailed wallabies), *Osphranter* (antilopine wallaroo, black wallaroo, common wallaroo, red kangaroo), *Petrogale* (rock wallabies), *Setonix* (quokka), *Thylogale* (pademelons), and *Wallabia* (swamp wallaby)

Family POTOROIDAE (bettongs, potoroos, and rat-kangaroos)

Small to medium-sized marsupials (body 150–420 mm, tail 120–390 mm, mass 0.4–3.5 kg) with elongate, partially prehensile, and well-furred tails (often ending in a brush), long hind legs and feet, and shortened forelegs and feet. Locomotion by pentapodal walk, faster quadrupedal bound, and at high speeds, bipedal hopping. Hind feet with four toes, lacking hallux, with dominant 4th toe transmitting thrust of hopping, as in macropodids. Inner digits of forefeet with long, well-developed claws used in digging for food. Tail prehensile in bettongs, used to carry nesting material; only weakly prehensile in potoroos; some bettongs have a terminal brush. Pelage short and dense, generally uniform gray-brown with paler underparts. Females with well-developed, anteriorly opening pouch; reproductive pattern includes embryonic diapause, as in macropodids. There is no defined breeding season. Food habits mycophagous or omnivorous, with fungi constituting major portion of diet, but plant material and invertebrates are also consumed; stomach is less elaborately sacculated than in macropodids, being unspecialized in some, and having a few simple chambers with bacterial fermentation in others. Primarily nocturnal, with activity often highest in early night; daytime spent in elaborate, well-constructed nests made of plant material. Most are solitary, foraging and nesting alone, except for mothers with young or during courtship. A few may form stable associations between a single male and one or two females, and small aggregations form to exploit concentrated food patches. Mating system ranges from promiscuity to serial monogamy.

RECOGNITION CHARACTERS:

1. small to medium-sized (mass up to 3.5 kg)
2. **tail haired, some with terminal brush; prehensile to semi-prehensile**
3. bipedal saltators
4. **digits 5/4** (hallux absent)
5. 1st upper incisor larger than 2nd and 3rd; arcade narrow and pointed, with 2nd and 3rd behind 1st
6. upper canine well developed
7. premolars 1/1, distinctly elongate blades (plagiaulacoid)
8. molars quadritubercular
9. molars stable in position throughout life, do not push out premolars with age

DENTAL FORMULA: $\frac{3}{1}\ \frac{1}{0}\ \frac{1}{1}\ \frac{4}{4} = 30$

TAXONOMIC DIVERSITY: Eight species in three extant genera in two subfamilies:
Bettonginae—*Aepyprymnus* (rufous bettong) and *Bettongia* (bettongs)
Potoroinae—*Potorus* (potoroos) [a fourth genus, *Caloprymnus*, the desert rat-kangaroo, recently extinct].

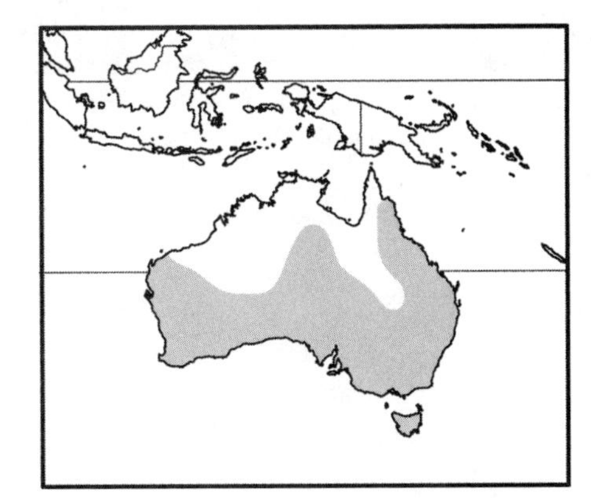

FIG. 92. Geographic range of family Potoroidae: open woodlands, subtropical and tropical forests, sclerophyll forests, and arid spinifex grasslands of eastern, southern, and western Australia and Tasmania.

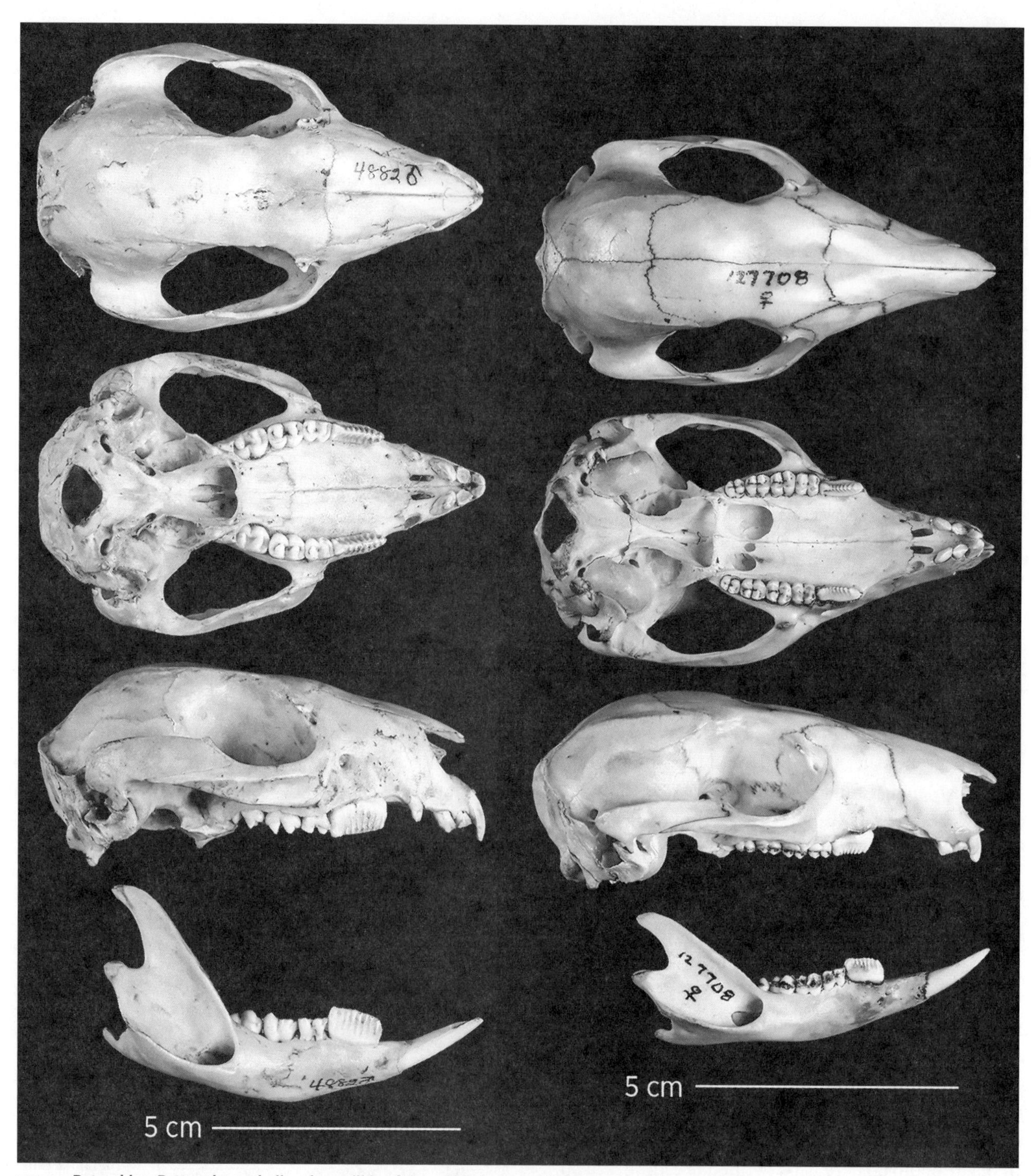

FIG. 93. Potoroidae: Bettonginae; skull and mandible of the rufous bettong, *Aepyprymnus rufescens* (*left*), and the eastern bettong, *Bettongia gaimardi* (*right*).

Infraclass EUTHERIA or PLACENTALIA

The terms Eutheria and Placentalia are often used synonymously, but in the phylogenetic classification of mammals, Eutheria is the clade that includes all mammals with the following set of characters, including those known only from the Jurassic and Cretaceous as fossils. The term Placentalia is used for the more exclusive "crown group" that contains all Recent eutherian mammals.

RECOGNITION CHARACTERS OF EUTHERIA:

1. epipubic bones not present
2. braincase large; cerebral hemispheres large
3. jugal not forming part of mandibular fossa
4. alisphenoid not contributing to auditory bulla
5. angular process generally not inflected
6. palatal vacuities usually not present
7. no stylar shelf on upper tribosphenic molars
8. entoconid, hypoconulid, and hypoconid equidistant on talonid

RECOGNITION CHARACTERS OF PLACENTALIA (IN ADDITION TO THOSE LISTED ABOVE FOR EUTHERIA):

1. pouch not present
2. single vagina
3. uteri medial to ureters; fully fused in some groups to form simple uterus
4. male phallus not divided

Supermatrix analyses of morphological and molecular characters have converged on the same general hierarchical groupings of living placental mammals, although relationships among a few remain uncertain (figs. 94, 95, and see Gatesy et al. 2017; Esselstyn et al. 2017). These analyses have corroborated most of the traditional ordinal groupings, with two major exceptions: first, the Insectivora of older literature has become widely recognized as an assemblage of unrelated animals now split into separate orders. Second, the even-toed ungulates (Artiodactyla) are now understood as paraphyletic, with whales (Cetacea) nested within the Artiodactyla as the sister lineage to the Hippopotamidae; this combined group is referred to as the Cetartiodactyla, although some recent classifications (e.g., Wilson and Reeder 2005) retain Artiodactyla and Cetacea as separate orders (see also Asher and Helgen 2010), as we do in this manual.

Moreover, the role that plate tectonics has played in mammalian diversification is evident when current understanding of higher mammalian relationships is superimposed on the historical breakup of Pangaea. Subsequent evolution on each resulting supercontinent produced the major groups of placental mammals: the Afrotheria and Xenarthra, loosely aligned in the Atlantogenata, hail from Gondwanaland, while the Boreoeutheria, including the Euarchontoglires and Laurasiatheria, arose on Laurasia.

The history of the "Insectivora" provides a compelling lesson in convergence and associated taxonomic challenges. As employed over much of the twentieth century, the order Insectivora included a number of families of largely dilambdodont or zalambdodont, small-bodied, insectivorous eutherians that were divided into two suborders: **Lipotyphla** (Soricidae [shrews], Talpidae [moles], Erinaceidae [hedgehogs], Solenodontidae [solenodonts], Tenrecidae [tenrecs], and Chrysochloridae [golden moles]) and **Menotyphla** (Macroscelididae [elephant shrews] and Tupaiidae [tree shrews]). Other classifications removed the elephant shrews, placing them in the order Macroscelidea, while retaining the remaining families within the Insectivora. While a suite of uniquely derived characters diagnosed each of these families, they were united in these early classificatory schemes only by shared-primitive (symplesiomorphic) features. Even Simpson, in his 1945 classification of mammals (see Simpson 1945, 175–176, for an overview of this history), recognized that this assemblage of "insectivorous" eutherians was unnatural (often referred to as a "waste-basket") and that the classificatory groups he proposed "improved the situation somewhat, but not much."

Fortunately, cladistic analyses of both morphological and molecular characters have resolved the conundrum of "insectivore" relationships, and thus the classification of these disparate families. The taxon Menotyphla is clearly paraphyletic, as elephant shrews (Macroscelidea) are strongly supported as the sister to a clade uniting tenrecs and golden moles (grouped now in the order Tenrecoidea). In turn, this pair of lineages is phylogenetically placed within the Afrotheria (along with the aardvark, hyraxes, elephants, and sirenians). Furthermore, the Afrotheria has its origins in Gondwanaland, but all remaining "insectivores" evolved from groups whose origins were in Laurasia; hence, these groups are only distantly related. The remaining menotyphlan group, the tree shrews, is recognized as

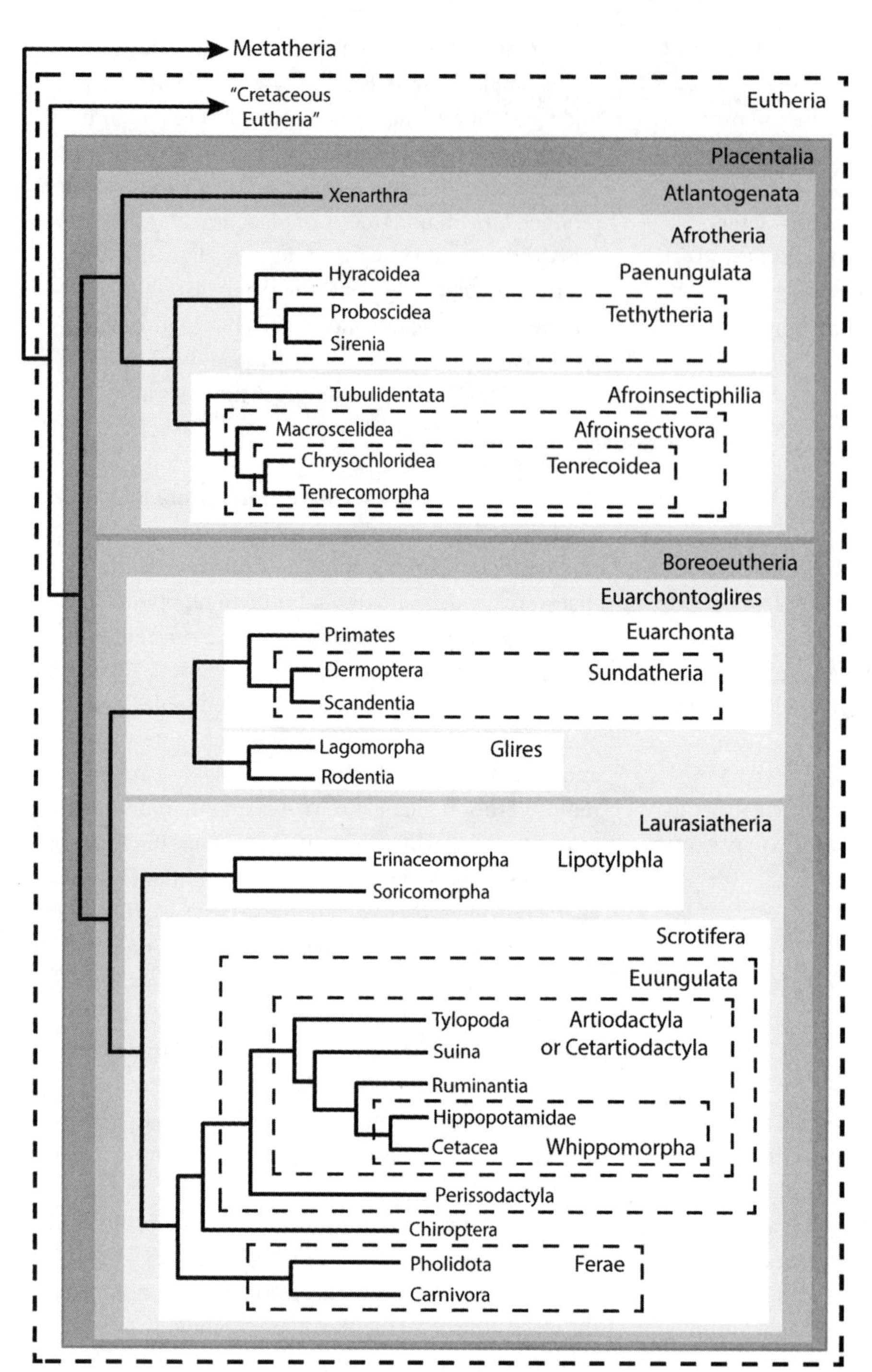

FIG. 94. High-level clades of the Placentalia, the crown group of living Eutheria. Commonly used clade names (bold) and the included taxa (orders, suborders, or families) are indicated. The tree drawn is fully resolved (after Asher and Helgen 2010), although some nodes remain questionable (see Fig. 95). Note that for practical purposes in this manual, we treat whales separately from hippopotamuses and remaining artiodactyls (or cetartiodactyls).

a member of the Euarchonta; they are a sister group to the colugos (Dermoptera), united with them in the clade Sundatheria and in turn sister to the Primates. All remaining lipotyphlan insectivores (the hedgehogs, shrews, moles, and solenodonts) group as basal to pangolins, bats, carnivorans, odd-toed ungulates, and the whales + even-toed ungulates in the second major Laurasian lineage, the Laurasiatheria. In summary, then, both lipotyphlan and menotyphlan families have been reallocated either to the Afrotheria or to one of the two clades within the Boreoeutheria. These lineages are not only phylogenetically distinct, they evolved on entirely different landmasses!

Clade ATLANTOGENATA

The clade Atlantogenata unites the two groups of living eutherian mammals that originated in the western

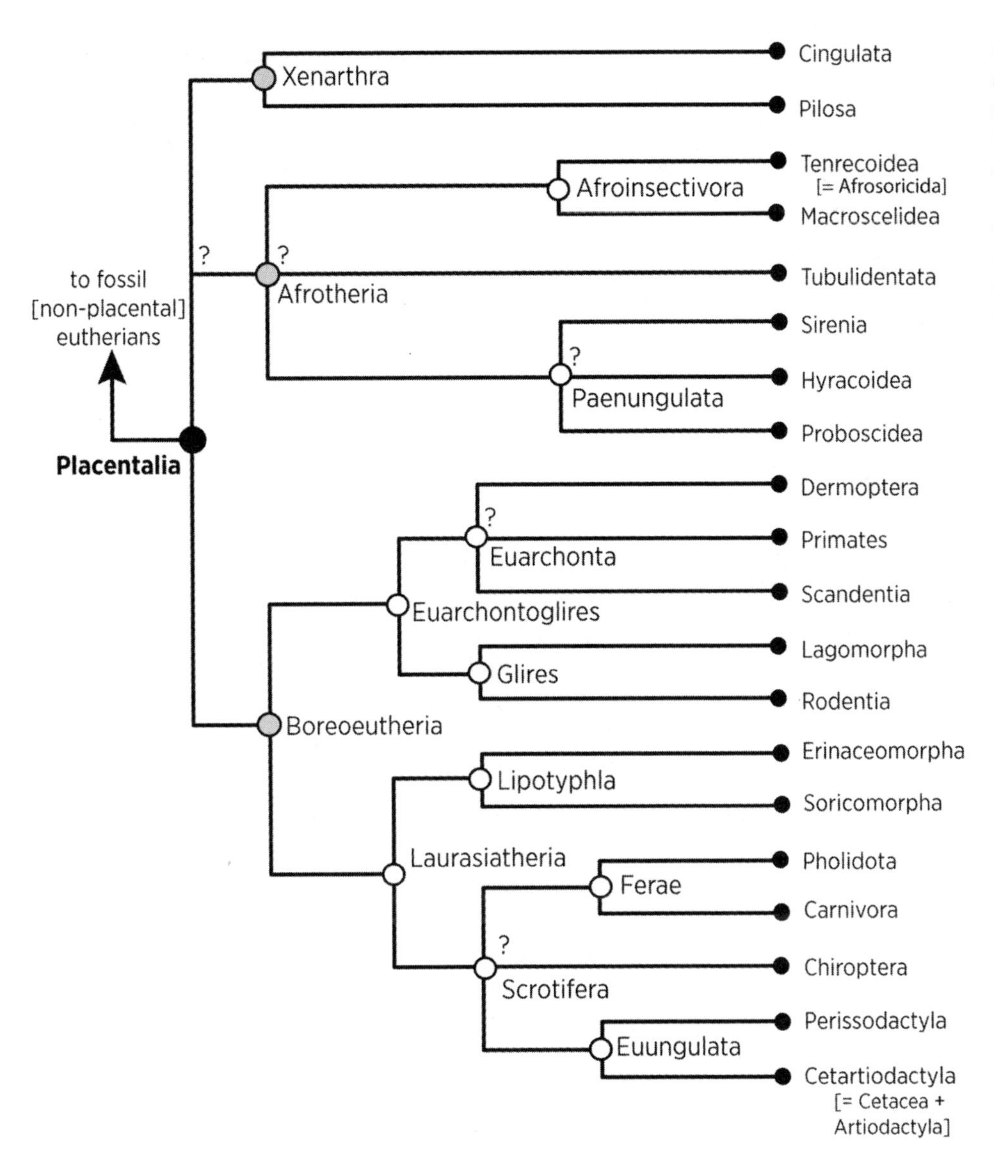

FIG. 95. Consensus tree depicting phylogenetic relationships among the Placentalia. The tree is derived from DNA sequences (Bininda-Emonds et al. 2007; dos Reis et al. 2012), amino acid sequences (Meredith et al. 2011), and morphological characters (O'Leary et al. 2013). Names commonly used for several internal nodes (white circles) grouping related orders are given. A "?" identifies those nodes that remain unresolved because of conflict between different data sets and/or analyses.

part of the southern supercontinent Gondwanaland in the mid- to late Cretaceous: the New World Xenarthra (the sloths, armadillos, and anteaters) and the Old World Afrotheria (the hyraxes, elephants, sirenians, tubulidentates, elephant shrews, golden moles, and tenrecs; see fig. 94). Although the Xenarthra and Afrotheria are united under Atlantogenata, the majority of their history has occurred on isolated landmasses; as such, these clades are more distantly related than the two clades that evolved on the northern supercontinent, Laurasia (see discussion under Boreoeutheria).

Clade XENARTHRA

Strictly New World in distribution, the clade Xenarthra is currently almost solely Neotropical, with the exception of one species of armadillo that continues to fare well in arid regions of North America. Older classifications treated xenarthrans as a single order, named the Edentata, with three families. The group is now divided into two orders (fig. 96): the Cingulata, containing only the armadillos (Dasypodidae), and the Pilosa, containing the tree sloths (suborder Folivora—Bradypodidae and Choloepodidae) and anteaters (suborder Vermilingua—Cyclopedidae and Myrmecophagidae).

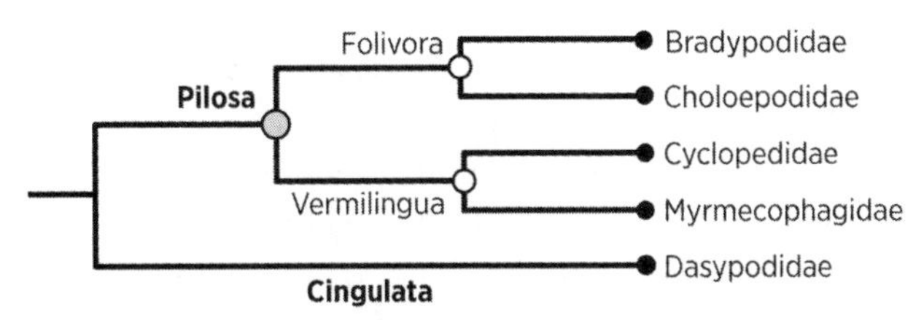

FIG. 96. Hypothesized phylogenetic relationships among families of xenarthrans; ordinal and subordinal ranks are indicated.

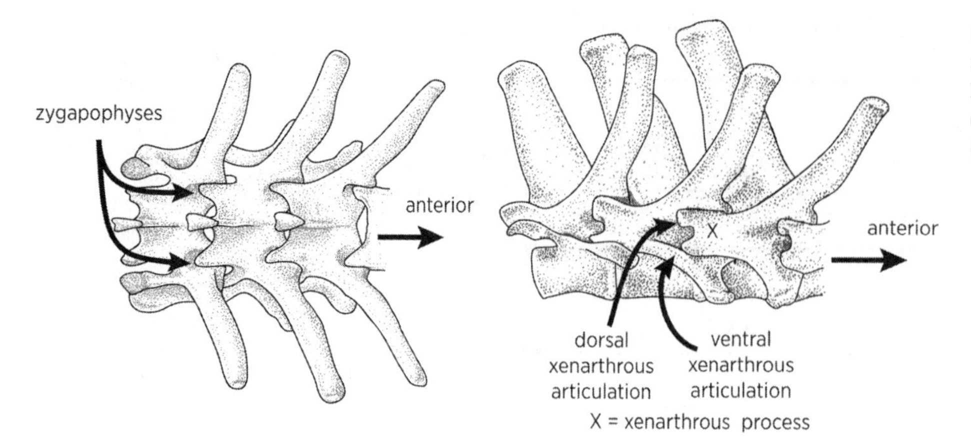

FIG. 97. Specialized, dual vertebral zygapophyses that unite all Xenarthra into a phylogenetic unit. Redrawn from Vaughan et al. (2015).

DIAGNOSTIC CHARACTER:

- **extra zygapophyses present on posterior thoracic and lumbar vertebrae** (fig. 97)

RECOGNITION CHARACTERS:

1. size small to medium-large (15–210 cm)
2. body covered with hair or scutes (the latter covered by scales)
3. **forefoot with two or three principal digits, each bearing long claws**; hind foot with two to five digits
4. no incisors or canines
5. **cheek teeth absent** (Myrmecophagidae) **or, if present, homodont and without enamel**
6. zygomatic arch complete (Cingulata) or incomplete (Pilosa)
7. **jugal, lacrimal, and interparietal bones present**
8. pterygoid bones variable, usually separate but meeting at midline to form part of palate in Myrmecophagidae

Order CINGULATA

The Cingulata includes all extant armadillos and extinct armored animals such as glyptodonts (†Glyptodontidae) and pampatheres (†Pampatheriidae). Delsuc et al. (2016), using ancient DNA technology, convincingly demonstrated that glyptodonts belong to a clade containing all extant armadillos except *Dasypus*; these authors thus divided the traditional Dasypodidae into two families, restricting the Dasypodidae to envelop *Dasypus* alone and elevating Chlamyphoridae to contain the remaining extant armadillos plus the extinct glyptodonts. Because all armadillos share such a distinctive and easily recognized set of morphological traits, for simplicity we treat all extant Cingulata together.

Families CHLAMYPHORIDAE and DASYPODIDAE (armadillos)

Size ranges from small (body length 150 mm, mass 120 g in fairy armadillos) to large (mass about 30 kg in giant armadillo); compact, heavy body; legs short, subequal in size; strong claws on forefeet and hind feet; armored dorsally, with ossified skin consisting of body (dermal, unlike pangolins) scutes, typically in five shields (head, scapular, dorsal, pelvic, and caudal); scutes covered with keratinous epidermal plates and scales; movable bands between plates; snout short to long; tail usually armored. Insectivorous or omnivorous; tongue long and protrusible. Fossorial, nesting in self-dug burrows. Occupy wide range of habitats, from forests, savannas, and plains and steppe to deserts, primarily in subtropics and tropics.

DIAGNOSTIC CHARACTER:

- **body covered with keratin scutes overlying bony plates**

RECOGNITION CHARACTERS:

1. **skull elongate, flattened; rostrum elongate**
2. zygomatic arch complete
3. **cheek teeth homodont**, subcylindrical; molars and premolars indistinguishable
4. premaxilla well developed
5. jugal well developed, lacrimal present, interparietal present but often indistinct
6. pterygoids usually separate, not forming part of palate (joined at midline in *Dasypus*)

DENTAL FORMULA: $\frac{0}{0}\ \frac{0}{0}\ \frac{7\text{–}25}{7\text{–}25} = 28\text{–}100$

(molars and premolars indistinguishable)

FIG. 98. Dasypodidae; nine-banded armadillo, *Dasypus novemcinctus*.

TAXONOMIC DIVERSITY: About 21 species in nine genera in two families, one with four subfamilies (including the extinct Glyptodontinae; see Delsuc et al. 2016):

Chlamyphoridae:

- Chlamyphorinae—*Calyptophractus* (fairy armadillo) and *Chlamyphorus* (fairy armadillo)
- Euphractinae—*Chaetophractus* (hairy armadillos), *Euphractus* (six-banded armadillos), and *Zaedyus* (pichi)
- Tolypeutinae—*Cabassous* (naked-tailed armadillos), *Priodontes* (giant armadillo), and *Tolypeutes* (three-banded armadillos)

Dasypodidae: *Dasypus* (long-nosed armadillos)

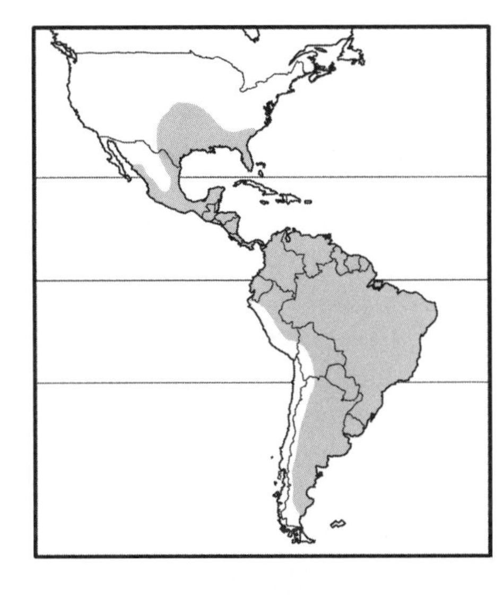

FIG. 99. Combined geographic range of families Dasyproctidae and Chlamyphoridae: central and southeastern United States south to southern South America.

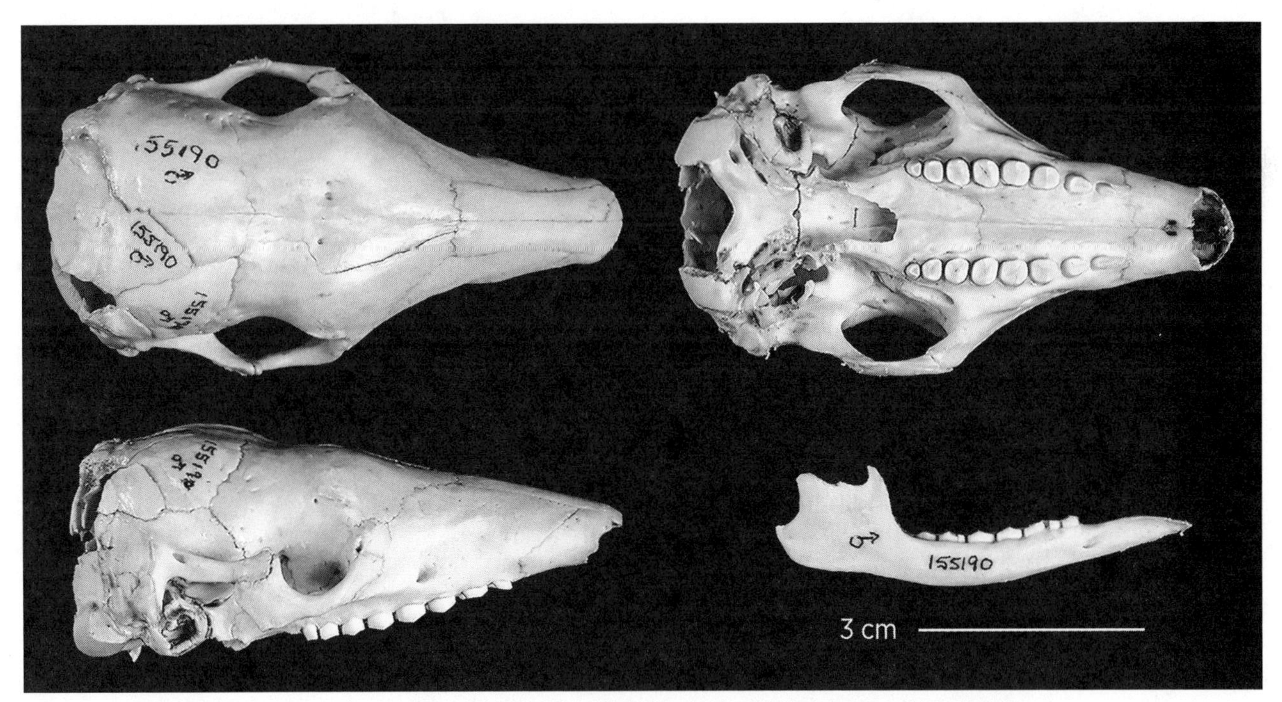

FIG. 100. Chlamyphoridae: Tolypeutinae; skull and mandible of a naked-tailed armadillo, *Cabassous unicinctus*.

Order PILOSA

The Pilosa includes all furred xenarthrans (hence the name), which morphologically and phylogenetically fall into two suborders, one containing extant sloths and extinct ground sloths (Folivora), and the other containing the anteaters (Vermilingua).

Suborder FOLIVORA (sloths)

The two living tree sloth genera (*Bradypus* and *Choloepus*) were traditionally placed in the same family, Bradypodidae, until character analyses suggested that two-toed sloths (*Choloepus*) were aligned with extinct ground sloths of the family Megalonychidae. Recent research, however, integrating ancient DNA from extinct sloth species, has clarified that *Choloepus* is more closely related to sloths in the Mylodontidae† than to those in the Megalodontidae (Delsuc et al. 2019; Presslee et al. 2019). As such, the Folivora now is treated as comprising three superfamilies: Megalocnoidea†, including two extinct families of Caribbean sloths; Mylodontoidea, including the Mylodontidae† and the extant Choloepodidae; and Megatherioidea, including Megatheriidae†, Megalonychidae†, Nothrotheriidae†, and the extant Bradypodidae. As we use it here, Folivora is equivalent to Phyllophaga of some other authors. Sloths are well known for their very slow movement and upside-down posture. Sloth fur may support red and green algae in longitudinal grooves, which presumably serve for camouflage. Some sloths host flightless moths (some of which are unique to sloths) that are dependent on the sloths for every step of their life cycle and which may serve, when dead, to fertilize the algae (Pauli et al. 2014).

DIAGNOSTIC CHARACTERS:

- **tail very small (long in other xenarthrans)**
- **skull blocky, snout blunt (both longer in other xenarthrans)**

RECOGNITION CHARACTERS:

1. **forelimbs much longer than hind limbs**
2. tongue unspecialized
3. zygomatic arch incomplete
4. premaxilla very small
5. jugal well developed
6. pterygoids separate, not forming part of palate
7. all cheek teeth homodont; molars and premolars indistinguishable

Family BRADYPODIDAE (three-toed sloths)

Body large (total length 520 mm; mass up to 4.5 kg); face small, with rounded head, very small ears, and long neck possessing eight or nine cervical vertebrae (most mammals have seven); forelimbs one and a half times longer than hind limbs; manus and pes with three digits, syndactylous except for long and sharp claws; stumpy tail very short but obvious; pelage shaggy, but hairs short and feathery in texture. Arboreal; folivorous, feeding solely on tree leaves; have color vision, poor hearing but acute sense of smell. Sexually dimorphic; males have a conspicuous speculum between the shoulder blades that lacks overhairs and therefore

FIG. 101. Bradypodidae; brown-throated three-toed sloth, *Bradypus variegatus*.

FIG. 102. Geographic range of family Bradypodidae: Neotropical forests from northern Argentina to Honduras.

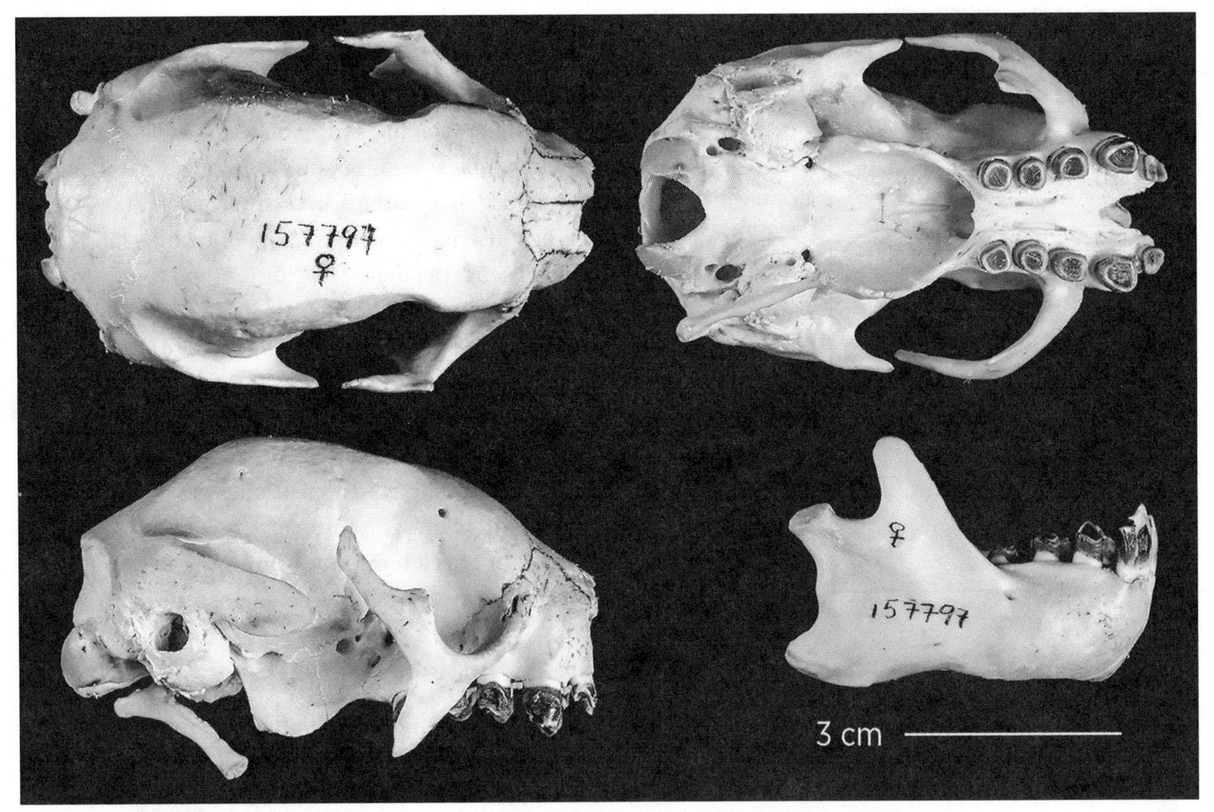

FIG. 103. Bradypodidae; skull and mandible of the brown-throated three-toed sloth, *Bradypus variegatus*.

shows black and white underfur and, in adults, yellow- or orange-stained hair. Restricted to closed-canopy tropical forests. Low but heterothermic metabolism, probably reflecting dietary restrictions.

DIAGNOSTIC CHARACTERS:

- tail short
- fur shaggy, but shorter and more feathery in texture relative to Choloepodidae
- **3 toes (with claws) on forefeet**
- **1st pair of cheek teeth not caniniform**
- **angular process enlarged, projecting posteroventrally**

DENTAL FORMULA: $\frac{0}{0}\ \frac{0}{0}\ \frac{5}{4\text{–}5} = 18\text{–}20$
(molars and premolars indistinguishable)

TAXONOMIC DIVERSITY: The family is monotypic, with four species in the single genus *Bradypus*.

Family CHOLOEPODIDAE (two-toed sloths)

Size larger than Bradypodidae (body length 660 mm; mass 5.7 kg); face larger but head less rounded, shorter

FIG. 104. Choloepodidae; Hoffmann's two-toed sloth, *Choloepus hoffmanni*.

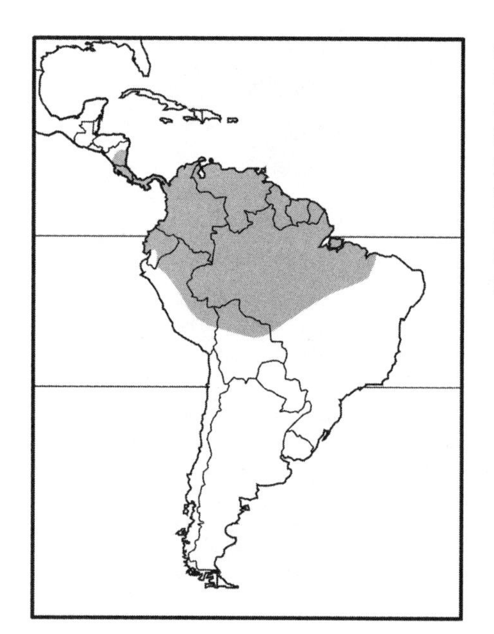

FIG. 105. Geographic range of family Choloepodidae: Neotropical forests from northern Argentina to Honduras.

and less flexible neck with five to seven vertebrae (occasionally eight); forelimbs only slightly longer than hind limbs; manus with two syndactylous digits, pes with three; each digit with long, curved claw; tail rudimentary; pelage shaggy but with longer hairs than in Bradypodidae, not feathery in texture. Arboreal folivore; nocturnal. Also have color vision, poor hearing, and well-developed sense of smell. Move faster than *Bradypus*. Also restricted to closed-canopy tropical forests, and also share low and heterothermic metabolic rate.

DIAGNOSTIC CHARACTERS:

- tail very short
- fur long and shaggy in appearance
- **2 toes (with claws) on forefeet**
- **1st pair of cheek teeth caniniform**
- **angular process small, directed posteriorly**

DENTAL FORMULA: $\frac{0}{0}\ \frac{0}{0}\ \frac{5}{4\text{–}5} = 18\text{–}20$

(molars and premolars indistinguishable)

TAXONOMIC DIVERSITY: The extant family is monotypic, with two species in the single genus *Choloepus*.

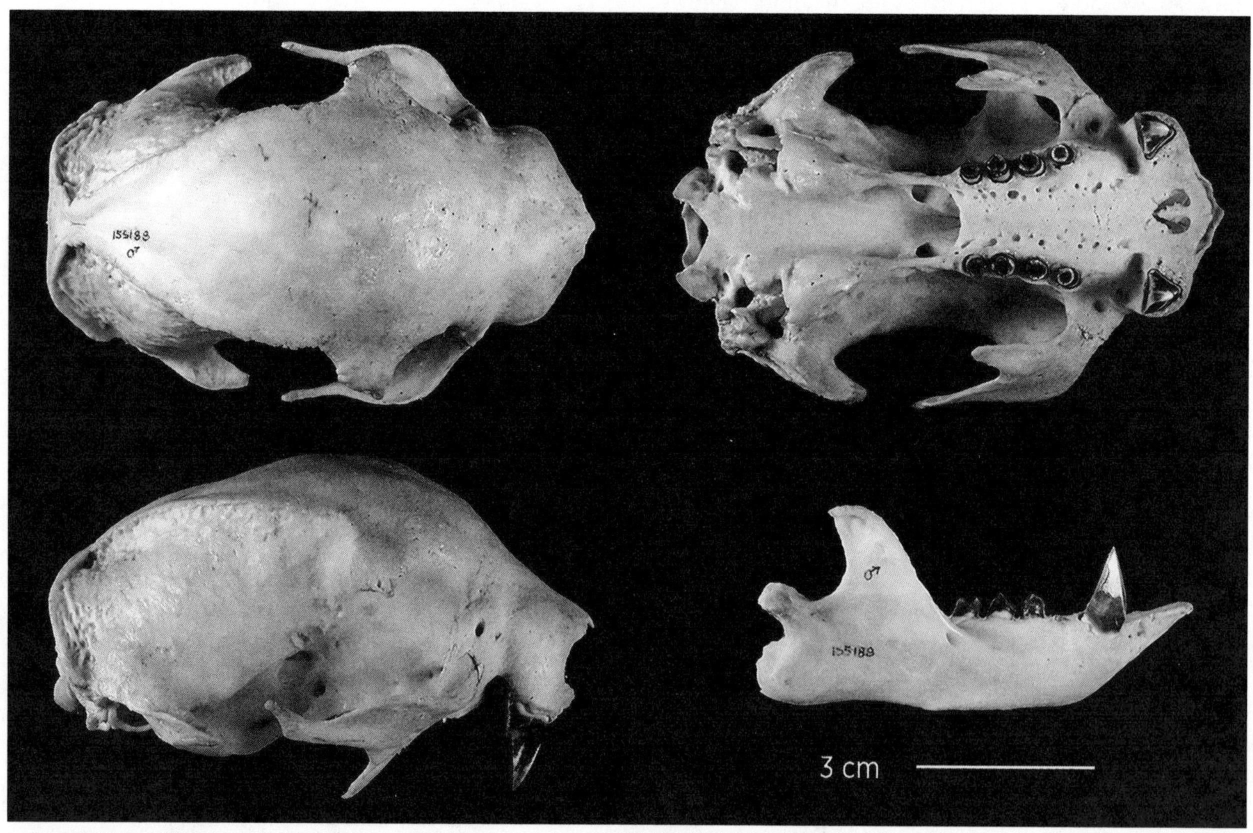

FIG. 106. Choloepodidae; skull and mandible of Hoffmann's two-toed sloth, *Choloepus hoffmanni*.

Suborder VERMILINGUA (anteaters)

The name Vermilingua reflects the long and wormlike tongues of this group, which are well adapted for extracting ants and termites from nests.

The anteaters have been traditionally placed in a single family (Myrmecophagidae), which was divided into two subfamilies: Cyclopodinae for the silky anteater, *Cyclopes*, and Myrmecophaginae for the tamandua and giant anteaters, *Tamandua* and *Myrmecophaga*. Gardner (2005) elevated these subfamilies to familial rank. We treat them together here, as the three genera share a number of synapomorphies relative to other xenarthrans.

Families CYCLOPEDIDAE and MYRMECOPHAGIDAE (anteaters)

Size varies extensively (*Cyclopes* mean total length 435 mm, mass 235 g; *Myrmecophaga* mean length 2 m, mass 33 kg); in *Tamandua* and *Cyclopes*, tail is long, partly naked at tip and ventrally for a longer stretch in *Cyclopes*, and prehensile; in *Myrmecophaga*, tail is rigid, non-prehensile, but crested dorsally and ventrally with long hair; forelimbs and hind limbs subequal in length; pelage either long and coarse (*Myrmecophaga*), short and coarse (*Tamandua*), or short, dense, and silky (*Cyclopes*); manus with greatly enlarged 3rd digit with long, heavy claw and other claws reduced; four to five subequal clawed pedal digits. Nocturnal, crepuscular, or diurnal; feed mostly on ants and termites. *Cyclopes* forages in trees, *Tamandua* both on ground and in trees, and *Myrmecophaga* is strictly terrestrial. Habitat ranges from savanna to subtropical woodlands to lowland tropical forests.

DIAGNOSTIC CHARACTER:

- **no teeth (= edentulous)**

RECOGNITION CHARACTERS:

1. **tongue very long, protrusible, wormlike**
2. **skull elongate; rostrum very long and curving downward**
3. zygomatic arch incomplete
4. premaxilla very small
5. **jugal present but small**, articulating with maxilla (not squamosal); **lacrimals and interparietals present**
6. **pterygoids extend posteriorly to auditory bullae; separate in Cyclopedidae but meet at midline along entire length in Myrmecophagidae**

DENTAL FORMULA: No teeth (= edentulous)

TAXONOMIC DIVERSITY: The New World anteaters are divided into two families, one monotypic and the other with two genera and a total of three species:

Cyclopedidae—*Cyclopes* (pygmy or silky anteater)

Myrmecophagidae—*Myrmecophaga* (giant anteater) and *Tamandua* (tamanduas or collared anteaters)

FIG. 107. Vermilingua; northern tamandua, *Tamandua mexicana* (Myrmecophagidae, *upper left*), silky anteater, *Cyclopes didactylus* (Cyclopedidae, *upper right*), and giant anteater, *Myrmecophaga tetradactyla* (Myrmecophagidae; *bottom*).

FIG. 108. Geographic range of family Cyclopedidae: Neotropics, from southern Mexico to eastern Bolivia and Amazonian Brazil.

FIG. 109. Geographic range of family Myrmecophagidae: Neotropics, from southern Mexico to northern Argentina and Uruguay.

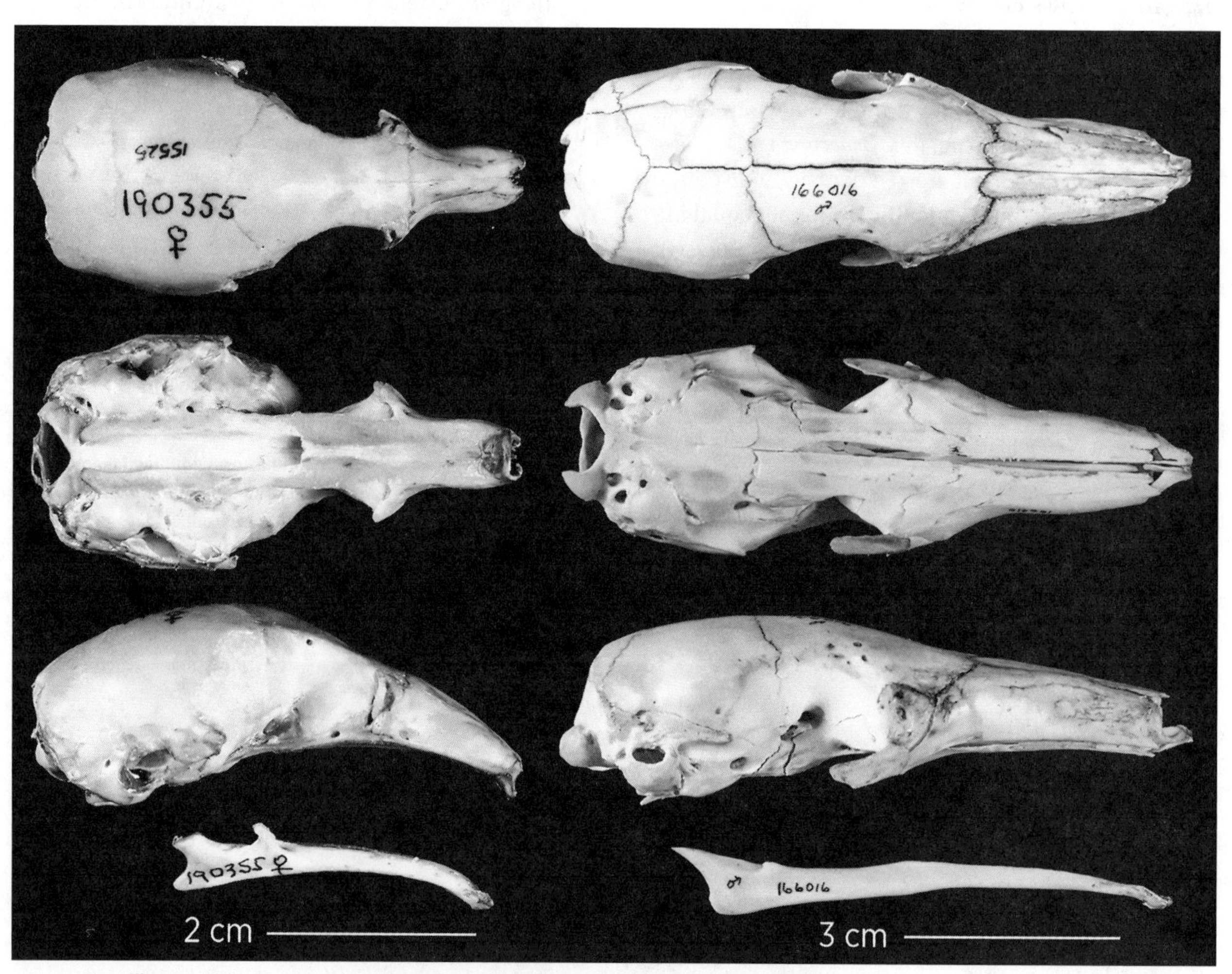

FIG. 110. Vermilingua; skull and mandible of the silky anteater, *Cyclopes didactylus* (Cyclopedidae; *left*), and the southern tamandua, *Tamandua tetradactyla* (Myrmecophagidae; *right*).

Clade AFROTHERIA

The Afrotheria is the second of the two major clades of eutherian mammals with southern (Gondwanaland) roots. As its name implies, the Afrotheria is rooted in Africa, but several lineages have radiated well beyond this region. The Afrotheria is further subdivided to two clades: the Afroinsectiphilia, which includes those lineages previously treated as members of the now-defunct Insectivora (Chrysochloridae, Macroscelididae, Tenrecidae) plus, somewhat oddly, the unique aardvark (Tubulidentata); and the Paenungulata (= Subungulata), which includes elephants (Elephantidae), sirenians (Sirenia), and hyraxes (Hyracoidea).

Clade AFROINSECTIPHILIA

Within the clade Afroinsectiphilia, the aardvark is basal to the afroinsectivoran lineages, the elephant shrews and tenrecs plus golden moles (see fig. 94).

Order TUBULIDENTATA

The order Tubulidentata includes a single extant family, Orycteropodidae, and one Recent species, the aardvark (*Orycteropus afer*). Found throughout much of sub-Saharan Africa outside of tropical rainforest, tubulidentates occupied Europe from the late Eocene to the early Oligocene, and most of the group's known history occurred in Europe and Africa. The phylogenetic placement of the tubulidentates has been problematic and varied. Simpson (1945) placed them in the superorder Protoungulata, a group that otherwise included four orders of primitive and now-extinct archaic ungulate groups. McKenna and Bell (1997) placed the aardvark in the grandorder Ungulata, a clade that also included the Cetacea, the true ungulates (Artiodactyla and Perissodactyla), and the so-called subungulates, composed of the closely related elephants (Proboscidea), hyraxes (Hyracoidea), and sirenians (Sirenia). Both of these views indicate a phyletic relationship between aardvarks and ungulates. However, more recent cladistic analyses of both morphological and molecular characters firmly place the tubulidentates within the Afrotheria, a clade that otherwise includes other African endemics (the elephant shrews [Macroscelidea], the tenrecs and golden moles [Tenrecoidea]) along with the elephants, hyraxes, and sirenians, rather than with any living ungulate group.

The common name "aardvark" is Afrikaans for "earth pig," and the ordinal name "Tubulidentata" refers to the characteristic "tube teeth"; both names identify obvious features of this enigmatic mammal. Superficial features include a slender head and snout, large closable ears, and muscular tail and limbs, with the forefeet well developed for digging. The structure of the teeth is unique: the permanent cheek teeth are ever-growing, consisting of vertical tubes of dentine in a matrix of pulp, with no enamel covering. Weighing as much as 80 kg, aardvarks are the largest termite specialists (myrmecophages) among mammals. They have a highly developed sense of smell, with enlarged olfactory lobes of the brain, and an elongate snout with elaborately scrolled turbinal bones that house extensive olfactory mucosae. The blunt end of the snout is soft and mobile; dense hairs and tentacle-like sensory vibrissae surround the nostrils.

Aardvarks are primarily nocturnal and solitary, and they are most common in grasslands and savannas where termites and ants are abundant. Their specialization on ants and termites is convergent on numbats (Myrmecobiidae), among the marsupials, and both Myrmecophagidae (Xenarthra: Pilosa) and Manidae (Pholidota), among eutherian mammals.

Family ORYCTEROPODIDAE (aardvark)

DIAGNOSTIC CHARACTER:

- **cheek teeth homodont, columnar, consisting of numerous vertical tubes of dentine in a matrix of pulp; no enamel present**

FIG. 111. Orycteropodidae; aardvark, *Orycteropus afer*.

RECOGNITION CHARACTERS:

1. **size medium-large (145–220 cm), piglike**
2. **body covered with bristly hairs**
3. **digits 4/5, with shovel-shaped nails**
4. no incisors or canines
5. zygomatic arch complete (unlike Myrmecophagidae, Pholidota)
6. **jugal and lacrimal present; no interparietal**
7. pterygoid bones separate, not forming part of palate (unlike Myrmecophagidae)
8. no extra zygapophyses on vertebrae (unlike xenarthrans)

DENTAL FORMULA: $\frac{0}{0}\ \frac{0}{0}\ \frac{2}{2}\ \frac{3}{3} = 20$

TAXONOMIC DIVERSITY: The family is monotypic, with a single species, *Orycteropus afer*.

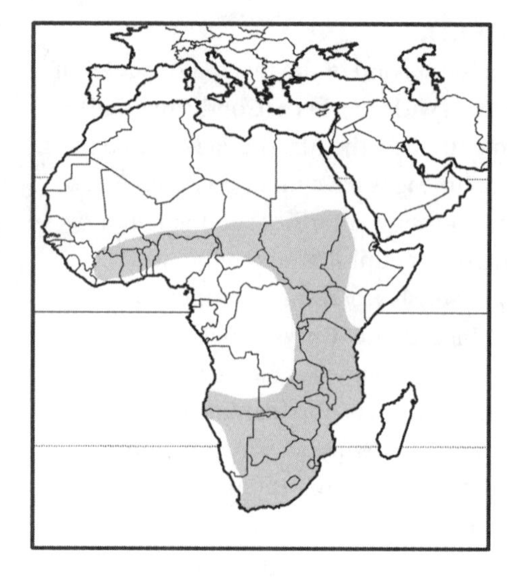

FIG. 112. Geographic range of family Orycteropodidae: sub-Saharan Africa, except Congo basin and Namib Desert.

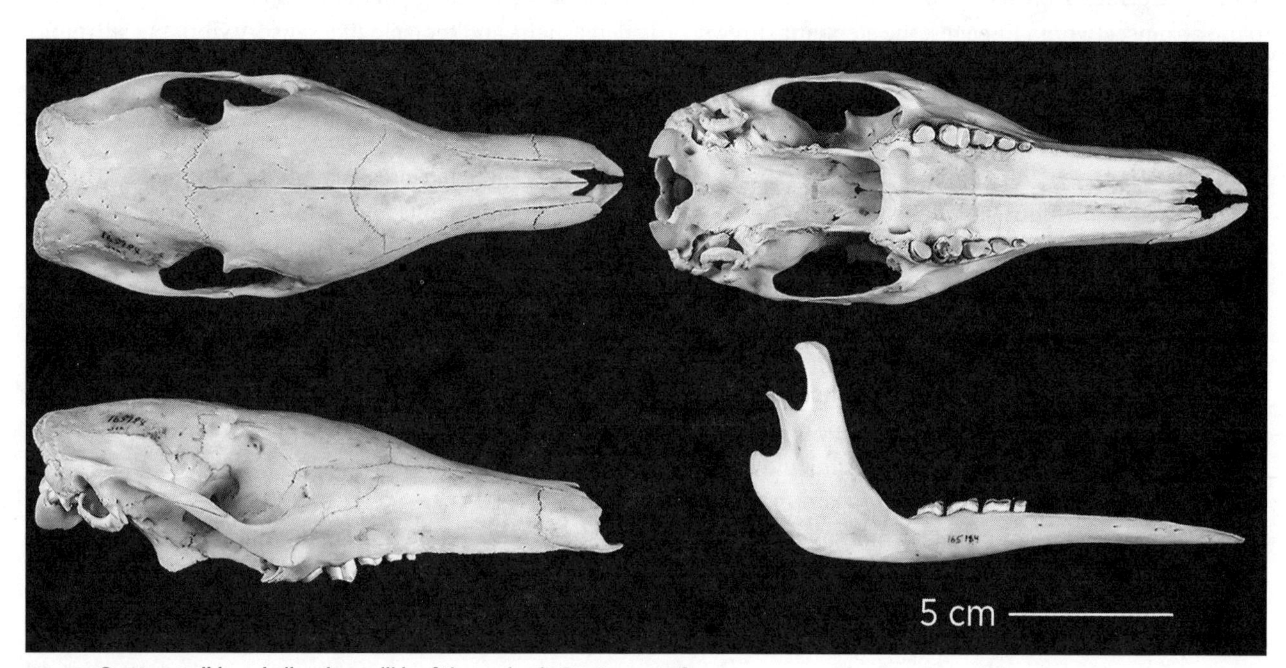

FIG. 113. Orycteropodidae; skull and mandible of the aardvark, *Orycteropus afer*.

Clade AFROINSECTIVORA

The Afroinsectivora is a clade that unites the order Macroscelidea and the order Tenrecoidea (see figs. 94 and 95). For discussion of the changing taxonomy of the Macroscelidea, see the introduction to infraclass Eutheria or Placentalia.

FIG. 114. Macroscelididae; short-nosed elephant shrew, *Elephantulus brachyrhynchus*.

Order MACROSCELIDEA

Family MACROSCELIDIDAE (elephant shrews)

Mouse- to rat-sized (body length 95–320 mm; mass 25–580 g); snout long, slender, and movable at base; ears and eyes large; hind limbs much longer than forelimbs, hind feet long and narrow, characters indicative of leaping quadrupedal gait; hind feet with four or five elongate digits, forefeet have five digits; tail long and slender, usually covered with bristles; pelage soft and lax in most. Diurnal, may be active on moonlit nights; insectivorous. Habitat varies from open sandy and gravelly plains to tall-grass savannas, dense thickets, woodlands, and forests.

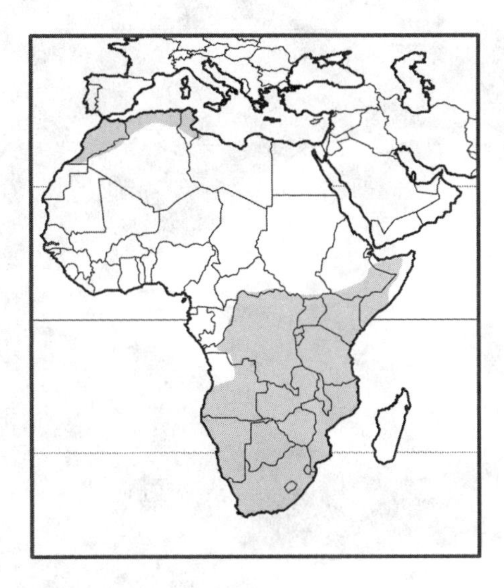

FIG. 115. Geographic range of family Macroscelididae: northwestern Africa; sub-Saharan Africa.

RECOGNITION CHARACTERS:

1. eyes large, vision acute
2. auditory bulla enlarged, greatly inflated in most genera
3. **palate contains a series of large perforations (palatal fenestrae)**, except in *Rhynchocyon*
4. postorbital process small or absent
5. zygomatic arch complete
6. molars quadritubercular, bunodont
7. lacrimal and palatine bones joined, not separated by maxilla
8. jugal well developed
9. innominate bones united in long symphysis

DENTAL FORMULA: $\frac{1-3}{3}\ \frac{1}{1}\ \frac{4}{4}\ \frac{2}{2-3} = 36–42$

TAXONOMIC DIVERSITY: About 15 species in four genera: *Elephantulus* (long-eared elephant shrews), *Macroscelides* (short-eared elephant shrew), *Petrodromus* (four-toed elephant shrew), and *Rhynchocyon* (giant elephant shrews).

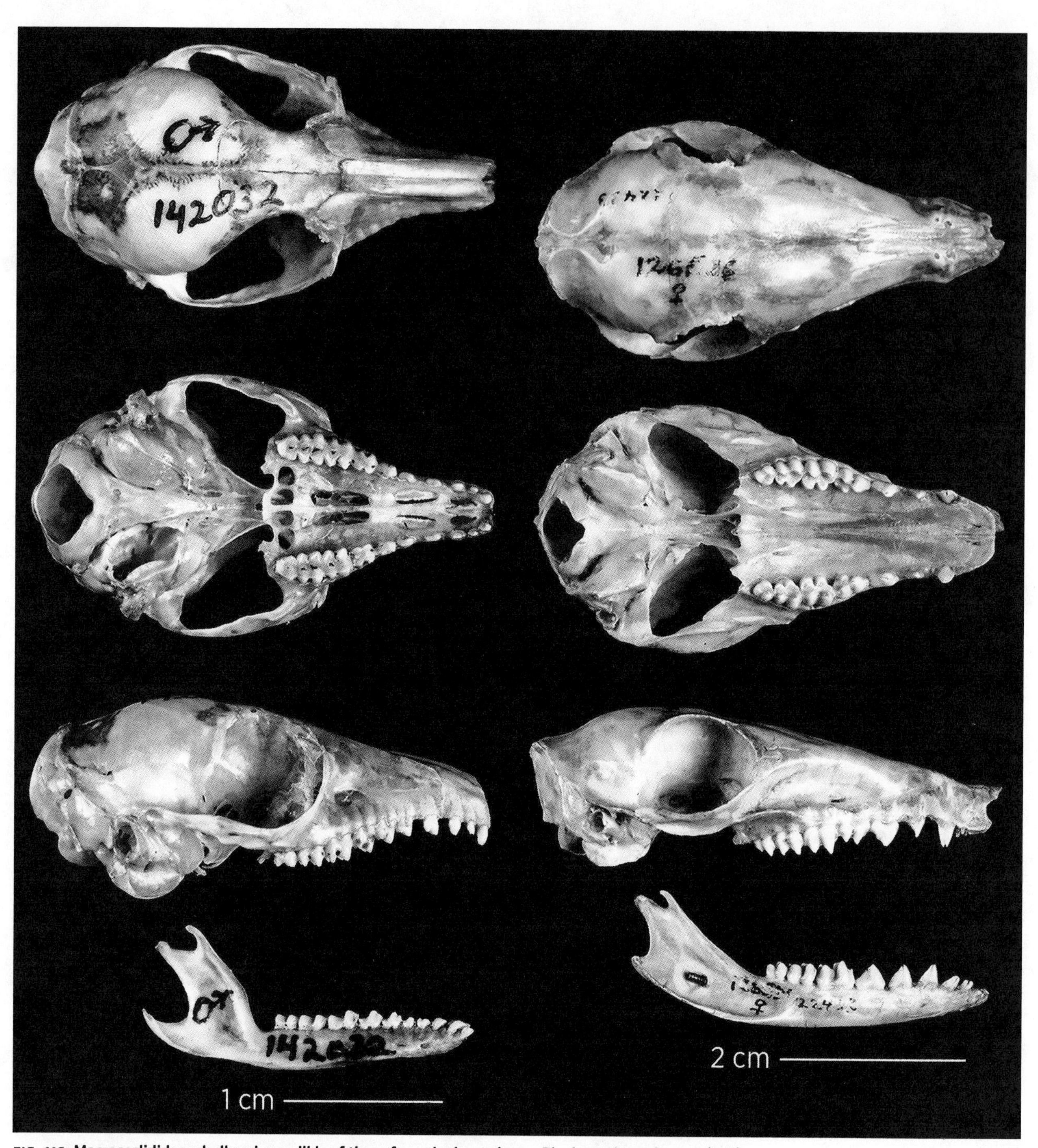

FIG. 116. Macroscelididae; skull and mandible of the rufous elephant shrew, *Elephantulus rufescens* (*left*), and the black and rufous elephant shrew, *Rhynchocyon petersi* (*right*).

Order TENRECOIDEA

Also called Afrosoricida (but see Asher and Helgen [2010], who argue that Tenrecoidea McDowell 1958 has priority over Afrosoricida Stanhope et al. 1998), the order Tenrecoidea includes two suborders, one for the golden moles (Chrysochloridea) and the other for the tenrecs (Tenrecomorpha).

Suborder CHRYSOCHLORIDEA

Family CHRYSOCHLORIDAE (golden moles)

Small (body length 75–240 mm; mass 15–100 g); all highly modified for subterranean existence; body short and stocky; forelegs and hind legs short and powerful; snout flattened and covered by leathery patch of skin; **eyes covered with skin; pinna absent; digits of manus with large, pointed claws** (3rd claw always largest; 2nd and 4th digits may have enlarged claws in some); hind feet with five digits and small claws; fur short, dense, soft, silky, and iridescent in some. Dig by powerful thrusts of forepaws and leathery snout; burrow networks extensive. Insectivorous to carnivorous; diurnal to nocturnal. Occupy friable, usually sandy soils in desert, grasslands, and forested areas.

DIAGNOSTIC CHARACTERS:

- **digits 4/5**
- **skull strongly triangular in shape in both dorsal and lateral views**
- **pair of extra bones (tabulars) present above and anterior to supraoccipital** (often fused with other bones in adults and thus not easily visible)

RECOGNITION CHARACTERS:

1. **zygomatic arch complete, formed by squamosal and long process of maxilla (no jugal)**
2. auditory bulla present or absent
3. 1st upper incisor enlarged
4. **upper molars tritubercular; crowns with V-shaped ectoloph** (zalambdodont)
5. pubic bones separate
6. no cloaca

DENTAL FORMULA: $\frac{3}{3}\ \frac{1}{1}\ \frac{3}{3}\ \frac{2\text{–}3}{2\text{–}3} = 36\text{–}40$

TAXONOMIC DIVERSITY: Over 20 species placed in nine genera in two subfamilies; common names for all are "golden moles."

Amblysominae—*Amblysomus*, *Calcochloris*, and *Neamblysomus*

Chrysochlorinae—*Carpitalpa*, *Chlorotalpa*, *Chrysochloris*, *Chrysospalax*, *Cryptochloris*, and *Eremitalpa*

FIG. 117. Chrysochloridae: Chrysochlorinae; Cape golden mole, *Chrysochloris asiatica*.

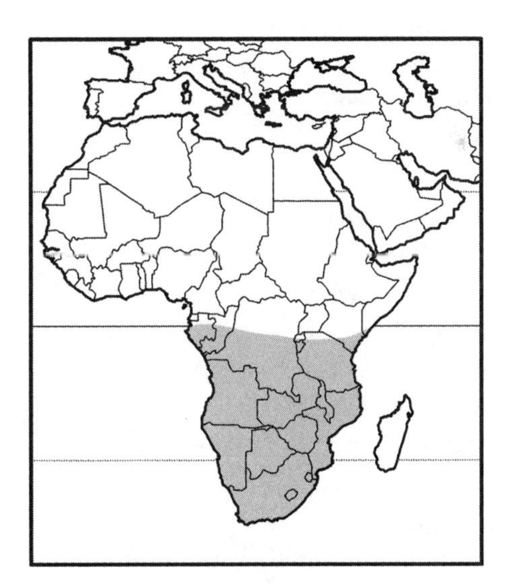

FIG. 118. Geographic range of family Chrysochloridae: sub-Saharan Africa.

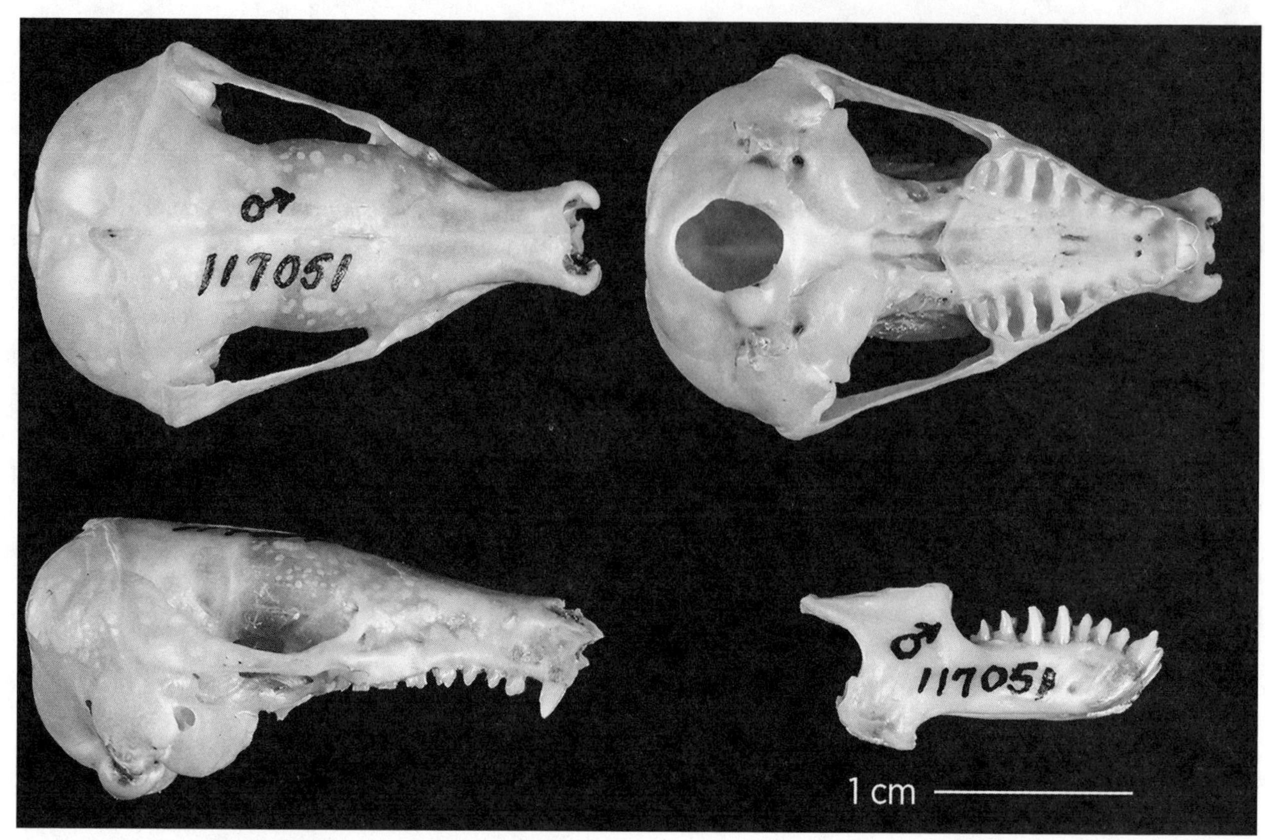

FIG. 119. Chrysochloridae: Amblysominae; skull and mandible of the Hottentot golden mole, *Amblysomus hottentotus*.

Suborder TENRECOMORPHA

The suborder Tenrecomorpha includes two families that were previously treated as subfamilies of the Tenrecidae; for simplicity, we treat both families together in the following account.

Families POTAMOGALIDAE (otter shrews) and TENRECIDAE (tenrecs)

Size varies from small (shrewlike; mass 5 g) to relatively large (mass up to 2 kg); body form also highly variable, with some resembling tailless opossums, with coarse fur and long muzzles (*Tenrec*); hedgehogs, covered dorsally with well-developed spines (*Echinops*, *Setifer*); fossorial moles, with strong digging claws and a flattened pad on the nose (*Oryzorictes*); aquatic rats with webbed feet (*Microgale mergulus*); shrews, some with very long, partially prehensile tails (*Microgale*); or otters, with laterally compressed tails for swimming (*Potamogale*); all have visible pinnae. Most are nocturnal omnivores and occupy a variety of habitats, from fast-flowing streams to leaf litter on the forest floor, with most found in wet tropical forest. *Hemicentetes* have special stridulating quills used in communication.

FIG. 120. Tenrecidae; lesser long-tailed shrew tenrec, *Microgale longicaudata* (Oryzorictinae; *top*), and lowland streaked tenrec, *Hemicentetes semispinosus* (Tenrecinae; *bottom*).

RECOGNITION CHARACTERS:

1. **zygomatic arch incomplete** (jugal absent)
2. no auditory bulla
3. **incisors usually small, relatively simple**
4. **upper molars tritubercular; crowns with V-shaped ectoloph** (zalambdodont); W-shaped ectoloph (dilambdodont) in *Potamogale*
5. pubic bones united in short symphysis in geogalines, oryzoryctines, and tenrecines; separate in potamogalids
6. cloaca present

DENTAL FORMULA: $\frac{2\text{–}3}{2\text{–}3}\ \frac{1}{1}\ \frac{2\text{–}3}{2\text{–}3}\ \frac{2\text{–}4}{2\text{–}3} = 28\text{–}42$

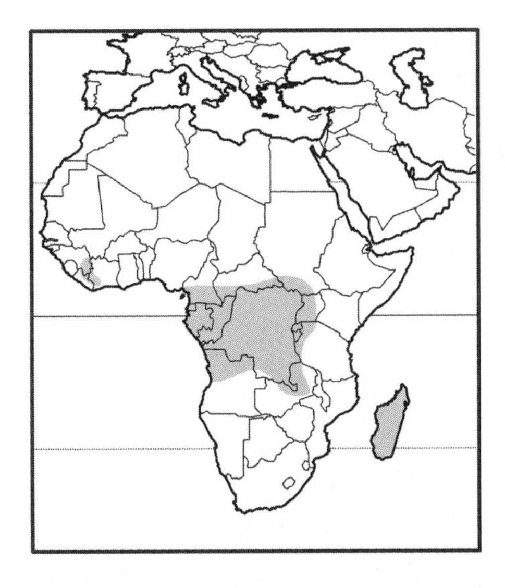

FIG. 121. Combined geographic range of families Potamogalidae (central Africa) and Tenrecidae (Madagascar).

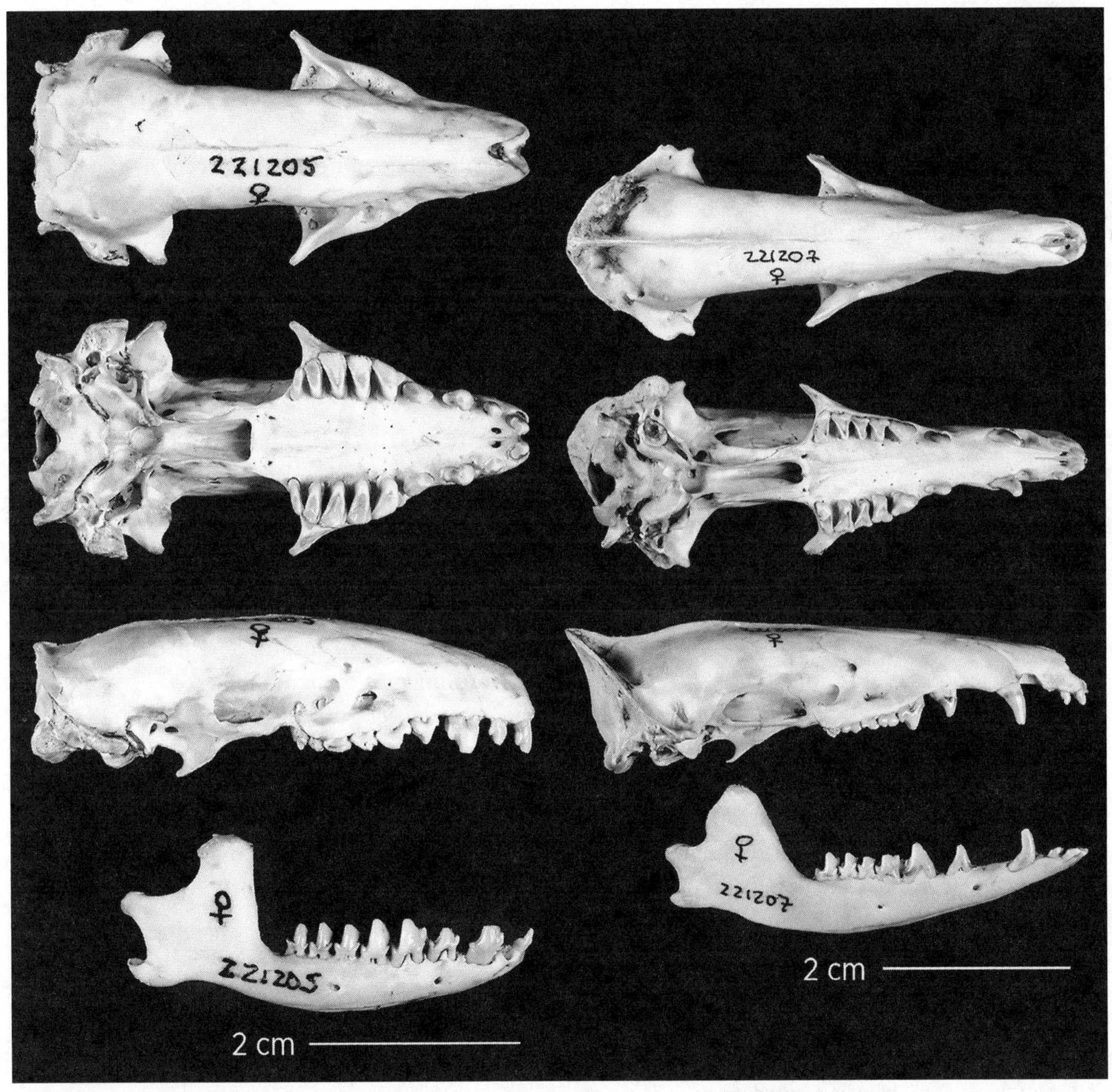

FIG. 122. Tenrecidae: Tenrecinae; skull and mandible of the greater hedgehog tenrec, *Setifer setosus* (*left*), and the tailless tenrec, *Tenrec ecaudatus* (*right*).

TAXONOMIC DIVERSITY: Approximately 35 species placed in 10 genera in two families, one with three subfamilies (Everson et al. 2016); the Potamogalidae are endemic to central Africa; the Tenrecidae are endemic to Madagascar.

Potamogalidae:

Micropotamogale (pygmy and Rwenzori otter shrews) and *Potamogale* (giant otter shrew)

Tenrecidae:

Geogalinae—*Geogale* (long-eared tenrec)

Oryzorictinae—*Microgale* (web-footed and shrew tenrecs), *Nesogale* (shrew tenrecs), and *Oryzorictes* (rice tenrecs)

Tenrecinae—*Echinops* (lesser hedgehog tenrec), *Hemicentetes* (streaked tenrecs), *Setifer* (greater hedgehog tenrec), and *Tenrec* (tailless tenrec)

Clade PAENUNGULATA (= Subungulata)

The Paenungulata is the clade that unites hyraxes, elephants, and sirenians (see fig. 94), a group long recognized as phyletically related. Simpson (1945) included these three orders along with a number of extinct taxa in the superorder Paenungulata within the cohort Ferungulata; McKenna and Bell (1997) grouped them together as the order Uranotheria within the grandorder Ungulata. While the elephants and sirenians are commonly united in the clade Tethytheria (see fig. 94), the most recent multi-gene molecular data (Esselstyn et al. 2017; Gatesy et al. 2017) failed to resolve phyletic relationships among the three included orders (see also fig. 95).

Order HYRACOIDEA

Members of the Hyracoidea are medium-sized (body length 300–650 mm), short-tailed, rodent-like creatures, commonly called hyraxes or dassies, that are phylogenetically related to elephants (Proboscidea) and sirenians (Sirenia). There is a single extant family, the Procaviidae, and one extinct family extending back into the Eocene. Living members, including five species in three genera, are distributed throughout most of Africa, except the arid northwestern part, and extend to parts of the Middle East. The roughly rabbit-sized procaviids of today have a short skull with a deep lower jaw; specialized incisors that are broadly separated, with the flattened posterior surfaces lacking enamel, and with lower incisors chisel-shaped and generally tricuspid. A broad diastema separates the incisors from the cheek teeth, which are either brachydont or hypsodont. The molars are lophodont, resembling those of rhinoceroses; the upper molars have an ectoloph and two transverse lophs, and the lower ones have two V-shaped lophs. The body is fairly compact, and the tail is tiny. The forefoot has four toes and the hind foot has three; the **digits are joined at the base of the last phalanges (e.g., syndactylous)**, and all bear flattened nails except for the innermost toe of the hind foot, which bears a large claw. The feet are mesaxonic, with the plane of symmetry passing through digit 3 (as in Perissodactyla). The feet are **plantigrade, with specialized elastic pads on the soles** that are kept moist by abundant skin glands and provide remarkable traction.

The stomach is simple, but digestion is aided by microbiota in a pair of pouches in the colon and the caecum. Hyraxes are mainly herbivorous and are nimble climbers and jumpers. They occur in a variety of habitats, from forests and scrub country to rock outcrops and lava beds in grasslands, and at elevations up to 5,000 m. The bush hyrax (*Heterohyrax*) and rock hyrax (*Procavia*) live in cliffs, ledges, and talus; the tree hyraxes (*Dendrohyrax*) are highly arboreal.

Family PROCAVIIDAE (hyraxes or dassies)

RECOGNITION CHARACTERS:

1. **interparietal bone well developed**
2. **postorbital process well developed, often forming complete postorbital bar**
3. nasal opening of skull located at end of rostrum
4. **jugal forming part of mandibular fossa**
5. **upper incisor ever-growing, tusklike, composed chiefly or entirely of dentine**
6. cheek teeth lophodont, relatively simple, becoming progressively more complex from front to back

DENTAL FORMULA: $\frac{1}{2}\ \frac{0}{0}\ \frac{4}{4}\ \frac{3}{3} = 34$

TAXONOMIC DIVERSITY: Five species in three genera: *Dendrohyrax* (tree hyraxes), *Heterohyrax* (bush hyrax), and *Procavia* (rock hyrax).

FIG. 123. Procaviidae; rock hyrax, *Procavia capensis*.

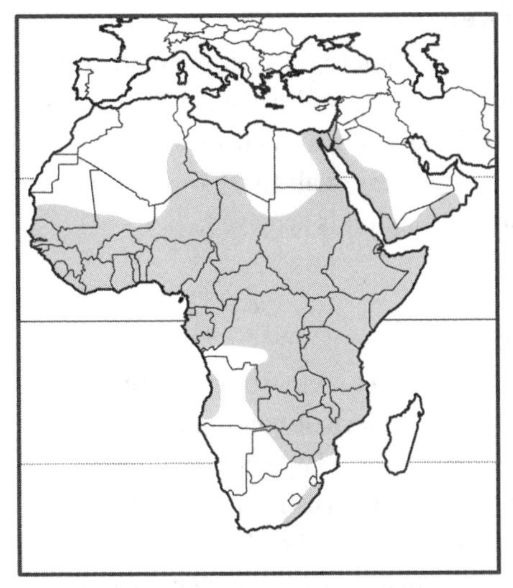

FIG. 124. Geographic range of family Procaviidae: Africa (except northwestern and southwestern arid lands and deserts) and Middle East.

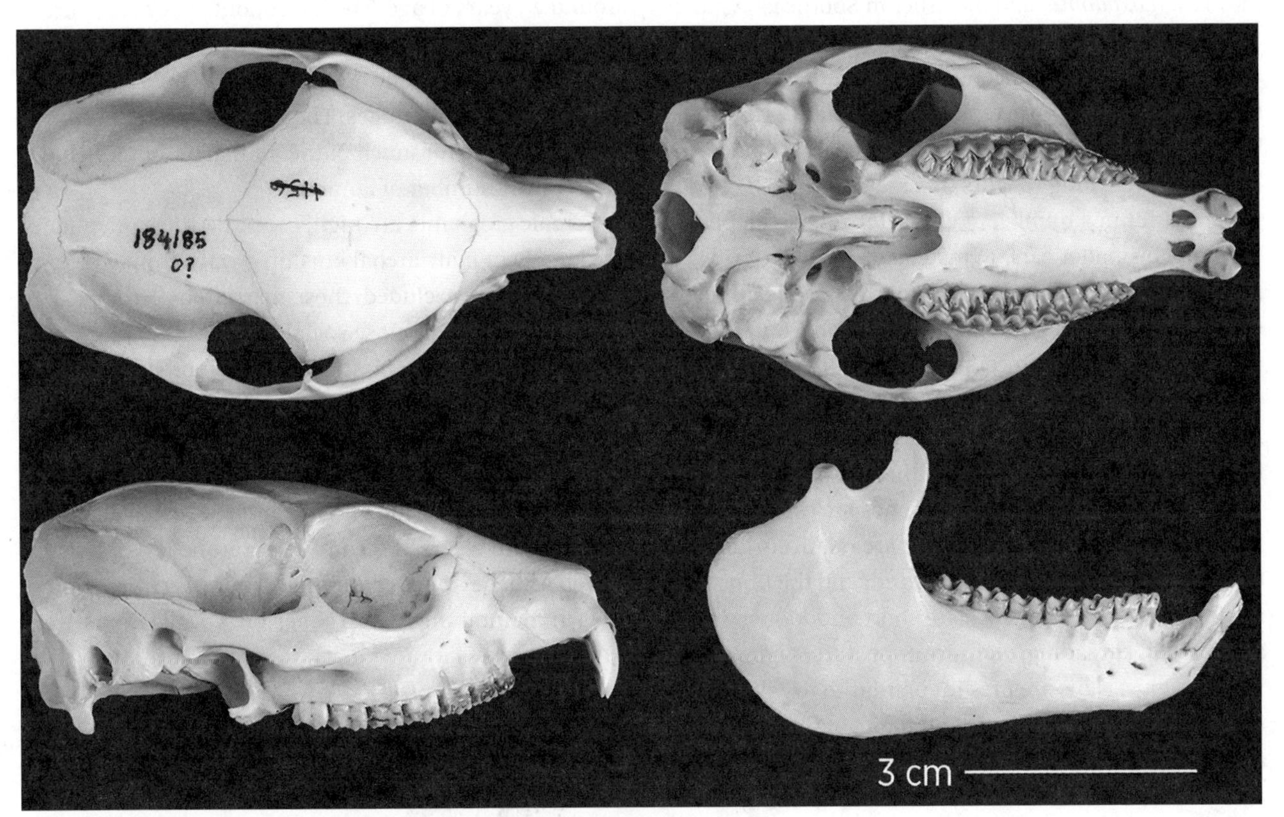

FIG. 125. Procaviidae; skull and mandible of the rock hyrax, *Procavia capensis*.

Clade TETHYTHERIA

As noted in the introduction to the Paenungulata, the Tethytheria unites elephants and sirenians. While this coupling currently lacks uniform cladistic support (Esselstyn et al. 2017; Gatesy et al. 2017), we continue its use until a final resolution is achieved.

Order PROBOSCIDEA

Elephants and their relatives stem from the late Paleocene, with the characteristic columnar (graviportal) limb posture and large body size well developed by the late Eocene. Distributed nearly worldwide (except Australia) by the Pleistocene, elephants are now restricted to a pair of genera, one in Africa (*Loxodonta*) and the other in Southeast Asia (*Elephas*). The extant family Elephantidae, which appears in the early Miocene, is characterized by molar teeth that erupt and are replaced in sequence from back to front. The living genera are the largest land mammals, reaching weights of 6 metric tons. The two genera differ in a number of characters: Asian elephants have much smaller ears, 19 instead of 21 pairs of ribs, and a flattened forehead; the top of the head is dome-shaped and is the highest point of the body; and there is a single fingerlike process at the tip of the trunk. In African elephants, the shoulders are generally the highest point, the ears are very large, and there are two fingerlike processes at the tip of the trunk.

In the Proboscidea, the limb bones are especially heavy, and the proximal segments are relatively long; the ulna and fibula are unspecialized; and the bones of the five-toed manus and pes are short and robust with an unusual, spreading, digitigrade posture; digits are syndactylous. Nails are present on four to five digits of the forefoot and on three to four digits of the hind foot. A heel pad of dense connective tissue braces the toes and largely supports the weight of the animal (fig. 126).

To help support this massive weight, the long axis of the pelvic girdle is oriented nearly at right angles to the vertebral column, and the acetabulum faces ventrally. There is little angulation between limb segments; that is, each segment is stacked on top of the others to form a single column bearing the weight of the animal (this arrangement is called graviportal). Elephants cannot run (a gait in which all four feet are off the ground at one time), but can only walk or employ a running walk in which at least two feet are always in contact with the ground.

The skull is unusually short and high, to provide the necessary mechanical advantage for the neck muscle to elevate the head, and is lightened by large and numerous air cells, particularly in the cranial roof. The highly specialized dentition consists of the tusks (each a second upper incisor; the first incisor is absent) and six cheek teeth in each half of each jaw. Tooth replacement is remarkable, with individual teeth erupting in sequence from front to rear, but with only a single tooth or a fragment of another in place and functional in each half of each jaw at one time (fig. 127). As a tooth becomes worn, it is replaced by the next posterior tooth. The first three cheek teeth erupt during an animal's youth; the fourth erupts at 4–5 years of age; the fifth at age 12–13; and the final tooth erupts at around 25 years of age. The hypsodont cheek teeth are formed of thin laminae, each consisting of an enamel band surrounding dentine, with cementum filling the spaces between the ridges. The last molar, the tooth that must serve for much of the animal's adult life, has the greatest number of laminae.

Female elephants are highly social. African elephants live largely in matriarchal kinship groups from which adult males are excluded; these matriarchal groups are held together by close social ties between adult females and between mothers and their offspring. Adolescent males leave the maternal herd and become solitary or form loose bachelor herds. Male elephants of both species experience episodes of musth, periods of increased aggression and heightened sexual activity coupled with elevated serum testosterone levels. Vocal, as well as behavioral, communications play important roles in maintaining group cohesion. Vocalizations are

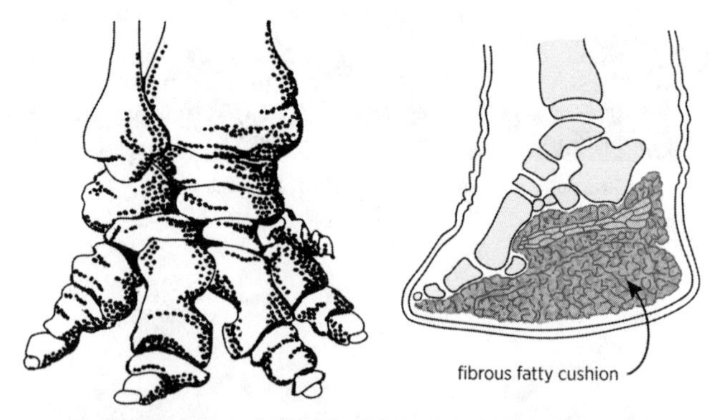

FIG. 126. Elephant feet. *Left*, bones of the right hind foot, frontal view, showing the spreading, digitigrade foot posture. Redrawn from Romer (1966). *Right*, lateral view of forefoot of an African elephant (*Loxodonta africana*), illustrating the heel pad that absorbs the shock of the animal's weight during locomotion. Redrawn from Feldhamer et al. (2015).

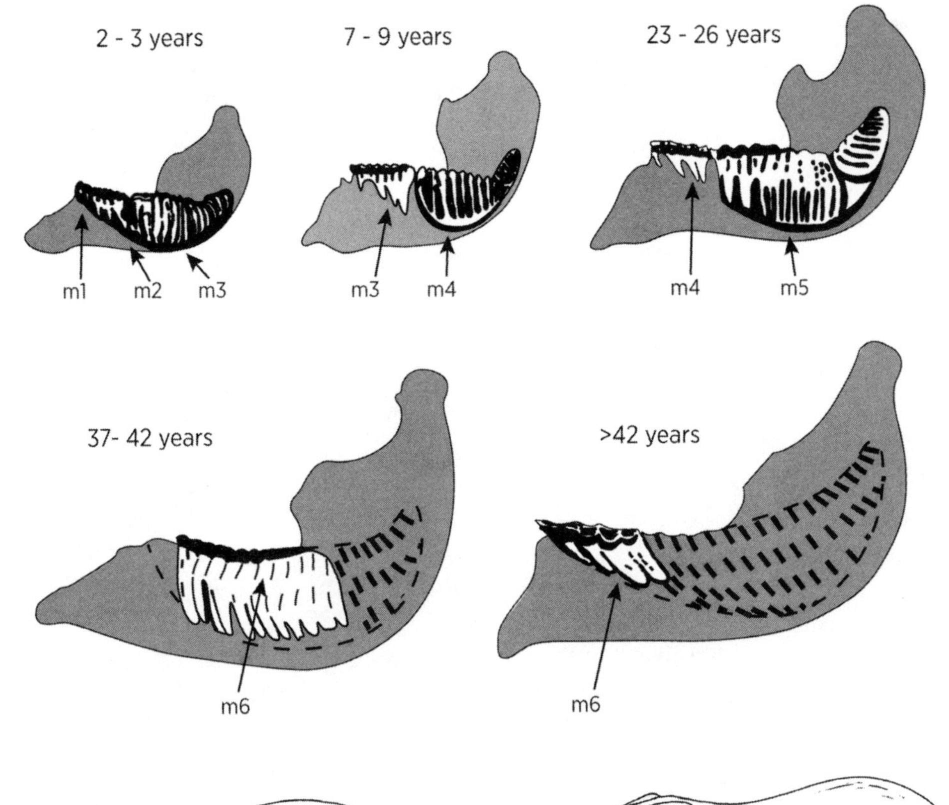

FIG. 127. Progression of molar tooth wear and tooth replacement in the lower jaw of the African elephant (*Loxodonta africana*) from birth to approximately 55 years of age (m1–m3 are considered by some to be true premolars; m4–m6 are molars). Redrawn from Kingdon (1979).

FIG. 128. Elephantidae; Asian elephant, *Elephas maximus* (*left*), and African elephant, *Loxodonta africana* (*right*).

diverse, with African elephants using infrasonic sounds (very low-frequency sounds of 14–24 Hz, often below human hearing thresholds) that can travel over vast distances.

Family ELEPHANTIDAE (elephants)

DIAGNOSTIC CHARACTERS:

- **long proboscis (trunk) present; nostrils and fingerlike projection at tip** (one projection in *Elephas*, two—dorsal and ventral—in *Loxodonta*)
- **pinna large, fanlike** (larger in *Loxodonta*)
- **limbs pillar-like; graviportal**
- **sole of foot with large elastic pad**
- **size enormous** (up to 4 m in height); body thickset

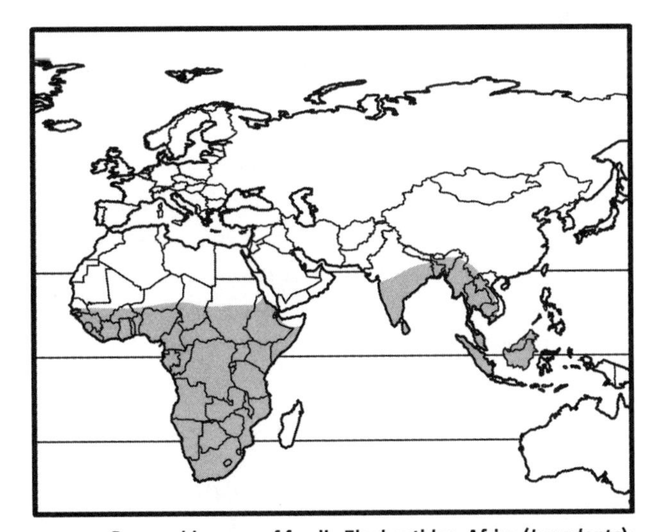

FIG. 129. Geographic range of family Elephantidae: Africa (*Loxodonta*); Southeast Asia (*Elephas*); current ranges of both genera considerably reduced.

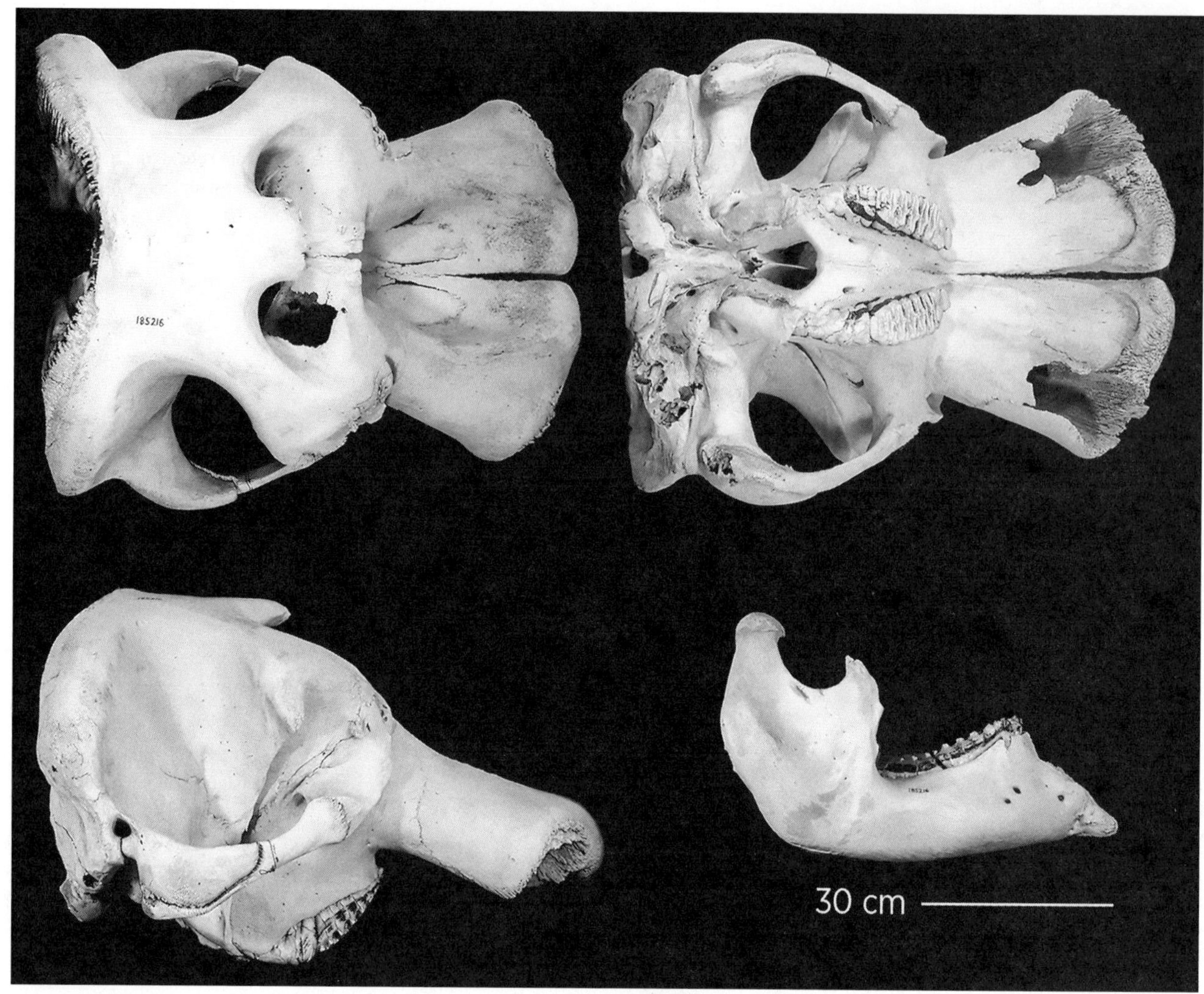

FIG. 130. Elephantidae; skull and mandible of the African elephant, *Loxodonta africana*.

RECOGNITION CHARACTERS:

1. no interparietal bone
2. no postorbital process or bar
3. nasal opening of skull located high on face
4. jugal not forming part of mandibular fossa
5. upper incisor ever-growing, tusklike (much larger in males), composed chiefly or entirely of dentine
6. cheek teeth lophodont, with many transverse ridges, replaced consecutively from rear (only one functional tooth or parts of two at a time on each jaw), becoming progressively more complex posteriorly

DENTAL FORMULA: $\frac{1}{0}\ \frac{0}{0}\ \frac{3}{3}\ \frac{3}{3} = 26$

(tusk usually absent in female *Elephas*)

TAXONOMIC DIVERSITY: Three species in two genera: *Loxodonta* (two species, Africa) and *Elephas* (one species, Southeast Asia).

Order SIRENIA

The sirenians are the only completely aquatic mammals that are herbivorous. There are four living species in two genera, each of which is placed in a separate family. All are largely confined now to tropical coastal marine and freshwater river systems. The genus *Hydrodamalis* (Steller's sea cow), rendered extinct by 1769 as a result of overhunting, occurred in the Bering Sea.

Living members are fusiform in shape and large, reaching weights in excess of 1.5 metric tons. They are nearly hairless, except for bristles on the snout, and have thick, rough, or finely wrinkled skin. The nostrils are valvular, the nasal opening extends posterior to the anterior borders of the orbits, and the nasals are reduced or absent. The skull is highly specialized: the dentary is deep, the tympanic bone is semicircular, and the external auditory meatus is small. The periotic bone has no bony attachments to the skull, but is instead attached by ligaments. Sirenians have very dense bone structure (a condition known as pachystosis), and their lungs are unusually long and oriented horizontally; in combination, these features allow the animal to use minor adjustments in lung volume to maintain a horizontal attitude while feeding at various depths. The heavy bones may also counterbalance added buoyancy from gas production in the gut. The forefoot is modified into a paddle-like flipper with five toes enclosed by skin; the pelvis is vestigial; and the **tail is a horizontal fluke.** There are no hind limbs.

The teeth of dugongs (Dugongidae) are large and columnar, lack enamel, are covered by cement, and are of a set number. They have open roots, and the occlusal surfaces are wrinkled and bunodont. The teeth of manatees (Trichechidae) are of indefinite number, with posterior teeth serially replacing anterior ones with wear and age; they are covered with enamel, lack cement, and each has two transverse ridges and closed roots. Five to eight teeth may be functional at any one time. Horny plates cover the front of the palate and the adjacent surface of the mandible in both living genera.

Sirenians are heavy-bodied, slow-moving animals that inhabit coastal seas, large rivers, and lakes. They graze while submerged for periods of up to about 15 minutes. Due to a low-nutrient diet, sirenians have very slow metabolism, resulting in the production of relatively little body heat for animals of their size; consequently, both living genera are confined to warm tropical waters.

Family DUGONGIDAE (dugongs)

DIAGNOSTIC CHARACTERS:

- **tail fluke dolphin-like, with pointed lateral projections; posterior margin deeply notched**
- **upper lip only slightly cleft**
- **no nails on flippers**

RECOGNITION CHARACTERS:

1. supraorbital process neither enlarged nor broadly expanded over orbit
2. no nasal bones
3. premaxilla large, bent sharply downward
4. **jugal broadened below orbit, in contact with premaxilla**
5. **palate narrow, distinctly elevated above tooth row, with small median ridge**
6. **lower jaw with coronoid process projecting upward**
7. upper incisor tusklike in males, small and often not protruding through gum in females
8. **cheek teeth simple columns**

DENTAL FORMULA: $\frac{1}{1}\ \frac{0}{0}\ \frac{2\text{–}3}{2\text{–}3} = 12\text{–}16$

(identity of cheek teeth uncertain)

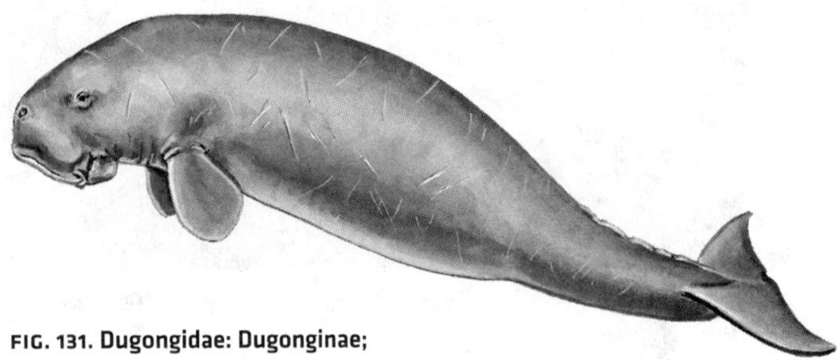

FIG. 131. Dugongidae: Dugonginae; dugong, *Dugong dugon*.

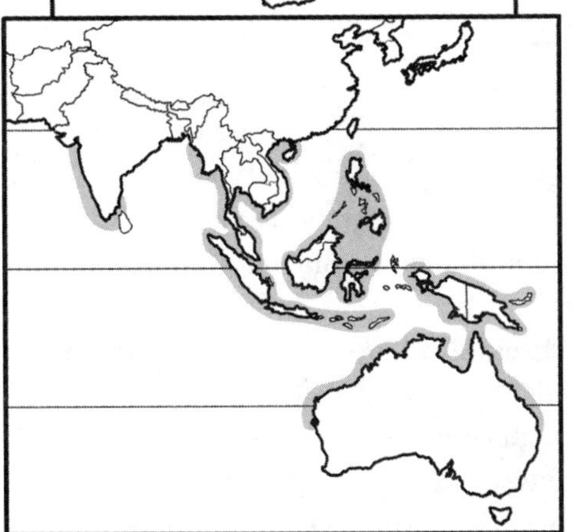

FIG. 132. Geographic range of family Dugongidae: coastal waters of Indo-Pacific region (extant range only).

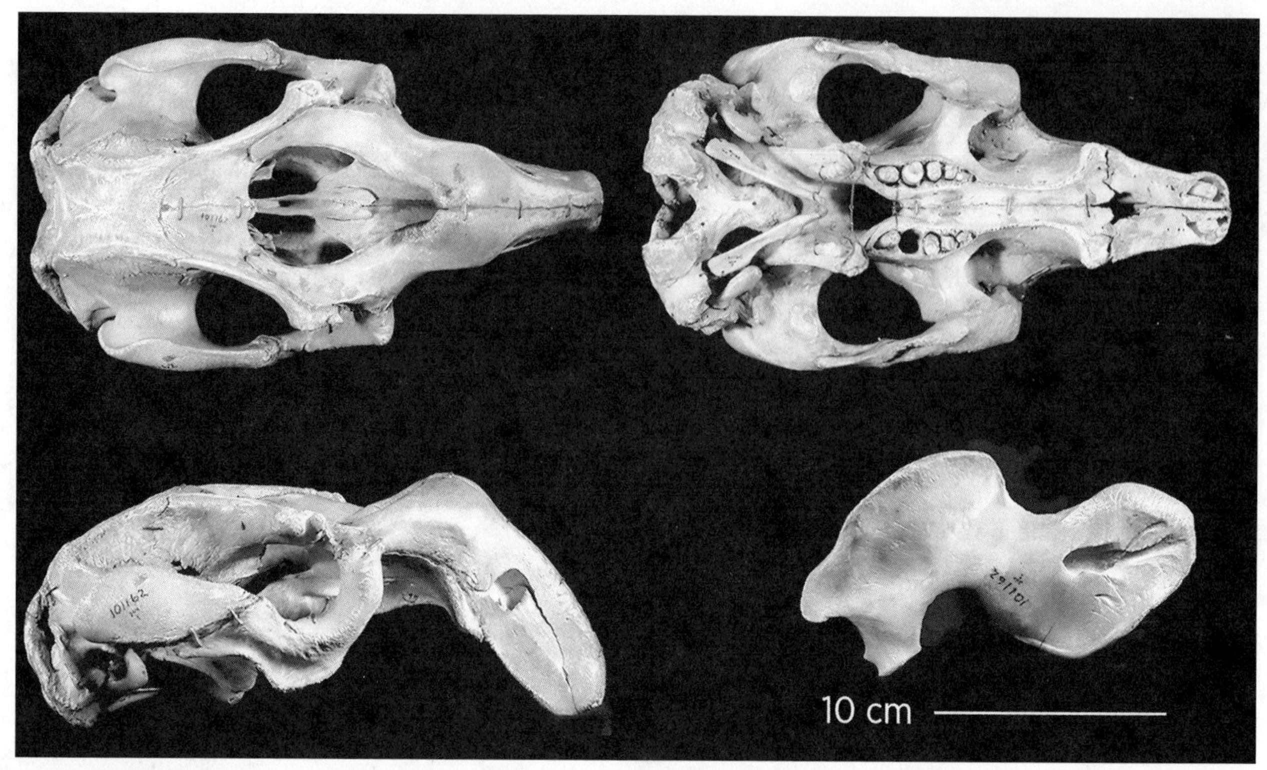

FIG. 133. Dugongidae: Dugonginae; skull and mandible of the dugong, *Dugong dugon*.

TAXONOMIC DIVERSITY: Two monotypic genera, each placed in separate subfamilies:
Dugonginae—*Dugong* (dugong)
†Hydrodamalinae—*Hydrodamalis* (Steller's sea cow)

Family TRICHECHIDAE (manatees)

DIAGNOSTIC CHARACTERS:

- **tail fluke evenly rounded, posterior margin not notched**
- **upper lip deeply cleft**
- **small nails usually present on flippers** (absent in *Trichechus inunguis*)

RECOGNITION CHARACTERS:

1. **supraorbital process large, broadly expanded over orbit**
2. nasal bones present, small
3. premaxilla small, only slightly bent downward
4. **jugal broadened behind orbit, not in contact with premaxilla**
5. **palate relatively broad, not elevated above tooth row, with distinct median ridge**
6. **lower jaw with coronoid process projecting forward**
7. **no upper incisors**
8. **cheek teeth each with two transverse ridges**

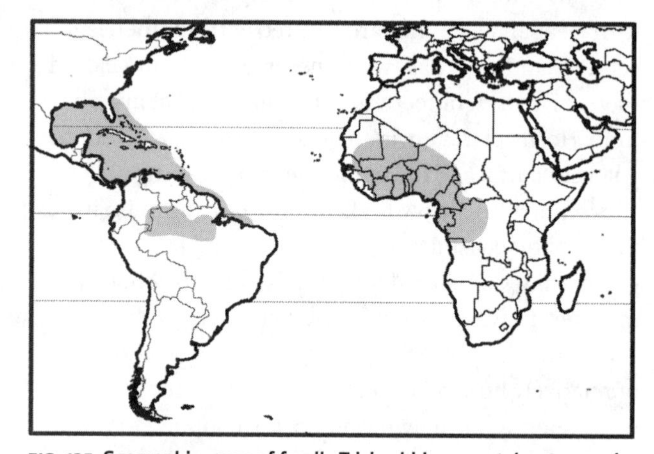

FIG. 135. Geographic range of family Trichechidae: coastal waters and rivers of tropical and subtropical Atlantic (including Florida south through Caribbean, coastal South America, and Amazon basin) and coastal West Africa and Congo basin.

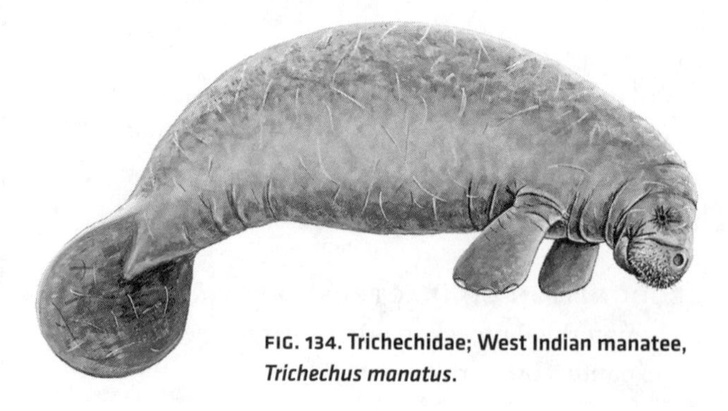

FIG. 134. Trichechidae; West Indian manatee, *Trichechus manatus*.

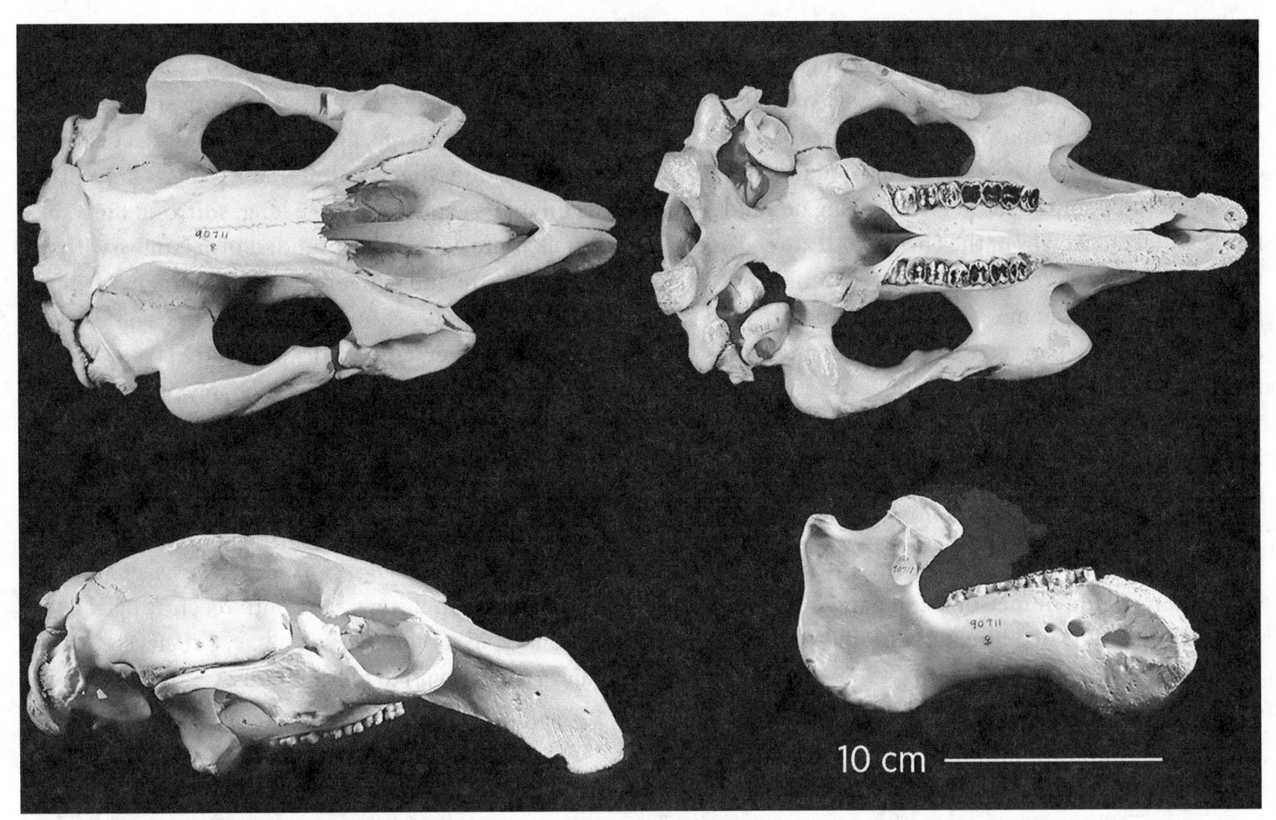

FIG. 136. Trichechidae; skull and mandible of the manatee, *Trichechus manatus*.

DENTAL FORMULA: $\frac{0}{0}\ \frac{0}{0}\ \frac{\text{indefinite}}{\text{indefinite}} = ??$
(identity of individual cheek teeth uncertain)

TAXONOMIC DIVERSITY: The family is monotypic, with three species in the single genus *Trichechus*.

Clade BOREOEUTHERIA

Clade EUARCHONTOGLIRES

Whereas the Xenarthra and Afrotheria were derived in southern continents during the Cretaceous, the northern supercontinent Laurasia spawned substantial diversification as well. Probably reflecting the greater connectivity among the landmasses comprised by Laurasia, the northern lineages are united in a single higher clade, Boreoeutheria, which subsequently diverged into two major lineages. One of these, the Euarchontoglires, takes its name from its two subordinate clades: Euarchonta, or sometimes just Archonta, which contains the Primates and the Sundatheria (colugos and tree shrews); and Glires, which combines the rabbits and rodents. For the curious student, the name Euarchonta itself means "the real, or true, archonta," and was proposed as a subset of a group called Archonta, which included bats as well as the euarchontans; current understanding, however, places bats in the second major boreoeutherian lineage, the Laurasiatheria, which also includes carnivorans, pangolins, ungulates and whales, and lipotyphlan insectivores (see the introduction to infraclass Eutheria or Placentalia and fig. 94).

Clade GLIRES

Rabbits and pikas are united with rodents in the cohort Glires in the Linnaean classification of Simpson (1945). In phylogenetic classifications, Glires is simply the node that joins the two orders Lagomorpha and Rodentia. The clade Glires is defined by the uniquely enlarged, single pair of ever-growing gnawing incisors, with enamel restricted to the anterior face so that differential wear of the softer, posterior dentine produces a chisel-like enamel cutting blade; this pair of incisors, above and below, is separated from the cheek teeth by a large diastema, and there are no canines.

Order LAGOMORPHA

Lagomorphs superficially resemble rodents, but have an additional, peg-like second upper incisor that rodents lack. All lagomorphs have a rudimentary or short tail and travel by a bounding, quadrupedal gait enabled by somewhat to greatly elongate hind limbs and feet. Crowns of the cheek teeth are relatively simple, composed of a pair of transverse basins separating enamel ridges; upper tooth rows are more widely separated than the lower ones, so that chewing takes place on only one side at a time. Testes in males are located anterior to the penis, as in marsupials—a position unique among eutherian mammals.

DIAGNOSTIC CHARACTERS (FIG. 137):

- **two pairs of upper incisors, 2nd pair small and peg-like and located directly behind the enlarged anterior pair**
- **fenestrated maxilla**

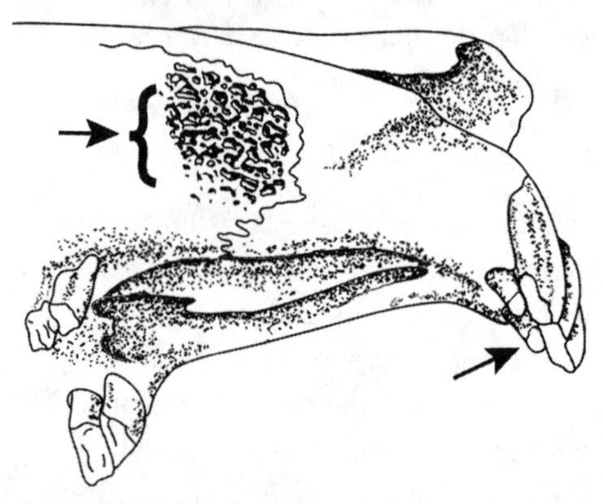

FIG. 137. Diagnostic characters of the Lagomorpha. Note peg-like second incisors located behind the enlarged anterior pair (right arrow) and the fenestration of the lateral side of the rostrum (upper left arrow). Redrawn from Feldhamer et al. (2015).

RECOGNITION CHARACTERS:

1. small to medium-sized
2. foot posture digitigrade
3. **tail indistinct or small**
4. **soles of feet largely or entirely covered with fur**
5. incisors and cheek teeth separated by large diastema
6. testes positioned anterior to penis

The order has limited taxonomic diversity, with only two families and about 12 genera.

Family OCHOTONIDAE (pikas)

Small (body length 160–260 mm; mass 150–200 g); body compact; eyes small; **ears haired on both surfaces, nearly circular in shape, and visible but project above body pelage only slightly**; **limbs short**; **digits 5/4, with pads on digits exposed**; pelage long, soft, and fluffy. Most pikas are active year-round; diurnal; move with a scampering gait and with great agility on rocky terrain. Chiefly herbivorous, eating leaves and stems of forbs and shrubs; gather, cure, and store food in large "hay piles" for winter use; all species probably coprophagic. Inhabit primarily rocky terrain of outcrops and talus in Arctic and alpine tundra, taiga, and mountains above and below tree line; individuals dig burrows in areas free of rocks. Very vocal, producing high-pitched trill. The extinct genus *Prolagus*, which survived until historical times in Europe, was until recently placed in a monotypic subfamily Prolaginae, but elevated by Erbajeva (1988, 1994) to family status.

FIG. 138. Ochotonidae; American pika, *Ochotona princeps*.

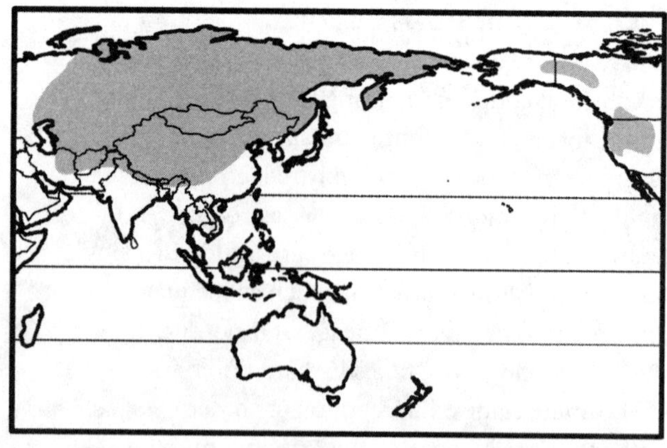

FIG. 139. Geographic range of family Ochotonidae: discontinuous distribution in mountains of western North America, Europe, and Asia, including Japan.

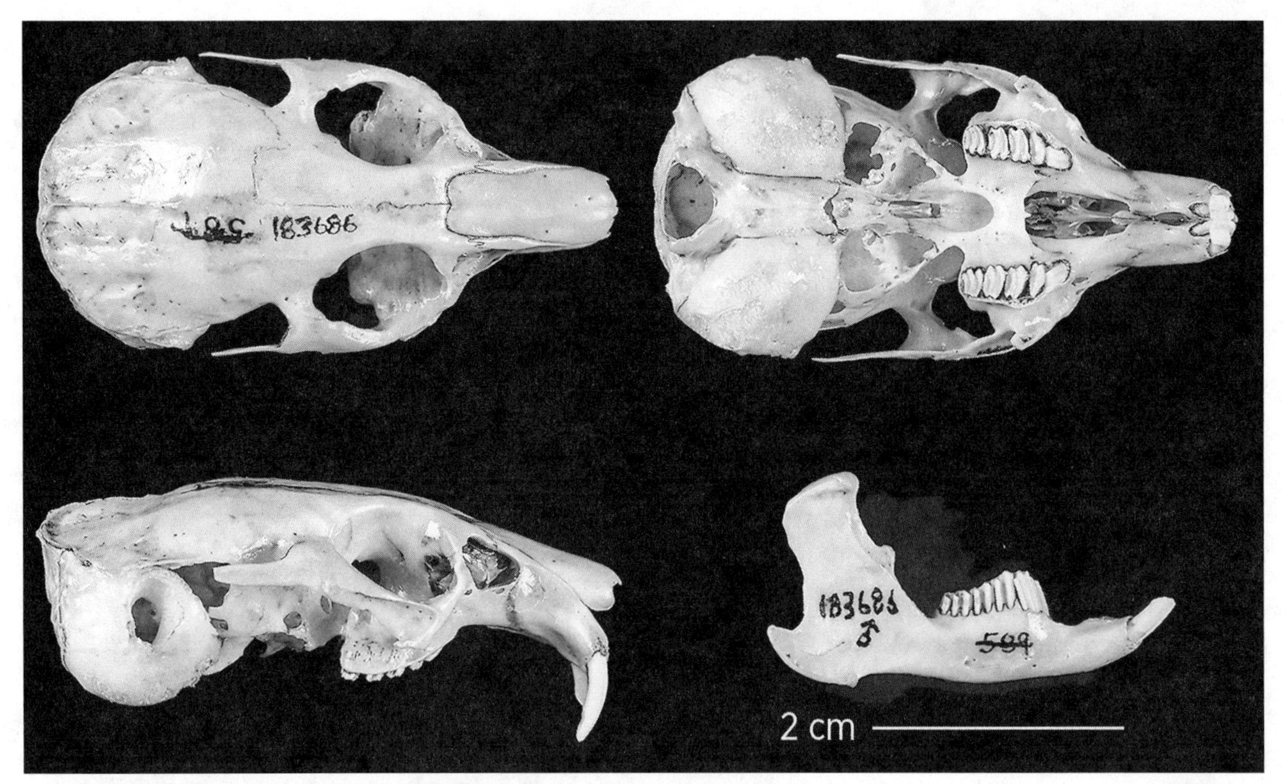

FIG. 140. Ochotonidae; skull and mandible of the collared pika, *Ochotona collaris*.

RECOGNITION CHARACTERS:

1. **no supraorbital process**
2. **maxilla with single (occasionally two or three) perforations**
3. **nasal widest anteriorly**
4. jugal projecting conspicuously beyond posterior margin of zygomatic arch
5. cutting edge of 1st upper incisor V-shaped
6. cheek teeth with simple transverse bilophodonty
7. baculum present in males (Weimann et al. 2014)

DENTAL FORMULA: $\frac{2}{1}\ \frac{0}{0}\ \frac{3}{2}\ \frac{2}{3} = 26$

TAXONOMIC DIVERSITY: The family is monotypic, with about 29 species in the single genus *Ochotona*.

Family LEPORIDAE (rabbits, cottontails, hares, jackrabbits)

Small to medium-sized (body length 250–700 mm; mass 300 g–4 kg [domesticated rabbits may exceed 7 kg]); muzzle elongate; **ears longer than wide, highly movable and can rotate**, proximal portion tubular with opening well above skull; hind limbs much longer than forelimbs, modified for quadrupedal saltatory locomotion; hind feet very long, **soles and digital pads covered by long, thick hair**; **digits functionally 4/4** (both pollex and hallux reduced); claws nearly straight and pointed; **tail short, recurved, and visible externally**. Generally nocturnal to crepuscular; do not hibernate or estivate; shelter in self-dug burrows, crevices, depressions at base of shrubs (called "forms"), or dense vegetation, replying on rapid movement to escape predators. Usually nonvocal. Strictly herbivorous, and do not store foods; coprophagic, producing two types of fecal pellets. Most are solitary and territorial, but some live in larger groups with a well-defined hierarchical system. Some species are subject to multiannual population fluctuations with a periodicity of about 10 years. Habitat highly variable, ranging from snowfields and Arctic tundra to alpine environments above tree line, grasslands, savannas, woodlands, brushlands, marshes and swamps, and boreal, temperate, and tropical forests.

RECOGNITION CHARACTERS:

1. **supraorbital process present, fan-shaped** (often fused to varying degrees with frontal bones)
2. **maxilla with numerous perforations, often highly fenestrated**
3. **nasal widest posteriorly**
4. jugal contained wholly within zygomatic arch
5. cutting edge of 1st upper incisor straight
6. cheek teeth with transverse bilophodonty with interior crenulated enamel ridges
7. baculum absent in males

DENTAL FORMULA: $\frac{2}{1}\ \frac{0}{0}\ \frac{3}{2}\ \frac{2\text{–}3}{3} = 26\text{–}28$

TAXONOMIC DIVERSITY: 11 genera, divided by some authors into two subfamilies. Most genera monotypic, but *Sylvilagus* and *Lepus* very speciose.

Leporinae—*Brachylagus* (pygmy rabbit of western North America), *Bunolagus* (riverine rabbit of South Africa), *Caprolagus* (hispid hare of Himalayan region), *Lepus* (jackrabbits and hares of North America, Eurasia, and Africa), *Nesolagus* (striped rabbits), *Oryctolagus* (European rabbit; widely introduced), *Poelagus* (Bunyoro rabbit of central Africa), and *Sylvilagus* (cottontail rabbits of North and South America)

Paleolaginae—*Pentalagus* (Amami rabbit of Japan), *Pronolagus* (rock hares of Africa), and *Romerolagus* (volcano rabbit of southern Mexico)

FIG. 141. Leporidae: Leporinae; eastern cottontail, *Sylvilagus floridanus* (*upper left*), and Cape hare, *Lepus capensis* (*lower right*).

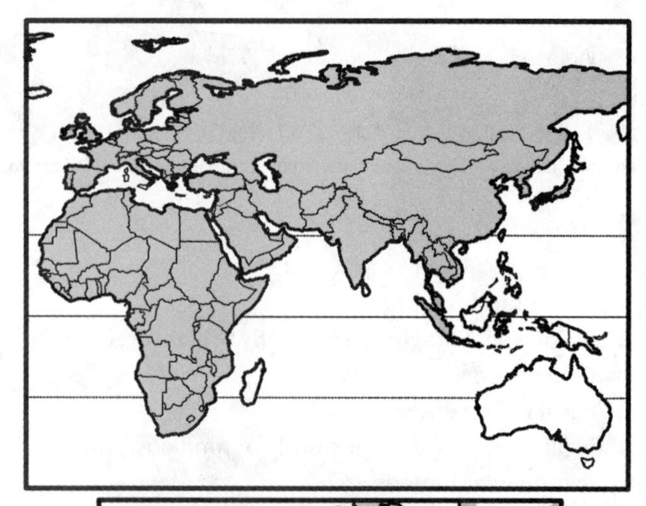

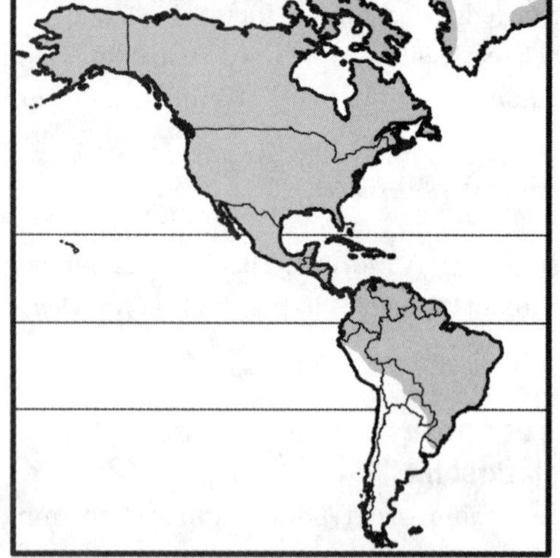

FIG. 142. Geographic range of family Leporidae: native to all continents except Australia (but introduced there) and Antarctica.

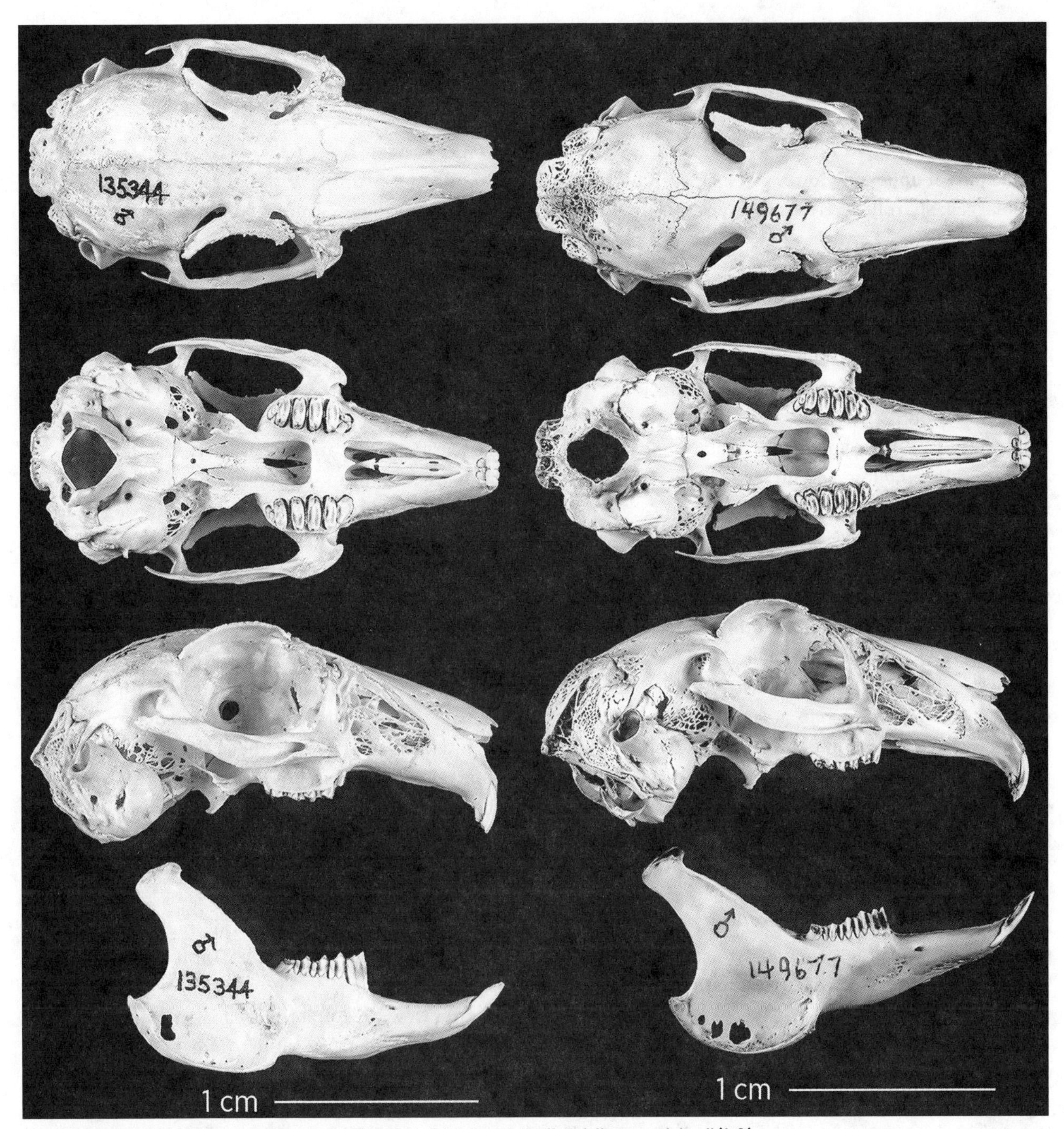

FIG. 143. Leporidae: Leporinae; skull and mandible of Audubon's cottontail, *Sylvilagus audubonii* (*left*), and black-tailed jackrabbit, *Lepus californicus* (*right*).

Order RODENTIA

UNIQUELY DERIVED CHARACTERS THAT DIAGNOSE A MONOPHYLETIC RODENTIA (FROM HARTENBERGER 1985):

- **one pair of upper and lower incisors; each tooth enlarged, sharply beveled, and ever-growing**
- broad diastema between incisors and cheek teeth in both upper and lower jaws
- incisor enamel restricted to anterior face only
- paraconid lost on lower cheek teeth
- orbital cavity lying just dorsal to cheek teeth
- zygomatic process of the maxilla lies anterior to the 1st cheek tooth
- mandibular (= glenoid) fossa is an anteroposterior trough allowing fore-and-aft movement of the mandible

Rodents as a group make up the largest order of living mammals The order Rodentia comprises nearly 45% of all extant taxa, currently organized into 34 living families (1 recently extinct family, the Heptaxodontidae, is not covered here), over 500 genera, and roughly 2,480 species. Nearly cosmopolitan in distribution, rodents exploit a wide range of foods, are important members of most terrestrial faunas, and often reach extremely high population densities. Repeated and parallel radiations have occurred on most continents, with spectacular examples of convergence especially in lineages of subterranean, bipedal jumping (= ricochetal), and arboreal taxa. As a result of the rodents' great diversity and parallel patterns of evolutionary divergence, their taxonomy is among the least understood for any living mammalian order, and numerous attempts to classify the group have been made.

Rodents range in size from very small (about 5 g) to very large (more than 50 kg). The occlusal surfaces of the cheek teeth are often highly complex, designed for vertical crushing of seeds and insects or horizontal grinding of plant materials. The dental formula never exceeds 1/1 0/0 2/1 3/3 = 22. No rodent possesses a canine in either the upper or lower jaw. The mandibular fossa of the squamosal bone is elongate and primarily allows anteroposterior (= fore-aft) jaw movement. The mandibular symphysis is typically unfused and has sufficient "give" in many species to enable the transverse mandibular muscles to pull the ventral borders of the rami together and spread the tips of the incisors. The masseter muscles are large and complexly subdivided, provide most of the power for mastication and gnawing, and, in all but one species (*Aplodontia rufa*: Aplodontiidae), have at least one division that originates on the rostrum.

PATTERNS OF THE SKULL, JAWS, AND MASSETER MUSCULATURE IN RODENTS

The terms protrogomorphy, sciuromorphy, myomorphy, and hystricomorphy refer to different patterns in the arrangement of the masseter muscles and the corresponding structures of the rostrum and zygomatic arch.

1. PROTROGOMORPHY

The pattern of jaw muscles in primitive rodents: large temporal (= temporalis) muscle and a masseter muscle with two parts (superficial and lateral), both originating entirely on the zygomatic arch (fig. 144). There is no zygomatic plate; the infraorbital foramen is very small (shaded in gray in anterior view) and serves only vessels and nerves (fig. 145). Among living rodents, found only in the family Aplodontiidae.

2. SCIUROMORPHY

Anterior portion of the lateral masseter originates on flat anterior surface of the zygomatic arch (the zygomatic plate) and superficial masseter originates from a masseteric tubercle on the lower side of the rostrum (fig. 146, *top*). A very small infraorbital foramen transmits only vessels and nerves. Temporal muscle is relatively large and coronoid process is moderately well developed.

Major force of jaw action is upward (thick arrow); secondary forces are up and forward (narrow arrows) (fig. 146, *bottom*).

In anterior view, the infraorbital canal of sciuromorphous rodents is small, often slit-like (indicated by arrow; fig. 147). Note well-developed zygomatic plate.

3. MYOMORPHY

Anterior portion of the lateral masseter originates on highly modified anterior extension of zygomatic arch (zygomatic plate and zygomatic spine; fig. 148, *top*), and anterior part of medial masseter originates on rostrum and passes through a somewhat enlarged infraorbital foramen (fig. 148, *middle*). Temporal muscle is reduced in size and coronoid process may be vestigial.

Direction of jaw movement (arrows) is equally upward and forward (fig. 148, *bottom*).

The infraorbital canal of myomorphous rodents tends to be larger than those of sciuromorphous rodents

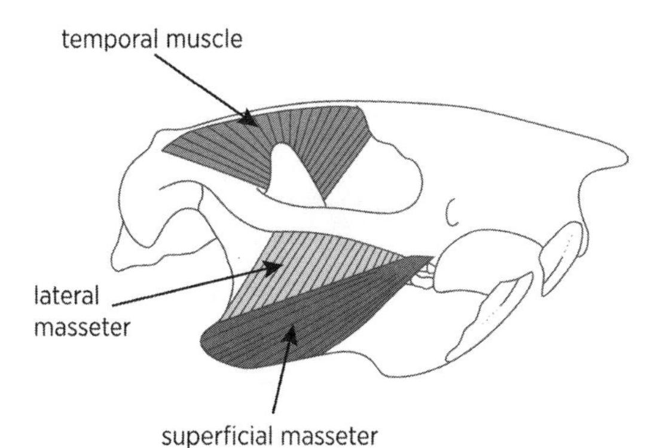

FIG. 144. Protrogomorphous masseter muscles. Redrawn from Vaughan et al. (2015).

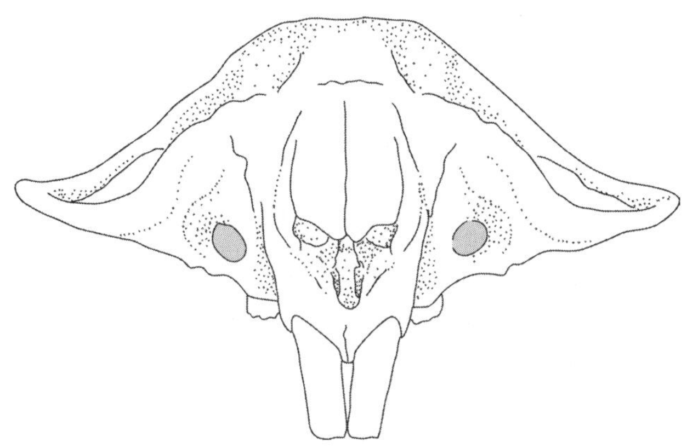

FIG. 145. Protrogomorphous infraorbital foramen, anterior view. Redrawn from Lawlor (1979).

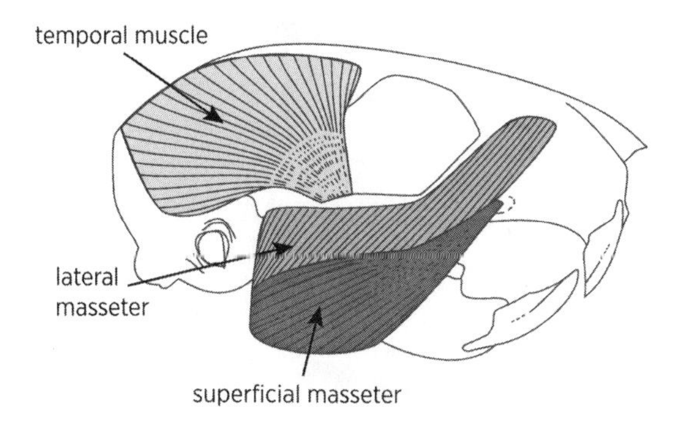

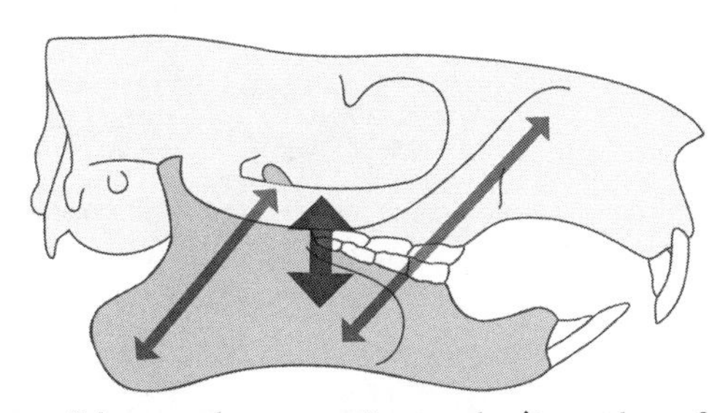

FIG. 146. Sciuromorphous masseter muscles (*top*; redrawn from Vaughan et al. 2015) and their force of action (*bottom*; redrawn from MacDonald 1984).

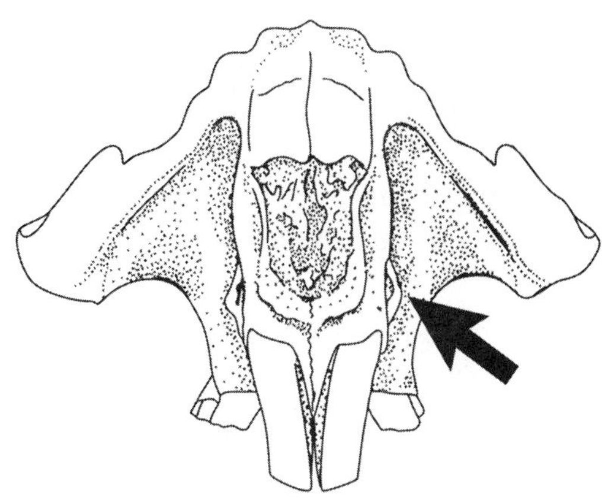

FIG. 147. Sciuromorphous zygomatic plate/infraorbital foramen, anterior view. Redrawn from Lawlor (1979).

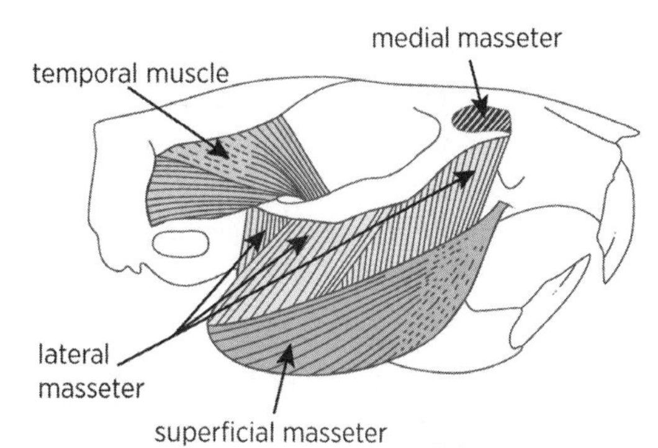

medial masseter

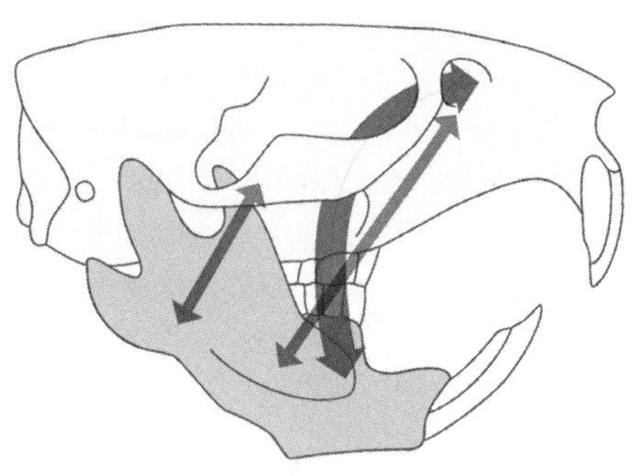

FIG. 148. Myomorphous masseter muscles (*top*, *middle*; redrawn from Vaughan et al. 2015) and their force of action (*bottom*; redrawn from MacDonald 1984).

and may be oval or V-shaped in anterior view (fig. 149). Note well-developed zygomatic plate.

4. HYSTRICOMORPHY

Superficial and lateral masseters originate from zygomatic arch (fig. 150, *top*); origin of medial masseter shifts from zygomatic arch to extensive area on side of rostrum, with muscle passing through greatly enlarged infraorbital foramen (fig. 150, *middle*). Temporal muscle may be large or small; if small, coronoid process is vestigial.

Direction of jaw movement up and strongly forward (fig. 150, *bottom*).

The large infraorbital canal of hystricomorphous rodents is generally oval and distinctive in anterior view; there is no zygomatic plate (fig. 151).

Not every rodent can be classified readily as sciuromorphous, myomorphous, or hystricomorphous, and even experts do not agree on which masseter type characterizes which families or how these conditions evolved. Earlier workers suggested a linear sequence in rodent jaw evolution, from protrogomorphous to sciuromorphous to myomorphous to hystricomorphous; alternatively, some recent workers believe that the myomorphous condition was derived from a hystricomorphous one. A well-resolved phylogenetic hypothesis will eventually allow us to determine the direction of jaw-muscle evolution and how many times each jaw-muscle system has evolved.

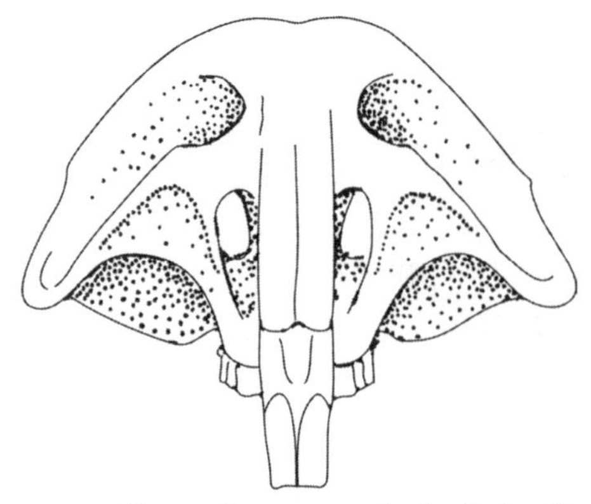

FIG. 149. Myomorphous zygomatic plate/infraorbital foramen, anterior view. Redrawn from Lawlor (1979).

SCIUROGNATHY VERSUS HYSTRICOGNATHY: Structure and Angle of the Lower Jaw

The spatial relationship between the alveolar root of the lower incisor and the angular process of the lower jaw subdivides rodents into two groups. In the sciurognathous condition the angular process lies more or less in line with the rest of the jaw. A derived condition is the hystricognathous jaw, in which the angular process is distinctly shifted laterally (fig. 152). The hystricognaths are uniformly recognized by most workers as a monophyletic lineage, uniquely defined by this jaw condition; the sciurognaths share the ancestral character state.

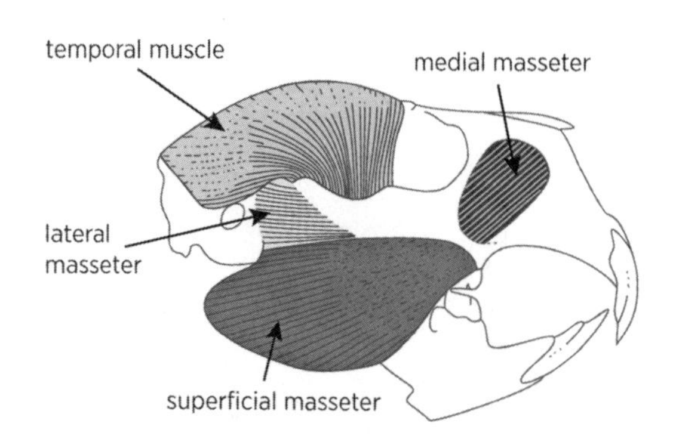

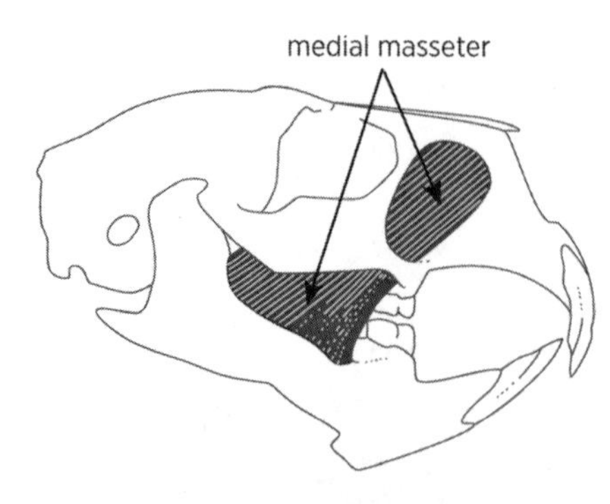

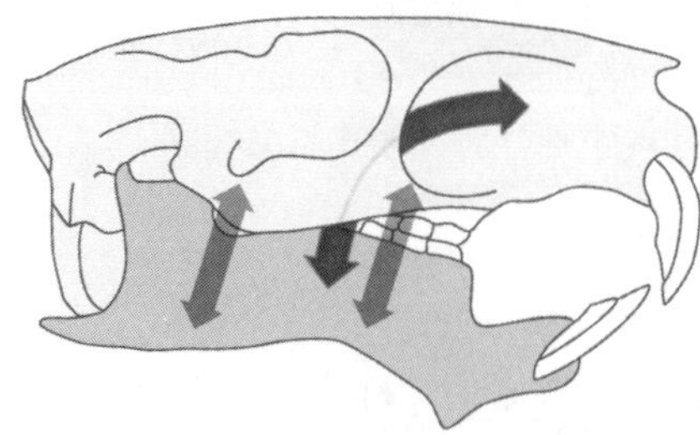

FIG. 150. Hystricomorphous masseter muscles (*top*, *middle*; redrawn from Vaughan et al. 2015) and their force of action (*bottom*; redrawn from MacDonald 1984).

PHYLOGENY AND CLASSIFICATION OF RODENTS

The traditional subdivision of rodents into four suborders, each based on a different jaw-muscle arrangement (Protrogomorpha, Sciuromorpha, Myomorpha, and Hystricomorpha; table 3), has been

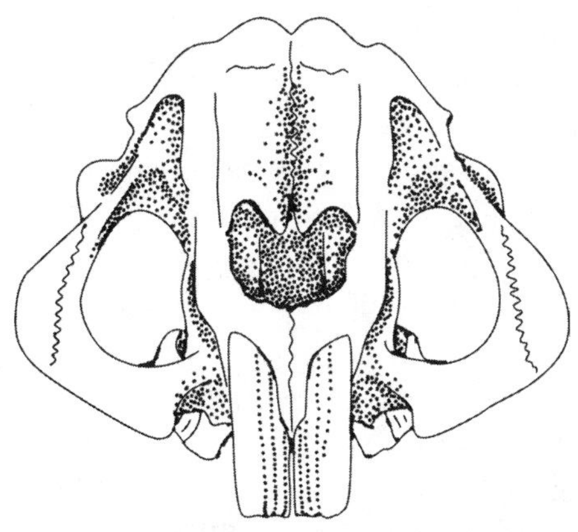

FIG. 151. Hystricomorphous infraorbital foramen, anterior view. Redrawn from Lawlor (1979).

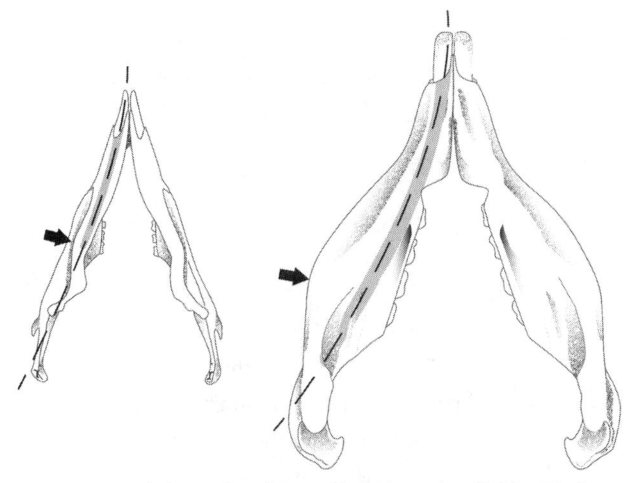

FIG. 152. Ventral views of a sciurognathous lower jaw (*left*), with the angular process (bold arrow) directly ventral to sheath of lower incisor (dashed line), and a hystricognathous lower jaw (*right*), with angular process (bold arrow) deflected lateral to sheath of lower incisor (dashed line). Redrawn from Lawlor (1979).

largely abandoned. Some combination of lower jaw structure and zygomasseteric structure is the basis for most current attempts to group rodents into phylogenetic units, such as those presented in recent textbooks (Feldhamer et al. 2015; Vaughan et al. 2015), although Landry (1999) attempted to reclassify rodents by characters other than those of the jaw-muscle system. While many problems remain, reassuring congruence in phylogenetic hypotheses based on cladistic analyses of morphological characters and DNA sequences has developed in recent years. The classification adopted by the account authors in Wilson and Reeder (2005), which divides rodents into five suborders, is generally corroborated by available DNA sequence analyses (see, for example, Fabre et al. 2012, 2015). The classification

Table 3. General groupings among Recent rodents evident in masseter formation and lower jaw condition.

	Sciurognathous	*Hystricognathous*
Protrogomorphous	Aplodontiidae	
Sciuromorphous	Sciuridae	Bathyergidae
	Castoridae	Heterocephalidae
Hystricomorphous	Anomaluridae	Hystricidae
	Zenkerellidae	Petromuridae
	Pedetidae	Thryonomyidae
	Ctenodactylidae	Erethizontidae
	Diatomyidae	Chinchillidae
	Dipodidae	Dinomyidae
		Caviidae
		Dasyproctidae
		Cuniculidae
		Ctenomyidae
		Octodontidae
		Abrocomidae
		Echimyidae
		†Heptaxodontidae
Myomorphous	Geomyidae*	
	Heteromyidae*	
	Platacanthomyidae	
	Spalacidae	
	Calomyscidae	
	Nesomyidae	
	Cricetidae	
	Muridae	
	Gliridae	

*Geomyoids (Geomyidae + Heteromyidae) are sometimes labeled as "pseudo-myomorphous."

†Heptaxodontidae—the Antillean giant hutias are known only from sub-fossil remains; the family is not covered in this manual.

we use herein (fig. 153) is based on Wilson and Reeder (2005), as modified by more recent molecular and other morphological studies cited in the accounts that follow.

Suborder Sciuromorpha
- Family Aplodontiidae (mountain beaver, sewellel)
- Family Gliridae (= Myoxidae) (dormice)
- Family Sciuridae (squirrels)

Suborder Castorimorpha
- Family Castoridae (beavers)
- Family Geomyidae (pocket gophers)

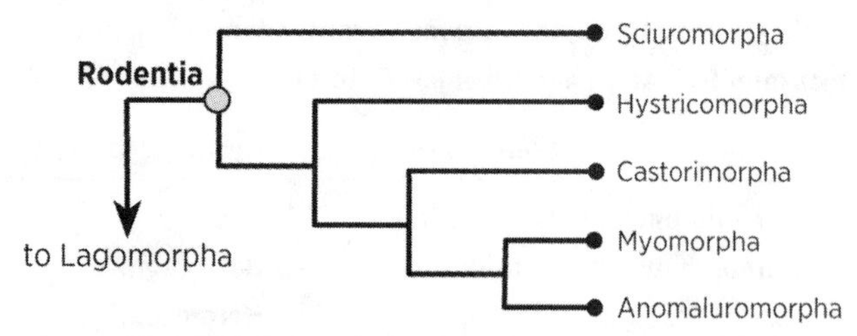

FIG. 153. Phylogenetic hypothesis of relationships among the five suborders of Rodentia, as recognized by Wilson and Reeder (2005), based on DNA sequences from mitochondrial and nuclear gene sequences (from Fabre et al. 2012, 2015). The suborder Hystricomorpha is equivalent to the Ctenohystrica as used by Fabre et al. (2012, 2015) and MacPhee (2011); the suborder Myomorpha is equivalent to the Myodonta as used by Fabre et al. (2015). The latter authors were unable to resolve relationships among the Castorimorpha, Anomaluromorpha, and Myomorpha.

Family Heteromyidae (pocket mice, kangaroo rats and mice)
Suborder Myomorpha (= Myodonta)
Superfamily Dipodoidea
Family Dipodidae (jerboas)
Family Sminthidae (birch mice)
Family Zapodidae (jumping mice)
Superfamily Muroidea
Family Platacanthomyidae (tree mice)
Family Spalacidae (zokors, bamboo rats, mole rats)
Family Calomyscidae (brush-tailed mice)
Family Nesomyidae (pouched rats, climbing mice, fat mice, and tufted-tailed rats)
Family Cricetidae (voles, New World mice and rats)
Family Muridae (gerbils, hamsters, Old World mice and rats)
Suborder Anomaluromorpha
Family Anomaluridae (scaly-tailed squirrels)
Family Zenkerellidae (scaly-tailed squirrels)
Family Pedetidae (spring hares)
Suborder Hystricomorpha (= Ctenohystrica)
Infraorder Ctenodactylomorphi
Family Ctenodactylidae (gundis)
Family Diatomyidae (Laotian rock rat)
Infraorder Hystricognathi
Family Bathyergidae (mole rats)
Family Heterocephalidae (naked mole rats)
Family Hystricidae (Old World porcupines)
Family Petromuridae (dassie rat or rock rat)
Family Thryonomyidae (cane rats)
Clade Caviomorpha
Superfamily Erethizontoidea
Family Erethizontidae (New World porcupines)
Superfamily Chinchilloidea
Family Chinchillidae (chinchillas, viscachas)
Family Dinomyidae (pacarana)
Superfamily Cavioidea
Family Caviidae (guinea pigs, Patagonian mara, capybara)
Family Cuniculidae (= Agoutidae) (pacas)
Family Dasyproctidae (agoutis, acouchis)
Superfamily Octodontoidea
Family Abrocomidae (chinchilla rats)
Family Echimyidae (spiny rats, coypu, hutias)
Family Ctenomyidae (tuco-tucos)
Family Octodontidae (octodonts, degu)
Family †Heptaxodontidae (key mouse, giant hutia)

Suborder SCIUROMORPHA

The suborder Sciuromorpha includes three living families (fig. 154). The apparent sister-group relationship of the Aplodontiidae and Sciuridae provides a good example in which the classical jaw-muscle morphology (protrogomorphic in the former, sciuromorphic in the latter) does not reflect phylogenetic propinquity (as discussed above).

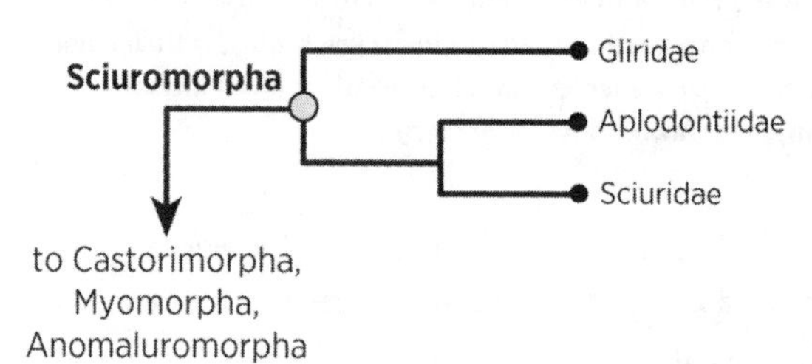

FIG. 154. Hypothesized relationships among the three families of sciuromorph rodents.

Family APLODONTIIDAE (mountain beaver, sewellel)

Medium-sized (total length 300–500 mm; mass 500–1,000 g) with stocky body, short ears that lack a tragus, **vestigial tail**, short and thick limbs, digits 5/5, not webbed, forefeet with enlarged claws; fossorial, digs complex burrows; quasi-colonial; herbivorous; confined to moist microenvironments. (The family name is commonly, but incorrectly, spelled Aplodontidae; see Helgen 2005a.)

FIG. 155. Aplodontiidae; mountain beaver, or sewellel, *Aplodontia rufa*.

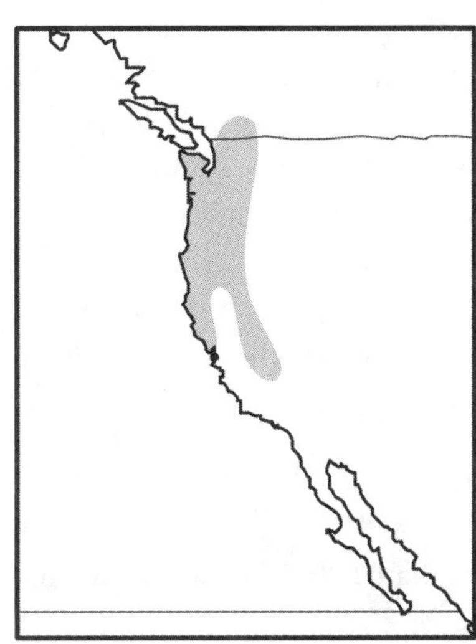

FIG. 156. Geographic range of family Aplodontiidae: western North America, from the northwestern United States (California, Oregon, and Washington) to southwestern British Columbia, Canada.

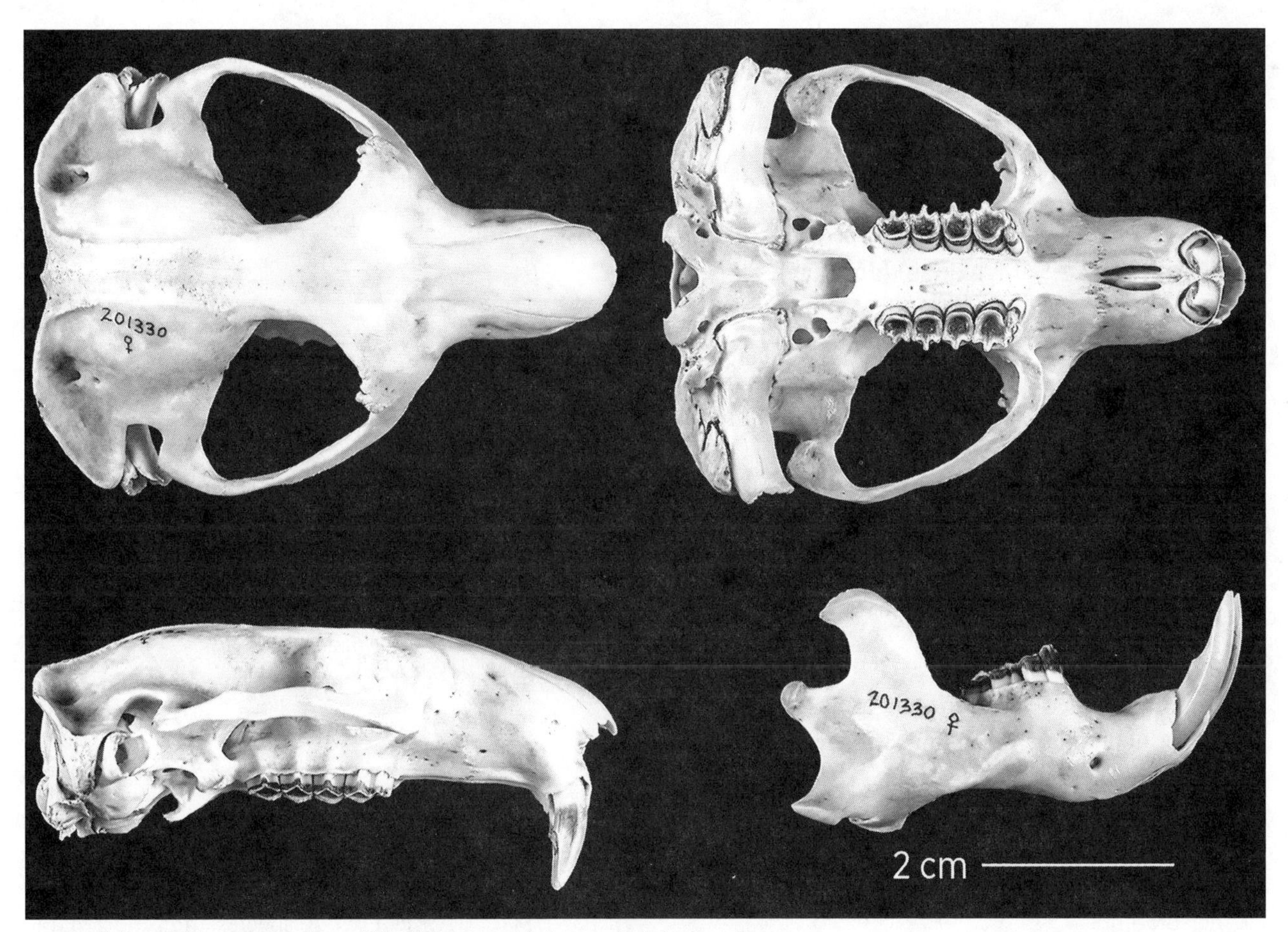

FIG. 157. Aplodontiidae; skull and mandible of the mountain beaver, *Aplodontia rufa*.

DIAGNOSTIC CHARACTERS:

- **skull flattened, greatly widened posteriorly**
- **jaw musculature protrogomorphous**, with all parts of masseter muscle having origin on zygomatic arch, none on the rostrum; **no zygomatic plate; small, round infraorbital foramen** (see fig. 145)
- **crowns of cheek teeth ringlike, each with a small projection** (located laterally on uppers, medially on lowers)

RECOGNITION CHARACTERS:

1. **no zygomatic plate**
2. **auditory bullae not inflated, flask-shaped, with long tympanic tube**
3. **angular process on lower jaw strongly inflected**

DENTAL FORMULA: $\frac{1}{1}\ \frac{0}{0}\ \frac{2}{1}\ \frac{3}{3} = 22$

TAXONOMIC DIVERSITY: The family is monotypic, with a single species, *Aplodontia rufa*.

FIG. 158. Gliridae: Graphiurinae; spectacled African dormouse, *Graphiurus ocularis*.

Family GLIRIDAE (= Myoxidae) (dormice)

Small to medium-sized (body length 100–400 mm), resembling squirrels or chipmunks; body compact; eyes and ears relatively large; **tail moderately long, usually bushy,** may autotomize and be regenerated in some genera; limbs short; hind feet broad, toes with curved claws; fur thick and soft, some species with distinct black facial markings; good climbers, arboreal species with well-developed toe pads; omnivorous; some species hibernate.

RECOGNITION CHARACTERS:

1. cranium rounded or flattened in profile
2. infraorbital foramen medium-sized, oval or V-shaped, opening anteriorly
3. zygomatic plate narrow and below infraorbital foramen (*Graphiurus*) or vertical
4. nasals extending anteriorly beyond level of upper incisors
5. foramen in angular process of lower jaw present (some *Graphiurus*), but usually absent
6. **occlusal surface of cheek teeth with series of cross ridges, or basin-shaped with indistinct ridges; brachydont**

DENTAL FORMULA: $\frac{1}{1}\ \frac{0}{0}\ \frac{0\text{–}1}{0\text{–}1}\ \frac{3}{3} = 16\text{–}20$

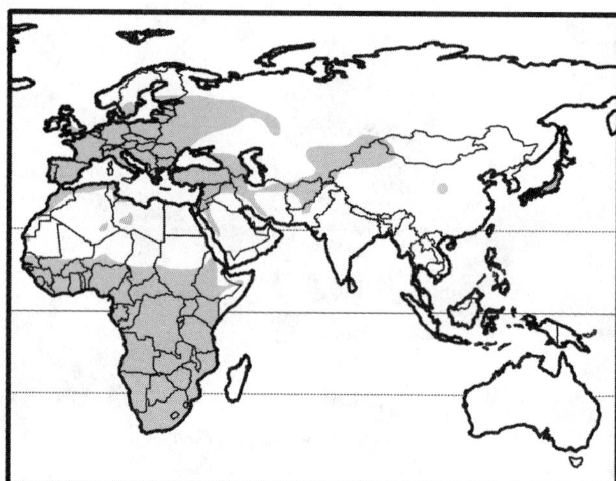

FIG. 159. Geographic range of family Gliridae: temperate, subtropical, and tropical forests, shrublands, and savannas in Africa, Europe, Middle East, southwestern Asia, Japan.

TAXONOMIC DIVERSITY: 29 species and nine genera allocated to three subfamilies:

Glirinae—*Glirulus* (Japanese dormouse) and *Glis* (= *Myoxus*; fat or edible dormouse)

Graphiurinae—*Graphiurus* (African dormice)

Leithiinae—*Chaetocauda* (Sichuan dormouse), *Dryomys* (forest dormice), *Eliomys* (garden dormice), *Muscardinus* (hazel dormouse), *Myomimus* (mouse-tailed dormice), and *Selevinia* (desert dormouse)

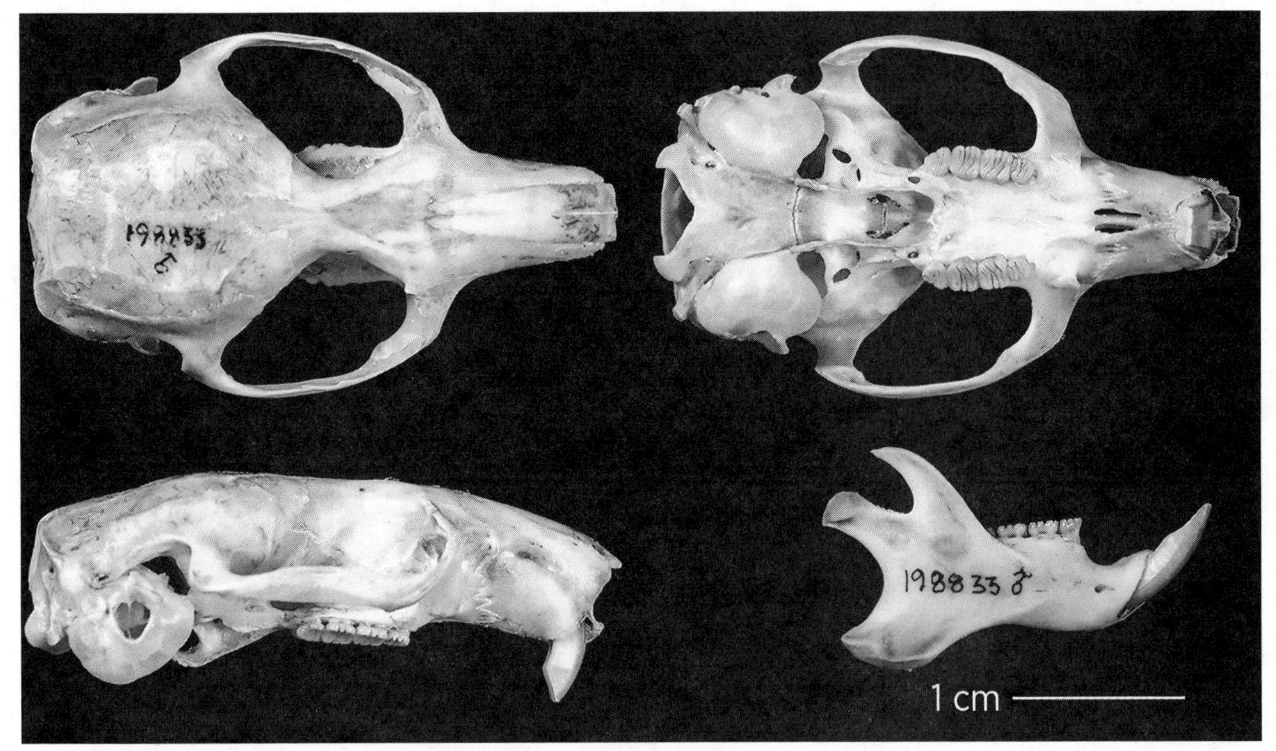

FIG. 160. Gliridae: Glirinae; skull and mandible of the fat dormouse, *Glis glis* (Glirinae).

Family SCIURIDAE

Very diverse group that includes typical squirrels, marmots, and prairie dogs. Tree squirrels have long, bushy tails; sharp, usually recurved claws; large ears, some with well-developed tufts; flying squirrels have furred gliding membranes (= patagia) extending from ankle to wrist, with a cartilaginous support (styliform cartilage) extending laterally from the wrist; ground squirrels have short, sturdy forelimbs with elongate claws for digging, and generally have short tails; arboreal members have specialized ankle allowing rotation of foot for head-down descent of tree trunks; body size ranges from mouse-sized (pygmy squirrels) to 2 kg or more (marmots, giant flying squirrels); all have ears without a tragus and **digits 4/5, not webbed**. Habitat breadth is extreme, from tropical and temperate forests and woodlands to open grasslands, deserts, and tundra. Most members are solitary but some (e.g., prairie dogs) live in large colonies with complex social organization. Food habits highly variable, but mostly herbivorous.

RECOGNITION CHARACTERS:

1. **postorbital process large**
2. auditory bulla moderately inflated, not flask-shaped and without tympanic tube
3. zygomatic plate well developed
4. **infraorbital foramen very small, slit-like**
5. angular process of lower jaw usually slightly inflected
6. **crowns of cheek teeth usually with distinct cusps or ridges and valleys, not flat**

DENTAL FORMULA: $\frac{1}{1}\ \frac{0}{0}\ \frac{1\text{–}2}{1}\ \frac{3}{3} = 20\text{–}22$

TAXONOMIC DIVERSITY: About 285 species in 58 genera allocated to five subfamilies, one with two tribes and another with three (following Thorington et al. 2012 and Patterson and Norris 2016). Helgen et al. (2009) revised the taxonomy of the paraphyletic *Spermophilus*, recognizing *Callospermophilus*, *Ictidomys*, *Otospermophilus*, *Xerospermophilus*, *Poliocitellus*, *Urocitellus*, and *Notocitellus*, thereby restricting *Spermophilus* to the Old World.

Callosciurinae: 67 species in 14 genera: *Callosciurus* (beautiful squirrels), *Dremomys* (Asian long-nosed squirrels), *Exilisciurus* (Asian pygmy squirrels), *Funambulus* (striped squirrels), *Glyphotes* (sculptor squirrel), *Hyosciurus* (Indonesian long-nosed squirrels), *Hypsciurus* (long-nosed squirrels), *Lariscus* (Indonesian striped ground squirrels), *Menetes* (Indochinese ground squirrel), *Nannosciurus* (black-eared squirrel), *Prosciurillus* (Sulawesi dwarf

FIG. 161. Sciuridae; fox squirrel, *Sciurus niger* (Sciurinae: Sciurini; *top*), and hoary marmot, *Marmota caligata* (Xerinae: Marmotini; *bottom*).

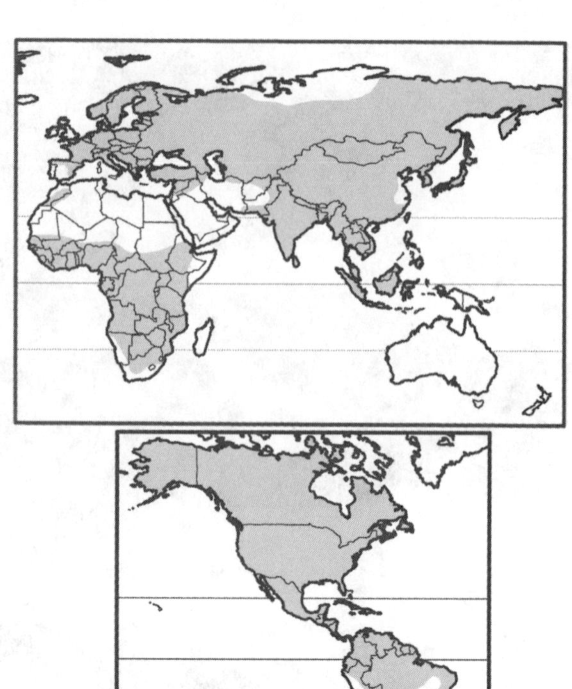

FIG. 162. Geographic range of family Sciuridae: tropical and temperate regions of all continents, except for Australasia and Antarctica.

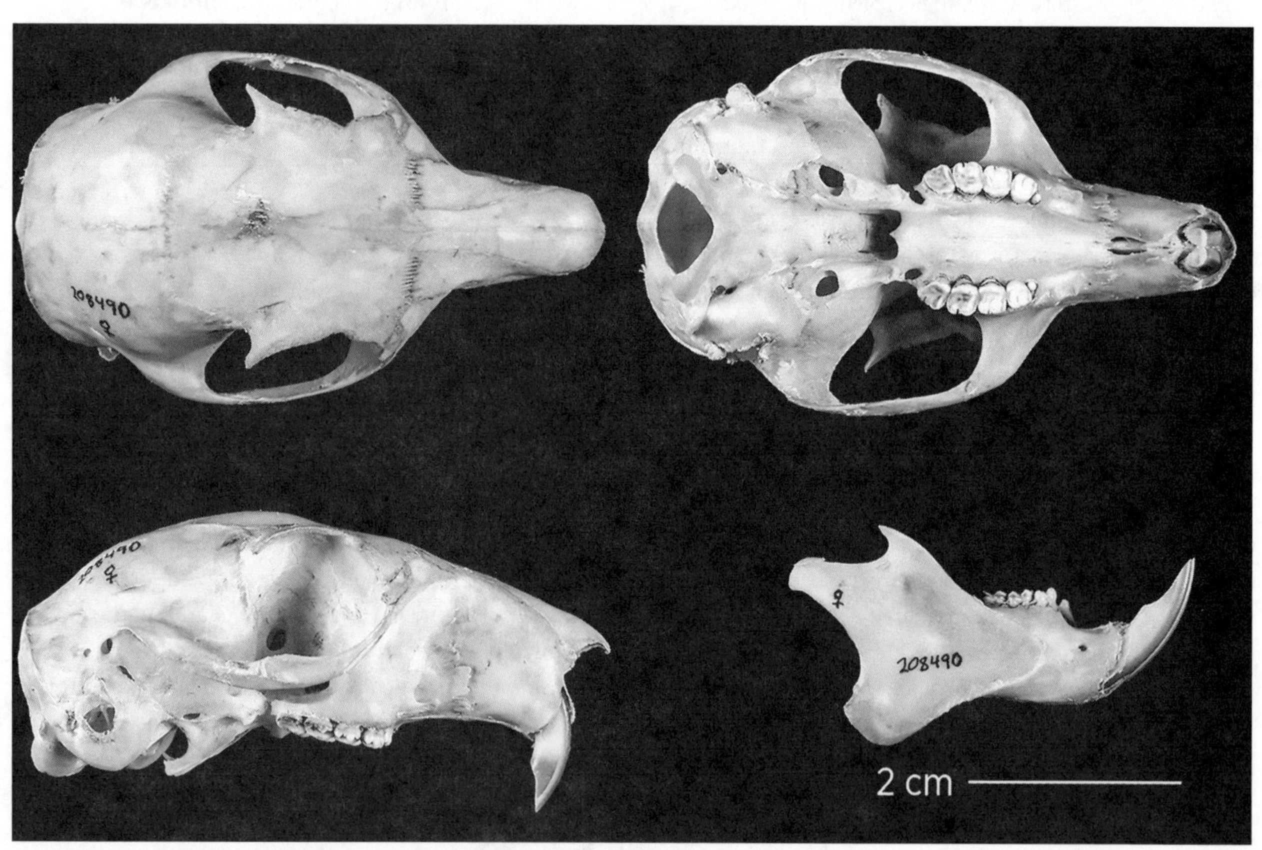

FIG. 163. Sciuridae: Sciurinae: Sciurini; skull and mandible of western gray squirrel, *Sciurus griseus*.

squirrels), *Rhinosciurus* (shrew-faced squirrel), *Rubrisciurus* (Sulawesi giant squirrel), *Sundasciurus* (Southeast Asian squirrels), and *Tamiops* (small striped tree squirrels)

Ratufinae: 4 species in the genus *Ratufa* (Oriental giant squirrels)

Sciurillinae: the monotypic genus *Sciurillus* (Neotropical pygmy squirrel)

Sciurinae:

Pteromyini—44 species in 15 genera: *Aeretes* (northern Chinese flying squirrel), *Aeromys* (Black and Thomas's flying squirrels), *Belomys* (hairy-footed flying squirrel), *Biswamoyopterus* (Namdapha and Laotian flying squirrels), *Eoglaucomys* (Kashmir flying squirrel), *Eupetaurus* (woolly flying squirrel), *Glaucomys* (North American flying squirrels), *Hyopetes* (Southeast Asian flying squirrels), *Iomys* (Javanese and Mentawai flying squirrels), *Petaurillus* (pygmy flying squirrels), *Petaurista* (giant flying squirrels), *Petinomys* (flying squirrels), *Pteromys* (Japanese and Siberian flying squirrels), *Pteromyscus* (smoky flying squirrel), and *Trogopterus* (complex-toothed flying squirrel).

Sciurini—37 species in 5 genera: *Microsciurus* (New World dwarf squirrels), *Reithrosciurus* (Borneo tufted ground squirrel), *Sciurus* (tree squirrels, 26 species in New World, only two in Eurasia), *Syntheosciurus* (Bang's mountain squirrel), and *Tamiasciurus* (North American chickarees)

Xerinae:

Marmotini—95 species in 15 genera of largely North American terrestrial squirrels: *Ammospermophilus* (antelope ground squirrels), *Callospermophilus* (mantled ground squirrels), *Cynomys* (prairie dogs), *Eutamias* (Siberian chipmunk), *Ictidomys* (little ground squirrels), *Marmota* (marmots), *Neotamias* (chipmunks of western North America), *Notocitellus* (tropical and ring-tailed ground squirrels), *Otospermophilus* (rock and Beechey ground squirrels), *Poliocitellus* (Franklin's ground squirrel), *Sciurotamias* (Asian rock squirrels), *Spermophilus* (Old World ground squirrels), *Tamias* (eastern chipmunk of North America), *Urocitellus* (Holarctic ground squirrels), and *Xerospermophilus* (pygmy ground squirrels)

Protoxerini—31 species in 6 genera: *Epixerus* (African palm squirrel), *Funisciurus* (African rope squirrels), *Heliosciurus* (African sun squirrels), *Myosciurus* (African pygmy squirrel), *Paraxerus* (African bush squirrels), and *Protoxerus* (slender-tailed and forest giant squirrels of Africa)

Xerini—6 species in 3 genera: *Atlantoxerus* (Barbary ground squirrel), *Euxerus* (striped ground squirrel of sub-Saharan Africa), *Geosciurus* (Damara and South African ground squirrels), *Spermophilopsis* (long-clawed ground squirrel of central Asia), and *Xerus* (unstriped ground squirrel of East Africa)

Suborder CASTORIMORPHA

The suborder Castorimorpha contains three families with quite different body morphologies (fig. 164). The Geomyidae and Heteromyidae are closely related sister groups (frequently united under the clade Geomyoidea) that share external fur-lined cheek pouches, a feature unique among living mammals. Recent molecular analysis (Fabre et al. 2012, 2015) suggests that the pocket gophers [Geomyidae] are nested within the Heteromyidae, a hypothesis originally advanced by Korth (1994) based on fossils. If so, Geomyidae would have date priority as the family name for both groups.

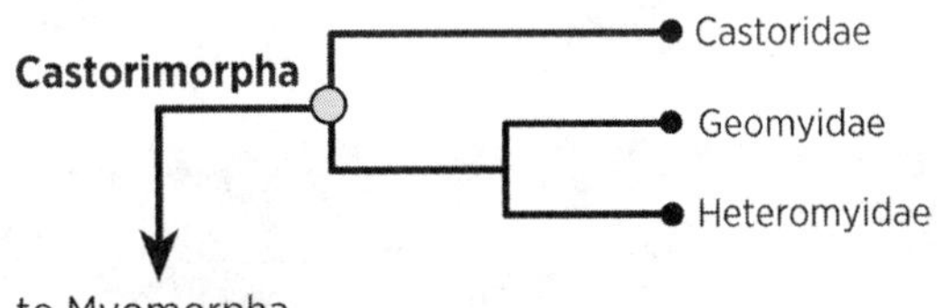

FIG. 164. Hypothesized relationships among the three families of castorimorph rodents.

Family CASTORIDAE (beavers)

Medium-sized to large (body length 950–1200 mm; mass up to 39 kg); body thickset; semiaquatic, with **5/5 digits on webbed feet**; small ears without a tragus; tail dorsoventrally flattened and scaly; dense underfur covered by elongate guard hairs; social, living in family groups; ecosystem engineers, building complex dams and lodges from sticks and mud; herbivorous.

DIAGNOSTIC CHARACTERS:

- **tail large, broad, and flat, mostly naked and scaly**
- **distinct depression in basioccipital**

RECOGNITION CHARACTERS:

1. postorbital process very small
2. auditory bulla not inflated, somewhat flask-shaped and with tympanic tube

FIG. 165. Castoridae; American beaver, *Castor canadensis*.

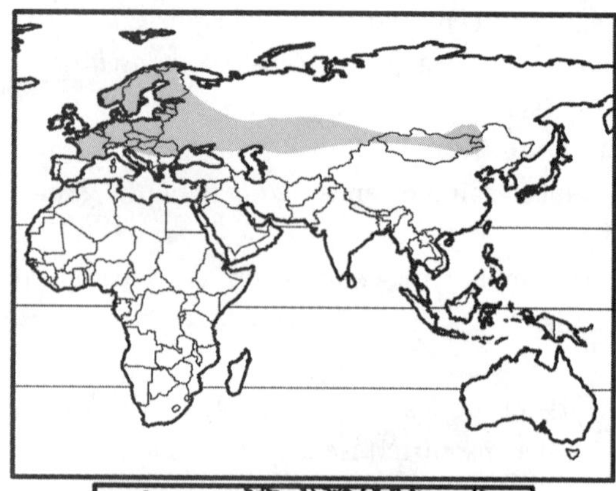

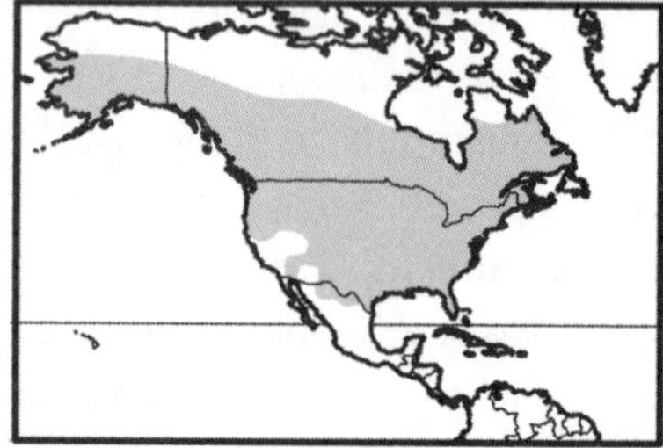

FIG. 167. Geographic range of family Castoridae: streams, rivers, and lakes in temperate North America and Eurasia. *Castor canadensis* has been introduced to Tierra del Fuego, where it is a major problem.

FIG. 166. Frontal view of infraorbital foramen of a beaver, *Castor canadensis*. Redrawn from Lawlor (1979).

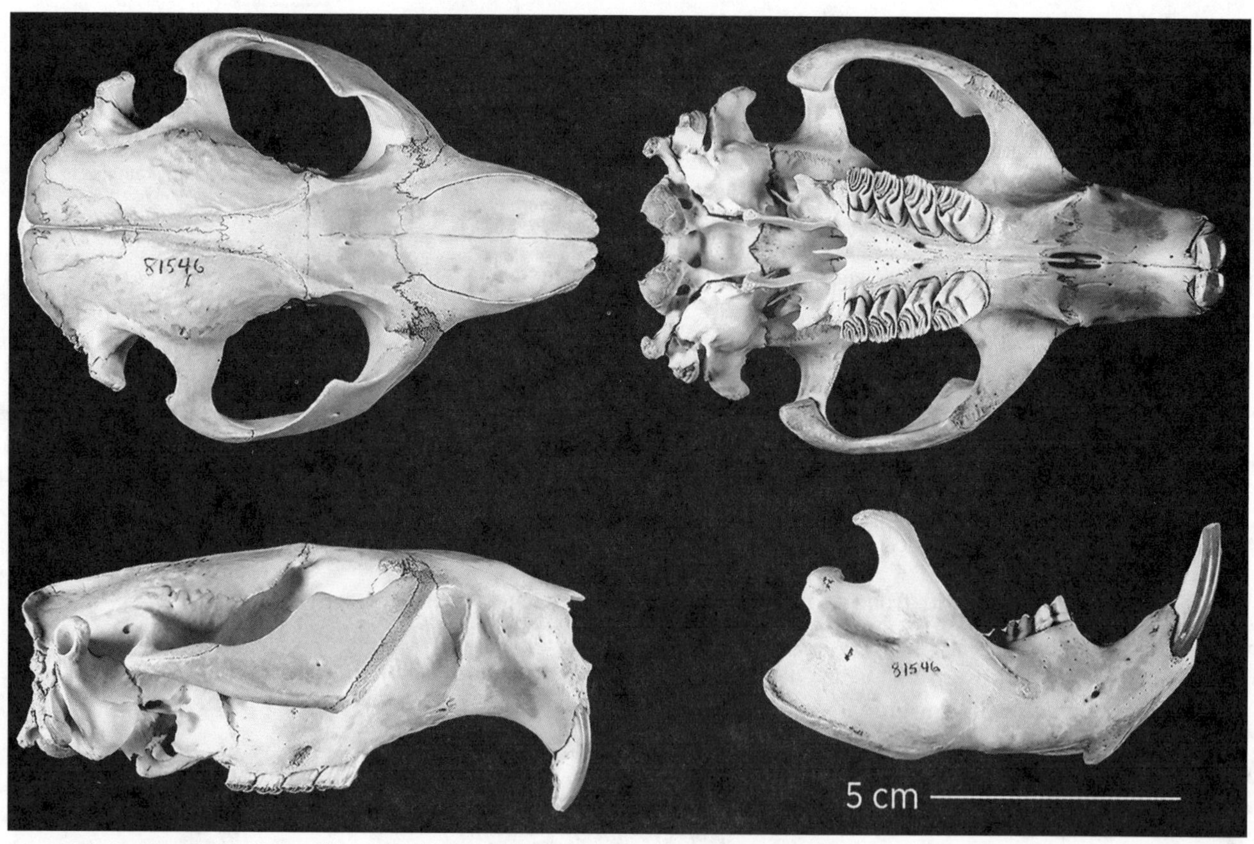

FIG. 168. Castoridae; skull and mandible of a beaver, *Castor canadensis*.

3. well-developed zygomatic plate
4. **infraorbital foramen very small, slit-like** (fig. 166, arrow)
5. angular process not inflected
6. crowns of cheek teeth flat, with narrow inner and outer lateral enamel folds

DENTAL FORMULA: $\frac{1}{1}\frac{0}{0}\frac{1}{1}\frac{3}{3} = 20$

TAXONOMIC DIVERSITY: The family has a single living genus, *Castor*, with one species each in North America and Eurasia; *C. canadensis* has been introduced to Tierra del Fuego and parts of northern Europe and the Russian Far East.

Family GEOMYIDAE (pocket gophers)

Subterranean rodents of small to medium body size (mass 100 g–2 kg), with tubular bodies, no discernible neck, small ears, short, nearly hairless (but not scaly) tail, and enlarged forefeet with powerful claws; burrow primarily by scratch-digging and secondarily by tooth-digging with enlarged incisors; burrow systems are long and complex; solitary; herbivorous; can be active at any time of day and all year long, even under snow; have many physiological adaptations for living in anoxic environment of closed tunnels.

DIAGNOSTIC CHARACTERS:

- **external fur-lined cheek pouches**
- **tubular body, short ears, small eyes, short forelegs and hind legs, short tail, manus with well-developed claws for digging**
- **two large pits in palate behind last molars**
- **premolar figure 8–shaped, larger than any molar**
- **infraorbital canal compressed against side of rostrum, opening anterolaterally**

RECOGNITION CHARACTERS:

1. cranium robust, flat in lateral profile
2. zygomatic plate broad, tilted strongly upward
3. zygomatic arches square and well developed
4. **nasals not extending anteriorly beyond level of upper incisors**
5. auditory bulla small
6. no foramen in angular process of lower jaw
7. **crowns of cheek teeth simple, those of molars ring-shaped**

DENTAL FORMULA: $\frac{1}{1}\frac{0}{0}\frac{1}{1}\frac{3}{3} = 20$

TAXONOMIC DIVERSITY: 41 species and seven genera, with all living pocket gophers in a single subfamily (Geomyinae) with two tribes; all share the common name of pocket gopher:

Geomyini—*Cratogeomys*, *Geomys*, *Heterogeomys*, *Orthogeomys*, *Pappogeomys*, and *Zygogeomys*

Thomomyini—*Thomomys*

FIG. 169. Geomyidae: Geomyinae: Thomomyini; Botta's pocket gopher, *Thomomys bottae*.

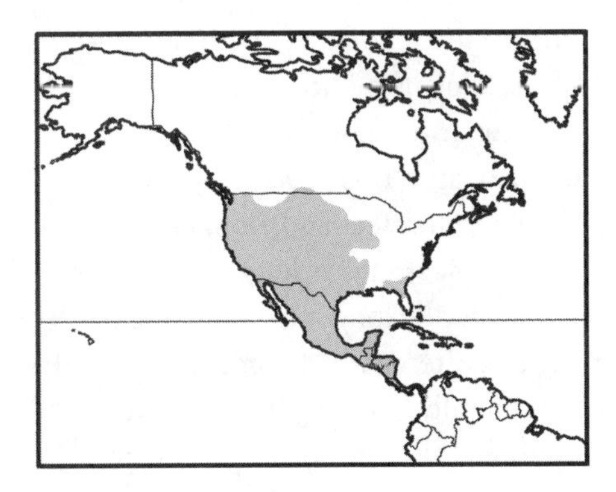

FIG. 170. Geographic range of family Geomyidae: North and Central America (southeastern United States, Great Plains to Pacific coast, southern Canada through Mexico) to extreme northwestern Colombia in South America.

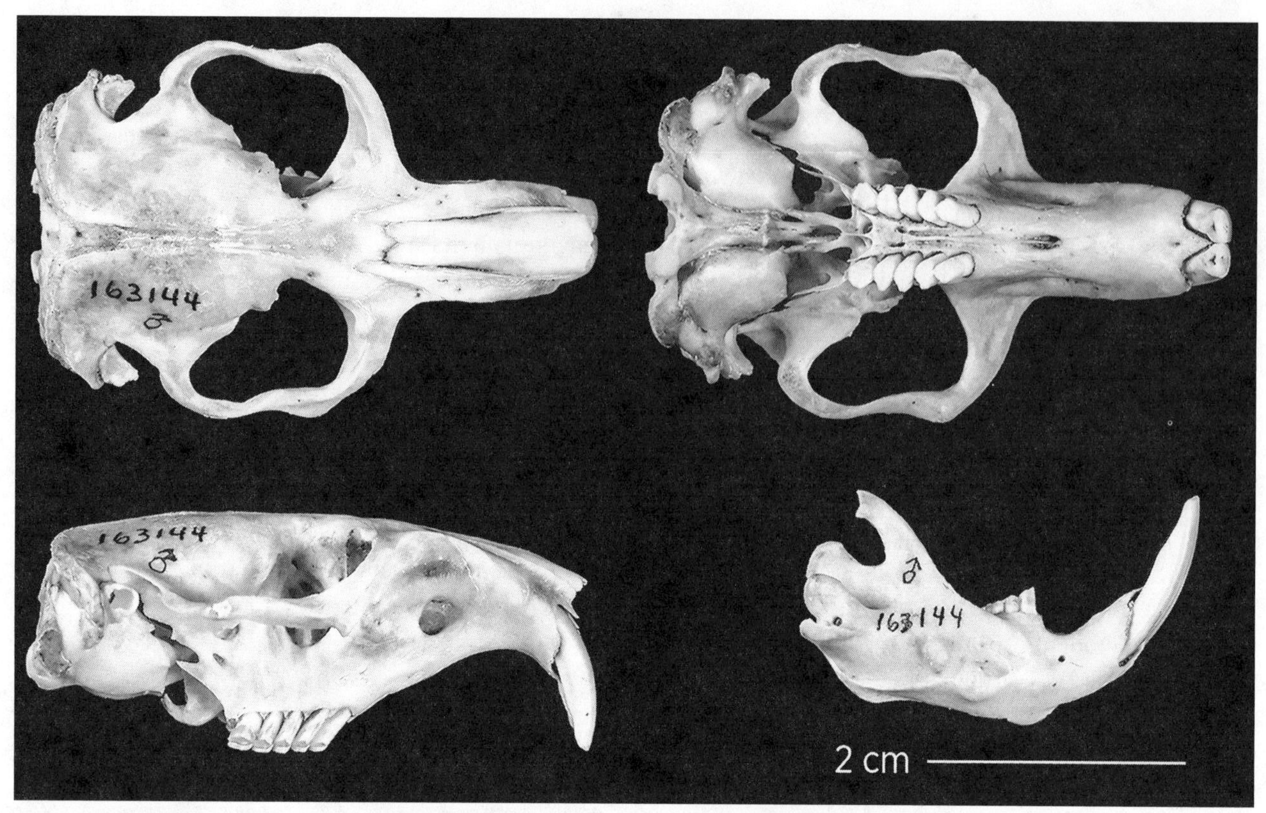

FIG. 171. Geomyidae: Geomyinae: Thomomyini; skull and mandible of Townsend's pocket gopher, *Thomomys townsendii*.

Family HETEROMYIDAE (pocket mice, kangaroo rats, kangaroo mice)

Small to medium-sized (body length 100–400 mm) mice and rats largely confined to arid environments but extending into tropical dry forests in North America and northern South America. Body form varies: pocket mice (*Chaetodipus*, *Heteromys*, and *Perognathus*) are quadrupedal bounders with narrow, naked-soled hind feet; kangaroo mice (*Microdipodops*) are quadrupedal saltators with elongate and hairy-soled hind feet; and kangaroo rats (*Dipodomys*) are truly ricochetal with short forefeet and elongate hind legs and hairy-soled hind feet; tail as long as head and body length, but substantially longer in some pocket mice and kangaroo rats, often with elongate tuft at end; all have short ears. Generally good diggers with well-developed claws on forefeet; some construct complex burrow systems; all species essentially granivorous, foraging above ground, with many hoarding seeds in surface or burrow caches; typically solitary, territorial, and nocturnal; some species estivate or undergo daily torpor to save energy.

DIAGNOSTIC CHARACTERS:

- **external fur-lined cheek pouches**
- **limbs modified for quadrupedal bounding or bipedal saltation**
- **infraorbital canal compressed against and entirely piercing the rostrum, opening laterally**

RECOGNITION CHARACTERS:

1. **cranium somewhat rounded in profile**
2. zygomatic plate broad and tilted upward
3. nasals extending anteriorly, well beyond level of upper incisors
4. **auditory bulla slightly inflated** (*Heteromys*), **moderately inflated** (*Chaetodipus*, *Perognathus*), **or greatly inflated** (*Dipodomys*, *Microdipodops*)
5. no foramen in the angular process of the lower jaw
6. **crowns of molars simple, bilophodont, rooted, and brachydont** (all pocket mice and kangaroo mice) **or simple ovals, unrooted, and hypsodont** (kangaroo rats)

DENTAL FORMULA: $\frac{1}{1}\ \frac{0}{0}\ \frac{1}{1}\ \frac{3}{3} = 20$

TAXONOMIC DIVERSITY: 66 species allocated to five genera in three subfamilies:

Dipodomyinae—*Dipodomys* (kangaroo rats) and *Microdipodops* (kangaroo mice)

Heteromyinae—*Heteromys* (including *Liomys*; spiny pocket mice)

Perognathinae—*Chaetodipus* (spiny pocket mice) and *Perognathus* (silky pocket mice)

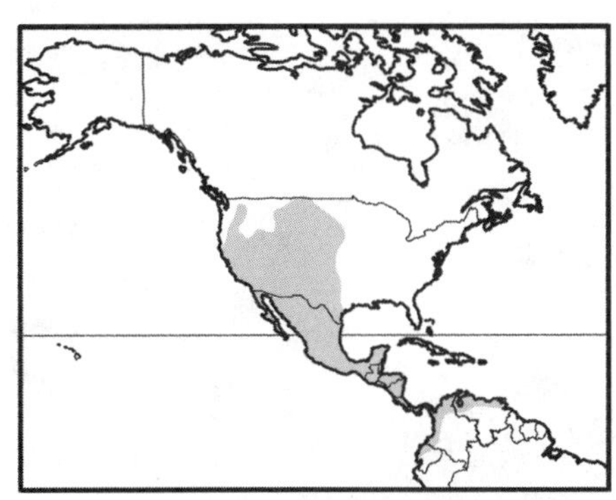

FIG. 173. Geographic range of family Heteromyidae: arid and semiarid zones of western North America, from southwestern Canada through deserts of United States and Mexico; subtropical forests of Mexico, Central America, and northern South America.

FIG. 172. Heteromyidae; giant kangaroo rat, *Dipodomys ingens* (Dipodomyinae; *left*), and San Joaquin pocket mouse, *Perognathus inornatus* (Perognathinae; *right*).

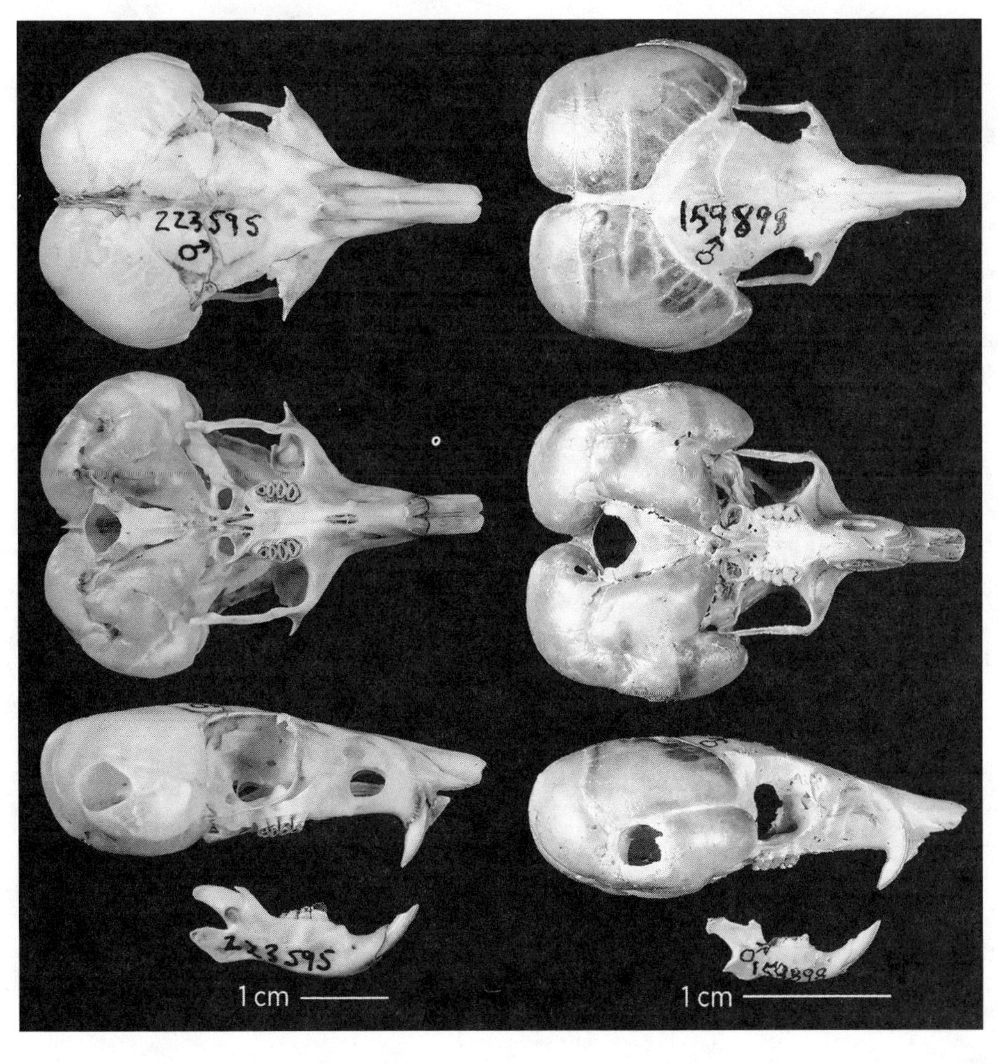

FIG. 174. Heteromyidae: Dipodomyinae; skull and mandible of the desert kangaroo rat, *Dipodomys deserti* (*left*), and the dark kangaroo mouse, *Microdipodops megacephalus* (*right*).

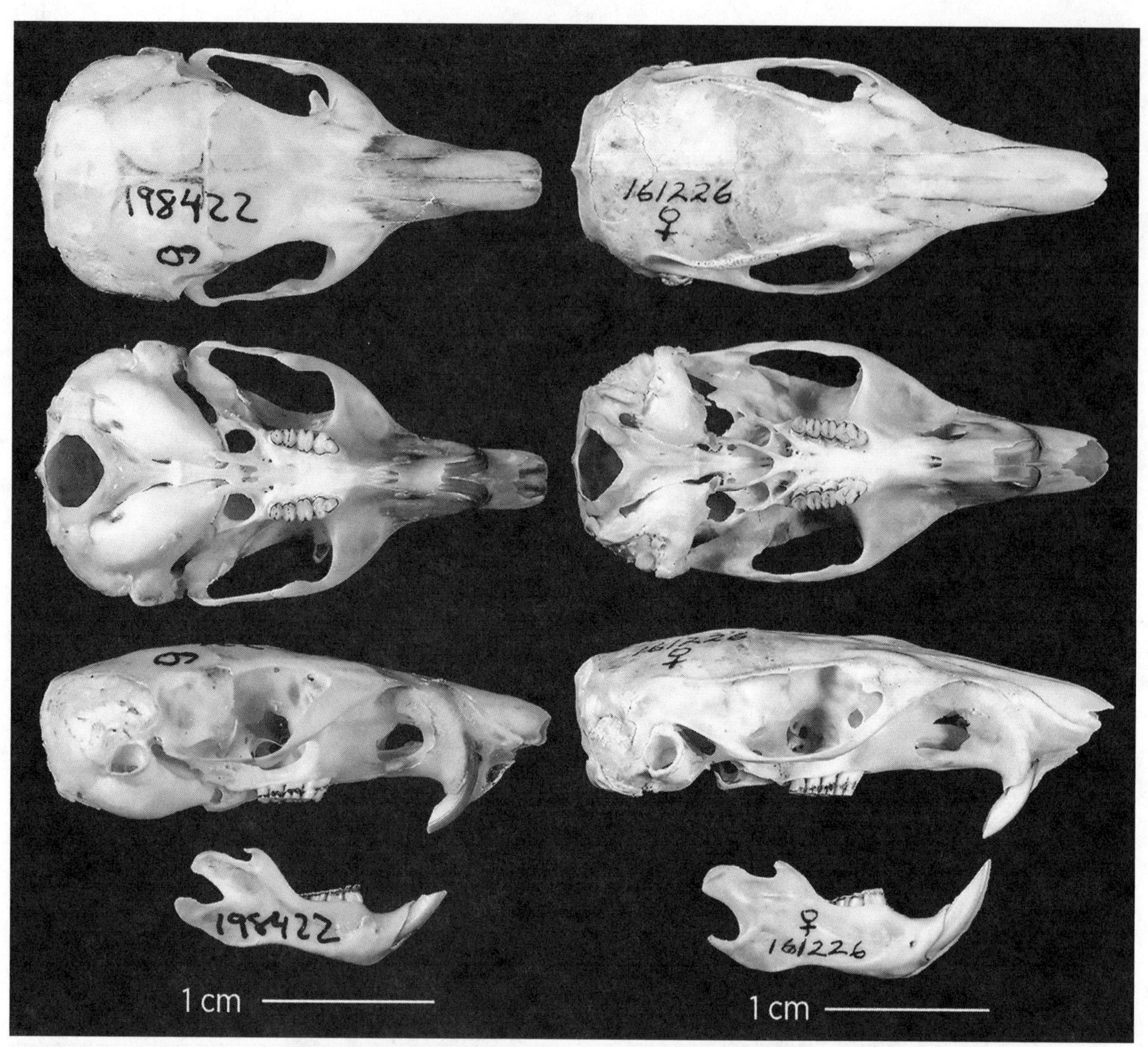

FIG. 175. **Heteromyidae; skull and mandible of the California pocket mouse, *Chaetodipus californicus* (Perognathinae; *left*), and the forest spiny pocket mouse, *Heteromys desmarestianus* (Heteromyinae; *right*).**

Suborder MYOMORPHA (= Myodonta)

The suborder Myomorpha comprises nine families allocated to two superfamilies: Dipodoidea and Muroidea (fig. 176). The Dipodoidea contains three families, the Dipodidae, Sminthidae, and Zapodidae, although until recently the Sminthidae was treated as a subfamily of the Dipodidae (see below). The Muroidea has six families, with the Platacanthomyidae basal to the remainder, followed successively by the Spalacidae, Calomyscidae, Nesomyidae, Cricetidae, and Muridae, the latter four of which have been collectively termed the Eumuroida (the true muroids; e.g., Schenk et al. 2013). The terminal pair of sister taxa (Cricetidae and Muridae) contain the vast majority of taxonomic diversity found in the suborder. As we use it here, Myomorpha is equivalent to Myodonta as employed by Schenk et al. (2013) and Fabre et al. (2015).

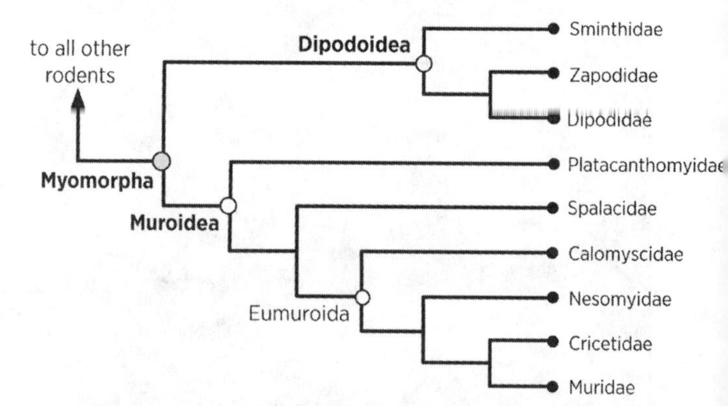

FIG. 176. **Hypothesized relationships among the seven families of myomorph rodents; nodes depicting subfamily groups are also identified.**

Superfamily DIPODOIDEA

Many earlier classifications placed members of the Dipodoidea into two families, Dipodidae (birch mice and jerboas) and Zapodidae (jumping mice), based on differences in body form, degree of specialization in the auditory bulla, and the occlusal surfaces of the cheek teeth, among other characters. More recently, molecular phylogenetic analyses (Lebedev et al. 2013; Pisano et al. 2015) have divided this group of taxa into three families, whose divergence appears to correspond with major tectonic changes associated with the collision of India with Asia in the Middle Eocene, followed by global cooling at the Eocene–Oligocene transition; in these classifications, the family Sminthidae (birch mice; Sicistidae of some authors) is basal to the Zapodidae (jumping mice) and Dipodidae (jerboas), with the latter group including four subfamilies. Because the three groups share common morphological attributes, despite their collective disparity, we follow Musser and Carleton (2005) and group them together here. Many genera in this family are highly convergent in body form and lifestyle with the kangaroo rats (*Dipodomys*) of the New World Heteromyidae.

DIAGNOSTIC CHARACTERS:

- **infraorbital canal large, oval in shape, and opening anteriorly** (fig. 178, upper arrow)
- lateral masseter originates from ventrolateral surface of the zygomatic arch, visible as a slight ridge below the infraorbital canal (fig. 178, lower arrow; see also Klingener 1964); while often referred to as a **narrow, horizontal zygomatic plate**, this condition is more analogous to that of hystricomorphous rodents

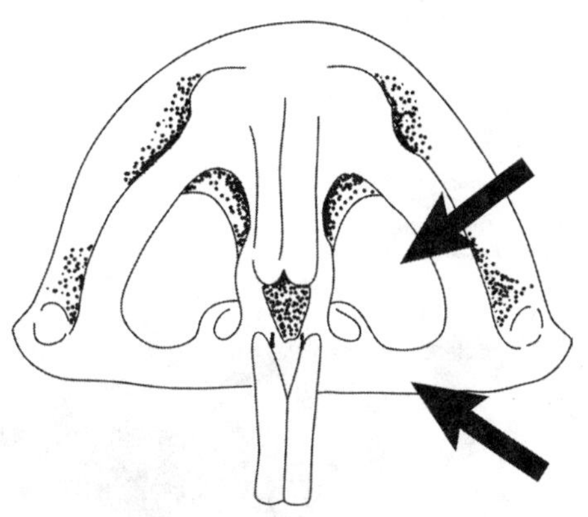

FIG. 178. Frontal view of infraorbital foramen and zygomatic plate of Dipodidae. Redrawn from Lawlor (1979).

Families DIPODIDAE, SMINTHIDAE, AND ZAPODIDAE (jerboas, birch mice, jumping mice)

Very diverse mouselike and ratlike rodents. Jumping mice have long and narrow hind feet and an elongate, slender tail, move by quadrupedal bounding, occupy moist habitats, and are solitary, nocturnal, and omnivorous. Birch mice have shorter tails and hind feet but also have bounding locomotion; they are nocturnal and occupy self-dug, shallow burrows. Jerboas are highly specialized for bipedal ricochetal saltation, with fused or nearly fused cervical vertebrate, shortened forelimbs, elongate hind limbs, elongate hind feet with metatarsal elements commonly fused into a single long cannon bone and with furred plantar surfaces in species that occupy sandy soils.

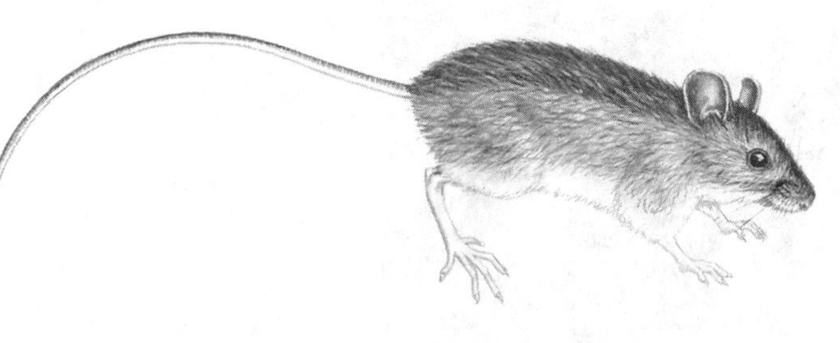

FIG. 177. Zapodidae; western jumping mouse, *Zapus princeps*.

RECOGNITION CHARACTERS:

1. body small, mouse- or kangaroo-like (100–400 mm)
2. **hind limbs elongate** in all genera but birch mice (*Sicista*), **greatly elongate with fused cannon bone with reduced or vestigial lateral toes in jerboas** (such as *Allactaga* and *Jaculus*)
3. soles of hind feet well haired, with long hairs on sides of toes (*Allactaga*, *Jaculus*)
4. **tail very long, either sparsely haired** (*Napaeozapus*, *Zapus*) **or well haired with long terminal tuft** (*Allactaga*, *Jaculus*)
5. eyes relatively large, visible
6. pinna varies from small (*Napaeozapus*, *Zapus*) to extremely elongate (*Allactaga*)
7. no external cheek pouches (cf. Heteromyidae)
8. cranium may be rounded or flat in lateral profile
9. **nasals may** (*Zapus*) **or may not** (*Allactaga*, *Jaculus*) **extend anteriorly beyond level of upper incisors**
10. **auditory bulla uninflated** (*Napaeozapus*, *Zapus*) **to greatly inflated** (*Allactaga*, *Jaculus*, *Salpingotus*)
11. foramen may be present in the angular process of lower jaw (*Allactaga*, *Jaculus*)
12. **cheek teeth modified quadritubercular** (*Napaeozapus*, *Zapus*) **or flattened with cusps separated by lateral enamel folds** (*Allactaga*, *Jaculus*)

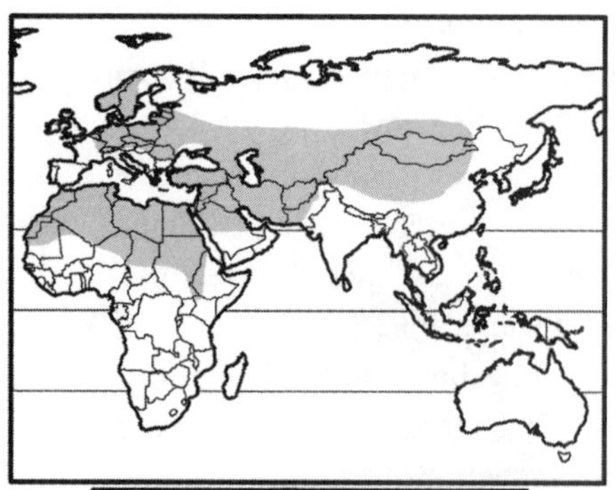

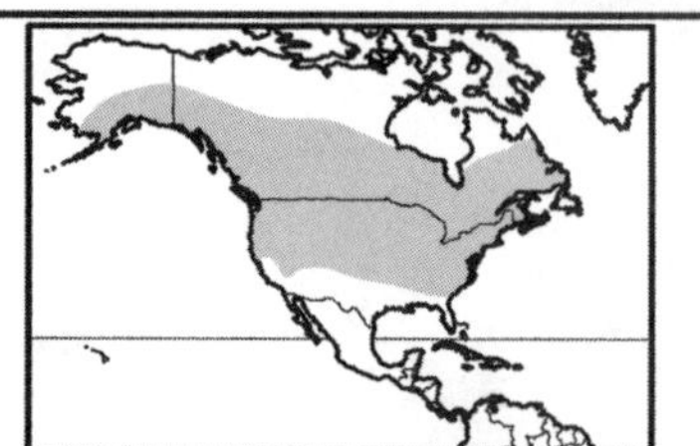

FIG. 179. Combined geographic range of families Dipodidae (North Africa and central Asia), Sminthidae (Eurasia), and Zapodidae (North America north of Mexico, plus one genus in China).

DENTAL FORMULA: $\frac{1}{1}\ \frac{0}{0}\ \frac{0\text{–}1}{0}\ \frac{3}{3} = 16\text{–}18$

TAXONOMIC DIVERSITY: About 50 species in 18 genera historically either split into two families (Zapodidae and Dipodidae) or placed within a single one (Dipodidae); now allocated to three families (one with four subfamilies, *sensu* Lebedev et al. 2013; Pisano et al. 2015; Holden et al. 2017; Michaux and Shenbrot 2017; Whitaker 2017).

Dipodidae:

Allactaginae—*Allactaga* (great and Severtzov's jerboas), *Allactodipus* (Bobrinski's jerboa), *Orientallactaga* (Siberian jerboas), *Paralactaga* (jerboas), *Pygeretmus* (fat-tailed jerboas), and *Scarturus* (jerboas)

Cardiocraniinae—*Cardiocranius* (five-toed pygmy

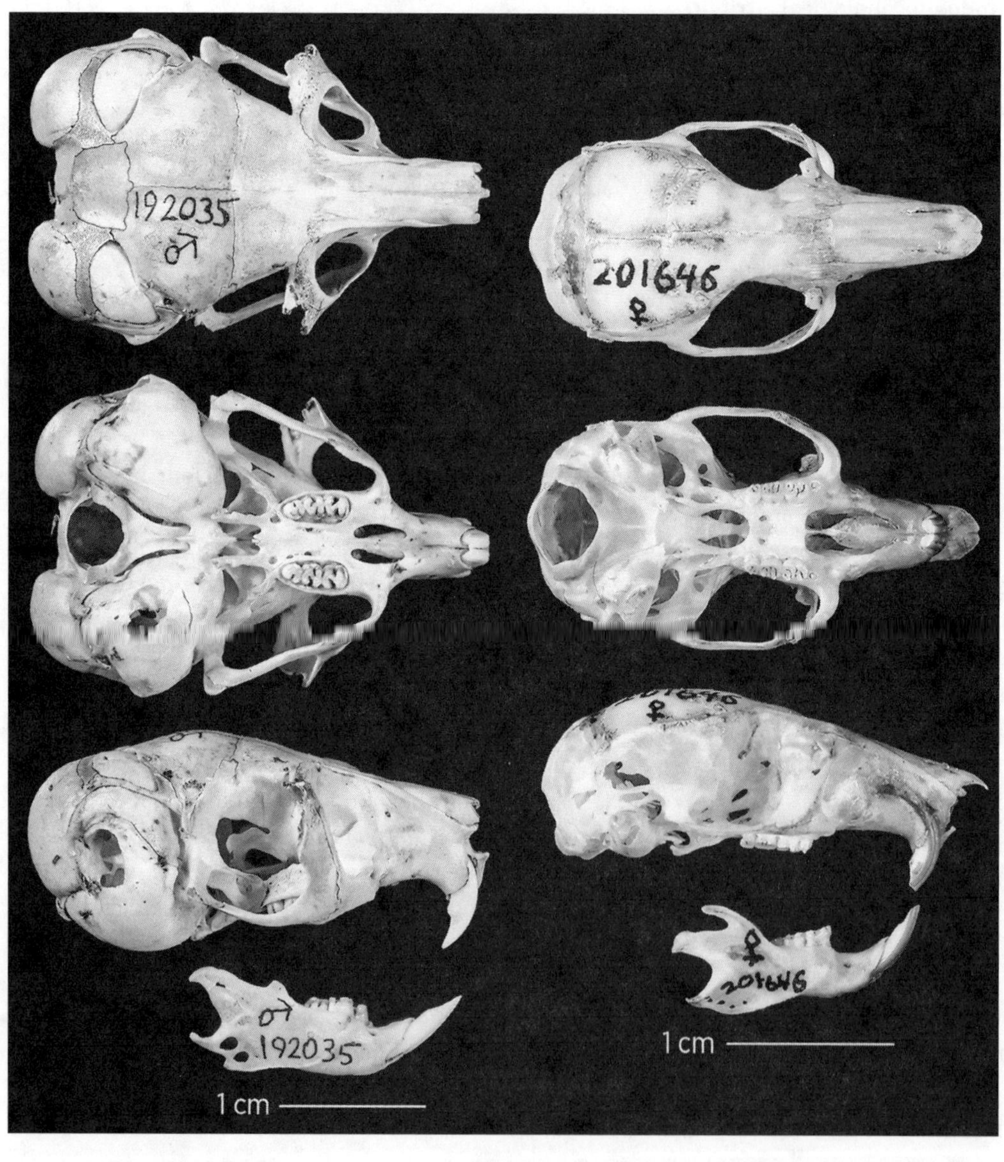

FIG. 180. Dipodoidea; skull and mandible of the three-toed jerboa, *Jaculus blanfordi* (Dipodidae: Dipodinae; *left*), and the western jumping mouse, *Zapus princeps* (Zapodidae; *right*).

jerboa) and *Salpingotus* (three-toed pygmy jerboas)
Dipodinae—*Dipus* (northern three-toed jerboa), *Eremodipus* (Lichtenstein's jerboa), *Jaculus* (desert jerboas), *Paradipus* (comb-toed jerboa), and *Stylodipus* (three-toed jerboas)
Euchoreutinae—*Euchoreutes* (long-eared jerboa)
Sminthidae: *Sicista* (birch mice)
Zapodidae: *Eozapus* (Chinese jumping mouse), *Napaeozapus* (woodland jumping mouse), and *Zapus* (jumping mice)

Superfamily MUROIDEA

The superfamily Muroidea is the largest group of mammals, comprising about 67% of all living rodents and more than 25% of all living species of mammals (>1,685 species; table 4). Body size ranges from small (50 mm) to quite large (1 m). Body form is diverse, with the generalized quadrupedal mouse (*Mus*) and rat (*Rattus*) the common theme, but with modifications for burrowing, climbing, hopping, or swimming exhibited by many. While generic assignments to suprageneric groups have become fairly well established, the

Table 4. Diversity and distribution of muroid taxa.

Taxon	*Approx. diversity (genera/species)*	*Common names*	*Range*
Superfamily Muroidea	327/1,690		
Family Platacanthomyidae	2/5	tree mice	India, Southeast Asia
Family Spalacidae	7/28		
Subfamily Myospalacinae	2/11	zokors	Siberia, northern China
Subfamily Rhizomyinae	2/4	bamboo rats	East Africa, southern Asia
Subfamily Spalacinae	2/11	blind mole rats	North Africa, eastern Mediterranean
Subfamily Tachyoryctinae	1/2	root rats	East Africa
Family Calomyscidae	1/8	brush-tailed mice	Middle East and southwestern Asia
Family Nesomyidae	21/68		
Subfamily Cricetomyinae	3/9	pouched rats and mice	sub-Saharan Africa
Subfamily Delanymyinae	1/1	swamp mouse	tropical Africa
Subfamily Dendromurinae	6/26	climbing mice, fat mice	sub-Saharan Africa
Subfamily Mystromyinae	1/1	white-tailed rat	South Africa
Subfamily Nesomyinae	9/27	Malagasy rats and mice	Madagascar
Subfamily Petromyscinae	1/4	pygmy rock mice	Africa
Family Cricetidae	142/765		
Subfamily Arvicolinae	29/162	voles, lemmings, muskrat	Holarctic
Subfamily Cricetinae	7/18	hamsters	Palearctic
Subfamily Neotominae	16/140	New World rats and mice	North, Central, and northwestern South America
Subfamily Sigmodontinae	86/434	New World rats and mice	southern North America, Central and South America
Subfamily Tylomyinae	4/10	vesper rats, climbing rats	Central and South America
Family Muridae	155/816		
Subfamily Deomyinae	4/57	spiny mice, brush-furred rats	Africa
Subfamily Gerbillinae	14/101	gerbils, jirds, sand rats	Africa, southern Asia
Subfamily Leimacomyinae	1/1	forest mouse	Africa (Togo)
Subfamily Lophiomyinae	1/1	maned rat	eastern Africa
Subfamily Murinae	136/625	Old World rats and mice	Old World (worldwide with introductions)
Subfamily Otomyinae	2/31	swamp or grass rats	Africa

Source: From Musser and Carleton (2005); updated from Wilson et al. (2017); classification reorganized to match the phylogenetic data presented in the text.

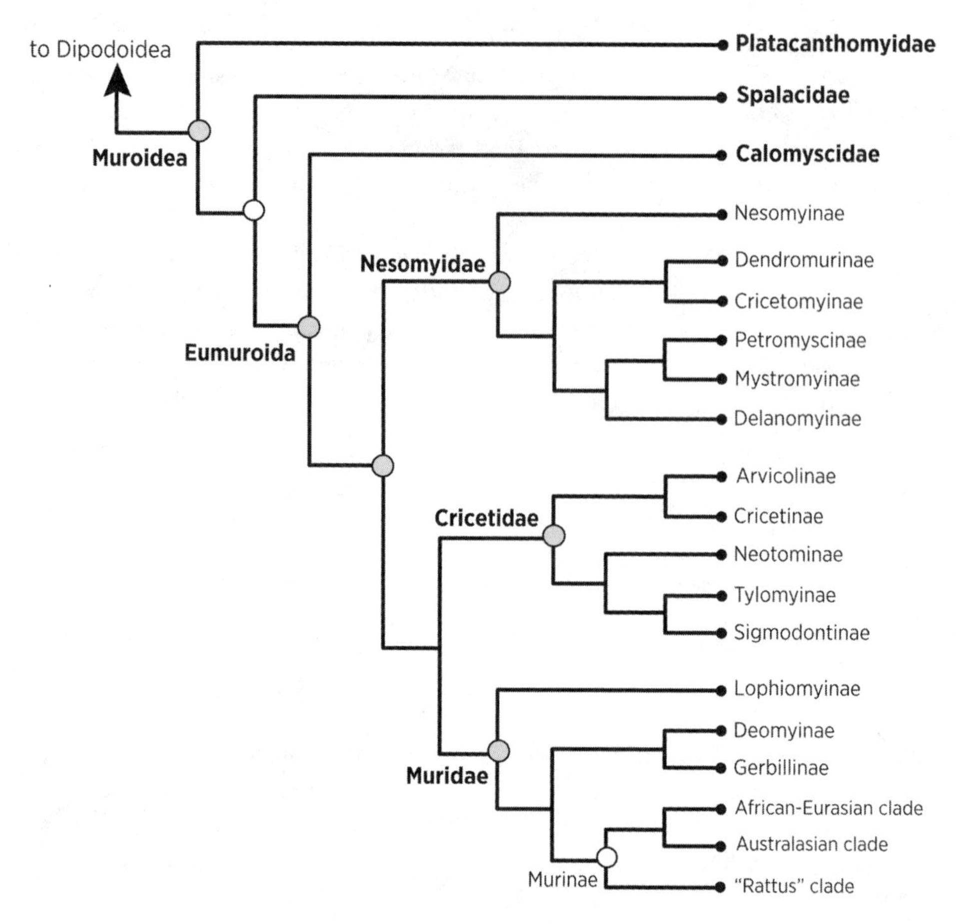

FIG. 181. Phylogenetic relationships among the families and subfamilies of Muroidea, based on mitochondrial and nuclear DNA sequences (consensus from Fabre et al. 2012, 2015; Schenk et al. 2013; Steppan and Schenk 2017). Note that this tree depicts a few different relationships among clades than inferred from Musser and Carleton's 2005 classification: molecular evidence places the Lophiomyinae as the basal clade of Muridae rather than as a member of the Cricetidae, and the Muridae subfamily Otomyinae is part of the African-Eurasian clade of Murinae. The Leimacomyinae is not included in the cladogram, as the single species has not as yet been included in any DNA-based phylogenetic study; morphologically it appears closest to the Gerbillinae (Denys et al. 1995).

hierarchy of division among these higher taxa has remained unclear. Steppan and Schenk (2017), however, provided a robustly supported tree of relationships within the superfamily based on molecular evidence, and in so doing solved many of the vexing problems that have plagued muroid classification for decades. The hierarchical taxa that these authors have defined are depicted in fig. 181. This tree is largely congruent with the morphologically based classification compiled by Musser and Carleton (2005).

This manual provides characters for the entire superfamily, and then separately for each family, expanding coverage for the highly diverse lineages.

MUROID RECOGNITION CHARACTERS:

1. mouse- to rat-sized (100–900 mm), unspecialized or adapted for swimming, climbing, digging, or hopping
2. limbs usually unspecialized (except for some that are modified for the above habits)
3. tail short to very long, scaly to well-haired
4. eyes relatively large, visible
5. pinna small to large
6. cheek pouches, if present, opening internally
7. cranium variable in shape, rounded in profile (most genera) or flat and triangular (subterranean arvicoline *Ellobius*, rhizomyines, myospalacines, spalacines)
8. **infraorbital canal medium-sized, usually V-shaped** (wider dorsally than ventrally) but oval in rhizomyines, **opening anteriorly** (fig. 182, arrows)

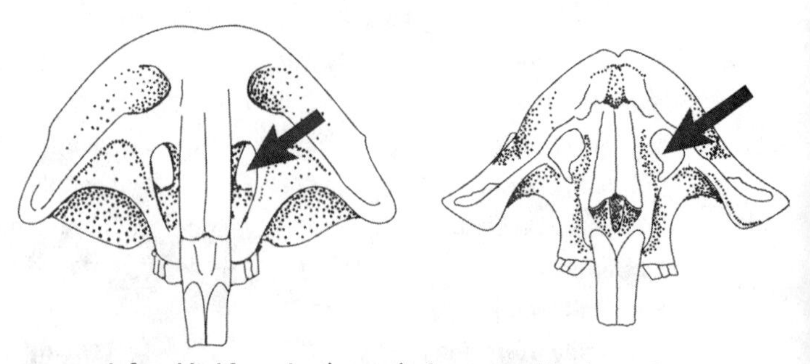

FIG. 182. Infraorbital foramina (arrows) of two muroid rodents, *Rattus* (Muridae: Murinae; *left*), and *Rhizomys* (Spalacidae: Rhizomyinae; *right*). Redrawn from Lawlor (1979).

9. **zygomatic plate broad** (rarely narrow), **tilted upward**
10. nasals usually extending anteriorly to or beyond level of upper incisors (not extending beyond incisors in rhizomyines)
11. auditory bulla usually not greatly inflated (except in most genera of gerbillines)
12. no foramen in angular process of lower jaw
13. crowns of molars cuspidate or flat, with angular (prismatic) or rounded lateral folds of enamel
14. three or fewer cheek teeth above and below

DENTAL FORMULA: $\frac{1}{1}\ \frac{0}{0}\ \frac{0}{0}\ \frac{0\text{–}3}{0\text{–}3} = 4\text{–}16$

Most muroid rodents have a "normal" dentition, with an upper and lower pair of incisors and three cheek teeth above and below (16 total teeth). Some of the Australasian shrew rats (Muridae, Murinae) that specialize on earthworms and other soft-bodied soil invertebrates have lost their 3rd molars, and one recently described extreme example (*Paucidentomys vermidax*) has no cheek teeth at all (Esselstyn et al. 2012).

Family PLATACANTHOMYIDAE (spiny tree mice, soft-furred tree mice)

Mouselike with short muzzle, prominent but sparsely furred ears, long vibrissae, medium-sized to small eyes, a long, brush-tipped tail, and unspecialized limbs; body length 70–210 mm, tail length 75–140 mm; forefeet with claws on 2nd through 5th digits, pollex rudimentary and with a nail; hind feet narrow and small; soles of all four feet naked, with six pads; arboreal; herbivorous; probably nocturnal. *Platacanthomys* has spines in fur; *Typhlomys* does not.

DIAGNOSTIC CHARACTERS:

- tail naked and scaly at base, brushed or bushy distally
- large foramina present in palate between upper tooth rows

RECOGNITION CHARACTERS:

1. cranium rounded in profile
2. infraorbital canal relatively large, V-shaped (wider dorsally than ventrally), opening anteriorly
3. zygomatic plate relatively narrow, tilted slightly upward
4. bony palate perforated by two enlarged posterior foramina
5. auditory bulla small; no transbullar septae
6. no foramen in angular process or lower jaw
7. molars rooted and brachydont
8. **crowns of molars flat, with a series of oblique or transverse ridges**

DENTAL FORMULA: $\frac{1}{1}\ \frac{0}{0}\ \frac{0}{0}\ \frac{3}{3} = 16$

TAXONOMIC DIVERSITY: Five species in two genera: *Platacanthomys* (Malabar spiny tree mouse) and *Typhlomys* (soft-furred tree mice).

FIG. 183. Platacanthomyidae; Malabar spiny dormouse, *Platacanthomys lasiurus*.

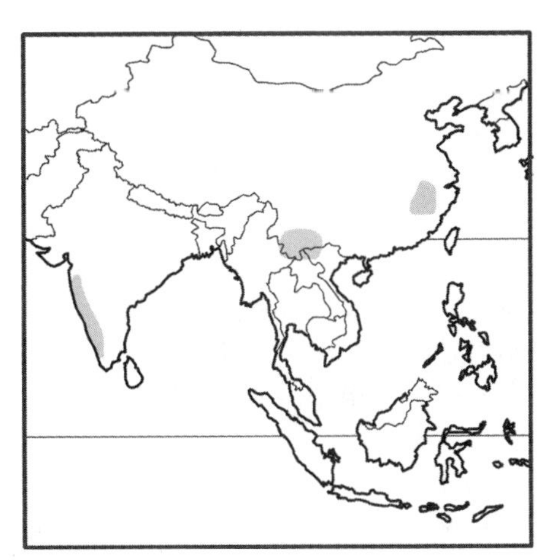

FIG. 184. Geographic range of family Platacanthomyidae: moist, rocky tropical and subtropical forests of southwestern India, southern China, northern Vietnam.

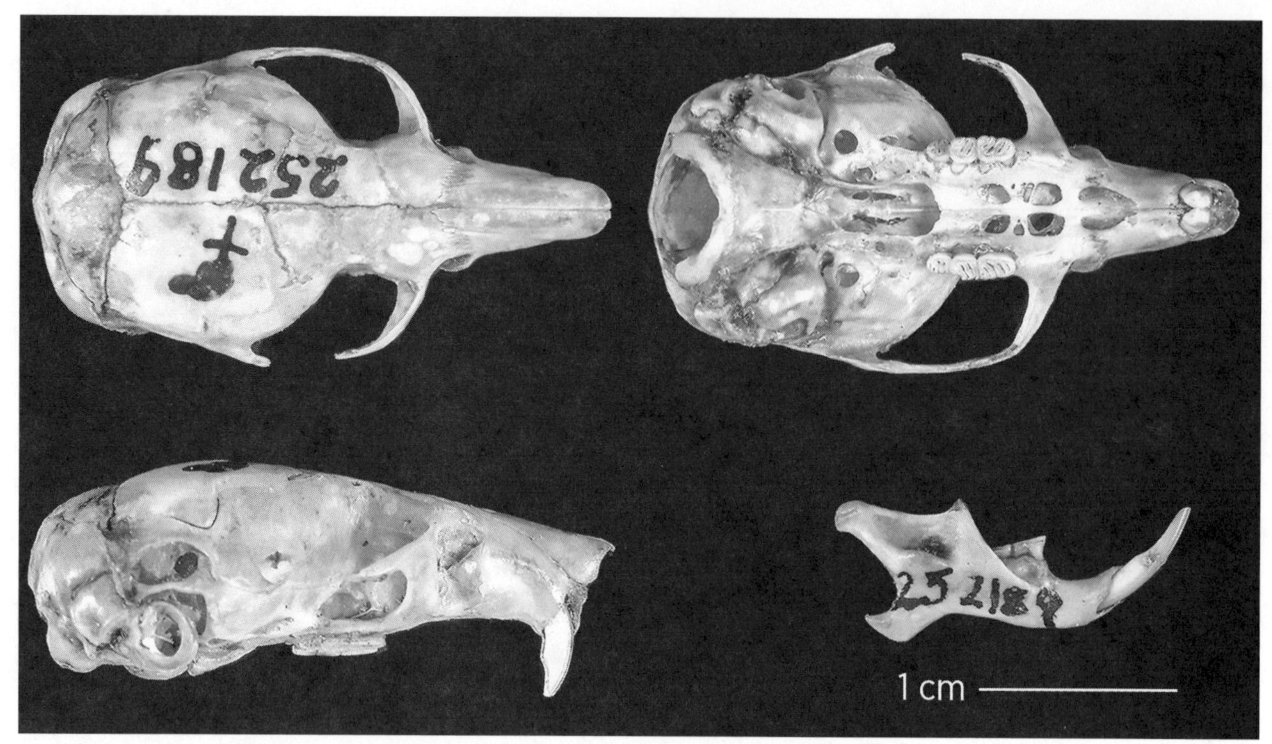

FIG. 185. Platacanthomyidae; skull and mandible of the Chinese pygmy dormouse, *Typhlomys cinereus*.

Family SPALACIDAE (zokors, bamboo rats, blind mole rats, root rats)

Mouse- to rat-sized (body length 130–500 mm; mass 100 g–4 kg); strongly adapted for subterranean existence, with tubular, stocky body, **short and powerful legs**, long claws on forefeet, **short ears, small to no visible eyes** (eyes present, small, usually visible in rhizomyines but covered with skin in other taxa), and **short tail** (<50% of body length; spalacines [blind mole rats] lack tails). Incisors large and wide, used for digging in all but myospalacines (zokors), which dig with forelimbs and enlarged claws; spalacines with very procumbent upper incisors and hardened patch of skin above nose, both used in digging; rhizomyines (bamboo rats) are least specialized for digging as they forage extensively on the surface. All are herbivorous, live in complex self-dug burrow systems, and are both solitary and territorial; some species are known to communicate by drumming heads against burrow walls and sensing vibrations of others. Habitat is moist or semimoist soils in grasslands, scrublands, forests, and agricultural areas.

The relationships among spalacids have a complex history, primarily because subfamilies exhibit different suites of morphological adaptations to a fossorial habit. As a result, some authors have viewed these subfamilies as reflecting different approaches to fossoriality from a common lineage, whereas others have argued that they are independent lineages subjected to convergent evolution (see Musser and Carleton 2005). Recent work suggests that both may be correct, and that the subfamilies, while constituting a monophyletic group, may have developed fossorial adaptations independently. Note the convergence of spalacids in body shape, ear length, tail length, and exposed incisors with North American subterranean pocket gophers (Geomyidae), African mole rats (Bathyergidae), and South American tuco-tucos (Ctenomyidae) and coruros (*Spalacopus*, Octodontidae).

RECOGNITION CHARACTERS:

1. **skull stout, flat in lateral profile** (rhizomyines, tachyoryctines) **or distinctly angular** (occipital region sloping far forward; myospalacines, *Spalax*), with strong zygomatic arches and short rostrum
2. broad parietal and occipital regions with strongly developed sagittal and lambdoidal crests
3. infraorbital foramen rounded, relatively large (*Spalax*) or medium-sized

4. zygomatic plate broad but mostly horizontal, **entirely below infraorbital canal** in *Spalax*
5. incisors highly procumbent (*Spalax*) or orthodont (other genera)
6. cheek teeth hypsodont
7. occlusal surface of cheek teeth flat, with deep lateral folds of enamel

DENTAL FORMULA: $\frac{1}{1}\frac{0}{0}\frac{0}{0}\frac{3}{3} = 16$

TAXONOMIC DIVERSITY: 28 species in seven genera divided into four subfamilies (Rhizomyinae and Tachyoryctinae are sometimes considered tribes of the single subfamily Rhizomyinae).

Myospalacinae—*Eospalax* (zokors) and *Myospalax* (zokors)

Rhizomyinae—*Cannomys* (lesser bamboo rat) and *Rhizomys* (bamboo rats)

Spalacinae—*Spalax* (blind mole rats) and *Nannospalax* (blind mole rats)

Tachyoryctinae—*Tachyoryctes* (African root rats)

FIG. 186. Spalacidae: Spalacinae; Middle East blind mole rat, *Spalax ehrenbergi*.

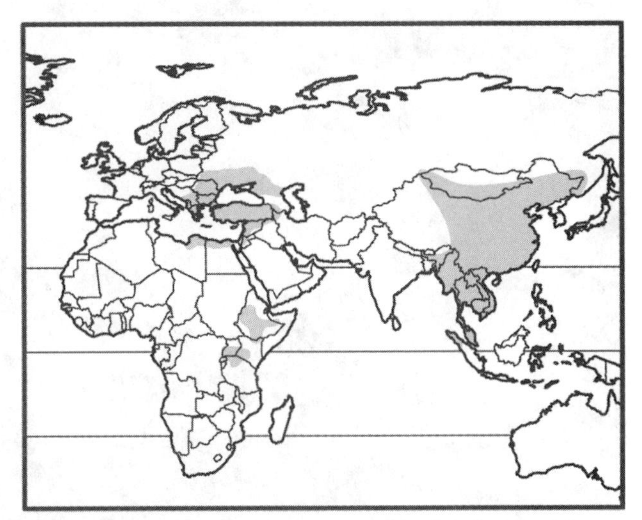

FIG. 187. Geographic range of family Spalacidae: Siberia, Mongolia, China (Myospalacinae); Indonesia (Sumatra), Southeast Asia to southern China, Nepal, and eastern India (Rhizomyinae); southeastern Europe, Middle East from Turkey to Israel, and North Africa (Libya) (Spalacinae); and eastern Africa (Tachyoryctinae).

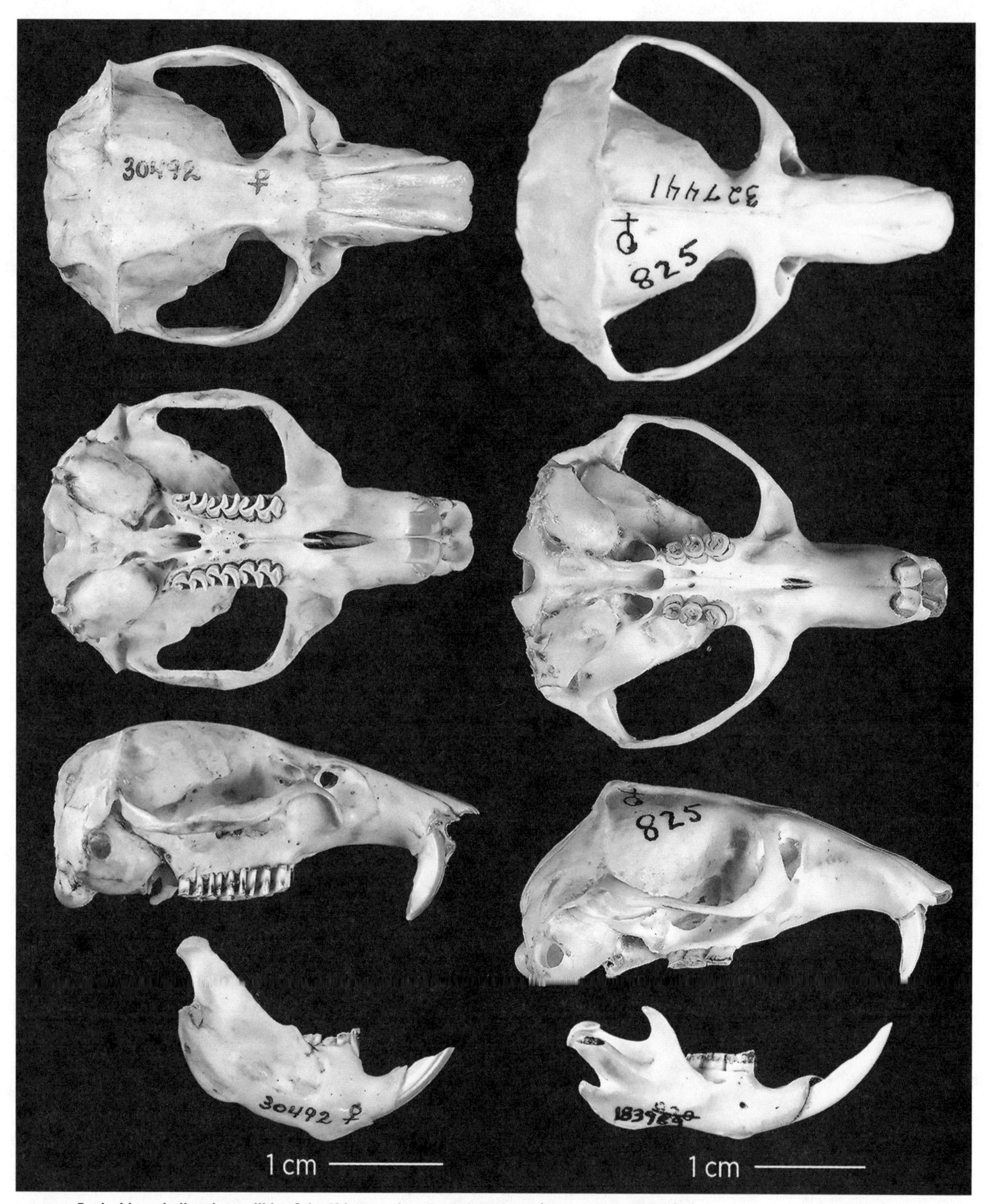

FIG. 188. Spalacidae; skull and mandible of the Chinese zokor, *Eospalax cansus* (Myospalacinae; *left*), and the lesser blind mole rat, *Spalax leucodon* (Spalacinae; *right*).

Family CALOMYSCIDAE (brush-tailed mice)

Small (15–30 g) and mouselike, with tail equal to or slightly longer than body (body length 60–100 mm, tail length 70–100 mm), prominent ears, long and lax fur, and tail well furred and slightly tufted at tip. Despite common name, and unlike true hamsters (Cricetidae, Cricetinae), calomyscids lack cheek pouches and sebaceous flank glands. Nocturnal to crepuscular; do not hibernate or estivate; feed primarily on seeds, but also eat flowers and leaves as well as insects and sometimes carrion; not social, but may huddle together. Prefer rocky, barren hillsides with sparse, shrubby vegetation in xeric regions, but extend in elevation to oak and juniper woodlands.

RECOGNITION CHARACTERS:

1. occlusal surfaces of cheek teeth relatively simple, without complex topography
2. 1st and 2nd molars with two longitudinal rows of cusps rather deeply divided by equal lateral and lingual folds
3. 3rd molar less than half the size of 2nd
4. well-developed, vertical zygomatic plate
5. V-shaped infraorbital foramen, widest at top
6. cranium smooth, non-ridged
7. auditory bulla small

DENTAL FORMULA: $\frac{1}{1}\ \frac{0}{0}\ \frac{0}{0}\ \frac{3}{3} = 16$

TAXONOMIC DIVERSITY: The family is monotypic, with eight species in the single genus *Calomyscus*.

FIG. 189. Calomyscidae; Great Balkan calomyscus, *Calomyscus mystax*.

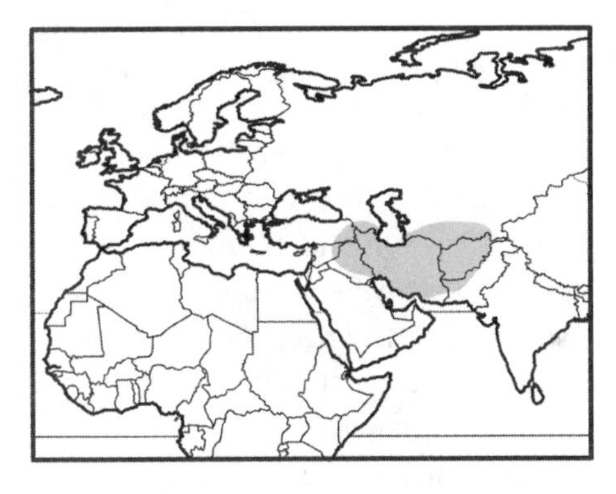

FIG. 190. Geographic range of family Calomyscidae: western Pakistan through Afghanistan and Iran to eastern Syria and Turkey, and north into southern Turkmenistan.

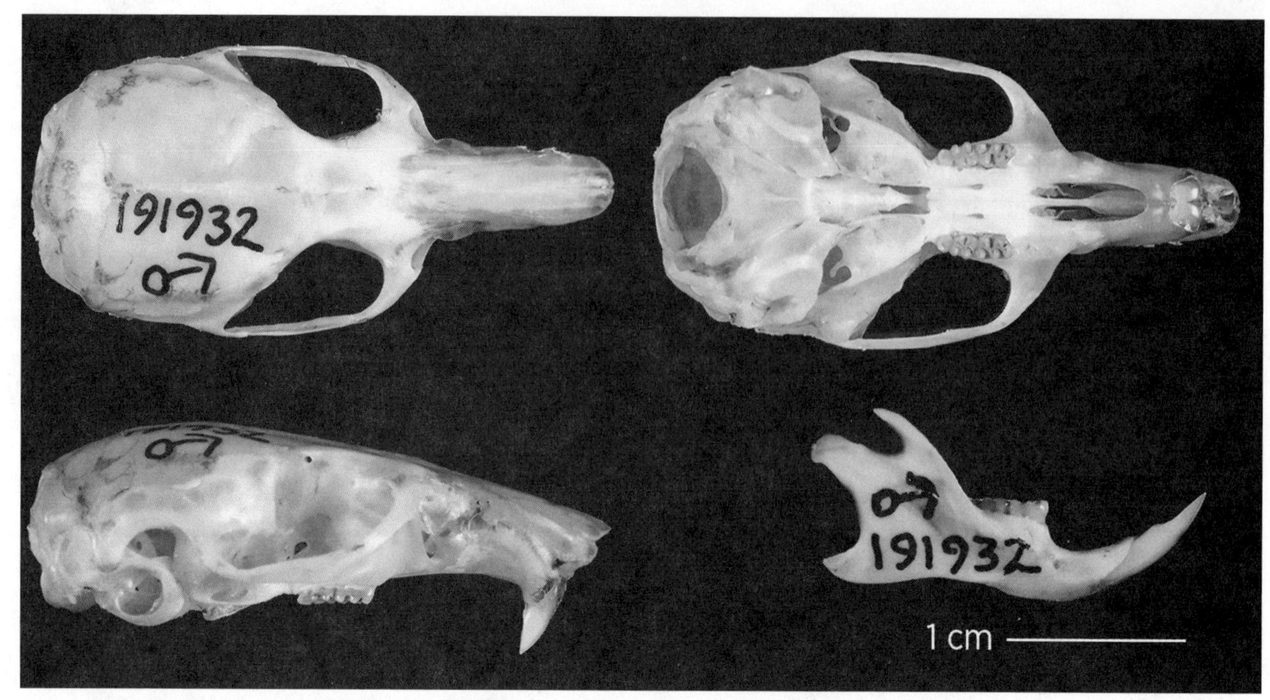

FIG. 191. Calomyscidae; skull and mandible of the Zagros Mountains mouselike hamster, *Calomyscus bailwardi*.

Family NESOMYIDAE

The family Nesomyidae comprises six subfamilies, including a radiation of endemic Malagasy rodents (Nesomyinae) and five subfamilies of limited diversity, mostly obscure genera restricted to sub-Saharan Africa. Nesomyids are physically diverse, with body forms ranging from squirrel-like to ratlike to vole-like to gerbil-like. Body sizes range very widely, from the tiny Delany's swamp mouse (*Delanymys*: Delanymyinae), at 5–6 g and with a head and body length of 50–65 mm, to the large African giant pouched rat (*Cricetomys*: Cricetomyinae), with a body length of 450 mm and a mass up to nearly 3 kg. Most nesomyids are thickly furred, although hair length varies considerably; some have nearly naked, prehensile tails, others have well-furred tails that are almost bushy and have tufted tips. Members of one subfamily (Cricetomyinae) have internal cheek pouches. All taxa have the full muroid dental formula, with three molars above and below. The individual subfamilies share no readily discernible diagnostic morphological features, and historically, each has been shuffled between the Cricetidae and Muridae or considered a family on its own. Members of all six subfamilies, however, form a well-supported clade in molecular analyses (Schenk et al. 2013). We provide general descriptions for each subfamily separately and illustrate the skulls and teeth of a few selected representatives.

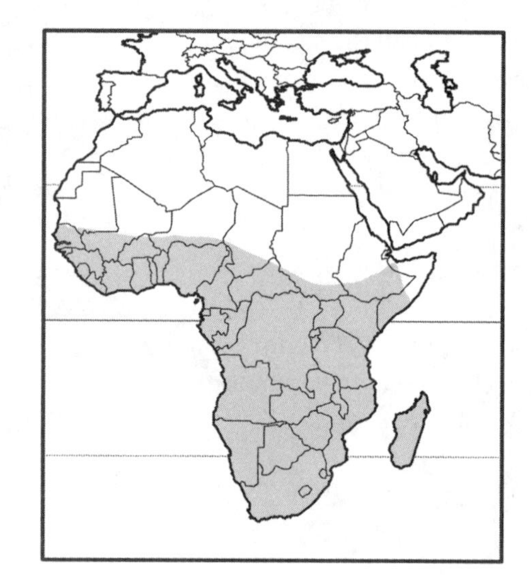

FIG. 192. Geographic range of family Nesomyidae: Madagascar (Nesomyinae) and sub-Saharan Africa (Cricetomyinae, Delanymyinae, Dendromurinae, Mystromyinae, and Petromyscinae).

Subfamily CRICETOMYINAE (pouched rats and mice)

Medium-sized to large relative to other muroids (body length 100–450 mm), body robust with large head, short limbs, stout hind feet with short toes; tail shorter than or nearly equal to body length, either naked and scaly or thinly haired; large internal cheek pouches present; mammae number from 4 to 6 pairs. Terrestrial to scansorial. Habitat includes forests, moist woodlands, and savannas. Skull with long rostrum (fig. 194). Upper incisors lack grooves; molars are rooted, cuspidate, and brachydont; main cusps form transverse laminae or

FIG. 193. Nesomyidae: Cricetomyinae; Gambian pouched rat, *Cricetomys gambianus*.

chevrons with isolated lingual cusplets, thus resembling the triserial condition (e.g., three longitudinal rows of cusps) of "typical" murids.

RANGE: Broadly distributed in sub-Saharan Africa.

TAXONOMIC DIVERSITY: Nine species in three genera: *Beamys* (greater and Hinde's long-tailed pouched rats), *Cricetomys* (giant pouched rats), and *Saccostomus* (pouched mice).

Subfamily DELANYMYINAE (Delany's swamp mouse)

Size small (body length 50–65 mm); body gracile with delicate limbs, long hind feet and toes, long tail twice body length and semi-prehensile; pelage dense and soft; four pairs of mammae. Arboreal. Habitat includes montane marshes and bamboo forests. Skull with high and globular braincase, short rostrum, narrow zygomatic plates, and short bony palate that terminates at end of M3s. Upper incisors lack grooves; molars are rooted, cuspidate, and brachydont; M1 and M2 subequal in size, M3 much smaller; occlusal surface with transverse laminae with lingual cusp attached by crest to posteromedial portion of protocone in M1 and M2.

RANGE: Restricted to east-central Africa.

TAXONOMIC DIVERSITY: The subfamily is monotypic, with a single species, *Delanymys brooksi* (Delany's swamp mouse).

Subfamily DENDROMURINAE (climbing mice, fat mice)

Small to medium-sized (body 50–90 mm), tail generally longer than body, either naked and scaly or moderately well haired, lacking a pencil or tuft, and semi-prehensile; number of digits on forefeet varies from 3 to 5; hind foot short and wide in arboreal genera, long and narrow in more terrestrial ones; plantar surfaces hairy; four pairs of mammae. Terrestrial to arboreal. Habitat variable, including mature tropical forest, woodlands, shrublands, and short grass. Skull with long rostrum, slender to wide; smooth braincase; narrow zygomatic plates; bony hard palate ending at or just behind 3rd molars. Upper incisors grooved in most genera; molars rooted, cuspidate, and brachydont; occlusal surface of M1 and M2 consisting of bicuspid laminae with a lingual circular cusp adjacent to middle lamina of M1 and anterolamina of M2; M1 molar large, M2 about half that size, and M3 noticeably small.

RANGE: Broadly distributed in sub-Saharan Africa.

TAXONOMIC DIVERSITY: 26 species in six genera: *Dendromus* (African climbing mice), *Steatomys* (African fat mice), and four monotypic genera, *Dendroprionomys* (velvet African climbing mouse), *Malacothrix* (large-eared African desert mouse), *Megadendromus* (Nikolaus's African climbing mouse), and *Prionomys* (Bates's African climbing mouse).

Subfamily MYSTROMYINAE (African white-tailed rat)

Medium-sized (body length 140–165 mm; mass 70–95 g); ears large; tail noticeably short, less than half of body length (60–70 mm), and thinly haired; body thickset with large head; fur soft and woolly. Terrestrial. Habitat includes shrublands and grasslands, often on stony or rocky soils. Skull with shortened braincase, moderately inflated bullae, and broad zygomatic plates (fig. 194). Molars with zig-zag enamel pattern resulting from opposite cusps and slanting folds; M1 with two inner and two outer folds; M2 with one inner and two outer folds.

RANGE: Restricted to southern Africa.

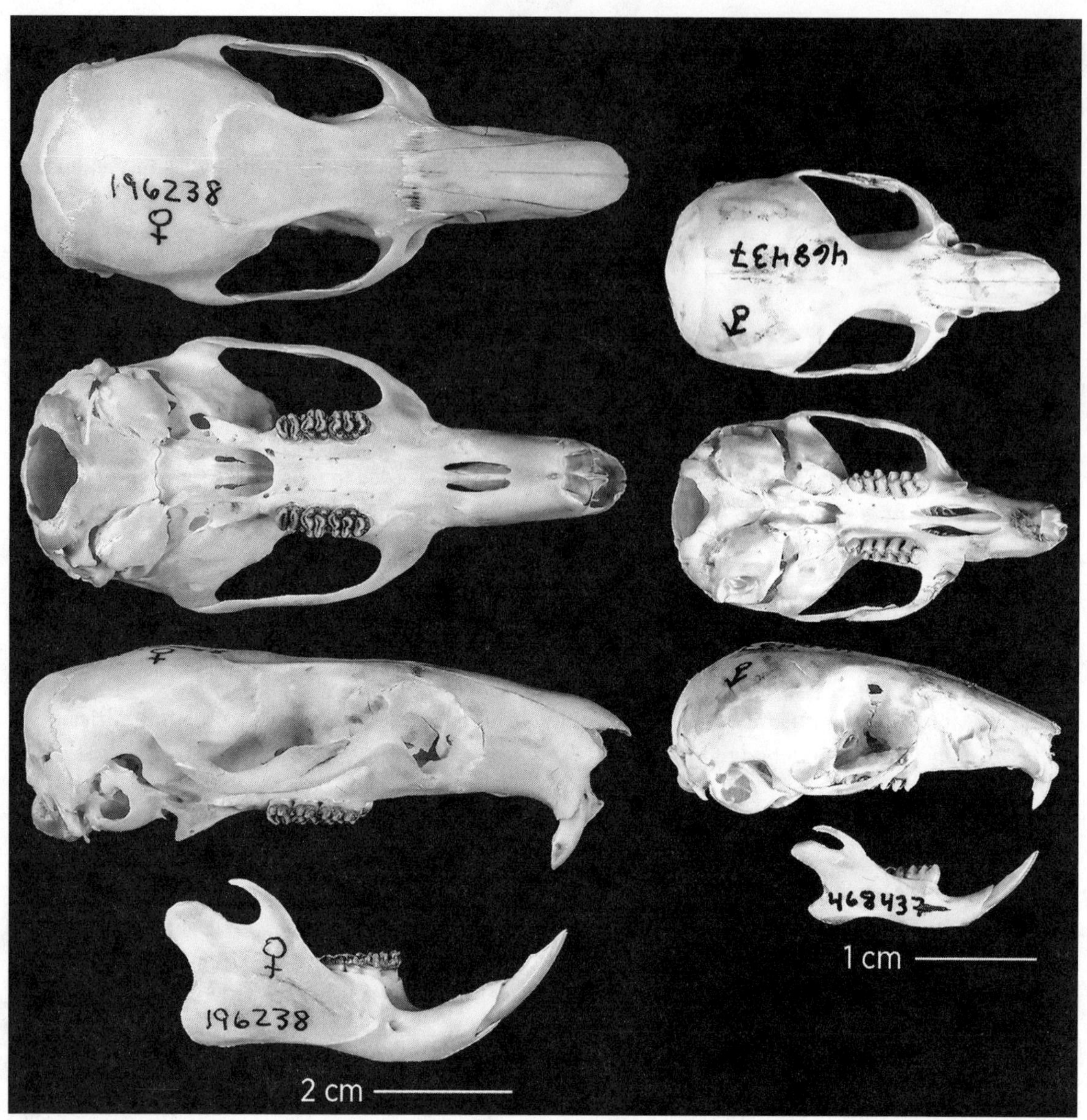

FIG. 194. **Nesomyidae; skull and mandible of the northern giant pouched rat, *Cricetomys gambianus* (Cricetomyinae; *left*), and the white-tailed rat, *Mystromys albicaudatus* (Mystromyinae; *right*).**

TAXONOMIC DIVERSITY: The subfamily is monotypic, with a single species, *Mystromys albicaudatus* (African white-tailed rat).

Subfamily NESOMYINAE (Malagasy endemic rodents)

Strikingly differentiated assemblage of genera, both morphologically and ecologically; medium-sized to large (length 80–300 mm); gerbil-like (*Macrotarsomys*), squirrel-like (*Brachytarsomys*), generalized ratlike (*Nesomys*), and vole-like (*Brachyuromys*); tail thus shorter than, subequal to, or longer than body, naked to well furred, and prehensile in some; hind feet long and narrow or short and broad; ears short and rounded to quite large; fur texture harsh to soft and woolly. Terrestrial and arboreal. Habitat includes moist tropical forest, dry forest, dry scrublands, grasslands, and moist meadows. Skull varies in size and shape, gracile to sturdy with short and broad to long rostrum (fig. 195). Incisors ungrooved; molars with biserial arrangement of cusps (two parallel longitudinal rows of cusps), rooted, crowns brachydont to hypsodont; occlusal pattern tuberculate, prismatic, planar, or laminate; M3 subequal in size to M2 in most genera.

TAXONOMIC DIVERSITY: 27 species in nine genera, all restricted to Madagascar: *Brachytarsomys*

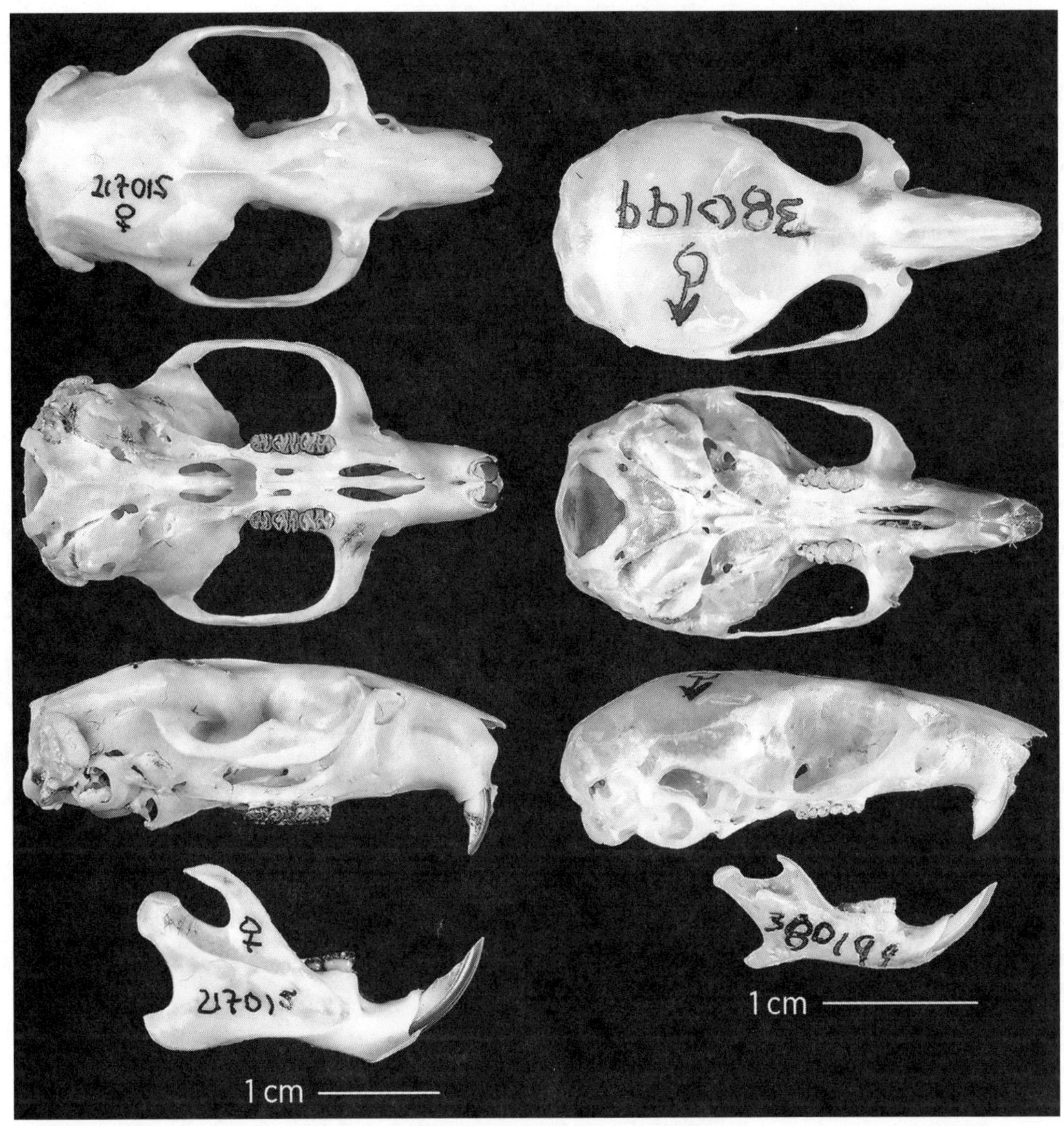

FIG. 195. Nesomyidae; skull and mandible of the white-tailed antsangy, *Brachytarsomys albicauda* (Nesomyinae; *left*), and Brukkaros pygmy rock mouse, *Petromyscus monticularis* (Petromyscinae; *right*).

(antsangys), *Brachyuromys* (short-tailed rats), *Eliurus* (tufted-tailed rats), *Gymnuromys* (voalavoanala), *Hypogeomys* (votsovotsa), *Macrotarsomys* (big-footed mice), *Monticolomys* (montane voalavo), *Nesomys* (nesomys), and *Voalavo* (naked-tailed voalavo).

Subfamily PETROMYSCINAE (pygmy rock mice)

Small (body length 75–115 mm); body chunky; ears large; limbs short; hind feet and digits short; tail shorter than or subequal to body; fur soft and silky; two pairs of mammae. Terrestrial. Habitat is predominantly arid regions with large boulders and rock outcrops. Skull is broad and flattened; rostrum moderately long and slender; zygomatic plates broad; bony palate long, extending past M3s as conspicuous shelf (fig. 195). Incisors are ungrooved; molars rooted, cuspidate, brachydont; M2 smaller than M1, M3 much smaller; occlusal surface has transverse laminae with a lingual cusp connected by a crest to the protocone in M1 and M2.

RANGE: Restricted to southwestern Africa.

TAXONOMIC DIVERSITY: The family is monotypic, with four species in the single genus *Petromyscus* (pygmy rock mice).

Family CRICETIDAE

The Cricetidae is a large and diverse family with five subfamilies: the Old World hamsters (Cricetinae), the Holarctic voles and lemmings (Arvicolinae), and the New World, largely North American Neotominae, largely Middle American Tylomyinae, and largely South American Sigmodontinae. As a group, cricetids are characterized by a biserial cusp arrangement (molars with two parallel longitudinal rows of cusps) that is uniformly evident despite a wide range in occlusal pattern, crown height, and degree of rootedness. Body size varies extremely, from tiny (pygmy mice, *Baiomys* [Neotominae] at 7–8 g) to large (woodrats, *Neotoma* [Neotominae], and swamp rats, *Kunsia* [Sigmodontinae], at nearly 700 g); most are intermediate-sized mice and rats. Body traits also vary extensively, as do the biomes occupied and habitat range within them. Genera in some subfamilies are relatively uniform with clear-cut diagnostic features; those in others are highly diverse and not easily characterized by common morphological attributes. Some (notably the Arvicolinae, Neotominae, and Sigmodontinae) are taxonomically very diverse, made up of dozens of genera and hundreds of species; other subfamilies are species and genus poor in comparison.

Subfamily ARVICOLINAE (voles, lemmings, muskrat)

Small (head and body length 75 mm, some *Myodes*) to large (length more than 300 mm, *Ondatra*); body form robust, with blunt, rounded muzzle, **small ears largely hidden by fur**, short legs, **tail always shorter than head and body** (barely evident in *Ellobius*; relatively long and laterally compressed in *Ondatra*); **body coloration generally uniform above and below**; sebaceous glands present on flank, hip, rump, or caudal region. **Skull relatively massive and rugose**, with same bauplan in all genera. Incisors typically broad; cheek teeth ever-growing in many genera, with **common enamel pattern of closed triangles or broadly confluent prisms**. Mainly terrestrial, but some are semiaquatic (*Arvicola*, *Neofiber*, some *Microtus*), aquatic (*Ondatra*), arboreal (*Arborimus*), or subterranean (*Ellobius* and *Prometheomys*). Primarily herbivores; some nocturnal, others polyphasic, few are diurnal. Many have the capacity to reach exceedingly high densities over multiannual cycles. Occur in nearly all habitats within temperate, boreal, montane, and Arctic biomes, including short- and tallgrass prairie, steppe, shrub-steppe, alpine and subalpine meadows, cloud forest, treeless tundra, deciduous woodlands, conifer forests, and marshes and other wetland areas.

DIAGNOSTIC CHARACTER:

- **occlusal surface of cheek teeth flat with sharp-edged lateral enamel folds producing characteristic "Christmas tree" prisms; usually strongly ever-growing and hypsodont with deep alveoli in maxilla and mandible; taxa with rooted molars exhibit same occlusal pattern, but crowns are also hypsodont.**

RECOGNITION CHARACTER:

1. skull with short and robust rostrum, broad and robust zygomatic arches, strongly constricted interorbital region, and postorbital crest or process on anterolateral margin of squamosal

FIG. 196. Cricetidae: Arvicolinae; muskrat, *Ondatra zibethicus* (Ondatrini; *left*), and brown lemming, *Lemmus trimucronatus* (Lemmini; *right*).

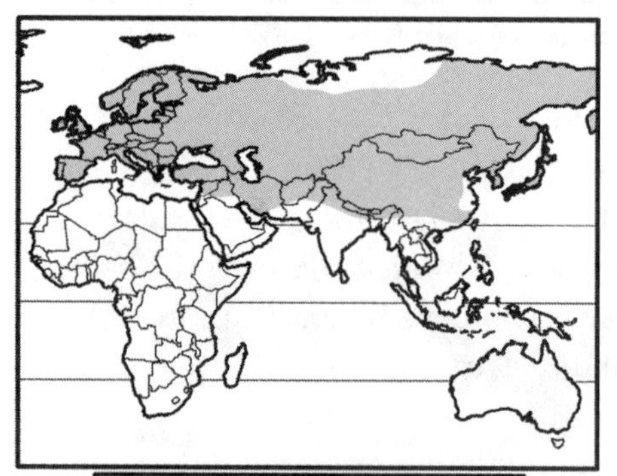

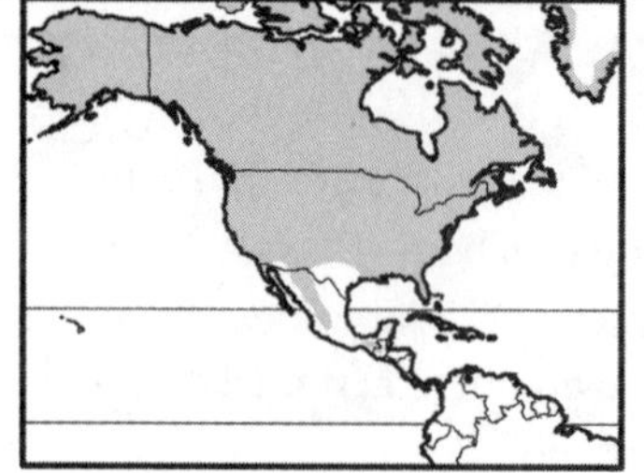

FIG. 197. Geographic range of subfamily Arvicolinae: Holarctic (North America, Eurasia).

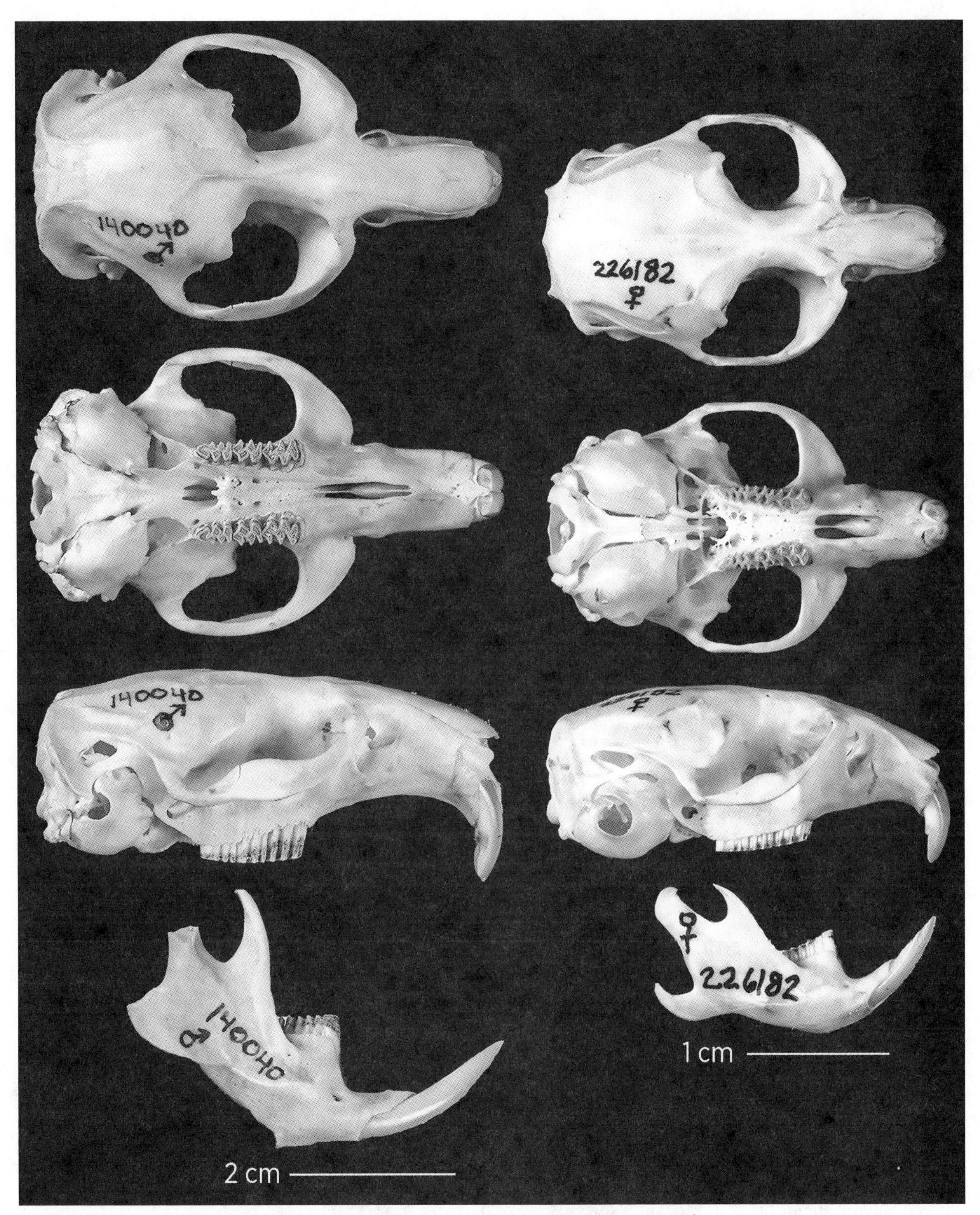

FIG. 198. Cricetidae: Arvicolinae; skull and mandible of the muskrat, *Ondatra zibethicus* (Ondatrini; *left*), and the California vole, *Microtus californicus* (Arvicolini; *right*).

DENTAL FORMULA: $\frac{1}{1}\ \frac{0}{0}\ \frac{0}{0}\ \frac{3}{3} = 16$

TAXONOMIC DIVERSITY: A large and highly successful group with more than 160 species and 29 genera grouped in 10 tribes, although suprageneric relationships remain ambiguous and in need of further study.

Arvicolini—*Alexandromys* (Asian voles), *Arvicola* (Eurasian water voles), *Chionomys* (snow voles), *Lasiopodomys* (Asian voles), *Lemmiscus* (sagebrush vole), *Microtus* (Holarctic voles), *Neodon* (mountain voles of Southeast Asia), *Proedromys* (Liangshan and Duke of Bedford's voles), and *Volemys* (Sichuan and Marie's voles)

Dicrostonychini—*Dicrostonyx* (collared lemmings)

Ellobiusini—*Ellobius* (mole voles)

Lagurini—*Eolagurus* (steppe lemmings) and *Lagurus* (steppe vole)

Lemmini—*Lemmus* (brown lemmings), *Myopus* (wood lemming), and *Synaptomys* (bog lemmings)

Myodini—*Alticola* (mountain voles), *Aschizomys* (lemming mountain vole), *Caryomys* (Gansu and Kolan red-backed voles), *Craseomys* (red-backed voles), *Eothenomys* (red-backed voles), *Myodes* (red-backed voles), and *Hyperacrius* (burrowing and Murree voles)

Ondatrini—*Neofiber* (round-tailed muskrat) and *Ondatra* (common muskrats)

Phenacomyini—*Arborimus* (white-footed and tree voles) and *Phenacomys* (heather voles)

Pliomyini—*Dinaromys* (Balkan snow vole)

Prometheomyini—*Prometheomys* (long-clawed mole vole)

Subfamily CRICETINAE (hamsters)

Size variable, from small (*Phodopus*, body length 50 mm) to large (*Cricetus*, body length >300); ears densely furred; hind feet short and broad; **tail much shorter than body** (<45%, stub-like in *Cricetus*, *Mesocricetus*, and *Phodopus*); claws of forefeet moderately well developed; fur thick and long, brightly colored and variegated in some, sometimes with distinct middorsal stripe; sebaceous flank glands present; **capacious internal cheek pouches present**. Terrestrial and scansorial; adept diggers; nocturnal to crepuscular; primarily granivores, but also eat other plant parts and use cheek pouches to cache food while foraging; construct extensive underground burrows; solitary; will enter torpor in cold months. Occupy dry, open biomes with xerophilous vegetation, including deserts, shrublands, forest and montane steppe, and brushy to rocky hillsides.

RECOGNITION CHARACTERS:

1. body form stocky, with short legs, small ears, and very short tail
2. **cheek teeth rooted, brachydont**, and cuspidate
3. **anterocone of M1 divided into prominent lingual and labial conules**
4. **mandible lacking capsular projection of incisor on lateral face**

DENTAL FORMULA: $\frac{1}{1}\ \frac{0}{0}\ \frac{0}{0}\ \frac{3}{3} = 16$

FIG. 199. Cricetidae: Cricetinae; common hamster, *Cricetus cricetus*.

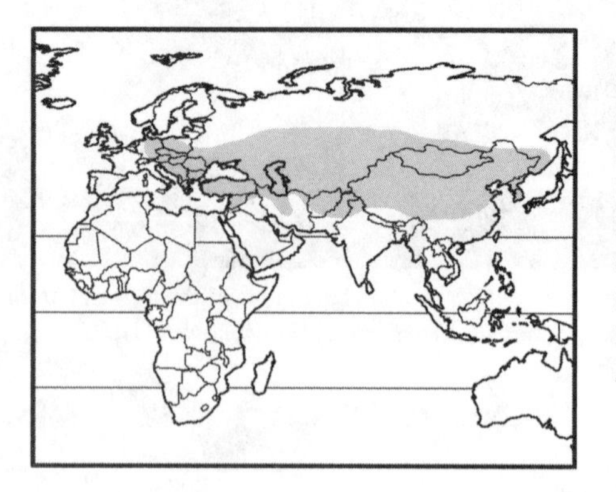

FIG. 200. Geographic range of subfamily Cricetinae: Eurasia (central Europe through central Asia to Mongolia and China, southwest Asia).

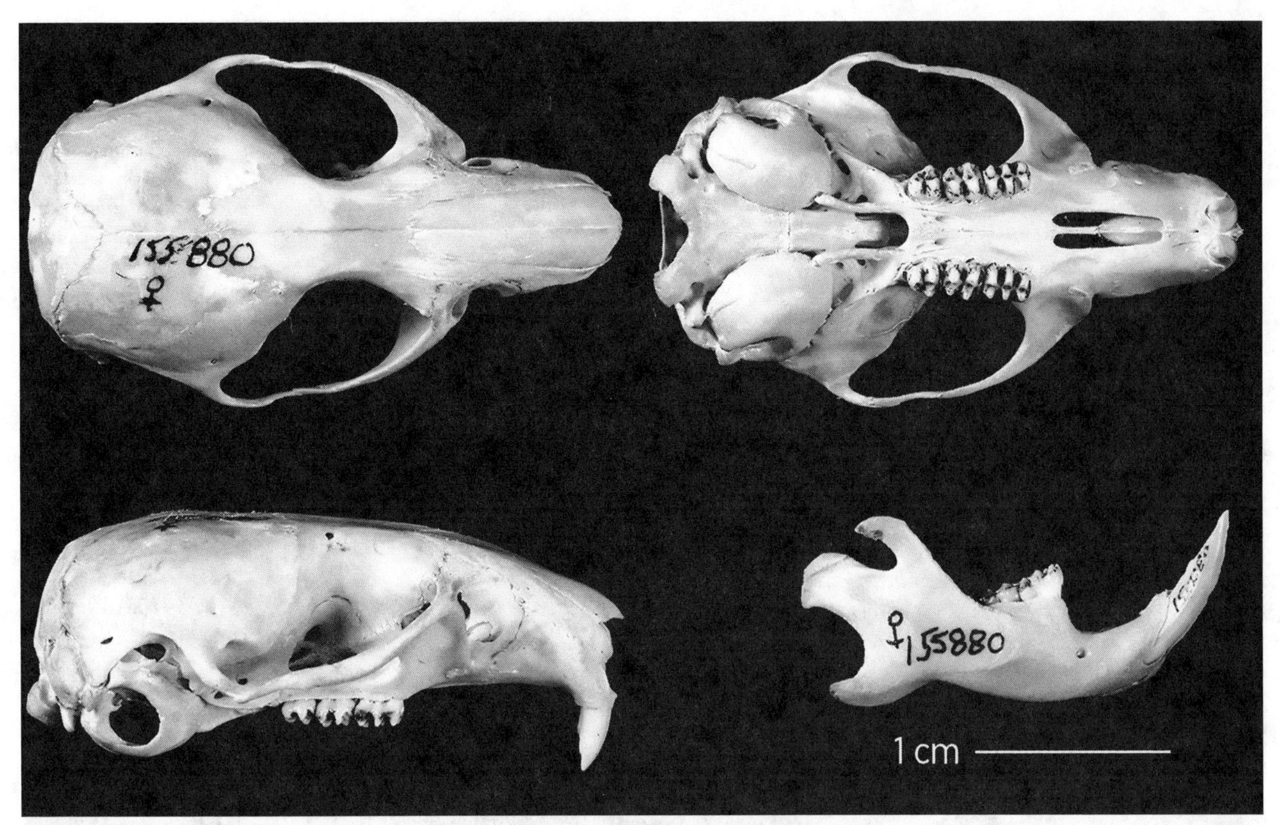

FIG. 201. Cricetidae: Cricetinae; skull and mandible of the European hamster, *Cricetus cricetus*.

TAXONOMIC DIVERSITY: 18 species in seven Old World genera, all known as various types of "hamster": *Allocricetulus* (Mongolian and Eversmann's hamsters), *Cansumys* (Gansu hamster), *Cricetulus* (dwarf hamsters), *Cricetus* (common hamster), *Mesocricetus* (golden hamster), *Phodopus* (desert hamsters), and *Tscherskia* (greater long-tailed hamster).

Subfamily NEOTOMINAE (North American [largely] rats and mice)

Mouse- and rat-sized rodents; body size diminutive to large (length 50–360 mm; mass 6–450 g); ears are large and conspicuous; tail is usually longer than the body but may be shorter than or subequal to the body in some, usually sparsely haired but may be bushy or with terminal short tuft; male *Neotoma* may have prominent ventral sebaceous glands, but all genera lack rump, hip, or flank glands; pelage is usually dense, ranges from short to long, and can be either relatively coarse or very soft. Terrestrial, arboreal, or semiarboreal, with feet modified for running (long and narrow) or climbing (short and broad with large plantar pads). Social organization varies widely; mating system varies from polygynous to monogamous, with males providing parental care. Most are herbivorous, eating a wide range of plant parts; some are capable of dietary specialization on plants heavy in plant secondary metabolites (e.g., woodrats, *Neotoma*); a few (like the grasshopper mice, *Onychomys*) are carnivorous. Some are notable for high-pitched calls (such as the singing mice, *Scotinomys*). Occur in virtually all biomes of North and Central America, including prairies, savannas, deserts, montane meadows, woodlands and scrublands, rainforest, conifer forest, and both temperate and tropical deciduous forest.

RECOGNITION CHARACTER:

1. cheek teeth always rooted; crowns either brachydont (most genera) and cuspidate, or hypsodont with flattened occlusal surface

DENTAL FORMULA: $\frac{1}{1}\ \frac{0}{0}\ \frac{0}{0}\ \frac{3}{3} = 16$

FIG. 202. Cricetidae: Neotominae; piñon mouse, *Peromyscus truei* (Reithrodontomyini; *left*), and eastern woodrat, *Neotoma floridana* (Neotomini; *right*).

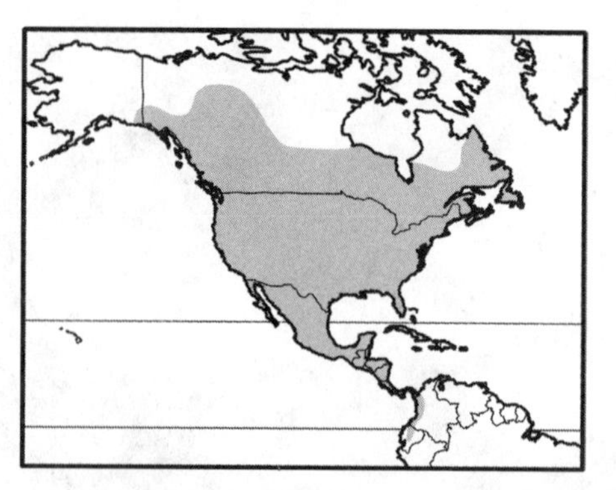

FIG. 203. Geographic range of subfamily Neotominae: throughout North America except extreme boreal regions, Central America, and northwestern South America.

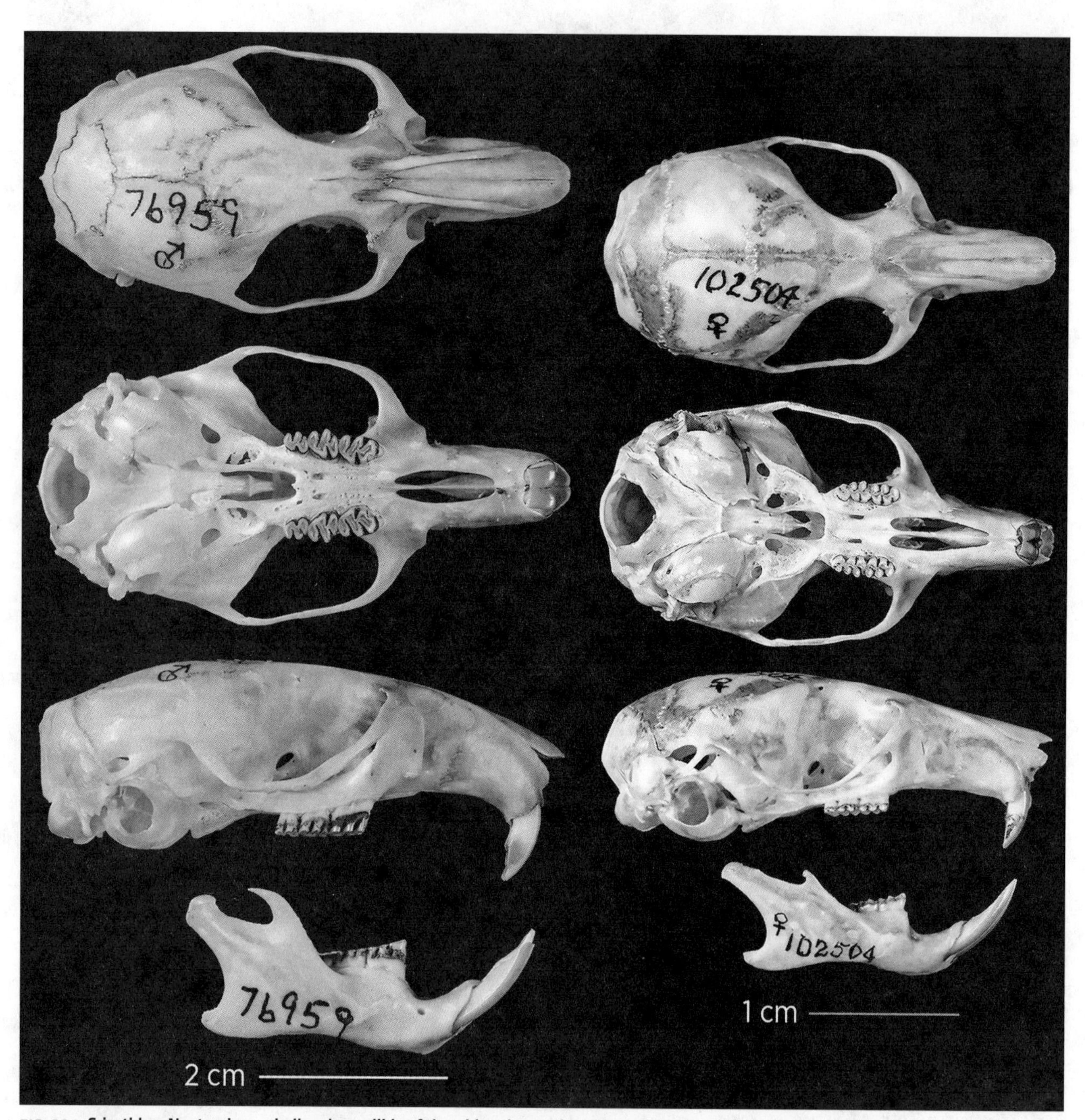

FIG. 204. Cricetidae: Neotominae; skull and mandible of the white-throated woodrat, *Neotoma albigula* (Neotomini; *left*), and the California mouse, *Peromyscus californicus* (Reithrodontomyini; *right*).

TAXONOMIC DIVERSITY: About 140 species allocated to 16 genera divided into four tribes:

Baiomyini—*Baiomys* (pygmy mice) and *Scotinomys* (singing mice)

Neotomini—*Hodomys* (Allen's woodrat), *Nelsonia* (diminutive woodrats), *Neotoma* (woodrats or pack rats), and *Xenomys* (Magdalena woodrat)

Ochrotomyini—*Ochrotomys* (golden mouse)

Reithrodontomyini—nine genera, including *Habromys* (Central American deer mice), *Megadontomys* (big-toothed deer mice), *Onychomys* (grasshopper mice), *Peromyscus* (deer mice), and *Reithrodontomys* (harvest mice)

Subfamily SIGMODONTINAE (South American [largely] rats and mice)

Exceedingly diverse in body form, habitat range, and ecology; small (15 g) to quite large (700 g); fossorial or semifossorial species (e.g., *Chelemys*, *Geoxus*, *Oxymycterus*) have small eyes, small ears, short tails, and well-developed claws on forefeet; grassland species (e.g., *Abrothrix*, *Akodon*, *Sigmodon*) are vole-like, with stout bodies, small eyes, short, rounded ears, short tails, and short limbs; arboreal species (*Oecomys*, *Rhipidomys*, *Irenomys*) vary in body size but have large eyes, short and broad feet with long toes, and elongate tails often terminating in a tuft; aquatic species (*Holochilus*, *Nectomys*, *Rheomys*) have dense underfur and often long guard hairs, webbed feet with natatory fringes of hair on outer digits and/or along edge of foot, and elongate, thickened tails; quasi-saltatorial forms inhabiting xeric, open communities (e.g., *Eligmodontia*, *Graomys*, *Phyllotis*) have large ears, elongate hind limbs and feet, and long tails also terminating in a pencil or tuft. Members occur in all biomes from the southern United States through Mexico, Central America, and South America to Tierra del Fuego, from deserts to lowland rainforest, and from deciduous woodlands to puna above tree line in the Andes. They include generalist, opportunistic feeders as well as members more specialized for granivory, herbivory, insectivory, or even piscivory.

FIG. 205. Cricetidae: Sigmodontinae; Azara's broad-headed oryzomys, *Hylaeamys megacephalus* (Oryzomyini; *left*), and hispid cotton rat, *Sigmodon hispidus* (Sigmodontini; *right*).

RECOGNITION CHARACTER:

1. cheek teeth always rooted, but with crowns varying from cuspidate and brachydont to flattened and hypsodont; cuspidate teeth typically more complex than those of similar members of Neotominae

DENTAL FORMULA: $\frac{1}{1}\ \frac{0}{0}\ \frac{0}{0}\ \frac{3}{3} = 16$

TAXONOMIC DIVERSITY: 86 living genera and over 430 species divided into 11 tribes; 5 genera remain as unique lineages of uncertain status (*incertae sedis*).

Abrotrichini—*Abrothrix* (soft-haired mice) and four genera of long-clawed mice (monotypic *Chelemys*, *Notiomys*, and *Paynomys*; and *Geoxus*, with four species)

Akodontini—16 genera, including *Akodon* (grass mice), *Bibimys* (swollen-nose mice), *Blarinomys* (burrowing mouse), *Brucepattersonius* (brucies), *Deltamys* (grass mice), *Juscelinomys* (Candago and Huanchaca burrowing mice), *Kunsia* (woolly giant rat), *Lenoxus* (Andean rat), *Necromys* (akodonts), *Oxymycterus* (hocicudos), and *Scapteromys* (swamp rats)

Andinomyini—*Andinomys* (Andean mouse) and *Punomys* (Puna mice)

Euneomyini—*Euneomys* (chinchilla rats), *Irenomys* (Chilean tree rat), and *Neotomys* (Andean swamp rat)

Ichthyomyini—collectively referred to as "water mice," including *Anotomys* (earless water mouse), *Chibchanomys* (Andean water mice), *Ichthyomys*

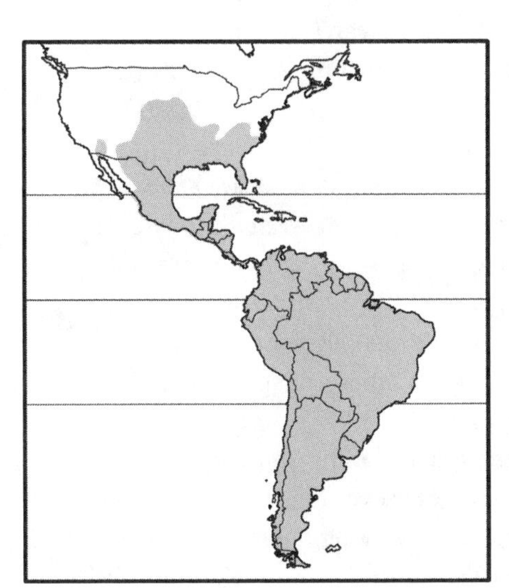

FIG. 206. Geographic range of subfamily Sigmodontinae: southern United States through Mexico, Central America, and throughout South America.

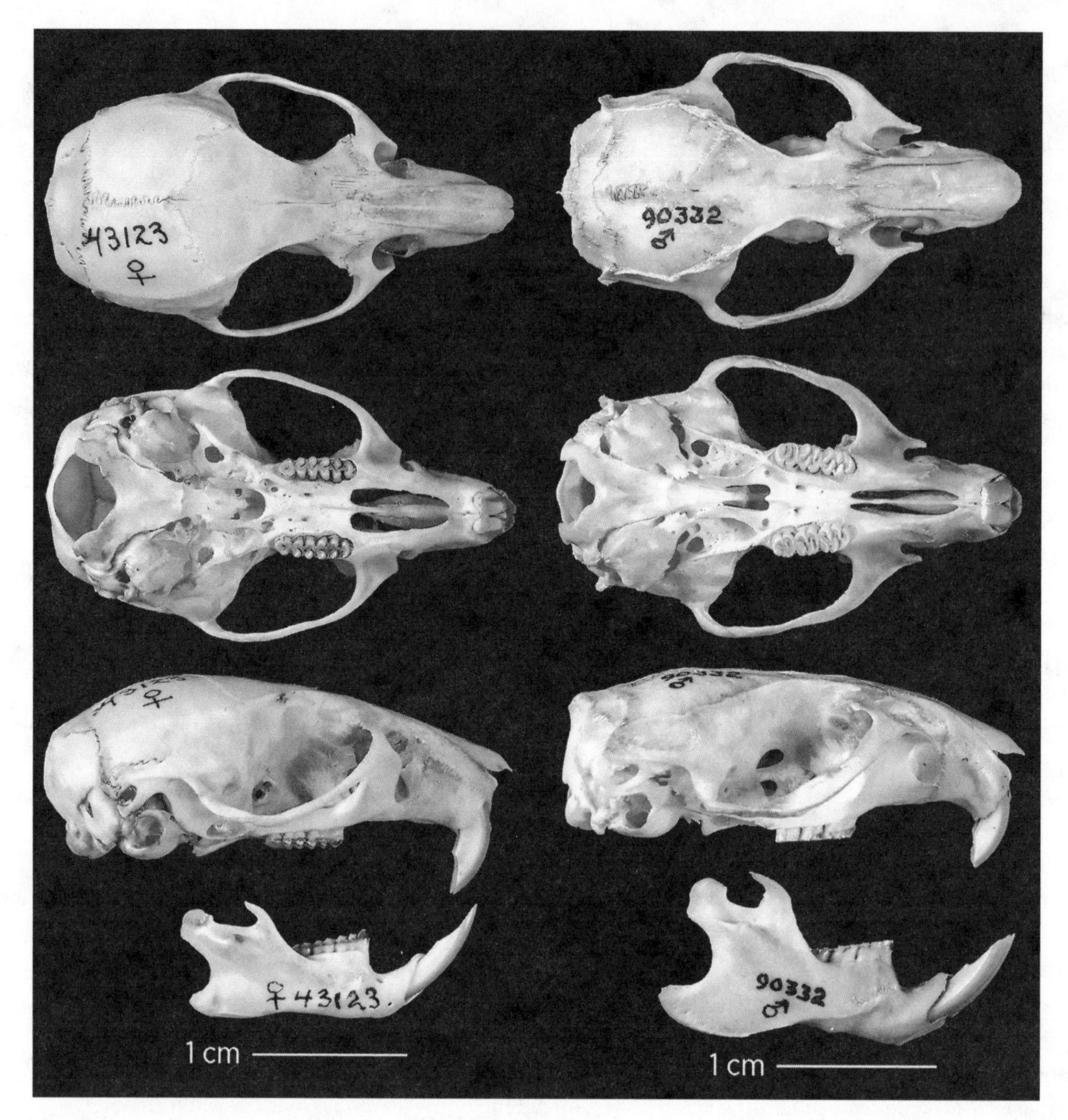

FIG. 207. Cricetidae: Sigmodontinae; skull and mandible of Coues's rice rat, *Oryzomys couesi* (Oryzomyini; *left*), and the hispid cotton rat, *Sigmodon hispidus* (Sigmodontini; *right*).

(crab-eating rats), *Neusticomys* (fish-eating rats), and *Rheomys* (Central American water mice)

Oryzomyini—29 genera, including *Aegialomys* (rice rats), *Cerradomys* (rice rats), *Euryoryzomys* (rice rats), *Handleyomys* (rice rats), *Hylaeamys* (rice rats), *Holochilus* (marsh rats), *Melanomys* (rice rats), *Neacomys* (spiny mice), *Nectomys* (water rats), *Nesoryzomys* (Galápagos mice), *Oecomys* (arboreal rice rats), *Oligoryzomys* (colilargos or pygmy rice rats), *Oryzomys* (rice rats), *Pseudoryzomys* (false rice rat), *Scolomys* (South American spiny mice), and *Zygodontomys* (cane mice)

Phyllotini—11 genera, including *Auliscomys* (pericotes), *Calomys* (lauchas or vesper mice), *Eligmodontia* (gerbil mice), *Graomys* (long-tailed pericotes), and *Phyllotis* (leaf-eared mice)

Reithrodontini—*Reithrodon* (conyrats)

Sigmodontini—*Sigmodon* (cotton rats)

Thomasomyini—*Aepeomys* (montane mice), *Chilomys* (forest mice), *Rhagomys* (arboreal mice), *Rhipidomys* (climbing rats), and *Thomasomys* (Thomas's mice or Oldfield mice)

Wiedomyini—*Phaenomys* (Rio de Janeiro arboreal rat), *Wiedomys* (red-nosed mice), and *Wilfredomys* (red-nosed tree mouse)

incertae sedis—*Abrawayaomys* (spiny mice), *Chinchillula* (Altiplano chinchilla mouse), *Delomys* (Atlantic Forest rats), *Juliomys* (tree mice), and *Neomicroxus* (Bogota and Ecuadoran grass mice)

Subfamily TYLOMYINAE (climbing rats)

Small to medium-sized (body length 95–255 mm; mass 25–280 g), largely arboreal rats; ears small to large but appear naked; vibrissae long; eyes large; tail slightly longer than the body, either naked with large scales visible or covered with hair and tufted at tip; hind feet modified for climbing. Solitary or live in family groups. Herbivorous, consuming seeds, fruits, and leaves. Habitat typically rocky areas in tropical evergreen and semideciduous forests.

FIG. 208. Cricetidae: Tylomyinae: Tylomyini; Watson's climbing rat, *Tylomys watsoni*.

RECOGNITION CHARACTERS:

1. skull with wedge-shaped (cuneate) interorbital region with prominent supraorbital shelves that continue posteriorly as pronounced temporal ridges
2. cheek teeth rooted and cuspidate, with major cusps lying opposite one another

DENTAL FORMULA: $\frac{1}{1}\ \frac{0}{0}\ \frac{0}{0}\ \frac{3}{3} = 16$

TAXONOMIC DIVERSITY: A group of limited species diversity, with only 10 species and four genera divided into two tribes:

Nyctomyini—*Nyctomys* (Sumichrast's vesper rat) and *Otonyctomys* (Yucatan vesper rat)

Tylomyini—*Ototylomys* and *Tylomys* (climbing rats)

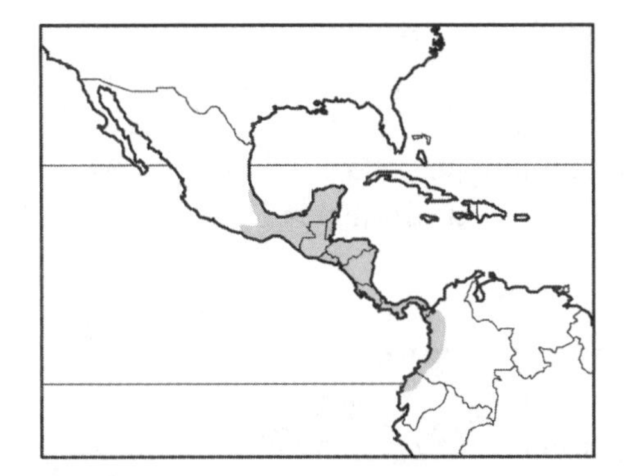

FIG. 209. Geographic range of subfamily Tylomyinae: Central America, from southern Mexico to northwestern Ecuador in South America.

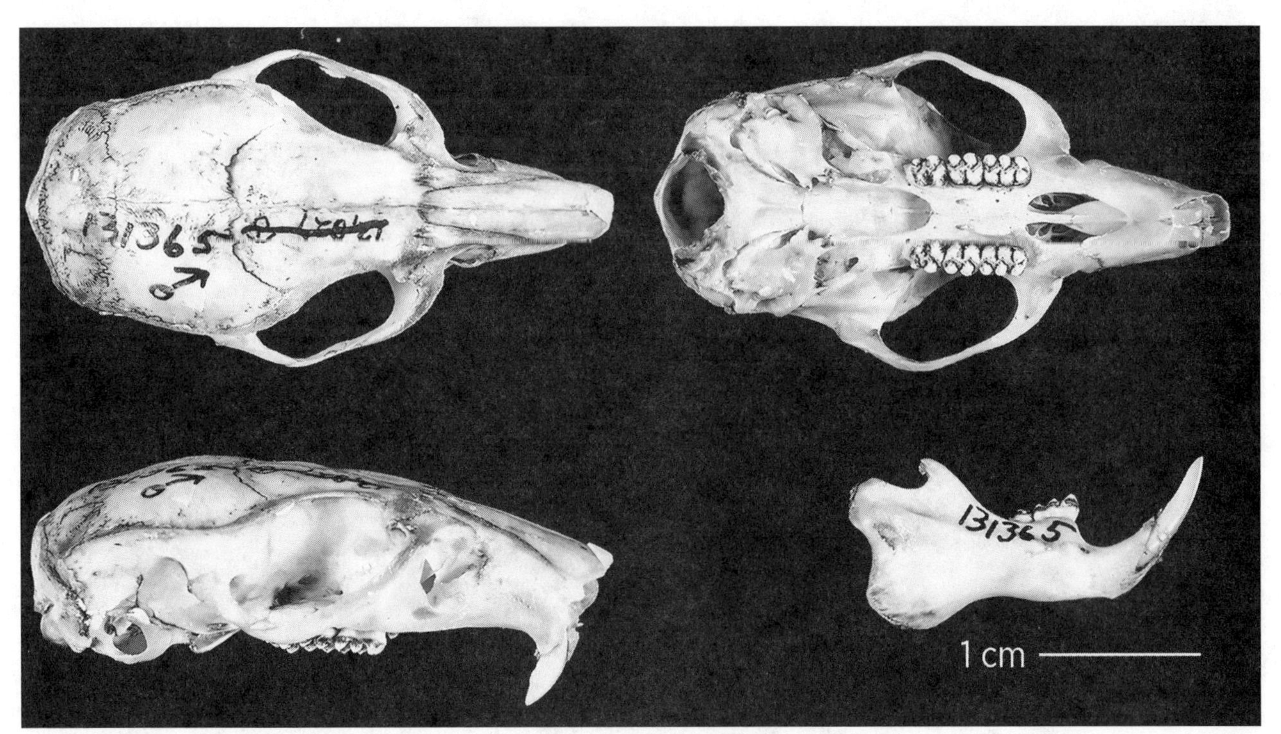

FIG. 210. Cricetidae: Tylomyinae: Tylomyini; skull and mandible of the climbing rat *Tylomys nudicaudatus*.

Family MURIDAE

Muridae is a very large family (about 155 genera and 816 species) that remains subject to revisions and descriptions of new species. It is currently divisible into five or six subfamilies. Members include the familiar mice and rats, but the family also encompasses an enormous array of diverse body forms and specializations. Size ranges from diminutive (body length 45 mm) to large (500 mm). Most have slender bodies, pointed muzzles, large ears, prominent vibrissae, and scaly tails, but these attributes vary widely. Natively, murids are geographically limited to the Old World, but commensal species of *Mus* and *Rattus* have spread worldwide. They occur in every biome, from dry deserts to wet lowland and montane cloud forests, from savanna to tundra, and throughout temperate woodlands and shrublands. Most are generalized terrestrial rats, but some exhibit aquatic adaptations, such as webbed feet, natatory fringes of stiffened hairs, and elongate, rudder-like tails; others are fossorial; and still others are highly adapted to life in the trees, with broad hind feet and prehensile or bushy tails. Food habits range from true omnivory to specialization on earthworms, fungi, and aquatic invertebrates. Some are extreme agricultural pests, others are vectors or reservoirs of a number of severe human diseases, and some are beneficial to humanity; several provide important ecosystem functions such as spreading mycorrhizal fungi or dispersing seeds.

All murids have three or fewer upper and lower cheek teeth, and most have a clearly evident triserial cusp arrangement (e.g., three longitudinal and parallel rows of cusps) on their molars.

Subfamily DEOMYINAE (spiny mice, Congo forest rat, brush-furred rats, bristle-furred rat)

Genera of the subfamily Deomyinae are phylogenetically united only by DNA sequences; they share no morphological characters that can be used to separate them from other murid subfamilies. All members are relatively small, mouselike rodents. Spiny mice of the genus *Acomys* have, as the name implies, dorsal hairs modified into broad, stiff spines, but are otherwise mouselike in general appearance, with short tails. The omnivorous *Deomys*, with a single species, is also mouselike but has an elongate tail, long hind limbs, large ears, and stiffened rump hairs. Members of the very speciose *Lophuromys* have very chunky bodies and shortened limbs, short tails, and fur composed of stiff hairs, but not spines, and are insectivorous or carnivorous. *Uranomys* resembles *Lophuromys* with its stiffened dorsal hairs, and is also insectivorous.

RANGE: Mostly sub-Saharan Africa (*Deomys*, *Lophuromys*, *Uranomys*), with *Acomys* distributed in southern Africa, eastern Africa, and the Middle East, from Turkey south through the Arabian Peninsula.

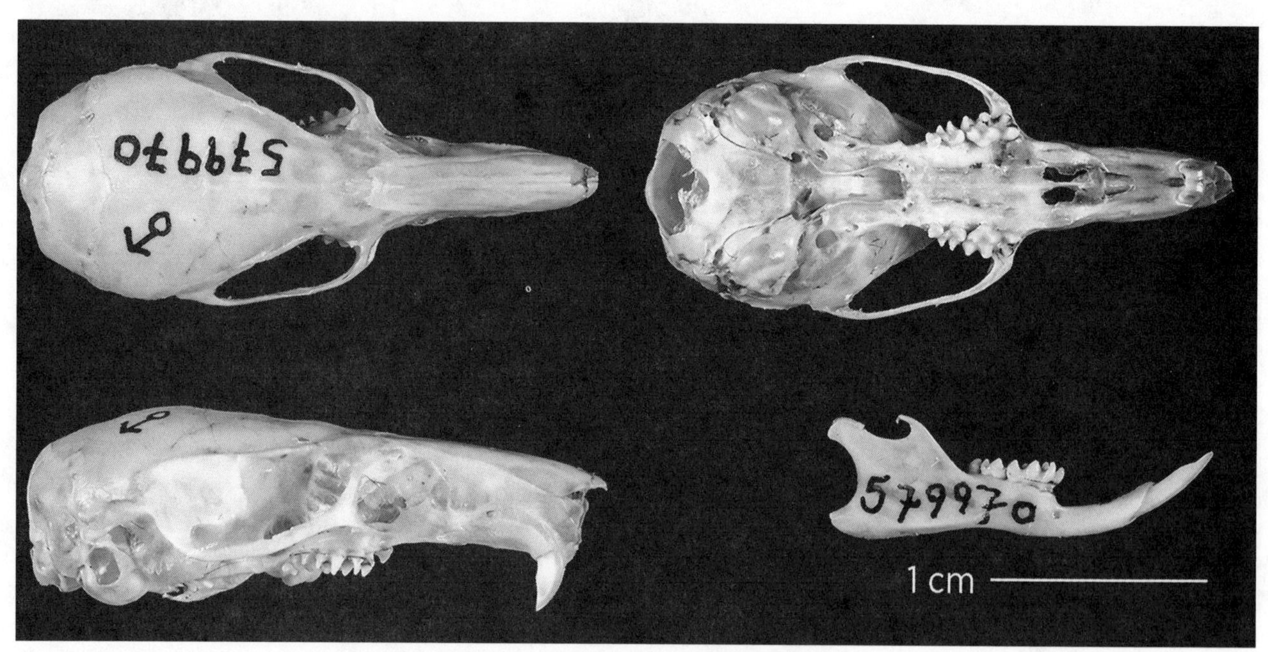

FIG. 211. Muridae: Deomyinae; skull and mandible of the Congo forest rat, *Deomys ferrugineus*.

DENTAL FORMULA: $\frac{1}{1}\ \frac{0}{0}\ \frac{0}{0}\ \frac{3}{3} = 16$

TAXONOMIC DIVERSITY: Four genera comprising nearly 60 species: *Acomys* (spiny mice), *Deomys* (Congo forest rat), *Lophuromys* (brush-furred rats), and *Uranomys* (Rudd's bristle-furred rat).

Subfamily GERBILLINAE (jirds, gerbils)

Small to medium-sized (body length 50–200 mm); body form stout and compact to slender and gracile; tail longer than, equal to, or shorter than body (extremely short and stubby in *Pachyuromys*, serving as site to store fat); tail usually densely haired, bushy in some, and can be black- or white-tipped; midventral sebaceous gland in males in many species; plantar surface either naked or moderately to well furred. Terrestrial, cursorial or highly saltatorial. Nocturnal or crepuscular, a few mainly diurnal. Primarily granivorous or herbivorous, but also consume insects; cache seeds in burrow chambers. Habitat is xeric areas of sparse vegetation, including sandy and clay deserts, gravelly plains, arid steppe, rocky slopes, and savanna woodlands.

RECOGNITION CHARACTERS:

1. supraorbital shelf (a distinctly flattened plate extending from the frontals and overhanging the orbits) well developed
2. zygomatic plate broad, producing a deep zygomatic notch, ventral portion of infraorbital foramen a narrow slit, dorsal portion normally rounded
3. lacrimal enlarged and forming a conspicuous ledge over the anterior margin of the orbit
4. tympanic portions of auditory bullae greatly inflated; mastoid portion enlarged in some genera
5. angular process of mandible deflected laterally
6. upper incisors grooved (except *Psammomys*)
7. occlusal pattern of cheek teeth lophate, planar, or prismatic
8. 3rd upper molar greatly reduced and cylindriform

DENTAL FORMULA: $\frac{1}{1}\ \frac{0}{0}\ \frac{0}{0}\ \frac{2\text{–}3}{2\text{–}3} = 14\text{–}16$

TAXONOMIC DIVERSITY: 14 genera and over 100 species divided into three tribes:

Ammodillini—*Ammodillus* (ammodile)

Gerbillini—*Brachiones* (Przewalski's jird), *Desmodilliscus* (pouched gerbil), *Gerbillus* (gerbils), *Meriones* (jirds), *Microdillus* (Somali pygmy gerbil), *Pachyuromys* (fat-tailed jird), *Psammomys* (sand rats), *Rhombomys* (great gerbil), and *Sekeetamys* (bushy-tailed jird)

Taterillini—*Desmodillus* (Cape short-tailed gerbil), *Gerbilliscus* (gerbils), *Tatera* (Indian gerbil), and *Taterillus* (taterils)

FIG. 212. Muridae: Gerbillinae: Gerbillini; lesser Egyptian gerbil, *Gerbillus gerbillus*.

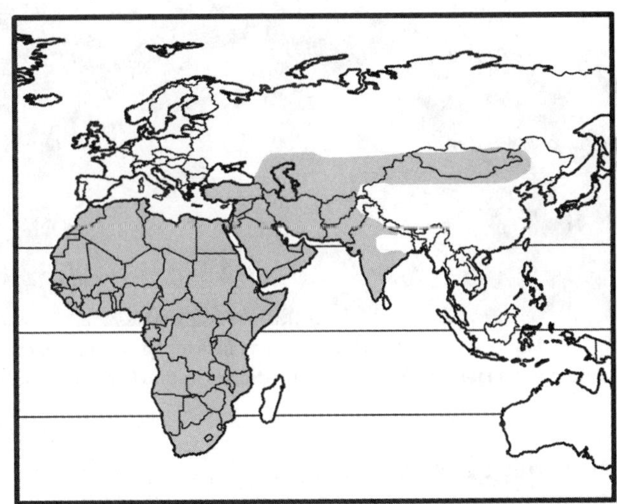

FIG. 213. Geographic range of subfamily Gerbillinae: Africa, Arabian Peninsula, Asia Minor, Middle East, central Asia to Mongolia, peninsular India, and island of Sri Lanka.

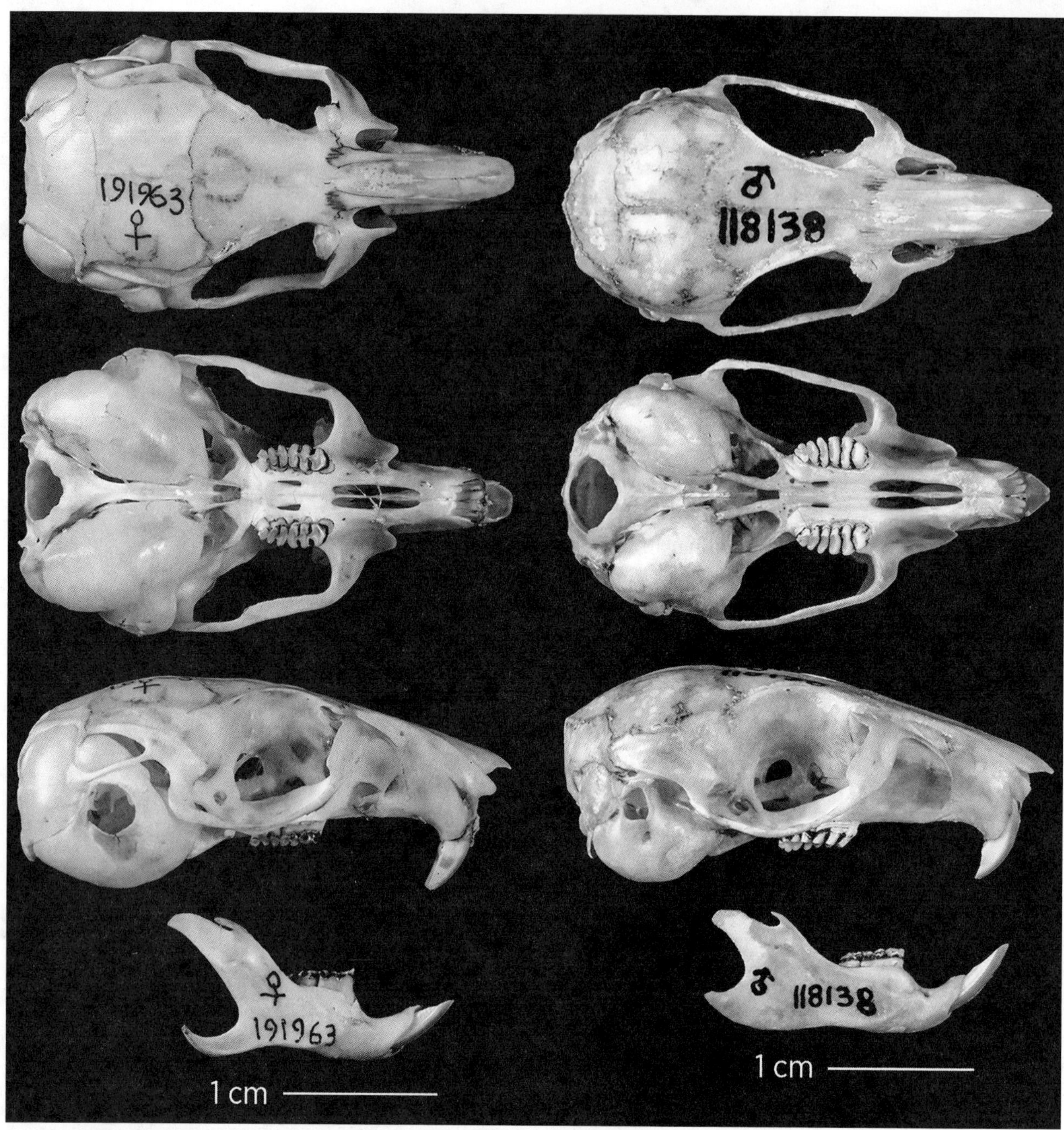

FIG. 214. Muridae: Gerbillinae; skull and mandible of the Libyan jird, *Meriones lybicus* (Gerbillini; *left*; note enlarged tympanic and mastoid portions of bullae), and the Cape gerbil, *Gerbilliscus afra* (Taterillini; *right*; note inflation of only tympanic portion of bullae).

Subfamily LEIMACOMYINAE (Togo or groove-toothed forest mouse)

The subfamily Leimacomyinae is monotypic, with a single species in the genus *Leimacomys* known from only two specimens collected in the African country of Togo in 1890. The mouse is small (head and body length 118 mm); its tail is short (37 mm) and slightly hairy; its fur is dark gray brown above and pale gray brown below; and its ears are small. Upper incisors have shallow grooves on their anterior face; the rostrum is long but wide; the interorbital region is broad; and the zygomatic plate is also large. It is believed to be insectivorous, based on its skull and tooth characteristics, but nothing is known about its behavior or ecology. Although it was previously grouped with the Dendromurinae, limited morphological comparisons suggest a close relationship with the Gerbillinae (Denys et al. 1995).

DENTAL FORMULA: $\frac{1}{1}\ \frac{0}{0}\ \frac{0}{0}\ \frac{3}{3} = 16$

TAXONOMIC DIVERSITY: The subfamily is monotypic, with a single species, *Leimacomys buettneri*.

Subfamily LOPHIOMYINAE (maned or crested rat)

Large (head and body length 300–360 mm); body stocky, with short muzzle and robust head, short limbs, and broad feet; trunk relatively elongate; tail shorter than body, bushy, and white-tipped; fur dense and long, with coarse silver- and black-tipped guard hairs and woolly underfur; strikingly colored black and white in stripes or variegated pattern; paired glandular tracts extend along sides and contain modified hairs; coarse middorsal hairs form erectile mane when the rat is alarmed, exposing broad white areas and the otherwise hidden glandular tracts; palms of forefeet and hind feet naked but with well-developed pads. Arboreal; facile climbers, with slow and deliberate movements; nest on ground in burrows, in fallen trees, or among boulders. Nocturnal and herbivorous, eating leaves, fruits, and young shoots. Inhabits montane forests and woodlands. Apparently extracts an ouabane-like toxin by chewing the bark of *Acokanthera schimperi* trees, which is slathered on the fur along the glandular tracts, and which is absorbed and retained in the modified hairs, hence the only vertebrate believed to employ toxicity by acquisition (Kingdon et al. 2012).

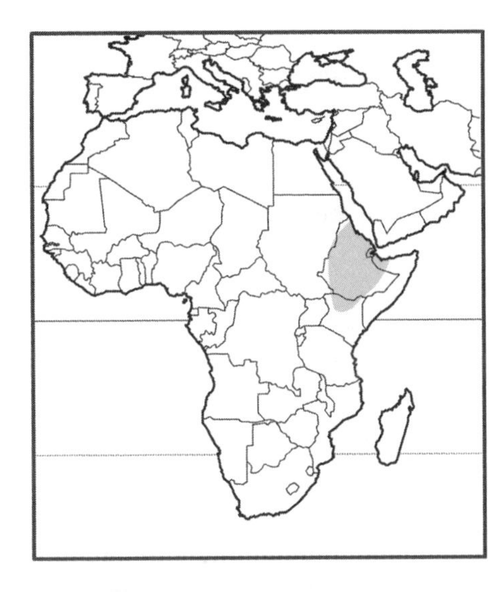

FIG. 215. Geographic range of subfamily Lophiomyinae: eastern Africa (Somalia, Ethiopia, Sudan, Kenya, and Tanzania).

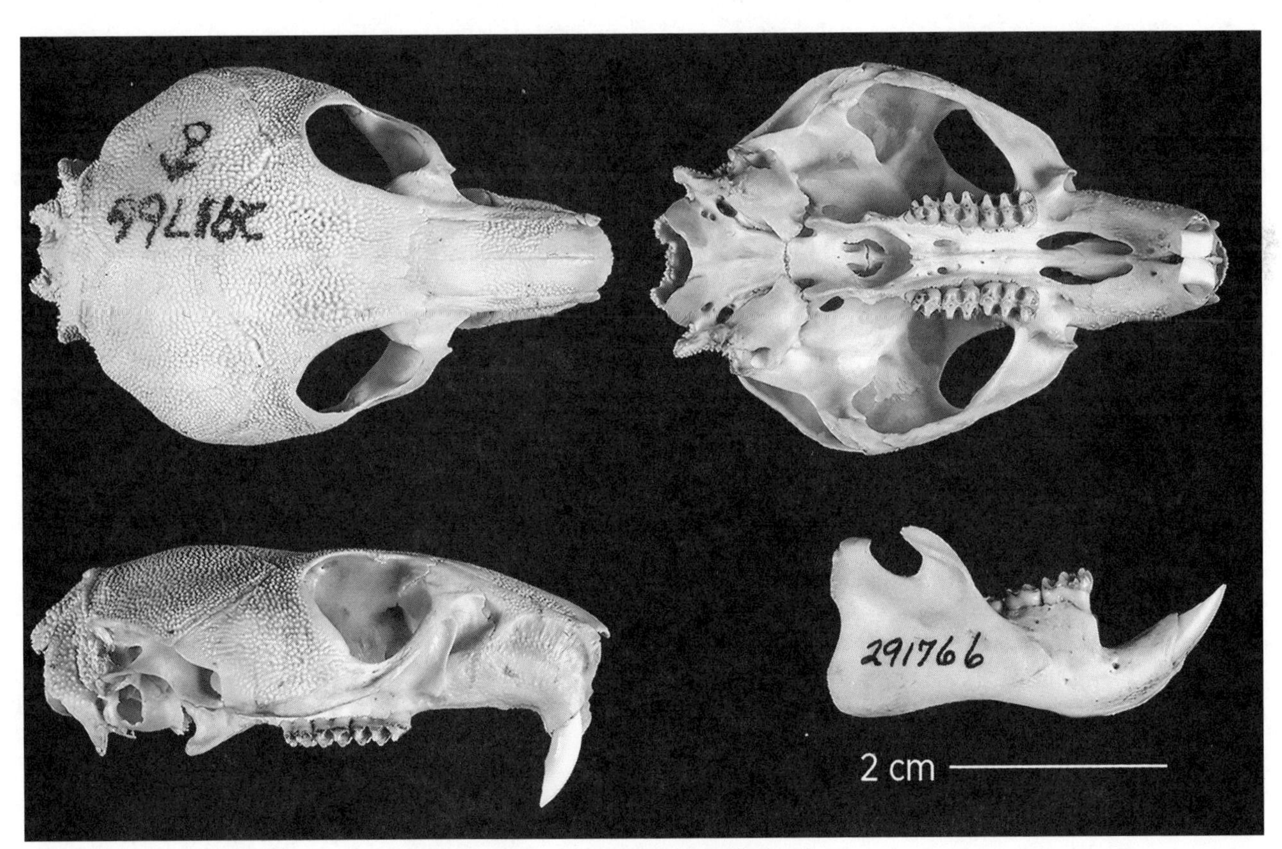

FIG. 216. Muridae: Lophiomyinae; skull and mandible of the maned rat, *Lophiomys imhausi*.

RECOGNITION CHARACTERS:

1. temporal fossa of skull roofed over by extensions of parietal, frontal, and jugal bones
2. dermal roofing bones of skull densely sculptured, giving granular texture and appearance
3. bony palate with deep grooves and median ridge
4. infraorbital canal ovoid, not notably constricted ventrally
5. cheek teeth rooted and cuspidate, with coronal pattern similar to that of Cricetinae

DENTAL FORMULA: $\frac{1}{1}\ \frac{0}{0}\ \frac{0}{0}\ \frac{3}{3} = 16$

TAXONOMIC DIVERSITY: The subfamily is monotypic, with a single species, *Lophiomys imhausi*.

Subfamily MURINAE (Old World rats and mice)

The Murinae is an extraordinarily varied group of rats and mice, with a diversity of body form similar to that of New World Sigmodontinae, ranging from gracile and small to robust and ponderous; their specializations in body form resemble those of water shrews and otter shrews (*Hydromys*), small-bodied terrestrial shrews (*Pseudohydromys*), large-bodied terrestrial shrews (*Tateomys*), elephant shrews (*Echiothrix*), kangaroo rats (*Notomys*), voles (*Arvicanthis* and *Mastacomys*), pocket gophers (*Nesokia*), squirrels (*Crateromys*), dormice (*Chiropodomys*), and mouse opossums (*Pithecheir*). Small to large (body length 55–400 mm); pinnae usually conspicuous, thinly to densely furred; tail shorter than, subequal to, or longer than head and body; tail appears naked in most genera, but may be densely haired; fur may be thick and velvety, coarse and thin, mixed with spinose hairs, or densely spinose. Diverse food habits, from granivorous to omnivorous; some feed on insects or soft-bodied invertebrates, some are fungus feeders, some eat foliage or roots, and the diets of some may include vertebrate flesh, small fish, and eggs. Skull and cheek tooth morphology is also highly variable, mirroring the broad range of body forms and ecologies—skulls are small and gracile to large and robust, sturdy and chunky to elongate and delicate. Most are terrestrial, some are amphibious to mostly aquatic, others arboreal, fossorial, or saxicolous. Most are nocturnal but some are crepuscular, others may be active throughout the day. Usually solitary, but some are quasi-colonial and occupy extensive burrow systems.

RECOGNITION CHARACTERS:

1. molars rooted, never ever-growing
2. molar crowns brachydont to hypsodont
3. upper molars typically with three lingual cusps, each forming the 3rd cusp of a transverse lamina, or with cusps on lingual sides of 1st and 2nd laminae but not on 3rd (triserial pattern)
4. occlusal pattern highly cuspidate in most, transversely laminar in a few, with chevron-shaped laminae in some, or basined

DENTAL FORMULA: $\frac{1}{1}\ \frac{0}{0}\ \frac{0}{0}\ \frac{0\text{–}3}{0\text{–}3} = 4\text{–}16$

The Sulawesi worm specialist *Paucidentomys vermidax* lacks cheek teeth altogether and has a bifid upper incisor designed to cut soft-bodied worms into small sections that can be swallowed whole; its dental formula is 1/1 0/0 0/0 0/0 = 4 (fig. 219).

TAXONOMIC DIVERSITY: About 630 species in 136 genera, divided into 14 tribes to reflect hypothesized relationships; three genera remain as unique lineages of uncertain status (*incertae sedis*). We list only the most common, best-known, or most diverse genera here.

Apodemini—*Apodemus* (field mice) and *Tokudaia* (spiny rats)

Arvicanthini—18 genera, including *Aetheomys* (rock rats), *Arvicanthis* (grass rats), *Dasymys* (shaggy rats), *Grammomys* (thicket rats), *Hybomys* (striped mice), *Lemniscomys* (striped grass mice), *Pelomys* (swamp rats), *Rhabdomys* (four-striped grass rats), *Thallomys* (acacia rats, thallomys), and *Thamnomys* (thicket rats)

Chiropodomyini—*Chiropodomys* (pencil-tailed tree mice)

Hapalomyini—*Hapalomys* (marmoset rats)

FIG. 217. Muridae: Murinae; Norway or brown rat, *Rattus norvegicus* (*Rattini*).

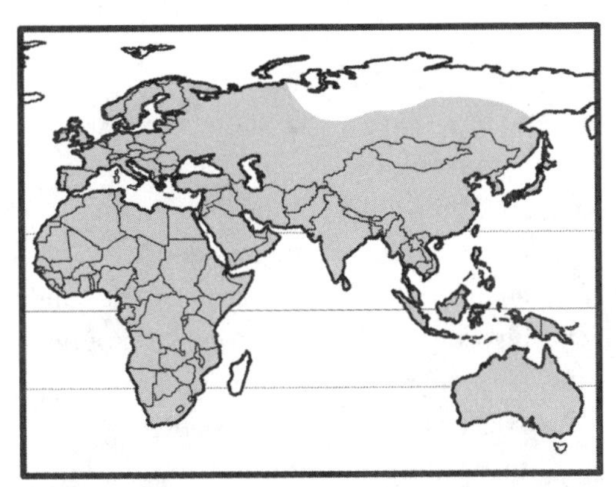

FIG. 218. Geographic range of subfamily Murinae: subarctic Old World including Australia, Wallacea, much of Eurasia, and Africa (excluding Madagascar, except as introduced); some genera (*Mus* and *Rattus*) introduced worldwide.

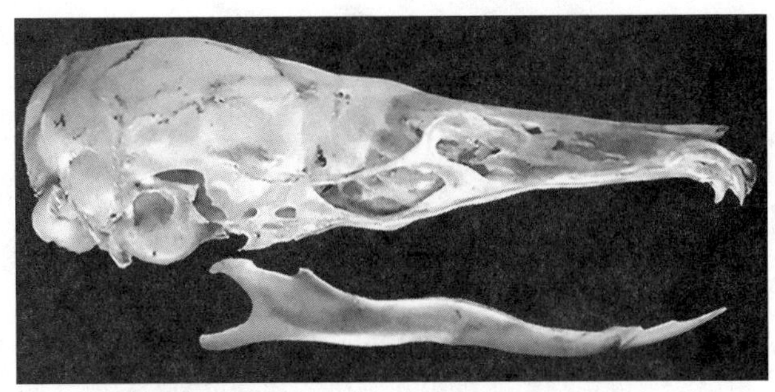

FIG. 219. Muridae: Murinae: Rattini; lateral view of skull and mandible of *Paucidentomys vermidax* (Rattini).

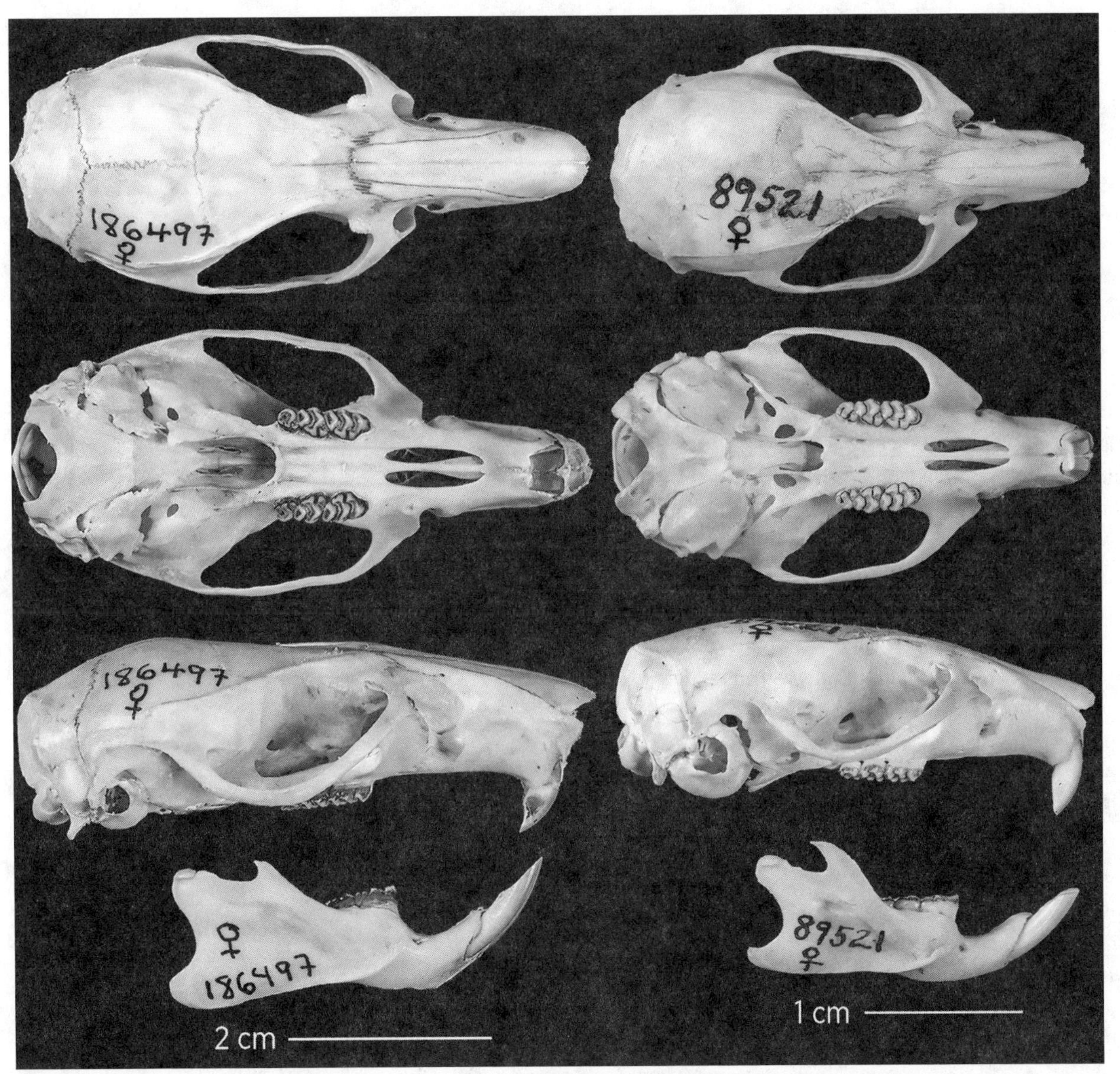

FIG. 220. Muridae: Murinae: Rattini; skull and mandible of the long-tailed giant rat, *Leopoldamys edwardsi* (*left*), and the Norway rat, *Rattus norvegicus* (*right*).

Hydromyini—42 genera, including *Apomys* (Philippine forest mice), *Archboldomys* (shrew mice), *Chrotomys* (Philippine striped shrew rats), *Crossomys* (earless New Guinea water rat), *Hydromys* (water rats), *Leggadina* (Forrest's and Lakeland Downs mice), *Leporillus* (greater stick-nest rat), *Leptomys* (New Guinea water rats), *Mallomys* (New Guinea woolly rats), *Mastacomys* (Australian broad-toothed rat), *Melomys* (mosaic-tailed rats, melomys), *Mesembriomys* (tree rats), *Notomys* (hopping mice), *Paramelomys* (mosaic-tailed rats, paramelomys), *Pogonomys* (New Guinea tree mice), *Pseudohydromys* (shrew mice), *Pseudomys* (Australian native mice), *Soricomys* (moss and shrew mice), *Uromys* (giant rats), and *Zyzomys* (Australian rock rats)

Malacomyini—*Malacomys* (swamp rats)

Micromyini—*Micromys* (Eurasian harvest mouse)

Millardiini—4 genera, including *Cremnomys* (cutch rat, Elvira rat) and *Millardia* (soft-furred rats)

Murini—*Mus* (mice)

Phloeomyini—*Batomys* (hairy-tailed rats), *Carpomys* (dwarf cloud rats), *Crateromys* (bushy-tailed cloud rats), *Musseromys* (tree mice), and *Phloeomys* (giant cloud rats)

Pithecheirini—*Pithecheir* (Malay tree rats) and *Pithecheirops* (Bornean tree rat)

Praomyini—8 genera, including *Hylomyscus* (wood mice), *Mastomys* (multimammate mice or rats), *Myomyscus* (meadow mice), *Praomys* (soft-furred mice), and *Stenocephalemys* (Ethiopian rats)

Rattini—43 genera, including *Bandicota* (bandicoot rats), *Berylmys* (white-toothed rats), *Bunomys* (hill rats), *Chiromyscus* (masked tree rats), *Hyorhinomys* (hog-nosed shrew rat), *Leopoldamys* (long-tailed giant rats), *Margaretamys* (Margareta rats), *Maxomys* (Asian spiny rats), *Nesokia* (bandicoot rats), *Niviventer* (Asian white-bellied rats), *Paruromys* (Sulawesi giant rat), *Paucidentomys* (edentate Sulawesi rat), *Rattus* (typical Old World rats), *Sundamys* (giant Sunda rats), *Taeromys* (Sulawesi rats), and *Tateomys* (Sulawesian shrew rats)

Vandeleurini—*Vandeleuria* (Asian long-tailed climbing mice)

incertae sedis—*Hadromys* (hadromys, bush rats), *Nilopegamys* (Ethiopian amphibious rat), and *Vermaya* (Vermay's climbing mouse)

Subfamily OTOMYINAE (African swamp or grass rats)

Medium-sized rats (body length 130–200 mm); body stocky, vole-like in appearance with small eyes, short ears, tail shorter than body, short limbs, palmar and plantar surfaces naked; fur long, soft, and dense. Primarily terrestrial and cursorial; can be active day and night. Mainly herbivorous. Nest in burrows, on ground, or in low shrubs; *Parotomys* constructs elaborate burrows, *Otomys* uses tunnels and runways through dense grass. Habitat ranges from xeric environments, such as desert flats and sandy veldts, to more mesic grasslands, alpine meadows, and stream and marsh edges. Treated as a tribe within Murinae by some authors (e.g., Denys et al. 2017).

FIG. 221. Muridae: Otomyinae; bush vlei rat, *Otomys unisulcatus*.

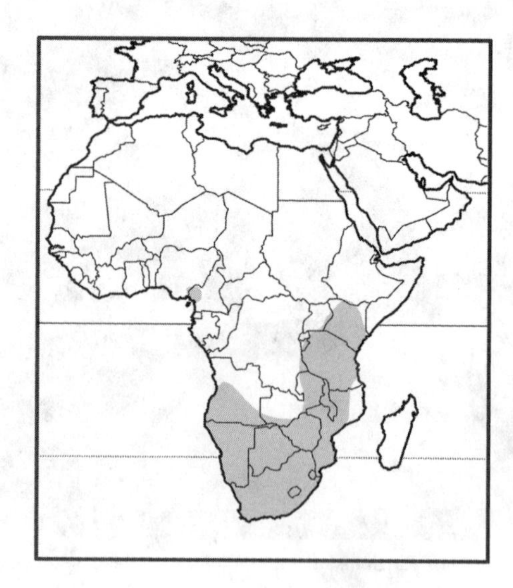

FIG. 222. Geographic range of subfamily Otomyinae: sub-Saharan Africa.

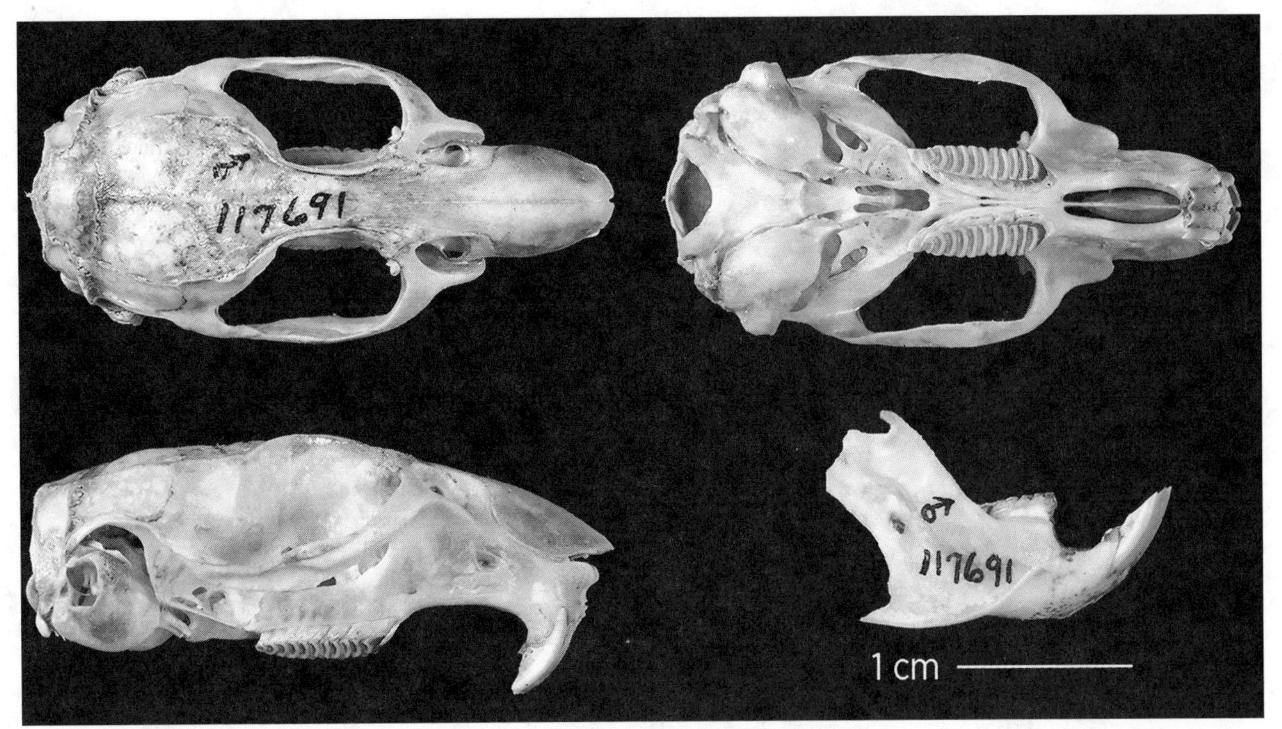

FIG. 223. Muridae: Otomyinae; skull and mandible of the Southern African vlei rat, *Otomys irroratus*.

RECOGNITION CHARACTERS:

1. cheek teeth hypsodont
2. occlusal surface of cheek teeth flat, with transversely oriented laminar folds of enamel

DENTAL FORMULA: $\frac{1}{1}\ \frac{0}{0}\ \frac{0}{0}\ \frac{3}{3} = 16$

TAXONOMIC DIVERSITY: Regarded by some as a tribe within the Murinae; over 30 species in two genera: *Otomys* (vlei rats) and *Parotomys* (whistling rats).

Suborder ANOMALUROMORPHA

The Anomaluromorpha includes three families restricted to sub-Saharan Africa. Two of these families were only recently distinguished. They share a hystricomorphous jaw musculature and sciurognathous lower jaw but present two very different bauplans; two of these resemble arboreal or even flying squirrels, whereas the third is semifossorial and exhibits bipedal locomotion.

Families ANOMALURIDAE and ZENKERELLIDAE (scaly-tailed squirrels or anomalures)

These two families resemble squirrels in external appearance, with body modified for arboreality; size small to large (body length 65–450 mm), with **tail longer than body** (75–460 mm), **tufted distally** (*Anomalurus*, *Zenkerella*) **or with long sparse hairs** (*Idiurus*); **gliding membrane (patagium) present** (except in *Zenkerella*) **between wrist and ankle, and extending to the basal part of the tail**, with front part supported by **cartilaginous rod extending from the olecranon process** of the ulna; ears enlarged. Herbivorous, feeding on all plant parts from seeds to flowers and leaves. Inhabit tropical and subtropical forests.

Although the anomalurids and zenkerellids were traditionally treated as a single family (Anomaluridae), Heritage et al. (2016) combined molecular and morphological analyses, with the latter including the rich assemblage of fossil taxa, to split *Zenkerella* into a family apart from the remaining anomalurids, *Amomalurus* and *Idiurus*. For convenience we present all three genera together, but we provide this new classification below.

FIG. 224. Anomaluridae: Idiurinae; long-eared flying squirrel, *Idiurus macrotis*.

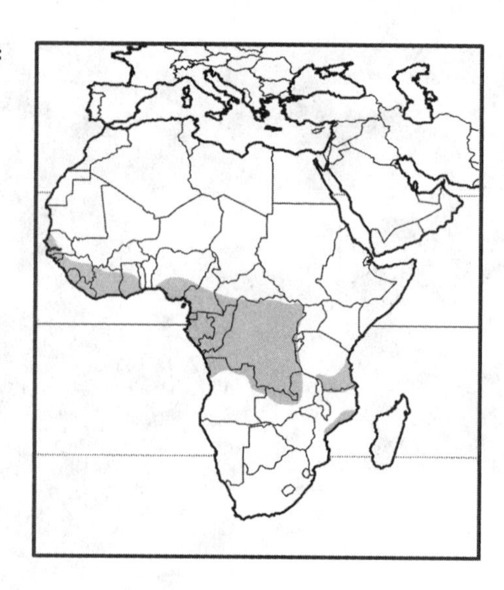

FIG. 225. Combined geographic range of families Anomaluridae and Zenkerellidae: tropical western and central Africa.

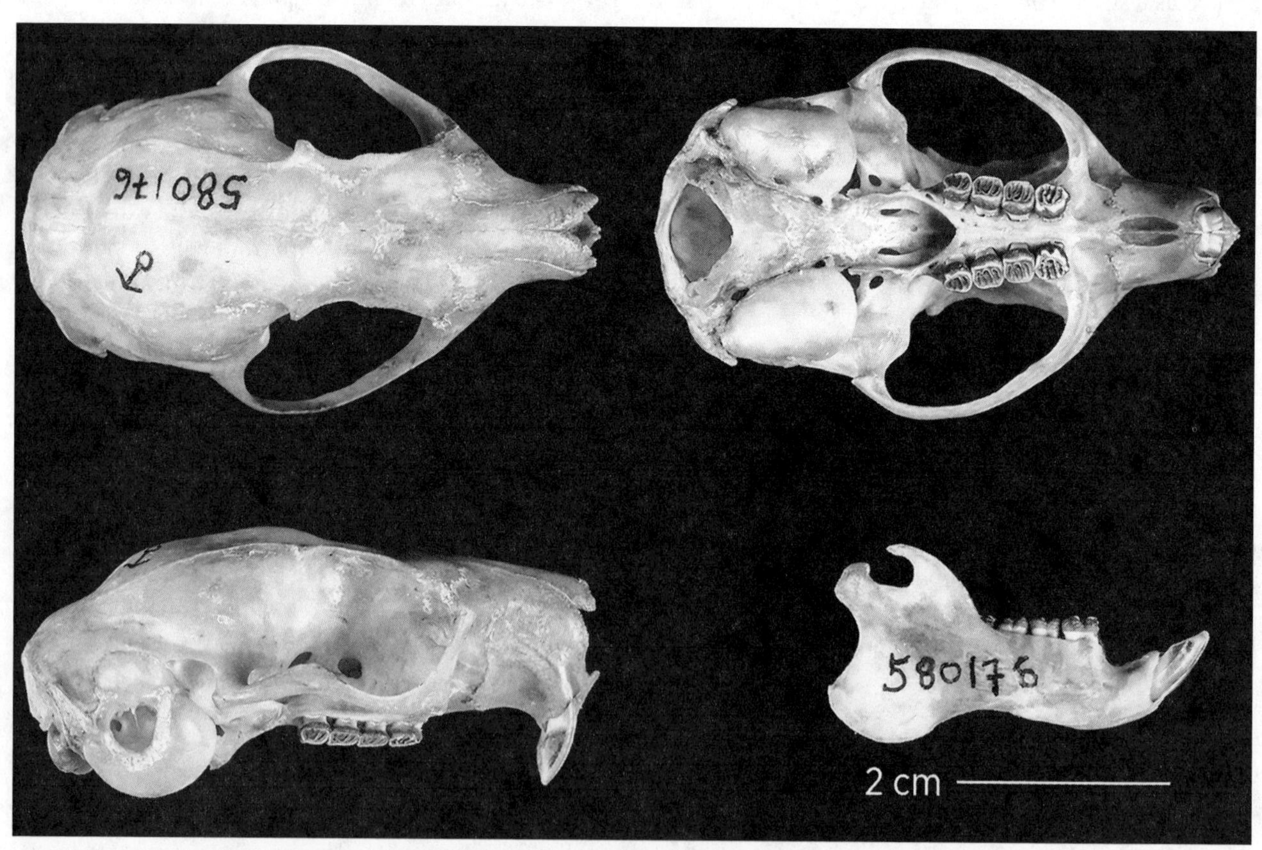

FIG. 226. Anomaluridae: Anomalurinae; skull and mandible of Lord Derby's scaly-tailed squirrel, *Anomalurus derbianus*.

DIAGNOSTIC CHARACTERS:

- **double row of scales on underside of tail at its base**
- **combination of large infraorbital canal with no zygomatic plate (hystricomorphous condition) but sciurognathous lower jaw**

RECOGNITION CHARACTERS:

1. postorbital process small
2. auditory bulla not inflated
3. **no zygomatic plate**
4. **infraorbital foramen large, oval**
5. angular process not inflated
6. occlusal surface of cheek teeth relatively flat, with distinct transverse ridges

DENTAL FORMULA: $\frac{1}{1}\ \frac{0}{0}\ \frac{1}{1}\ \frac{3}{3} = 20$

TAXONOMIC DIVERSITY: Seven species in three genera in two families, one with two subfamilies (see Heritage et al. 2016):

Anomaluridae:

Anomalurinae—*Anomalurus* (scaly-tailed squirrels)

Idiurinae—*Idiurus* (pygmy scaly-tailed flying squirrels)

Zenkerellidae: *Zenkerella* (Cameroon flightless scaly-tailed squirrels)

Family PEDETIDAE (spring hares or springhaas)

Size about that of a large rabbit, with mass up to 4 kg and total length 900 mm; externally modified for bipedal saltatorial locomotion, with heavy, brushy tail (500 mm) longer than head and body, shortened forelimbs, elongate hind limbs with enlarged thigh muscles, and elongate hind feet; **digits 5/4, brushy at tip**; 3rd digit of hind foot longest; well-developed digging claws on forefeet, hind claws flattened. **Pinnae large and presenting a distinct tragus.** Nocturnal and herbivorous. Semifossorial, excavating complex burrow system with forefeet. Live in arid or semiarid regions with deep, sandy soils.

RECOGNITION CHARACTERS:

1. **hind limbs much longer than forelimbs**
2. **frontal and nasal bones very broad**
3. **infraorbital foramen extremely large, oval**
4. **lower jaw sciurognathous**
5. postorbital process very small or absent
6. **auditory bulla greatly inflated,** with mastoid portion extending onto dorsal side of cranium
7. no zygomatic plate
8. angular process not inflected
9. **occlusal surface of cheek teeth flat, simple, bilophodont, with a single fold of enamel on the side of each tooth**

DENTAL FORMULA: $\frac{1}{1}\ \frac{0}{0}\ \frac{1}{1}\ \frac{3}{3} = 20$

TAXONOMIC DIVERSITY: The family is monotypic, with two species in the single genus *Pedetes.*

FIG. 227. Pedetidae; southern African spring hare, *Pedetes capensis.*

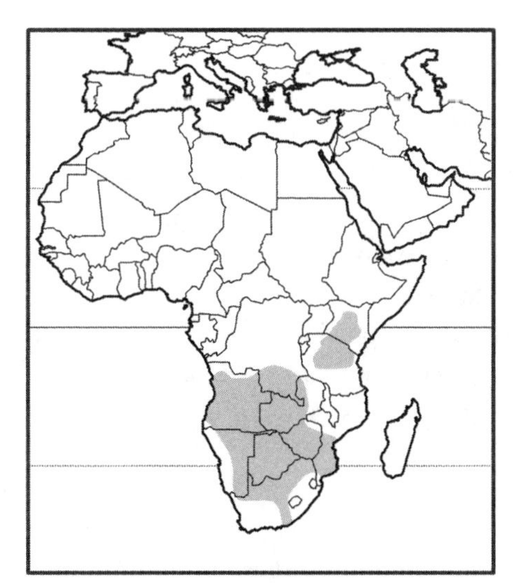

FIG. 228. Geographic range of family Pedetidae: east-central and southern Africa.

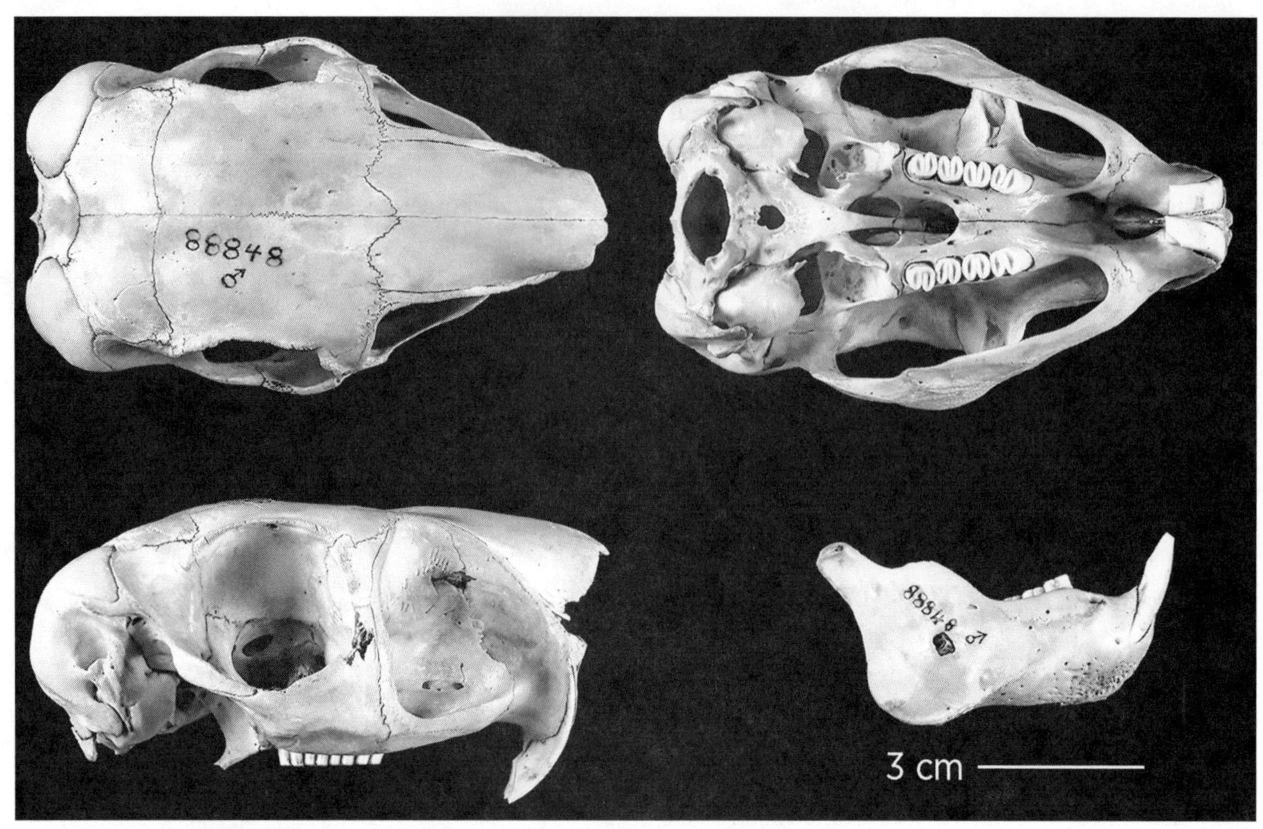

FIG. 229. Pedetidae; skull and mandible of a spring hare, *Pedetes capensis*.

Suborder HYSTRICOMORPHA (= Ctenohystrica)

RECOGNITION CHARACTERS:

1. infraorbital foramen very large (except in Bathyergidae), accommodating much of the medial masseter muscle (i.e., hystricomorphous) (fig. 230, arrow)
2. no zygomatic plate
3. cheek teeth usually 4/4 (one premolar, three molars), with four or five transverse crests primitively, or reduction to a minimum of two crests

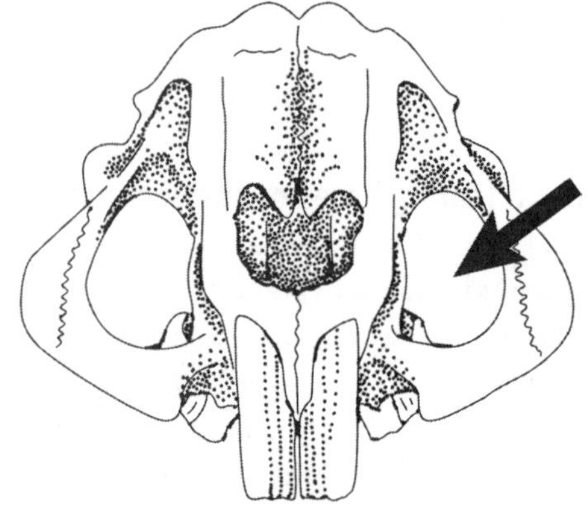

FIG. 230. Hystricomorph infraorbital foramen. Redrawn from Lawlor (1979).

PHYLOGENY AND CLASSIFICATION OF HYSTRICOMORPHA

The suborder Hystricomorpha is a monophyletic lineage, well supported by a number of morphological and molecular characters. DNA sequence analyses group the sciurognathous Ctenodactylidae and Diatomyidae as sister families in a hystricomorphous clade (Ctenodactylomorphi), which in turn is sister to the remaining hystricomorph families, all of which have hystricognathous lower jaws (Hystricognathi) (fig. 231). The Hystricognathi, in turn, have been traditionally divided into two lineages, the Old World Phiomorpha and the New World Caviomorpha. DNA sequence data, however, while strongly supporting the monophyly of the New World Caviomorpha, suggest that the Old World Phiomorpha is a paraphyletic group basal to the Caviomorpha. Relationships of New World porcupines (Erethizontoidea) to other caviomorph superfamilies are unclear. The tree depicted in figure 231 represents the most recent hypothesis of relationships among the Hystricomorpha.

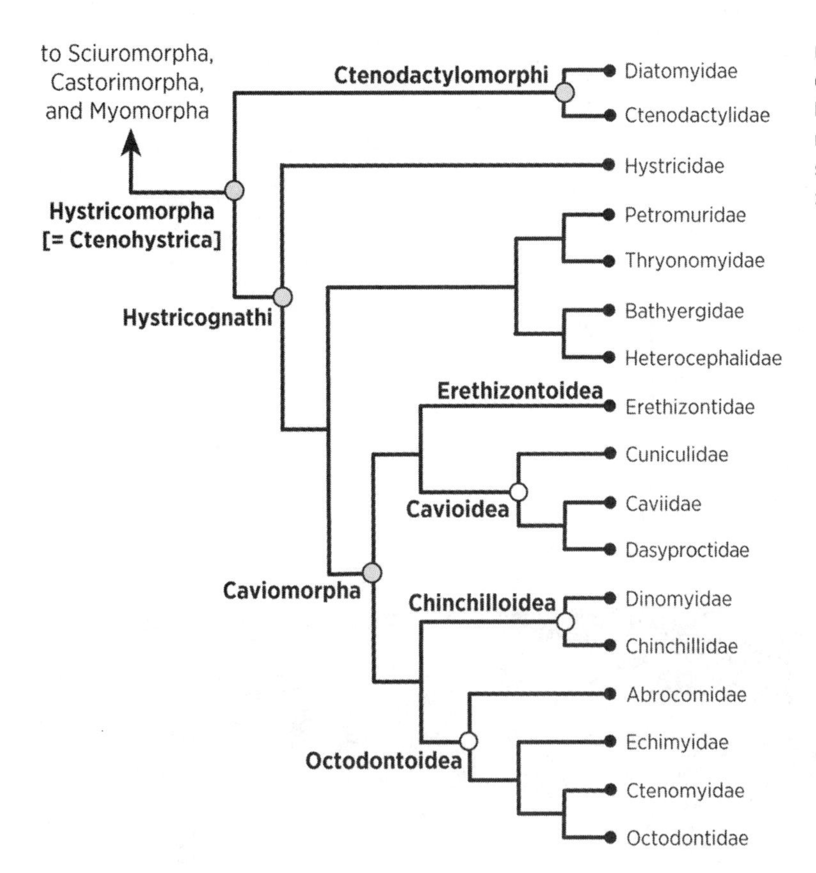

FIG. 231. Phylogenetic relationships among families of hystricomorph rodents; major clades and subfamily lineages are identified (based on the fully resolved mammalian tree presented in Gatesy et al. 2017, fig. 6; see also Fabre et al. 2012, 2015; Upham and Patterson 2015).

Infraorder CTENODACTYLOMORPHI

DIAGNOSTIC CHARACTER:

- **hystricomorphous jaw musculature but sciurognathous lower jaw**

Family CTENODACTYLIDAE (gundis)

Medium-sized (body length 160–240 mm); body form compact, with relatively large head, blunt nose, long vibrissae, short and rounded ears, short fluffy tail; four toes on both forefeet and hind feet; claws short, curved; inner two digits with comblike rows of specially shaped bristles; hind feet relatively narrow; hairless soles with ball-like friction pads on toes and both palmar and plantar surfaces; pelage soft, silky, and exceedingly dense. Resemble pikas (Lagomorpha: Ochotonidae) in general appearance, habits, and habitats occupied. Strictly herbivorous; diurnal, but seek shelter during hottest periods of the day; live in small colonies, and whistle when startled. Found in rocky outcrops in semiarid and arid regions with sparse vegetation.

RECOGNITION CHARACTERS:

1. skull broad, swollen, not ridged
2. infraorbital foramen very large, without distinct groove for maxillary nerve passage
3. lacrimal canal not opening on side of rostrum (an oblique canal is present on maxilla)
4. auditory bulla large, inflated
5. **paroccipital process broad, curving under and in contact with bulla**
6. **jugal in broad contact with lacrimal**
7. upper tooth rows roughly parallel
8. occlusal surface of cheek teeth relatively simple; upper teeth kidney- or figure 8–shaped, lower teeth with one or two inner enamel folds
9. **lower jaw with small ridge lateral and ventral to cheek teeth; angular process elongate but not deflected; no coronoid process**

DENTAL FORMULA: $\frac{1}{1}\ \frac{0}{0}\ \frac{1\text{–}2}{1\text{–}2}\ \frac{3}{3} = 20\text{–}24$

TAXONOMIC DIVERSITY: Four genera, one with two species (*Ctenodactylus*) and three that are monotypic (*Felovia*, *Massoutiera*, and *Pectinator*).

FIG. 232. Ctenodactylidae; common gundi, *Ctenodactylus gundi*.

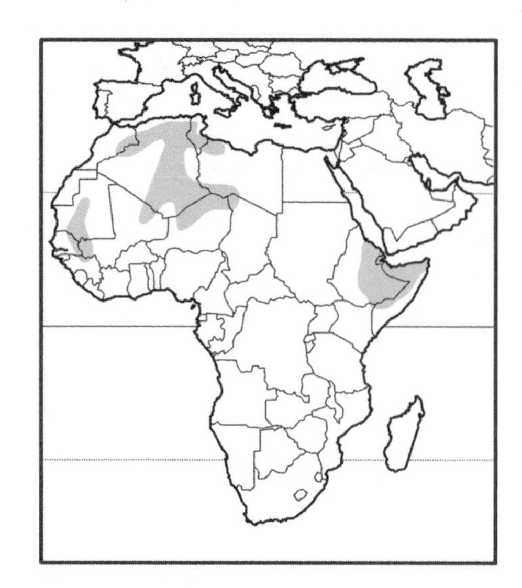

FIG. 233. Geographic range of family Ctenodactylidae: northern Africa, from Horn of Africa in the east to Senegal and Morocco in the west.

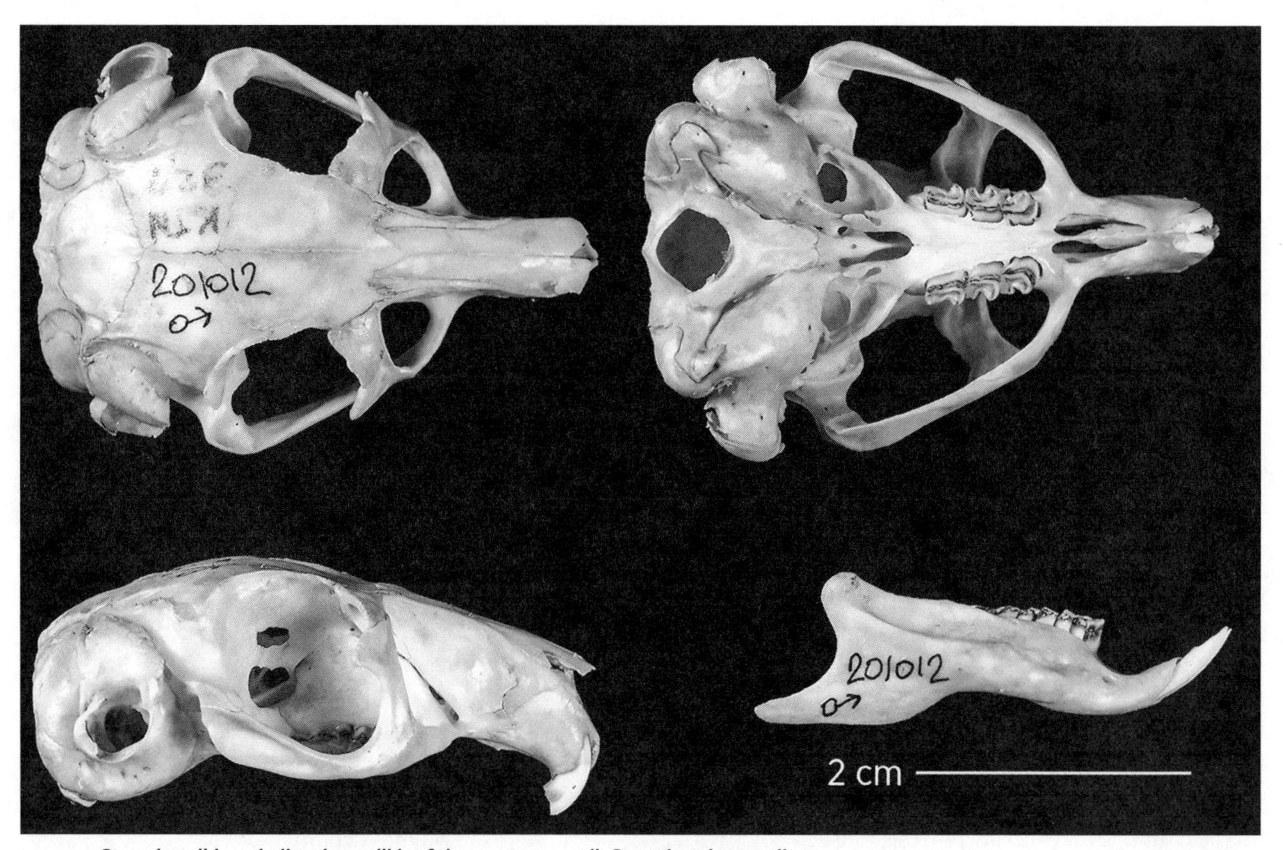

FIG. 234. Ctenodactylidae; skull and mandible of the common gundi, *Ctenodactylus gundi*.

Family DIATOMYIDAE (Laotian rock rat)

Medium-sized (body length 260 mm; mass 340 g) and ratlike in general appearance; ears rounded, medium-sized, with fringe of long hairs around entire margin; vibrissae robust and very long, extending to level of pectoral girdle if laid back; tail shorter than body (135 mm); hind feet short and broad; both forefeet and hind feet with noticeably large palmar-plantar pads; forefeet with four clawed toes and pollex with nail; hind feet with five toes, all with claws. Described in 2005 and initially placed in monotypic family Laonastidae (Jenkins et al. 2005); subsequently recognized as a living member of a widespread Asian family of otherwise extinct rodents, the Diatomyidae (Dawson et al. 2006).

DIAGNOSTIC CHARACTER:

- **pterygoid fossa secondarily reduced, foramina evident in young individuals, fused in older ones**

RECOGNITION CHARACTERS:

1. separate infraorbital neurovascular canal present
2. deep groove on ventral surface of maxillary part of zygomatic arch
3. upper cheek teeth diverge posteriorly
4. cheek teeth rooted, hypsodont
5. occlusal surface of cheek teeth moderately simple, each tooth with a pair of broad transverse laminae
6. upper and lower 3rd molars as large as 2nd

FIG. 235. Diatomyidae; Laotian rock rat, *Laonastes aenigmamus*.

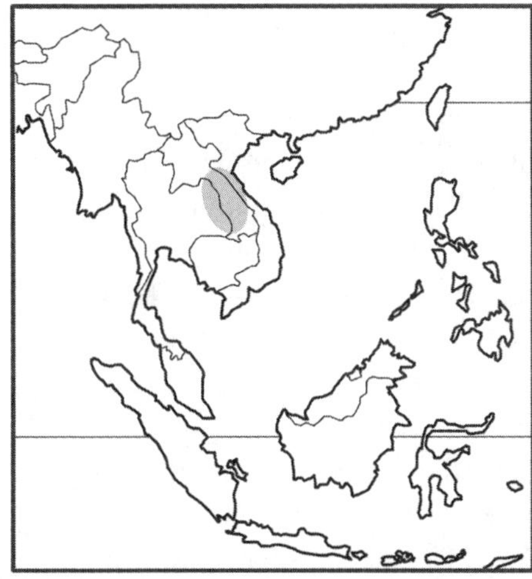

FIG. 236. Geographic range of family Diatomyidae: known only from a single region in central Laos in Southeast Asia.

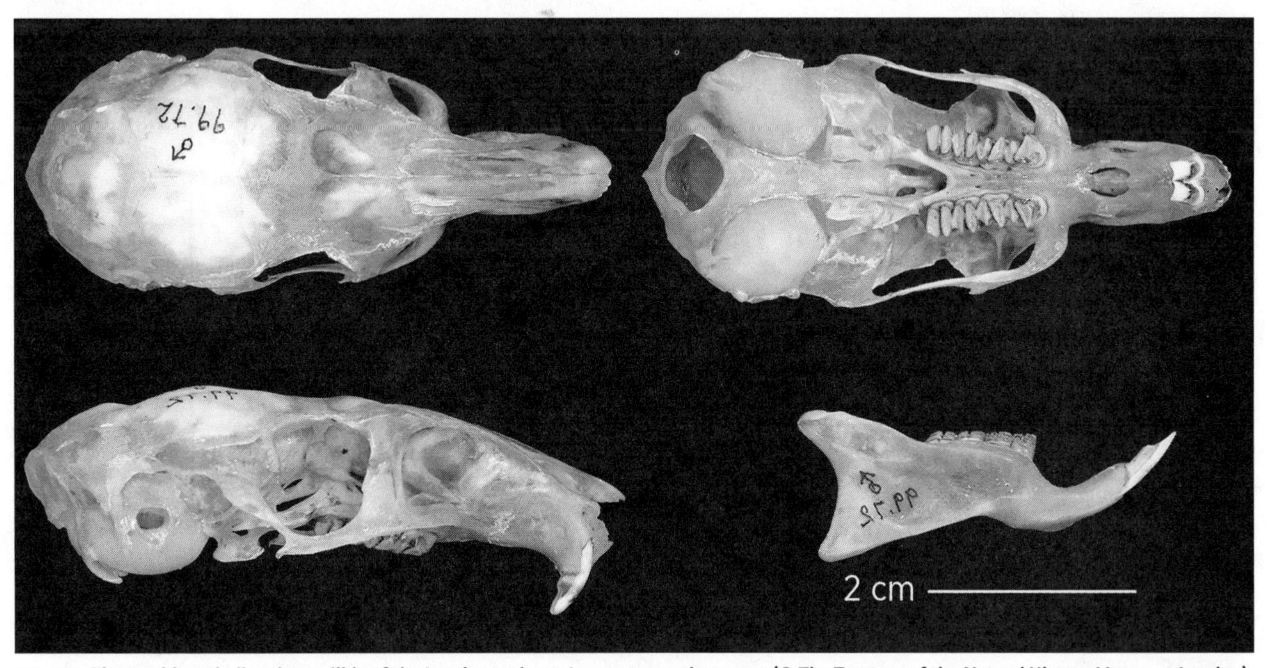

FIG. 237. Diatomyidae; skull and mandible of the Laotian rock rat, *Laonastes aenigmamus* (© The Trustees of the Natural History Museum, London).

DENTAL FORMULA: $\frac{1}{1}\ \frac{0}{0}\ \frac{1}{1}\ \frac{3}{3} = 20$

TAXONOMIC DIVERSITY: The sole living member of the family is *Laonastes aenigmamus*.

Infraorder HYSTRICOGNATHI

DIAGNOSTIC CHARACTER:

- **hystricomorphous jaw musculature** (except Bathyergidae and Heterocephalidae) **and hystricognathous lower jaw** (although it is less well developed in some)

Families BATHYERGIDAE and HETEROCEPHALIDAE (mole rats, blesmols, naked mole rat)

Small to medium-sized (body length 90–200 mm); body form typical of subterranean existence, with tubular body, small eyes, small ears, enlarged incisors, short to virtually absent tail, and short legs; most genera dig with enlarged, very procumbent incisors and hence have **small claws on forefeet** (*Bathyergus* is unique, with enlarged foreclaws and orthodont incisors); **digits 5/5**; fur short and soft, or reduced to a few scattered hairs in the naked mole rat *Heterocephalus*. Subterranean, excavating extensive burrow systems. Herbivorous, with social species tending to feed on individually very large but widely scattered geophytes. Some (*Bathyergus*, *Georhychus*, and *Heliophobius*) are solitary, with individuals occupying exclusive-use burrow systems; others (especially *Heterocephalus* but also *Cryptomys* and *Fukomys*) have a highly complex eusocial system comprising multiple individuals but a singular male-female mating pair. Inhabit sandy and loamy to very hard soils in arid regions, most common in dry steppes and savannas.

Although this group has been traditionally treated as a single family divided into two subfamilies, Patterson and Upham (2014) recently elevated the Heterocephalinae to family status due to the long temporal separation between *Heterocephalus* and the five genera of Bathyerginae. For convenience we present all six genera together, but we provide this new classification below.

DIAGNOSTIC CHARACTER:

- **small infraorbital foramen transmitting little or no muscle** (i.e., non-hystricomorphous)

RECOGNITION CHARACTERS:

1. skull blocky, usually heavily ridged
2. lacrimal canal not opening on side of rostrum
3. **auditory bulla relatively large, flattened**
4. paroccipital process short
5. **jugal widely separated from lacrimal**
6. lower jaw without ridge lateral and ventral to cheek teeth; **angular process strongly deflected**; coronoid process small to large
7. upper tooth rows approximately parallel
8. cheek teeth extremely hypsodont, but not open-rooted or ever-growing
9. **occlusal surface of cheek teeth simple, ringlike or figure 8–shaped**

DENTAL FORMULA: $\frac{1}{1}\ \frac{0}{0}\ \frac{2\text{–}3}{2\text{–}3}\ \frac{0\text{–}3}{0\text{–}3} = 12\text{–}28$

FIG. 238. Bathyergidae; Cape mole rat, *Georychus capensis*.

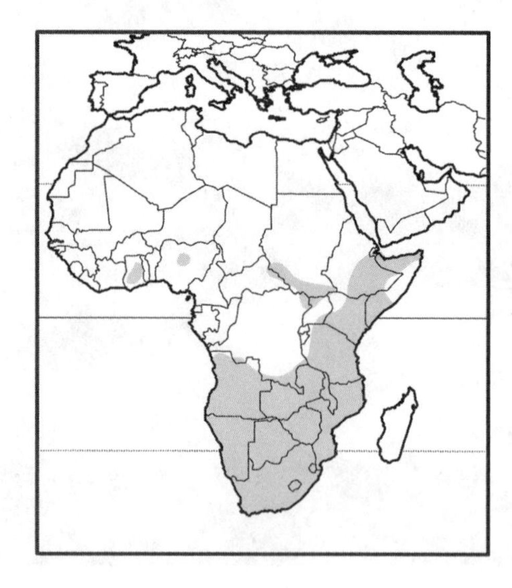

FIG. 239. Combined geographic range of families Bathyergidae and Heterocephalidae: Africa south of the Sahara, excluding Congo basin.

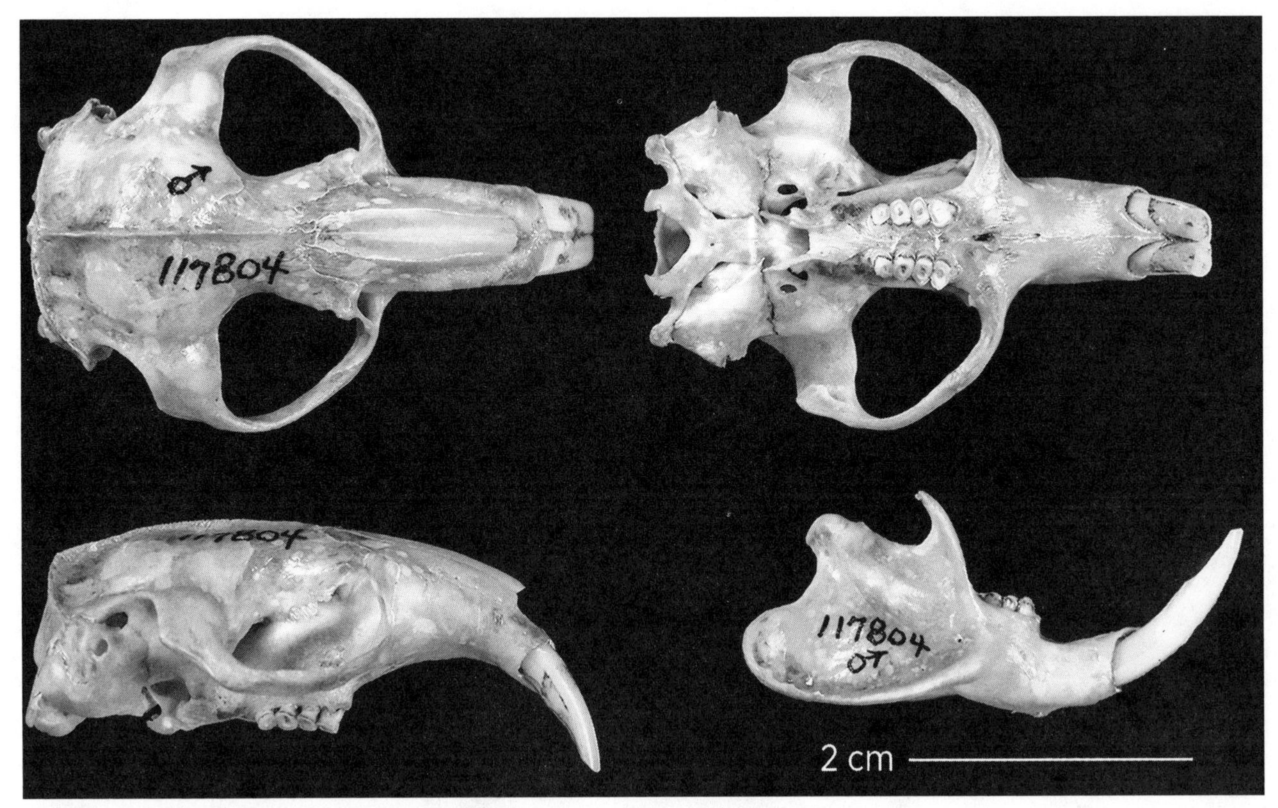

FIG. 240. Bathyergidae; skull and mandible of the common mole rat, or blesmol, *Cryptomys hottentotus*.

TAXONOMIC DIVERSITY: About 18 species in six genera now divided into two families:

Bathyergidae—*Bathyergus* (dune mole rats), *Cryptomys* (blesmol, common mole rat), *Fukomys* (common mole rats), *Georychus* (Cape mole rat), and *Heliophobius* (silvery mole rat)

Heterocephalidae—this family is monotypic, with a single species, *Heterocephalus glaber* (naked mole rat)

Family HYSTRICIDAE (Old World porcupines; brush-tailed porcupines)

Large, heavyset body (500–1,000 mm); tail varies from long, with terminal tuft of spines (*Atherurus*), to short, with broad, flat, still hairs (*Hystrix*); **pelage conspicuously spiny**, spines often modified into hollow, barbless quills; quills grouped into clusters of four to six quills; limbs short, feet plantigrade, **digits functionally 5/5, not webbed**. Terrestrial, semifossorial, nocturnal to crepuscular. Herbivorous, feeding on a variety of plant parts. Habitat ranges from deserts, savannas, and steppes to both temperate and tropical forests.

RECOGNITION CHARACTERS:

1. skull elongate, nasals greatly inflated and smooth (*Hystrix*), or ridged but not inflated (*Atherurus*)
2. enlarged infraorbital canal without distinct groove for nerve passage
3. lacrimal canal not opening on side of rostrum
4. auditory bulla relatively small
5. paroccipital process short
6. jugal variable in extent, but never meeting lacrimal
7. lower jaw without ridge or groove on lateral surface; coronoid process small (*Atherurus*) or prominent (*Hystrix*)
8. upper tooth rows approximately parallel
9. **occlusal surface of cheek teeth flat, with transverse folds of enamel forming islands with wear**

DENTAL FORMULA: $\frac{1}{1}\ \frac{0}{0}\ \frac{1}{1}\ \frac{3}{3} = 20$

TAXONOMIC DIVERSITY: 11 species in three genera: *Atherurus* (brush-tailed porcupines), *Hystrix* (porcupines), and *Trichys* (long-tailed porcupine).

FIG. 241. Hystricidae; crested porcupine, *Hystrix cristata*.

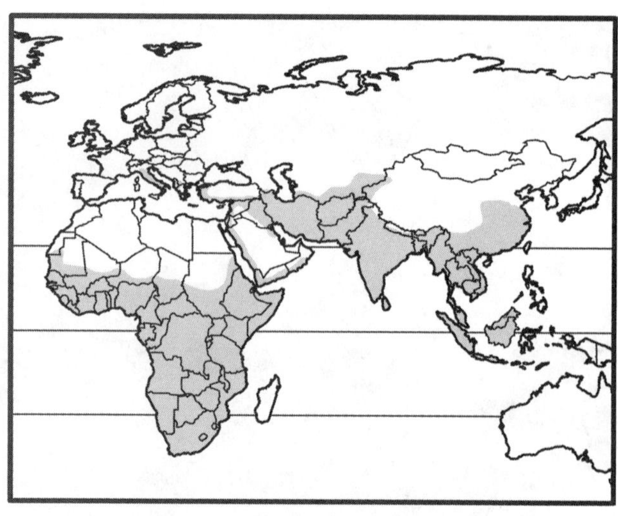

FIG. 242. Geographic range of family Hystricidae: southern Europe, sub-Saharan Africa, southwestern and southern Asia, Indonesia, Philippines.

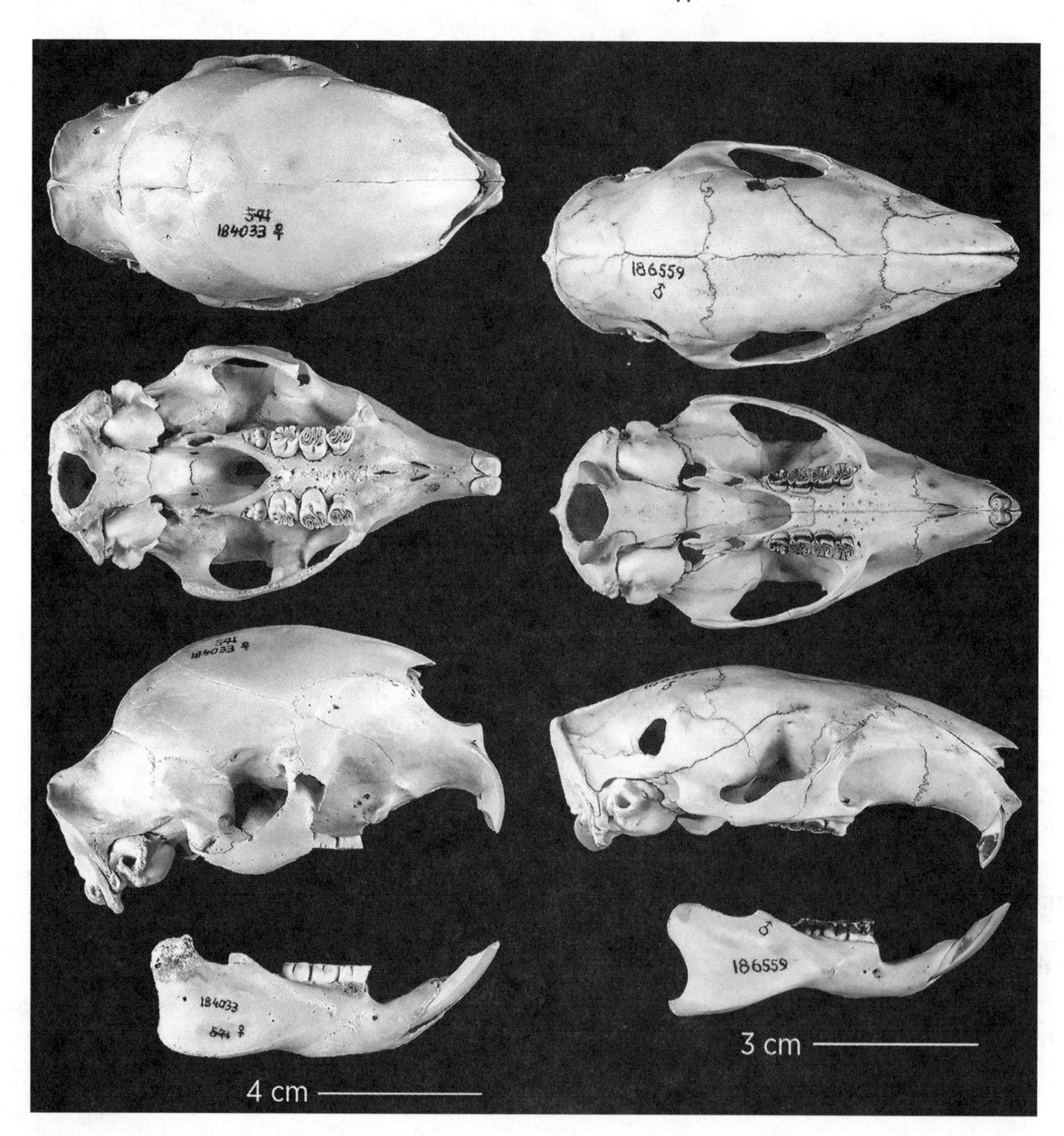

FIG. 243. Hystricidae; skull and mandible of the crested porcupine, *Hystrix cristata* (*left*), and the Asiatic brush-tailed porcupine, *Atherurus macrourus* (*right*).

Family PETROMURIDAE (dassie rat or rock rat)

Small (body length 145–190 mm) and **squirrel-like**, with moderately long limbs and well-developed but narrow feet with short claws and four functional digits (pollex vestigial, hallux short); stiff bristles on hind toes; **tail long and bushy**; ears short and rounded; **pelage soft and silky**. Live in groups and inhabit rocky areas in stony desert of Upper Karoo Plateau; herbivorous, feeding primarily on grasses, highly coprophagous.

FIG. 244. Petromuridae; dassie rat, *Petromus typicus*.

RECOGNITION CHARACTERS:

1. skull relatively broad, only slightly ridged
2. **large infraorbital foramen with distinct groove for nerve passage at inner base**
3. lacrimal canal not opening on side of rostrum
4. **auditory bulla large**
5. paroccipital process elongate, curved under bulla
6. jugal not approaching lacrimal
7. lower jaw without ridge or groove on lateral surface; coronoid process relatively small
8. upper tooth rows approximately parallel
9. cheek teeth strongly hypsodont but rooted
10. **occlusal surface of cheek teeth terraced: inner side of upper teeth and outer side of lower teeth elevated over rest of tooth surface; pattern simple, with one outer and one inner enamel fold per tooth**

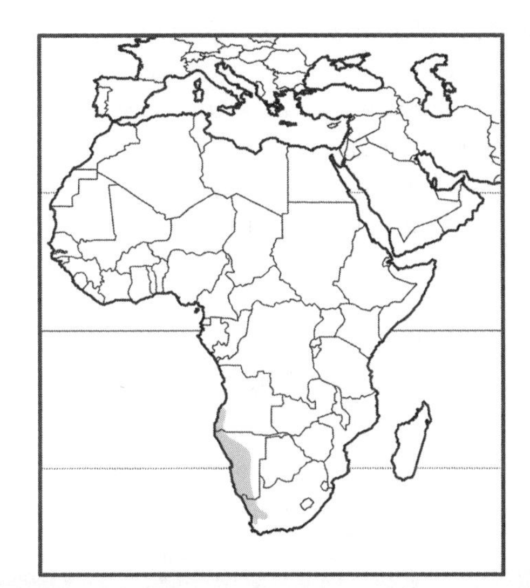

FIG. 245. Geographic range of family Petromuridae: southwestern Africa.

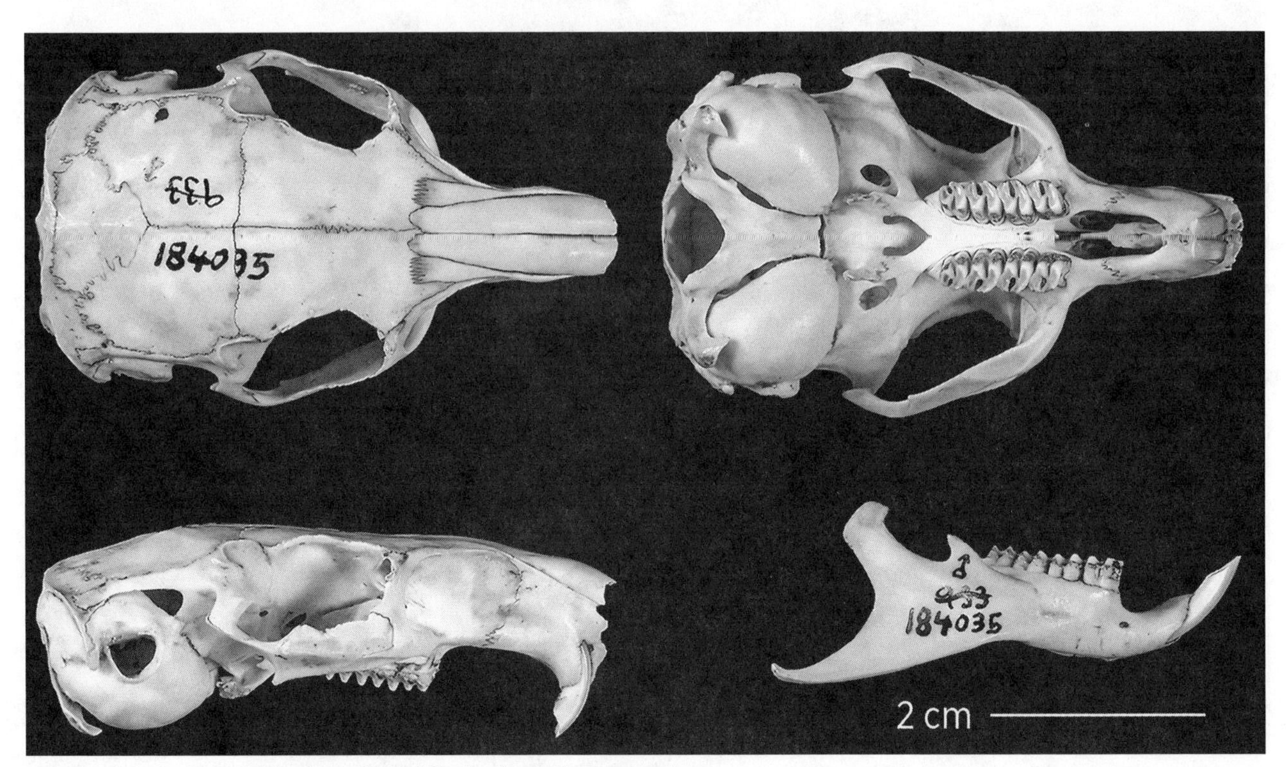

FIG. 246. Petromuridae; skull and mandible of the dassie rat, *Petromus typicus*.

DENTAL FORMULA: $\frac{1}{1}\ \frac{0}{0}\ \frac{1}{1}\ \frac{3}{3} = 20$

TAXONOMIC DIVERSITY: The family is monotypic, with a single species, *Petromus typicus* (rock rat or dassie rat).

FIG. 247. Thryonomyidae; cane rat, *Thryonomys swinderianus*.

Family THRYONOMYIDAE (cane rats or grass-cutters)

Large (body length 400–850 mm) and thickset; limbs short; ears short and rounded; tail short, scaly and bristly; **both forefeet and hind feet with three main digits**, hallux and pollex minute or lacking, 5th digit greatly reduced and lacks claws; claws thick and heavy; soles of feet naked; **pelage coarse and bristly.** Herbivorous; coprophagous. Live in small groups, foot thump and vocalize for communication. Occur in moist or swampy savannas and along coastal areas with dense grasslike vegetation and usually near water.

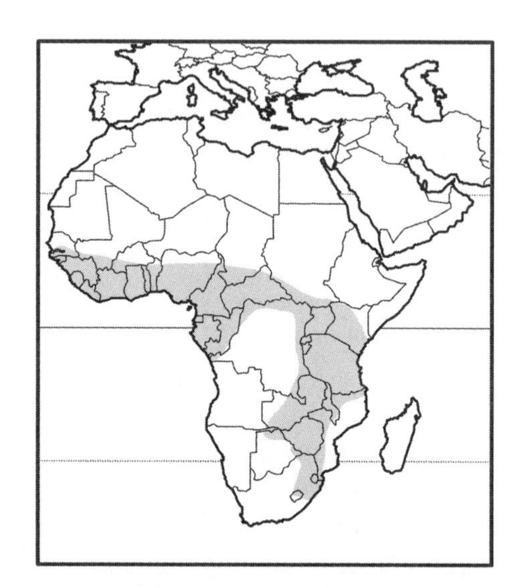

FIG. 248. Geographic range of family Thryonomyidae: sub-Saharan Africa.

DIAGNOSTIC CHARACTER:

- **upper incisors with three deep longitudinal grooves**

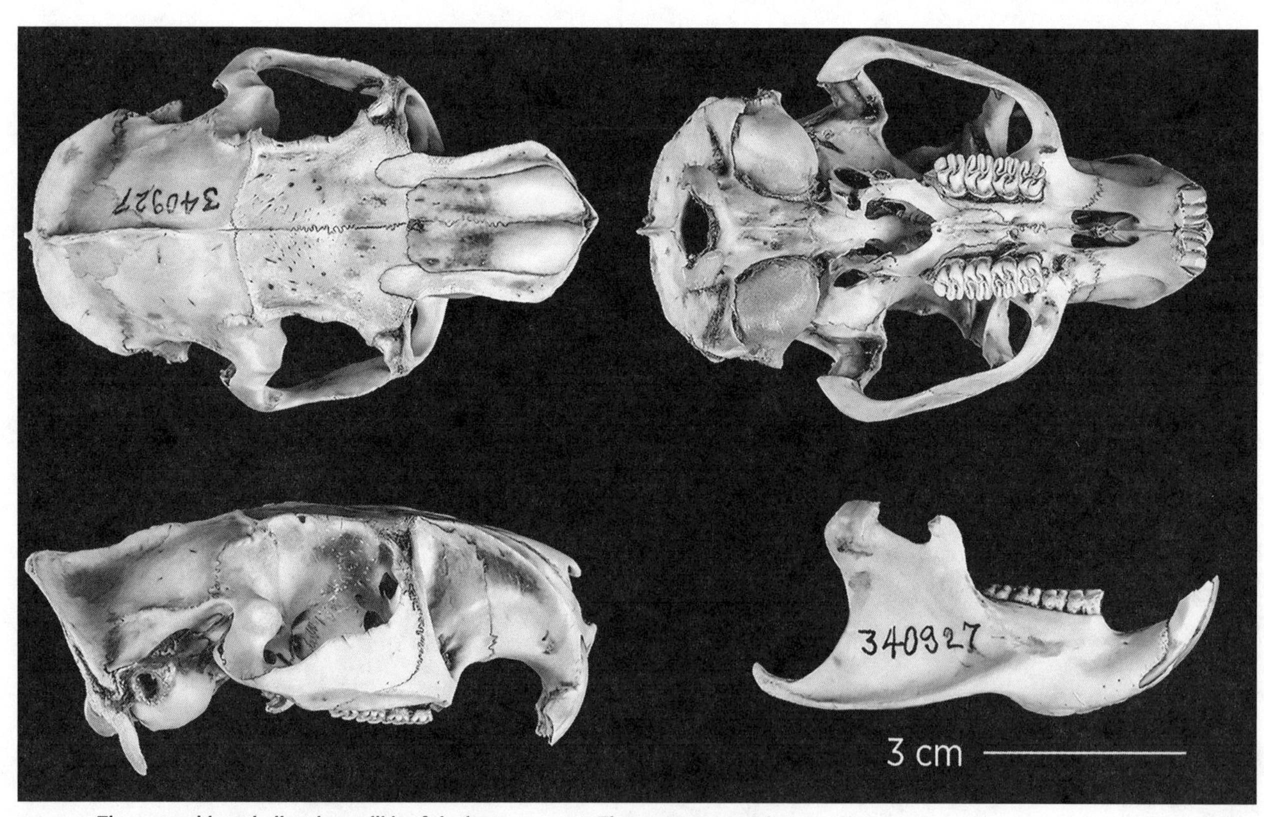

FIG. 249. Thryonomyidae; skull and mandible of the lesser cane rat, *Thryonomys gregorianus*.

RECOGNITION CHARACTERS:

1. **skull massive, strongly ridged**
2. **large infraorbital foramen with distinct groove for nerve passage at inner base**
3. lacrimal canal not opening on side of rostrum
4. **auditory bulla relatively small**
5. **paroccipital process relatively large and straight**
6. **jugal nearly in contact with lacrimal**
7. lower jaw without ridge or groove on lateral surface; angular process elongate; coronoid process prominent
8. upper tooth rows approximately parallel
9. **occlusal surface of cheek teeth with thick enamel folds that remain separate with wear**

DENTAL FORMULA: $\frac{1}{1}\ \frac{0}{0}\ \frac{1\text{–}2}{1\text{–}2}\ \frac{3}{3} = 20\text{–}24$

TAXONOMIC DIVERSITY: The family is monotypic, with two species in the single genus *Thryonomys.*

Clade CAVIOMORPHA (New World hystricognathous rodents)

The clade Caviomorpha, which includes the 10 families of New World hystricognathous rodents, is divided into four superfamilies. There are four hypotheses for the geographic origin of the New World Caviomorpha from an Old World Phiomorpha ancestral stock (fig. 250):

1. origin in Africa with dispersal across the mid-Atlantic to South America
2. origin in the Palearctic with dispersal across Eurasia into North America and then to South America
3. origin in Asia with dispersal across Africa and the mid-Atlantic to South America
4. origin in Asia with dispersal through Australia and Antarctica to South America

The phylogenetic hypothesis depicted in figure 231 supports the first alternative, but the ultimate origin of all hystricognathous rodents appears to have been in Asia.

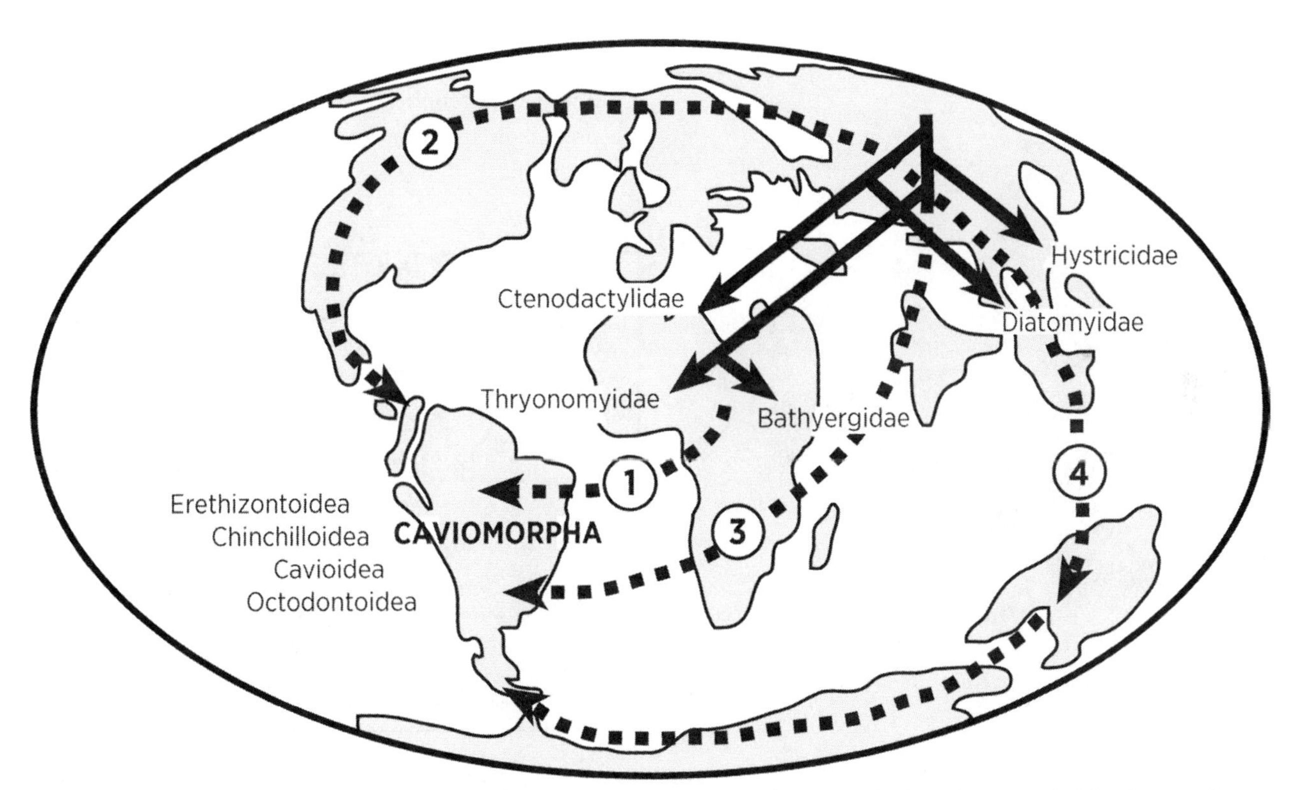

FIG. 250. Hypothesized pathways (numbered as in text) for the dispersal of New World Caviomorpha from the center of origin of ancestral Old World Phiomorpha. Solid lines illustrate dispersal routes of Old World families from putative area of origin in Asia; dashed lines are hypothesized alternative routes for origin of New World Caviomorpha. Updated and redrawn from Huchon and Douzery (2001).

Superfamily ERETHIZONTOIDEA
Family ERETHIZONTIDAE
(New World porcupines)

Medium-sized to large (body length 450–900 mm; mass of *Erethizon* up to 18 kg); heavyset; legs short; tail variable, from short (*Erethizon*, some *Coendou*) to long and prehensile (*Chaetomys*, most *Coendou*); **four functional digits on forefeet**, pollex reduced; **four or five digits on hind feet**, hallux reduced in some *Coendou* but present and clawed in *Erethizon*; **claws well developed** and soles of feet naked; **pelage coarse and long**, underfur usually present; **thick, barbed quills individually embedded in skin**; elongate guard hairs cover quills (*Erethizon*, some *Coendou*) or guard hairs absent (other *Coendou*). Arboreal and terrestrial (*Erethizon*) to highly arboreal (*Coendou*). Herbivorous, consuming wide range of plant parts; nocturnal, either solitary or found in pairs. Inhabit mixed-conifer forests and riparian communities in grasslands and deserts (*Erethizon*); lowland to montane tropical forests (*Coendou* and *Chaetomys*)

FIG. 251. Erethizontidae: Erethizontinae; North American porcupine, *Erethizon dorsatum*.

RECOGNITION CHARACTERS:

1. skull usually blocky, moderately ridged
2. infraorbital foramen very large and without distinct groove for nerve passage
3. lacrimal canal not opening on side of rostrum
4. auditory bulla relatively large
5. paroccipital process short
6. jugal not approaching lacrimal
7. lower jaw without ridge or groove on lateral surface; coronoid process prominent
8. upper tooth rows slightly convergent anteriorly
9. occlusal surface of cheek teeth flat, with narrow (*Chaetomys*) or wide enamel folds not usually separating from each other to form islands with wear

DENTAL FORMULA: $\frac{1}{1}\ \frac{0}{0}\ \frac{1}{1}\ \frac{3}{3} = 20$

TAXONOMIC DIVERSITY: 17 species in three genera allocated to two subfamilies:

Chaetomyinae—the monotypic *Chaetomys* (bristle-spined porcupine)

Erethizontinae—the monotypic *Erethizon* (North American porcupine) and the polytypic *Coendou* (prehensile-tailed, stump-tailed, and dwarf porcupines)

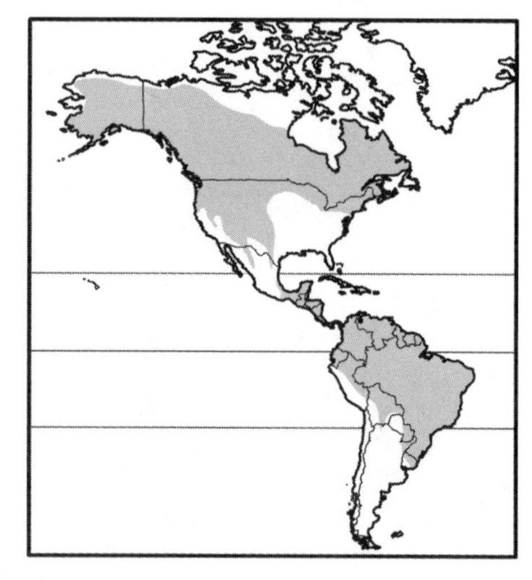

FIG. 252. Geographic range of family Erethizontidae: temperate North American forests (*Erethizon*); tropical forests of Central and South America (*Chaetomys, Coendou*).

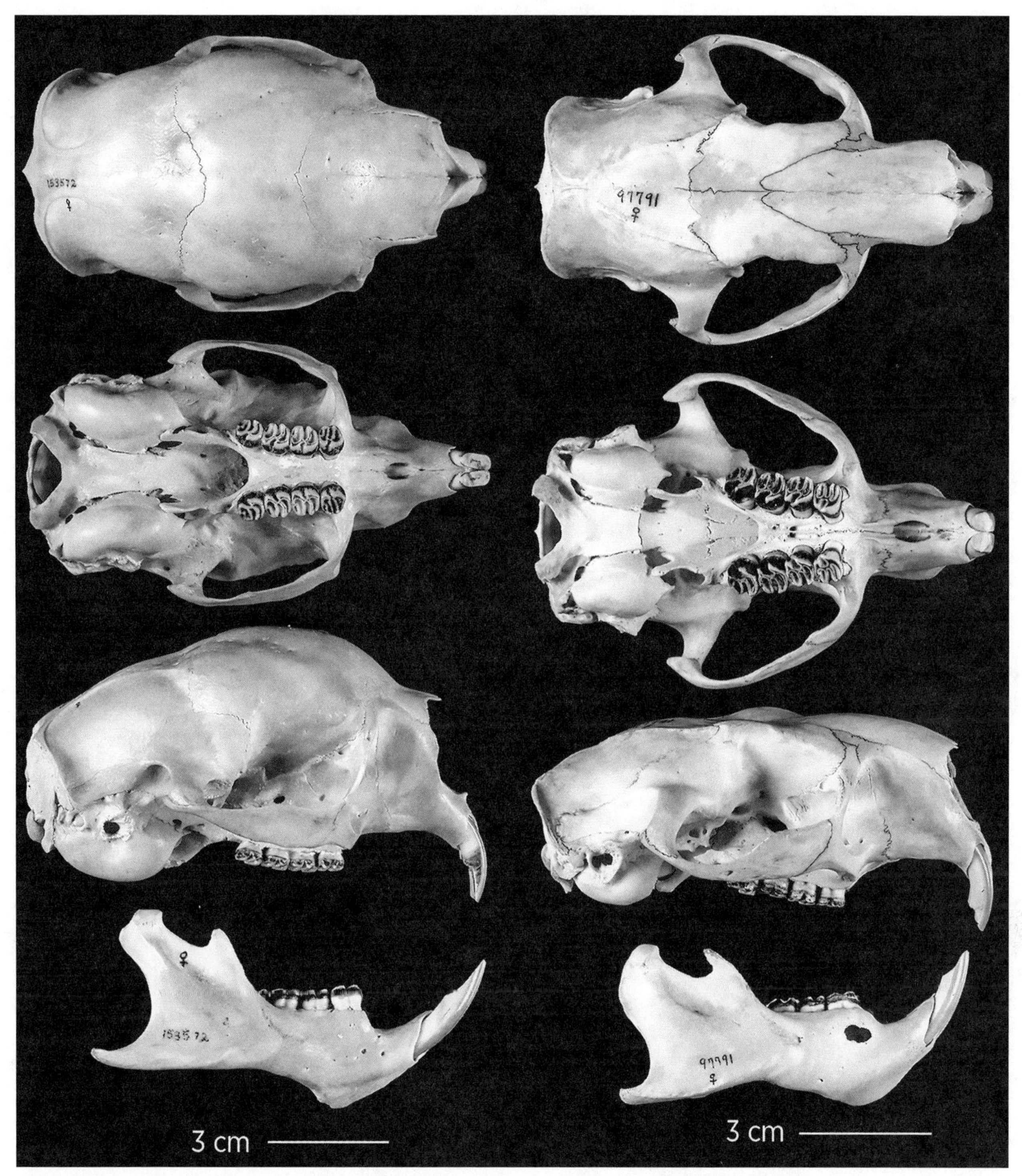

FIG. 253. Erethizontidae: Erethizontinae; skull and mandible of the Brazilian porcupine, *Coendou prehensilis* (*left*), and of the North American porcupine, *Erethizon dorsatum* (*right*).

Superfamily CHINCHILLOIDEA

Family CHINCHILLIDAE (chinchillas, viscachas)

Rabbitlike, with short forelimbs and elongate hind limbs, ears and eyes large; size variable, from small (*Chinchilla*, body length 380 mm, mass 500 g) to large (*Lagostomus*, body length 660 mm, mass up to 9 kg); tail variable, from short (*Chinchilla*) to long (*Lagidium*), always well furred; feet with fleshy pads; forefeet with four long digits; hind feet with three (*Lagostomus*) or four digits (*Chinchilla*, *Lagidium*); *Chinchilla* and *Lagostomus* with stiff bristles on hind toes; **pelage soft and dense**. Colonial, living in rocky outcrops, nesting among boulders (*Chinchilla*, *Lagidium*) or construct elaborate burrow systems occupied by family groups (*Lagostomus*). Diel activity variable; nocturnal (*Lagostomus*), crepuscular (*Chinchilla*), or diurnal (*Lagidium*). Occupy open habitats in the Andes, coastal plains, and grasslands.

RECOGNITION CHARACTERS:

1. skull elongate and ridged (*Lagostomus*) or relatively broad and with little or no ridging
2. infraorbital foramen very large, with (*Lagostomus*) or without distinct groove for nerve passage
3. **lacrimal canal with large opening on rostrum at front end of infraorbital canal**
4. auditory bulla small (*Lagostomus*) to very large
5. paroccipital process long (*Lagostomus*) or short and bound to bulla
6. **jugal approaching** (*Chinchilla*) **or in contact with lacrimal**
7. **lower jaw without ridge or groove on lateral surface; angular process elongate and not deflected; coronoid process prominent**
8. upper tooth rows convergent anteriorly
9. **occlusal surface of cheek teeth flat, consisting of a series of transverse plates**

DENTAL FORMULA: $\frac{1}{1}\ \frac{0}{0}\ \frac{1}{1}\ \frac{3}{3} = 20$

TAXONOMIC DIVERSITY: Six species in three genera allocated to two subfamilies:

Chinchillinae—five species in two genera, *Chinchilla* (common chinchillas) and *Lagidium* (mountain viscachas)

Lagostominae—the monotypic *Lagostomus maximus* (plains viscacha)

FIG. 254. Chinchillidae: Lagostominae; plains viscacha, *Lagostomus maximus*.

FIG. 255. Geographic range of family Chinchillidae: South America, high Andes (from southern Ecuador to southern Chile and Argentina) and lowland plains of Argentina.

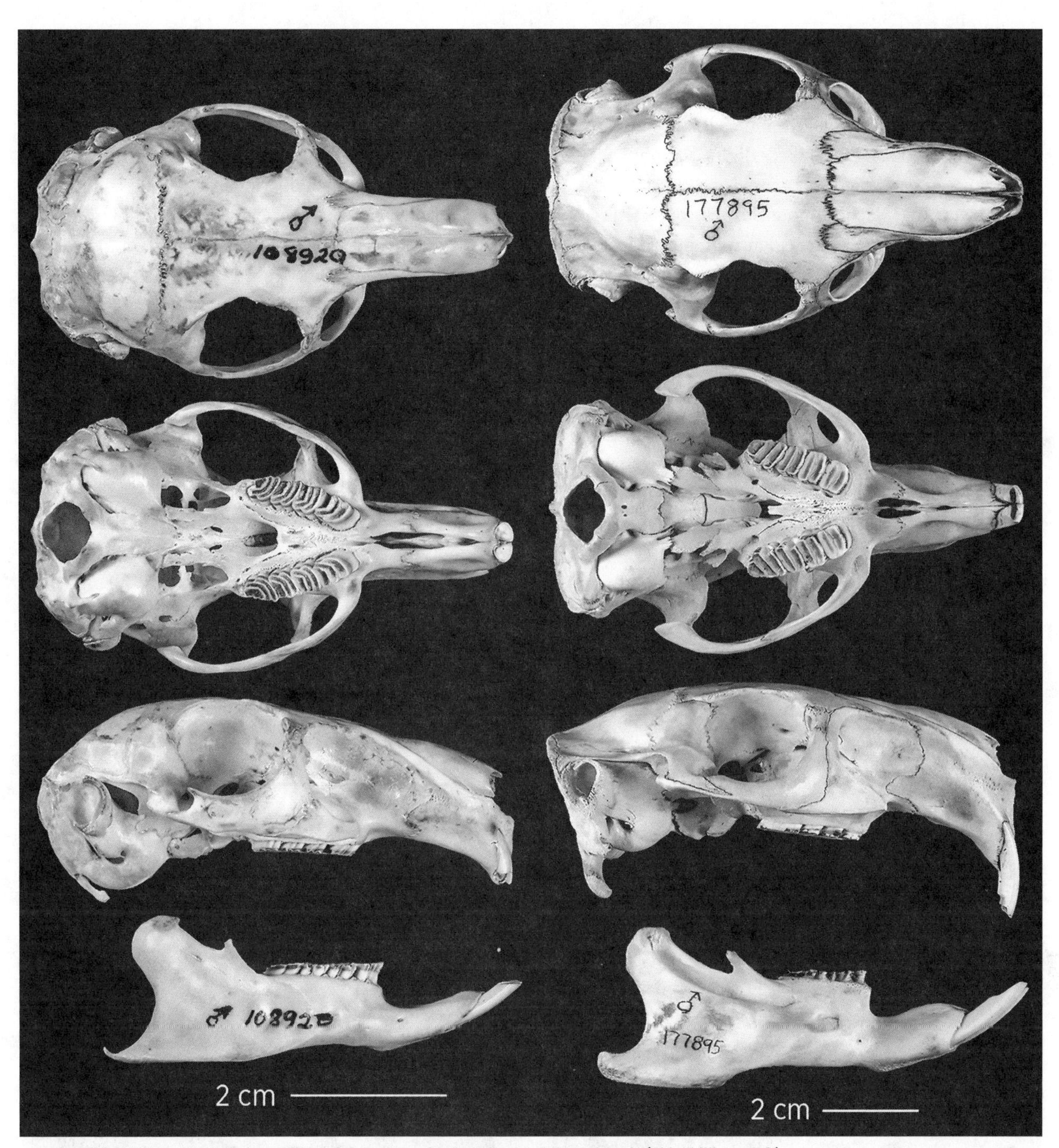

FIG. 256. Chinchillidae; skull and mandible of the mountain viscacha, *Lagidium viscacia* (Chinchillinae; *left*), and the plains viscacha, *Lagostomus maximus* (Lagostominae; *right*).

Family DINOMYIDAE (pacarana)

Large and heavyset (body length up to 1,000 mm; mass to 15 kg); ears short and curved; upper lip with deep cleft; vibrissae long; tail short and well haired; feet plantigrade, broad, soles naked; **digits 4/4**, not webbed, long claws on each digit; fur coarse and short, color pattern with two discontinuous stripes on back and series of spots on each side. Nocturnal; herbivorous; capable of climbing but primarily terrestrial; communicate via forefoot thumping, tooth chattering, and variety of vocalizations. Inhabit montane tropical forest on Andean slopes and in western Amazon basin.

RECOGNITION CHARACTERS:

1. skull blocky, not heavily ridged
2. infraorbital foramen very large, without distinct groove for nerve passage

FIG. 257. Dinomyidae; pacarana, *Dinomys branickii*.

FIG. 258. Geographic range of family Dinomyidae: eastern Andean slopes and western Amazon basin of tropical South America.

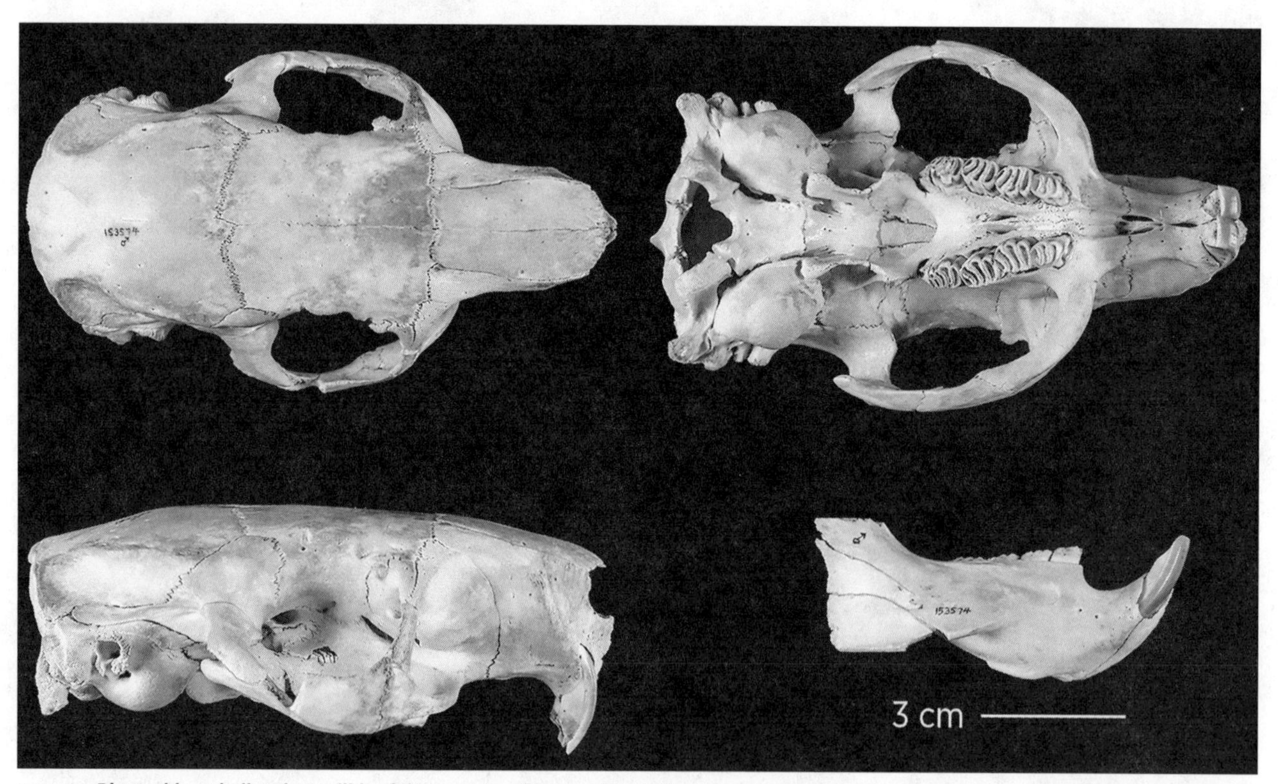

FIG. 259. Dinomyidae; skull and mandible of the pacarana, *Dinomys branickii*.

3. lacrimal canal not opening on side of rostrum
4. auditory bulla moderately large
5. paroccipital process short
6. jugal not approaching lacrimal
7. lower jaw without ridge or groove on lateral surface; **angular process strongly deflected; coronoid process very small or absent**
8. upper tooth rows strongly convergent anteriorly
9. **occlusal surface of cheek teeth flat, consisting of a series of four transverse plates**

DENTAL FORMULA: $\frac{1}{1}\ \frac{0}{0}\ \frac{1}{1}\ \frac{3}{3} = 20$

TAXONOMIC DIVERSITY: The family is monotypic, with a single species, *Dinomys branickii.*

FIG. 260. Caviidae; Brazilian guinea pig, *Cavia aperea* (Caviinae; *upper left*), Patagonian mara, *Dolichotis patagonum* (Dolichotinae; *upper right*), and capybara, *Hydrochoerus hydrochaeris* (Hydrochoerinae; *bottom*).

Superfamily CAVIOIDEA

Family CAVIIDAE (guinea pigs or cavies, Patagonian mara, moco, and capybara)

Size variable, from small (cavies up to 600 g) to large (*Dolichotis* up to 16 kg) to huge (*Hydrochoerus*, largest living rodent, up to 50 kg); body form variable (cavies stocky with short limbs, *Dolichotis* hare-like with elongate limbs, *Hydrochoerus* relatively short for size of body); all with **short tails;** ears short in caviines and *Hydrochoerus*, long in *Dolichotis*; digits 4/3; claws present in cavies, otherwise with blunt (*Kerodon*) or hooflike nails (*Dolichotis*, *Hydrochoerus*); feet partially webbed in semiaquatic *Hydrochoerus*. Typically diurnal or crepuscular; most are colonial or occur in large assemblages. Habitat of most ranges from open low-elevation grasslands and high Andean puna, temperate steppes, rocky outcrops; *Hydrochoerus*, convergent on African hippopotamids (Cetartiodactyla; Hippopotamidae), is found in a variety of mesic conditions, including riverbanks, streams, lakes, and swamps.

FIG. 261. Geographic range of family Caviidae: temperate South America, including high Andes but excluding Amazon basin and most of Chile (Caviinae and Dolichotinae); tropical lowlands from Panama south on both sides of the Andes to Argentina and Uruguay (Hydrochoerinae).

RECOGNITION CHARACTERS:

1. skull elongate (*Dolichotis*) or relatively short and broad, with moderate to no ridging; **massive** in Hydrochoerinae
2. infraorbital foramen very large, without distinct groove for nerve passage
3. **lacrimal canal with small opening on side of rostrum at front edge of infraorbital canal** (no opening in *Dolichotis*)
4. auditory bulla relatively large

5. paroccipital process prominent, slightly curved, free from bulla; **very long (longest of all rodents)** and slightly curved anteriorly in Hydrochoerinae
6. jugal not approaching lacrimal
7. **upper tooth rows strongly to weakly convergent anteriorly**
8. **lower jaw with prominent ridge and groove on lateral surface; angular process elongate but not deflected**; coronoid process prominent
9. cheek teeth hypsodont
10. occlusal surface of cheek teeth flat and prismatic with single major sharp, angular fold of enamel (*Cavia*, *Dolichotis*, *Kerodon*), or with **multiple angular folds, especially on 3rd molars** (*Hydrochoerus*)

DENTAL FORMULA: $\frac{1}{1}\ \frac{0}{0}\ \frac{1}{1}\ \frac{3}{3} = 20$

TAXONOMIC DIVERSITY: 20 species in six genera divided into three subfamilies:

Caviinae—*Cavia* (guinea pigs, cavies), *Galea* (yellow-toothed cavies), and *Microcavia* (mountain cavies)

Dolichotinae—*Dolichotis* (Patagonian and Chacoan maras or hares)

Hydrochoerinae—*Hydrochoerus* (capybaras) and *Kerodon* (mocos or rock cavies)

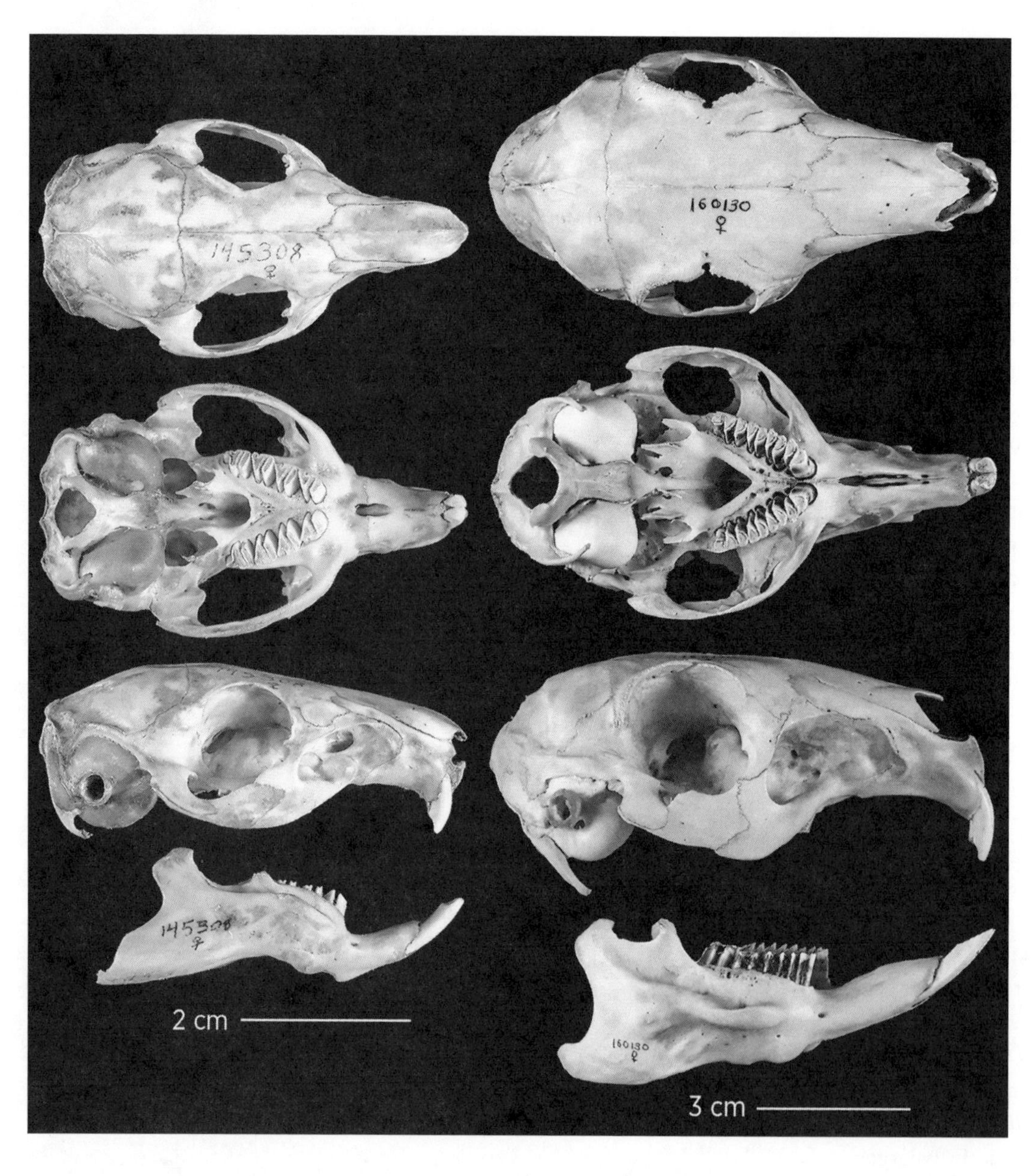

FIG. 262. Caviidae; skull and mandible of the Brazilian guinea pig, *Cavia aperea* (Caviinae; *left*), and the Patagonian mara, *Dolichotis patagonum* (Dolichotinae; *right*).

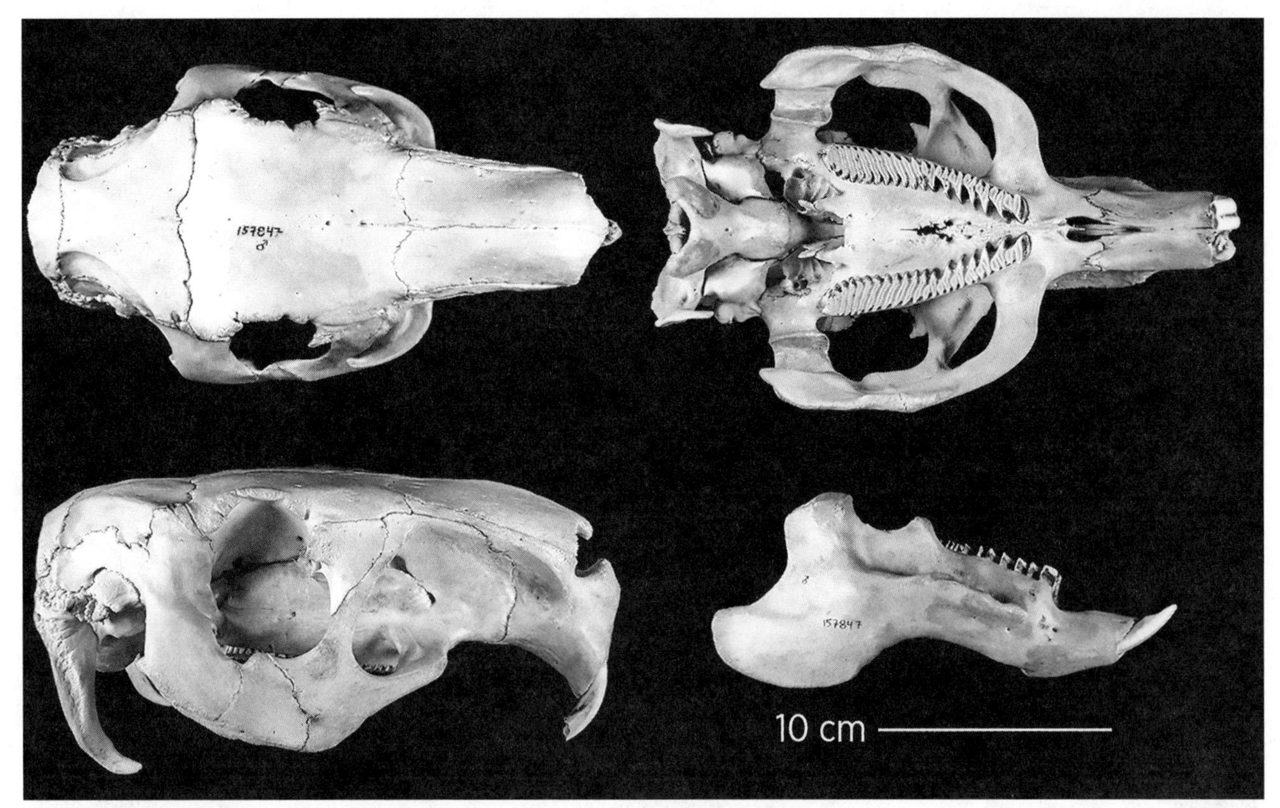

FIG. 263. Caviidae: Hydrochoerinae; skull and mandible of the capybara, *Hydrochoerus hydrochaeris*.

Family CUNICULIDAE (= Agoutidae) (pacas)

Large, with heavyset body, **short limbs**, and broad head (body length to 850 mm, mass to 12 kg); eyes large and situated dorsally; tail extremely short; **digits 4/5 but functionally 4/3** (reduced hallux and 5th toe); claws thick and hooflike; toes not webbed. Pelage coarse, without underfur, with **four longitudinal rows of pale spots**; internal cheek pouches associated with extremely enlarged zygomatic arches forming cheekplates (fig. 266). Terrestrial, nocturnal, semifossorial. Inhabit heavily wooded areas along tropical streams and rivers; live in pairs in self-dug burrows. Herbivorous, feeding on fruits, nuts, stems, and leaves.

DIAGNOSTIC CHARACTER:

- **hypertrophied zygomatic arch with corrugated jugal and maxilla; skull extremely robust**

RECOGNITION CHARACTERS:

1. **infraorbital foramen very large, with a distinct groove for nerve passage at inner base**
2. lacrimal canal not opening on side of rostrum
3. auditory bulla relatively small
4. paroccipital process relatively long
5. jugal not approaching lacrimal
6. lower jaw without ridge or groove on lateral surface; angular process deflected; coronoid process moderately prominent
7. upper tooth rows more or less parallel
8. occlusal surface of cheek teeth flat, consisting of transverse folds of enamel that separate from each other to form islands in adults

DENTAL FORMULA: $\frac{1}{1}\ \frac{0}{0}\ \frac{1}{1}\ \frac{3}{3} = 20$

TAXONOMIC DIVERSITY: The family is monotypic, with two species in the single genus *Cuniculus* (lowland and mountain pacas).

FIG. 264. Cuniculidae; paca, *Cuniculus paca*.

FIG. 265. Geographic range of family Cuniculidae: tropical Americas, from central Mexico to southern Brazil.

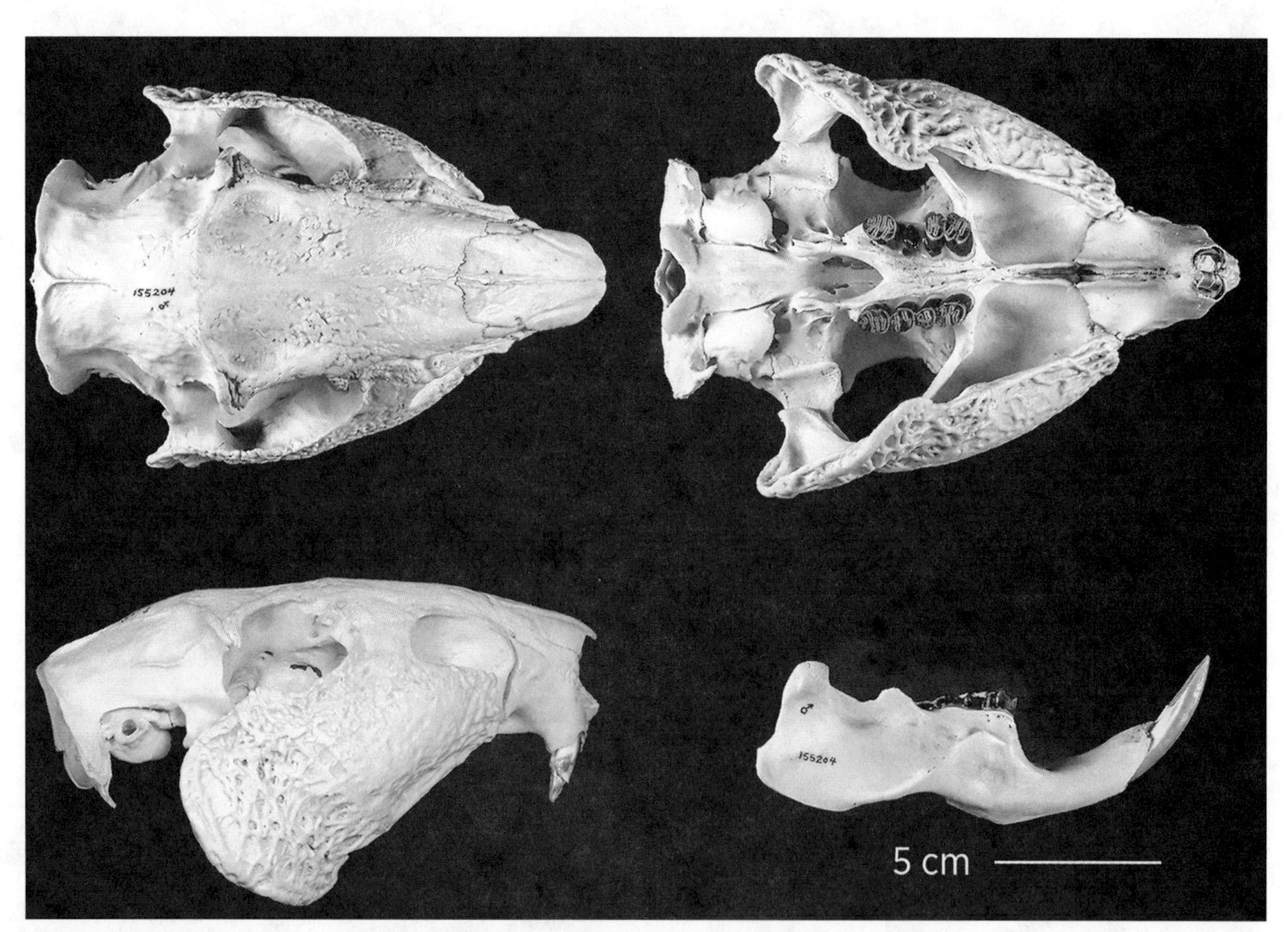

FIG. 266. Cuniculidae; skull and mandible of the lowland paca, *Cuniculus paca*.

Family DASYPROCTIDAE (agoutis, acouchis)

Small (*Myoprocta*, with body length 300–390 mm, mass up to 1.5 kg) to medium-sized (*Dasyprocta*, with body length 450–760 mm, mass up to 6 kg); body slender; limbs elongate with thin distal parts; tail very short and barely visible (*Dasyprocta*) or somewhat longer and clearly visible (*Myoprocta*). Digits functionally 4/3 (pollex knob-like), but 5th front digit reduced; middle digit on hind feet longest; primary digits on both feet with hooflike claws; locomotion digitigrade; soles of feet naked. Pelage coarse, thick, and appears glossy, with hairs over rump elongate. Diurnal; cursorial, run fast and jump well; generally solitary, constructing burrows along banks or under rocks or root masses. Herbivorous, specializing on fruits and large nuts, leaves, and stems. Habitat includes grassy stream banks and thick brush in lowland rainforest, gallery forest, and most Andean slopes.

FIG. 267. Dasyproctidae; Central American agouti, *Dasyprocta punctata*.

FIG. 268. Geographic range of family Dasyproctidae: tropical North (from southern Mexico), Central, and South America and Lesser Antilles.

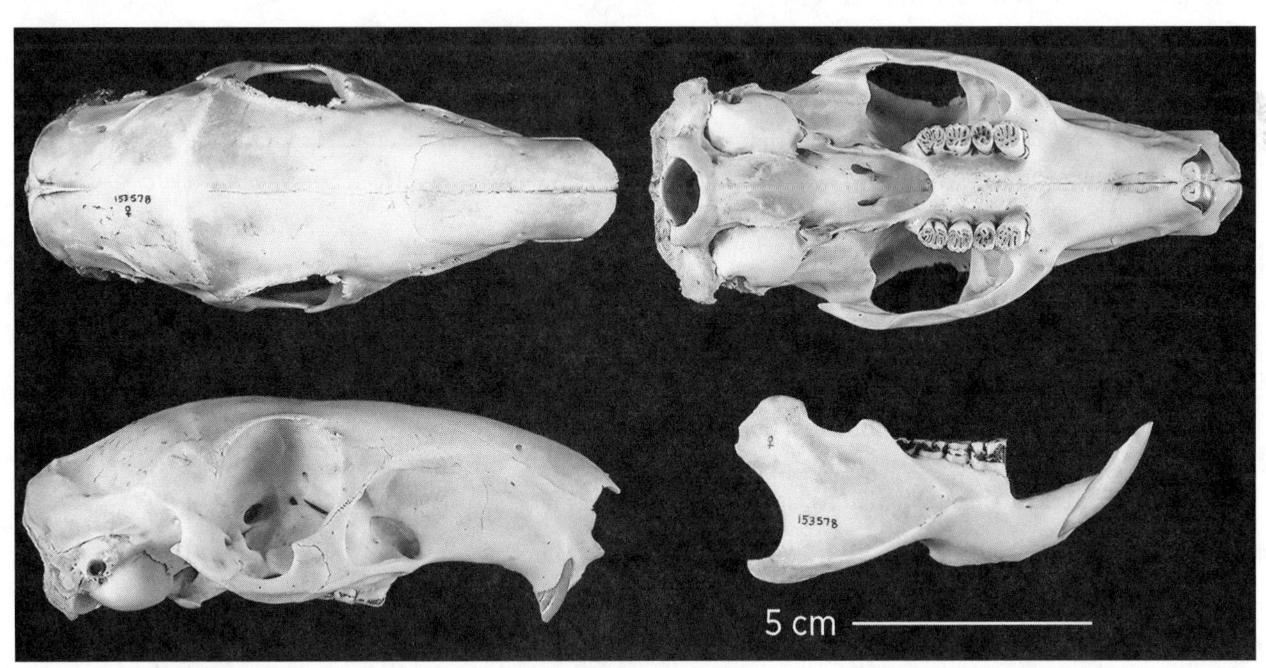

FIG. 269. Dasyproctidae; skull and mandible of the black agouti, *Dasyprocta fuliginosa*.

RECOGNITION CHARACTERS:
1. **skull elongate, not heavily ridged**
2. infraorbital foramen very large, without a distinct groove for nerve passage
3. **lacrimal canal with large opening on rostrum at front edge of infraorbital foramen**
4. auditory bulla relatively large
5. paroccipital process relatively short
6. jugal not approaching lacrimal
7. lower jaw without ridge or groove on lateral surface; angular process deflected; coronoid process prominent
8. upper tooth rows more or less parallel
9. cheek teeth semirooted but noticeably hypsodont
10. occlusal surface of cheek teeth flat, consisting of transverse folds of enamel that separate to form islands with age

DENTAL FORMULA: $\frac{1}{1}\ \frac{0}{0}\ \frac{1}{1}\ \frac{3}{3} = 20$

TAXONOMIC DIVERSITY: Two genera: *Dasyprocta* (agoutis), with about 13 species, and *Myoprocta* (acouchis), with 2 species.

Superfamily OCTODONTOIDEA

This caviomorph superfamily unites four families that span much of the Neotropical realm and include a correspondingly high diversity of species.

Family ABROCOMIDAE (chinchilla rats)

Body ratlike, with short limbs, large head with pointed muzzle, large eyes, large and rounded ears; *Cuscomys* larger (length >300 mm) than *Abrocoma* (150–250 mm); tail moderate in length, well furred; forefeet with four digits, hind feet with five; claws weak; soles naked and with small tubercles; **stiff hairs project over claws** of middle three digits of hind feet; pelage long and soft, underfur dense. Live in burrows or in rock crevices; *Abrocoma* may be colonial. Mostly nocturnal but may be active in daytime. Inhabit coastal hills and montane thickets, rocky areas in Andean highlands, and cloud forest edge.

RECOGNITION CHARACTERS:
1. **skull elongate, not ridged**
2. infraorbital foramen very large, without distinct groove for nerve passage
3. **lacrimal canal with large opening on side of rostrum at front edge**
4. **auditory bulla large**
5. paroccipital process short, curved under and bound to bulla
6. **jugal widely separated from lacrimal**
7. lower jaw without ridge or groove on lateral surface; **angular process elongate and only slightly deflected; coronoid process small**
8. upper tooth rows subparallel
9. **occlusal surface of cheek teeth flat, upper teeth with one inner and one outer angular enamel fold, lower teeth with two inner and one outer folds**

DENTAL FORMULA: $\frac{1}{1}\ \frac{0}{0}\ \frac{1}{1}\ \frac{3}{3} = 20$

TAXONOMIC DIVERSITY: About 10 species divided into two genera: *Abrocoma* (chinchilla rats) and *Cuscomys* (arboreal chinchilla rats or Inca rats).

FIG. 270. Abrocomidae; Bennett's chinchilla rat, *Abrocoma bennettii*.

FIG. 271. Geographic range of family Abrocomidae: mostly high Andes of South America, from southern Peru to central Chile and Argentina; *Abrocoma bennettii* to coastal north-central Chile.

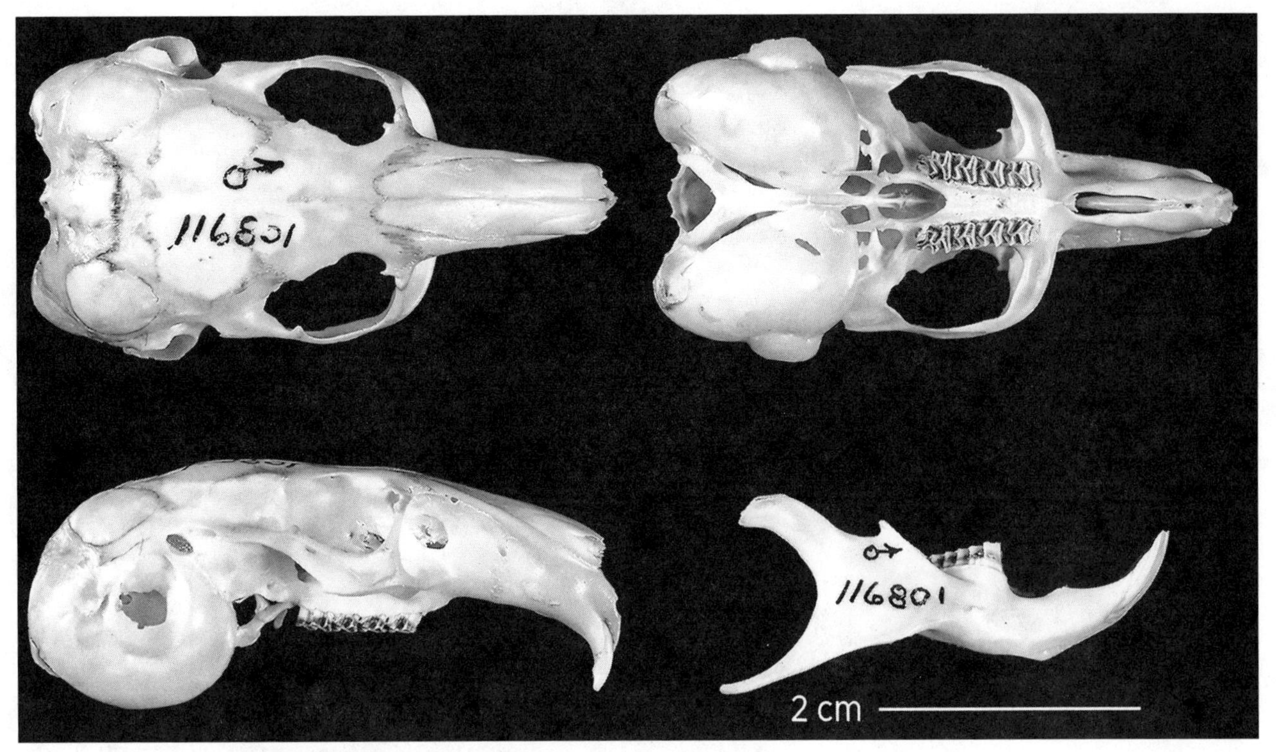

FIG. 272. Abrocomidae; skull and mandible of the ashy chinchilla rat, *Abrocoma cinerea*.

Family ECHIMYIDAE (hutias, spiny rats, tree rats, bamboo rats, coypu)

Echimyids are highly variable in body form and ecology. Most species are small to medium-sized (body length varies from 100 to 700 mm; mass up to 800 g), but two are large (the coypu, *Myocastor*, is >5 kg, and the hutia, *Capromys*, ranges up to 7 kg). Body form ratlike, but members vary depending on habitat and locomotory mode. Arboreal specialists (such as *Dactylomys* and *Echimys*) have elongate and either brushy or naked tails, short and broad hind feet with long toes, and naked palmar and plantar surfaces with or without large pads; terrestrial (*Proechimys*, *Trinomys*) to semifossorial species (*Clyomys* and *Euryzygomatomys*) have tails shorter than body length and narrow, elongate hind feet. All members have forefeet with four digits and hind feet with five. Many species have heavily spinose fur, others have non-spinose coarse to soft fur. *Myocastor* is large, robust, and aquatic, with a long, round, and sparsely haired tail; hind feet have four webbed digits, and dense underfur is covered by long guard hairs. All are herbivorous; some arboreal members are specialized folivores; terrestrial forms eat mostly seeds, nuts, and even some insects. Some, like the bamboo rat (*Dactylomys*), are highly vocal, emitting loud and distinctive calls carrying long distances in the tropical forest. All are nocturnal. Most live in lowland tropical forests, either on the ground or in trees; some extend to upper montane forests; several genera extend across the drier savanna regions of central Brazil, Paraguay, and Bolivia, and some of these genera specialize on rocky habitats. The Antillean hutias (Capromyinae) range from coastal mangroves to high-elevation forests.

The family is now divided into four subfamilies with the formal placement of the monotypic genus *Carterodon* into its own subfamily (Carterodontinae; Courcelle et al. 2019). Two subfamilies are further subdivided into tribal units. The inclusion of the capromyines and mycastorines, both traditionally considered separate families, is confirmed by comprehensive molecular phylogenetic analyses (e.g., Fabre et al. 2012, 2016; Upham and Patterson 2015; Courcelle et al. 2019).

Subfamily CAPROMYINAE (hutias)

Living members vary from rat-sized (*Mesocapromys nanus*, body length 200 mm) to the size of a large house cat (*Capromys pilorides*, up to 7 kg). Head is large; eyes and ears small, body stout; limbs short; tail nearly vestigial (*Geocapromys*) to long and prehensile (*Mysateles prehensilis*); feet plantigrade and pentadactyl (pollex may be reduced in size); fur thick, coarse to soft, but not spiny. Habits are variable—*Geocapromys* are totally terrestrial and feed on roots, tubers, and leaves; *Plagiodontia* feed on ground and in trees on a range of plant parts; some *Capromys* are arboreal folivores that live in tree cavities; terrestrial forms live in rock crevices. Most nocturnal; usually solitary. Habitat is forested areas with rocky soils and mangrove swamps on the edges of some islands.

FIG. 273. Echimyidae: Capromyinae: Plagiodontini; Hispaniolan hutia, *Plagiodontia aedium*.

RECOGNITION CHARACTERS:

1. skull blocky, ridged
2. infraorbital foramen very large, without distinct groove for nerve passage
3. **lacrimal canal not opening on side of rostrum**
4. auditory bulla moderately large
5. **paroccipital process relatively long, well separated from bulla**
6. jugal not approaching lacrimal
7. lower jaw without ridge or groove on lateral surface; angular process deflected; coronoid process prominent

FIG. 274. Geographic range of subfamily Capromyinae: Lesser and Greater Antilles; extant species known only from Cuba, Hispaniola, Jamaica, and the Bahamas (arrows); extirpated from Cayman Islands, Virgin Islands, and Puerto Rico.

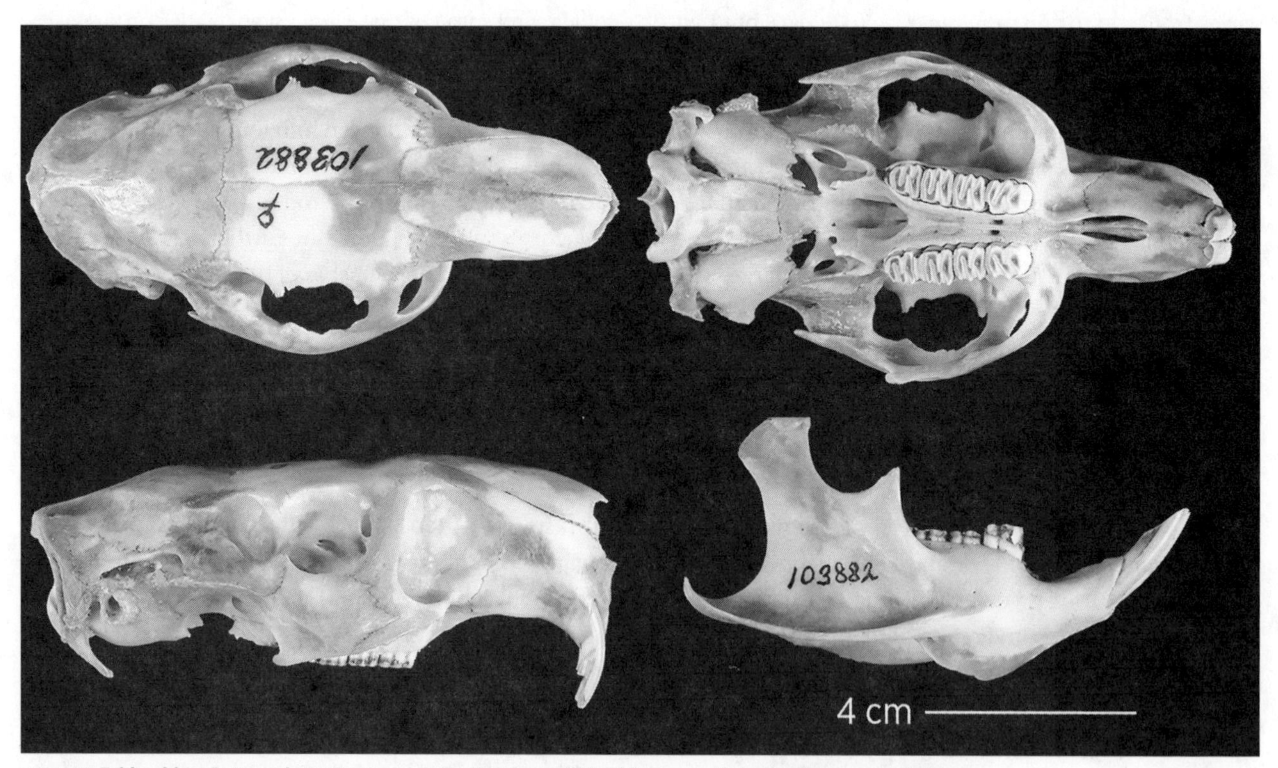

FIG. 275. Echimyidae: Capromyinae: Capromyini; skull and mandible of Desmarest's hutia, *Capromys pilorides*.

8. upper tooth rows convergent anteriorly
9. occlusal surface of cheek teeth flat, with single large inner and two outer enamel folds that separate from each other to form islands with age

DENTAL FORMULA: $\frac{1}{1}\frac{0}{0}\frac{1}{1}\frac{3}{3} = 20$

TAXONOMIC DIVERSITY: Nine species in seven genera, three of which are now extinct, divided into four tribes, two of which are extinct:

Capromyini—*Capromys* (one species in Cuba), *Geocapromys* (two species, one in Jamaica and one in the Bahamas), *Mesocapromys* (five species in Cuba), and *Mysateles* (three species in Cuba)

†Hexolobodontini—*Hexolobodon* (Haiti, one species, extinct)

†Isolobodontini—*Isolobodon* (Haiti, two species, extinct)

Plagiodontini—*Plagiodontia* (Hispaniola, four species, three extinct) and *Rhizoplagiodontia* (Haiti, one species, extinct)

Subfamily ECHIMYINAE (bamboo rats, armored rat, coypu, punarés, spiny rats, tree rats, toros)

The Echimyinae includes all non-Caribbean echimyid lineages except the three members of the Euryzygomatomyinae and Carterodontinae; it includes all arboreal specialists, all terrestrial forms (except *Trinomys*), and the semiaquatic *Myocastor*. This large and highly diverse assemblage is difficult to diagnose, and its membership has been substantially modified by recent molecular phylogenetic studies (Fabre et al. 2016).

RECOGNITION CHARACTERS:

1. **skull elongate, slightly to heavily ridged**
2. infraorbital foramen very large, with or without distinct groove on floor for nerve passage
3. **lacrimal canal not opening on side of rostrum**
4. auditory bulla moderately large
5. **paroccipital process elongate, curving under bulla**
6. jugal not approaching lacrimal
7. lower jaw without ridge or groove on lateral surface; angular process slender and deflected; coronoid process relatively small (e.g., *Dactylomys*) to moderately large (e.g., *Proechimys*)
8. upper tooth rows more or less parallel, except in *Dactylomys* or *Myocastor*, in which they diverge posteriorly
9. cheek teeth rooted but crowns hypsodont
10. occlusal surface of cheek teeth flat, variable in pattern, having rounded enamel folds that separate from each other to form islands with wear (e.g., *Proechimys*), sharply angular enamel folds forming prisms (e.g., *Dactylomys*), or transverse plates (e.g., *Makalata*)

FIG. 276. Echimyidae: Echimyinae; Atlantic bamboo rat, *Kannabateomys amblyonyx* (Echimyini; *upper left*), armored rat, *Hoplomys gymnurus* (Myocastorini; *upper right*), and coypu or nutria, *Myocastor coypus* (Myocastorini; *bottom*).

FIG. 277. Geographic range of subfamily Echimyinae: tropical and subtropical forests of Central America (southern Nicaragua) to central South America (southern Brazil); *Myocastor* in wetland habitats in south-central Chile and Argentina, northward to Bolivia, Paraguay, Uruguay, and southern Brazil (introduced to northern South America, North America, Europe, central Asia, and eastern Africa).

DENTAL FORMULA: $\frac{1}{1}\ \frac{0}{0}\ \frac{1}{1}\ \frac{3}{3} = 20$

TAXONOMIC DIVERSITY: Over 75 species in 18 genera divided into two tribes, each with multiple distinct lineages:

Echimyini—*Dactylomys* (bamboo rats), *Diplomys* (rufous tree rats), *Echimys* (tree rats), *Isothrix* (brush-tailed rats), *Kannabateomys* (Atlantic bamboo rat), *Lonchothrix* (tufted-tailed spiny tree rat), *Makalata* (armored tree rats), *Mesomys* (spiny tree rats), *Olallamys* (Olalla rats); *Pattonomys* (speckled tree rats), *Phyllomys* (Atlantic tree rats), *Santamartamys* (red-crested tree rat), and *Toromys* (Amazonian tree rats)

Myocastorini—*Callistomys* (painted tree rat), *Hoplomys* (armored rat), *Myocastor* (coypu or nutria), *Proechimys* (terrestrial spiny rats), and *Thrichomys* (punaré)

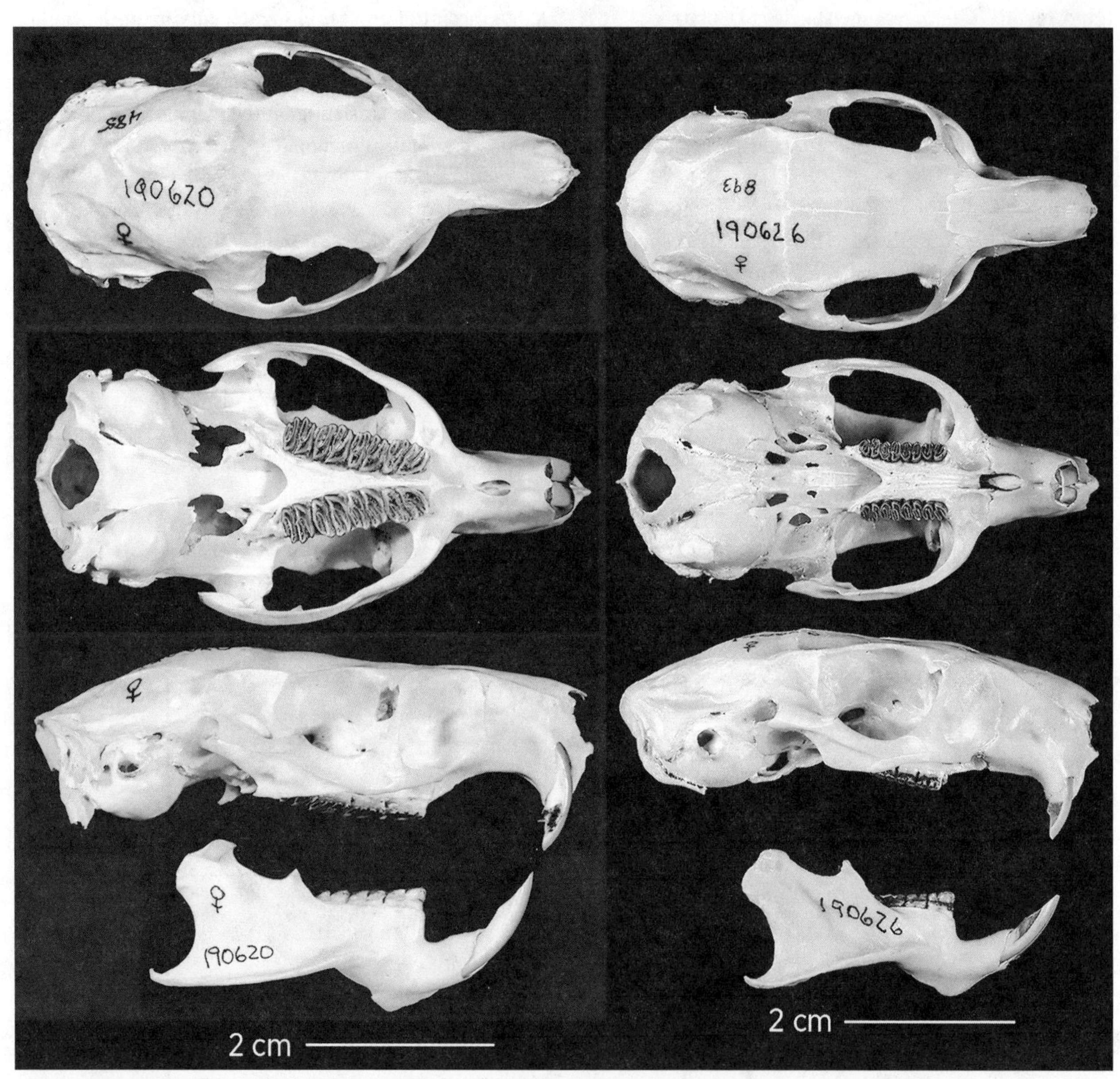

FIG. 278. Echimyidae: Echimyinae: Echimyini; skull and mandible of the Amazon bamboo rat, ***Dactylomys dactylinus*** **(*left*)**, and the brush-tailed rat, ***Isothrix bistriata*** **(*right*)**.

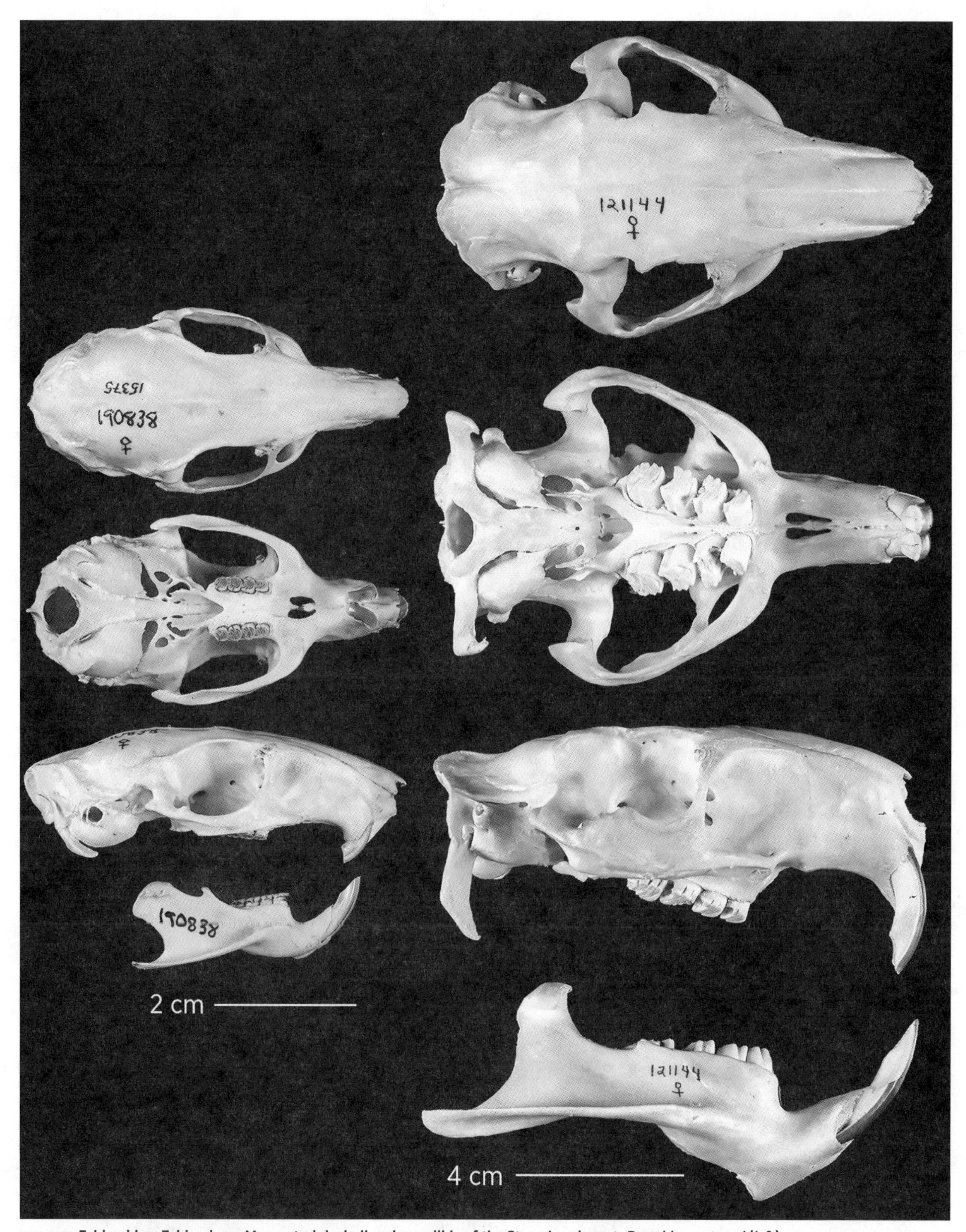

FIG. 279. Echimyidae: Echimyinae: Myocastorini; skull and mandible of the Steere's spiny rat, *Proechimys steerei* (*left*), and the coypu or nutria, *Myocastor coypus* (*right*).

Subfamily EURYZYGOMATOMYINAE (burrowing spiny rat, guiara, Atlantic spiny rats)

The Euryzygomatomyinae includes two semifossorial monotypic genera (*Clyomys* and *Euryzygomatomys*, both with coarse and spinose fur, short limbs, long and powerful claws, especially on the manus, and short tails) and the speciose *Trinomys*, comprising nine species of terrestrial spiny rats distributed within the dry Caatinga and moist Atlantic Forest of eastern and southern Brazil. *Trinomys* was long considered closely related to *Proechimys*, and members of both genera share many superficial external and cranial characters. The body is elongate; ears are enlarged; tails are slightly shorter to longer than the head and body, sparsely haired, but some terminate in a hairy tuft; the hind feet are narrow and elongate; and the fur varies from coarse to spiny.

Subfamily CARTERODONTINAE (groove-toothed spiny rat) (*incertae sedis*)

Conspicuously grooved upper incisors distinguish this monotypic and phylogenetically enigmatic genus from other echimyids; known from the Cerrado of western Brazil; previously grouped with the euryzygomatomine genera *Clyomys* and *Euryzygomatomys*, but now recognized as an independent lineage by molecular character analyses (Courcelle et al. 2019).

Family CTENOMYIDAE (tuco-tucos)

Small to medium-sized (body length 150–400 mm, mass 100–700 g); modified for subterranean existence, with heavy, tubular body, large head, thick neck, short limbs, small ears, short tail, and forefeet with enlarged claws (resemble North American pocket gophers, Geomyidae, but lack external cheek pouches). Digits functionally 4/5 with reduced pollex, not webbed, with **row of bristles extending beyond each claw; stiff hairs extend from sides of hind feet.** Dig with enlarged claws of forefeet but push dirt from burrows with hind feet; use teeth to cut through roots. Typically solitary, with each individual in exclusive tunnel system; a few species are colonial, sharing a single burrow system. Herbivorous. Occupy open biomes, such as grasslands, steppes, Andean puna, and shrublands.

RECOGNITION CHARACTERS:

1. skull blocky with strong zygomatic arches
2. infraorbital canal very large, with distinct groove for nerve passage at inner base
3. lacrimal canal not opening on side of rostrum
4. auditory bulla relatively large
5. **paroccipital process short, bound to bulla**
6. jugal not approaching lacrimal
7. lower jaw without ridge or groove on lateral surface; angular process deflected; coronoid process prominent
8. upper tooth rows slightly convergent anteriorly
9. **occlusal surface of cheek teeth simple, kidney-shaped**

DENTAL FORMULA: $\frac{1}{1}\frac{0}{0}\frac{1}{1}\frac{3}{3} = 20$

TAXONOMIC DIVERSITY: The family is monotypic, with about 70 species in the single genus *Ctenomys.*

FIG. 280. Ctenomyidae; white-bellied tuco-tuco, *Ctenomys colburni.*

FIG. 281. Geographic range of family Ctenomyidae: temperate South America, from southern Peru to Tierra del Fuego, and from sea level to the Altiplano.

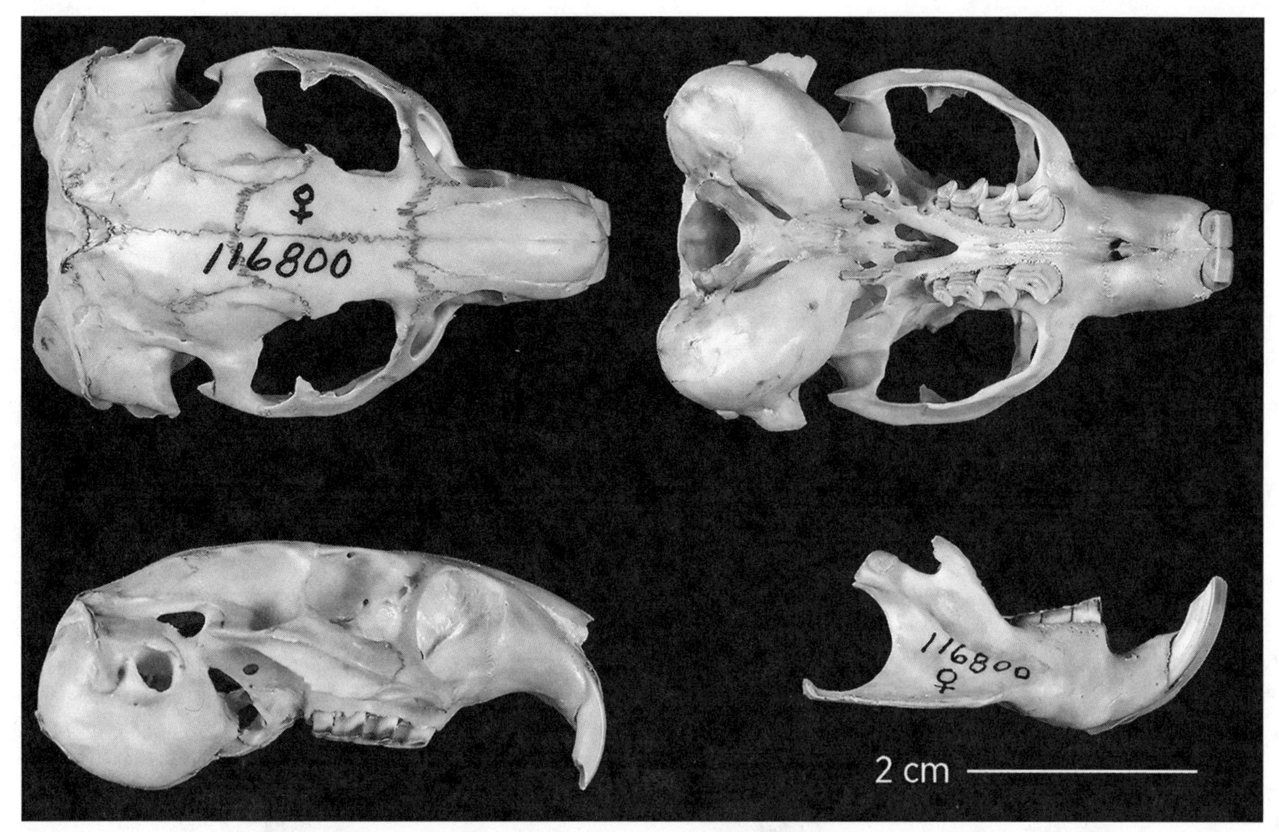

FIG. 282. Ctenomyidae; skull and mandible of the tawny tuco-tuco, *Ctenomys fulvus*.

Family OCTODONTIDAE (degus, chozchori, viscacha rats, coruros, rock rats)

Small (body length 170–370 mm), generally ratlike, with short limbs and moderately large eyes; ears ranging from large (*Octodon*, *Octodontomys*) to small (*Spalacopus*); tail either long and often tufted (*Octodon*, *Octodontomys*, *Octomys*, *Tympanoctomys*) or much shorter (*Aconaemys*, *Spalacopus*); *Spalacopus* is subterranean, with tubular body, short neck, but feet without extremely developed digging claws. Digits functionally 4/4 or 4/5 (5th digit on hind foot often reduced), not webbed, with **row of bristles extending beyond claw on each**; fur generally long, either soft or coarse. Some diurnal (*Octodon*), others nocturnal; some colonial, including subterranean *Spalacopus*, which constructs complex, interconnected burrow systems. Habitat ranges from desert scrub to shrublands to beech forests on Andean slopes.

RECOGNITION CHARACTERS (NEARLY IDENTICAL TO THOSE OF CTENOMYIDAE, BUT MOST NOT MODIFIED FOR SUBTERRANEAN LIFESTYLE):

1. skull blocky or elongate, slightly to strongly ridged
2. infraorbital foramen very large, with or without (*Spalacopus*) distinct groove for nerve passage at inner base
3. lacrimal canal not opening on side of rostrum
4. auditory bulla relatively small (*Spalacopus*) to large and inflated (*Octomys*)
5. **paroccipital process short, bound to bulla**
6. jugal not approaching lacrimal
7. lower jaw without ridge or groove on lateral surface; angular process deflected; coronoid process prominent
8. upper tooth rows slightly convergent anteriorly
9. **occlusal surface of cheek teeth simple, figure 8–shaped or kidney-shaped**

DENTAL FORMULA: $\frac{1}{1}\ \frac{0}{0}\ \frac{1}{1}\ \frac{3}{3} = 20$

TAXONOMIC DIVERSITY: About 14 species in six genera, with 1 to 4 species in each: *Aconaemys* (rock rats), *Octodon* (degus), *Octodontomys* (chozchori or mountain degu), *Octomys* (common viscacha rat), *Spalacopus* (coruro), and *Tympanoctomys* (viscacha rats; now including *Pipanactomys* and *Salinoctomys*; see Díaz et al. 2015).

FIG. 283. Octodontidae; mountain degu or chozchori, *Octodontomys gliroides* (*left*), and degu, *Octodon degus* (*right*).

FIG. 284. Geographic range of family Octodontidae: southern temperate South America (Bolivia, Chile, and Argentina).

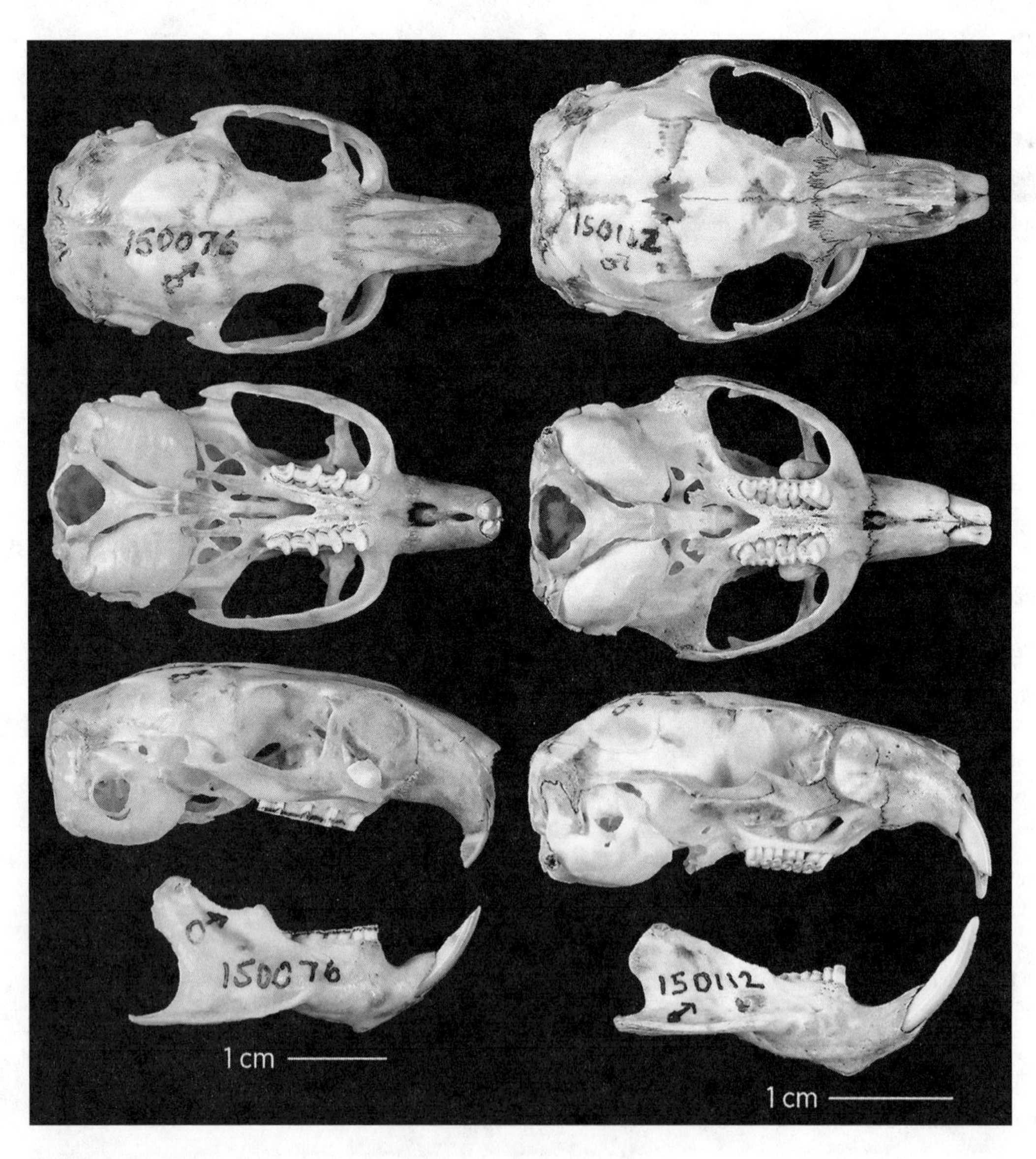

FIG. 285. Octodontidae; skull and mandible of the degu, *Octodon degus* (*left*), and coruro, *Spalacopus cyanus* (*right*).

Clade EUARCHONTA

The Euarchonta includes the Dermoptera (colugos), Primates (lemurs, galagos, lorises, tarsier, monkeys, and great apes), and Scandentia (tree shrews) (see fig. 94). Relationships among these three lineages, however, are not completely understood (see Esselstyn et al. 2017; Gatesy et al. 2017). Even the connection of Scandentia to other Placentalia has been historically ambiguous. Some early classifications placed the tree shrews, along with the elephant shrews (Macroscelidea), in the Menotyphla (see the introduction to infraclass Eutheria or Placentalia). Others regarded tree shrews as basal to the Primates. While molecular analyses have clearly shown that a Macroscelidea + Scandentia link is untenable, these studies have yet to resolve the relationships of tree shrews to Dermoptera and Primates. In some analyses (e.g., Meredith et al. 2011; Gatesy et al. 2017), tree shrews actually appear to be basal to rodents, and thus part of the larger Euarchontoglires clade; in others (e.g., Roberts et al. 2011), tree shrews are grouped with dermopterans and primates within the Euarchonta. Lin et al. (2014) concluded that these conflicting results were at least partly due to rapid nucleotide substitution rates in both Scandentia and Glires (rodents plus lagomorphs); their analyses supported the monophyly of Scandentia and Primates (unfortunately, they did not include dermopterans in their analyses). Esselstyn et al. (2017) reviewed these different hypotheses by using new genome-wide data for all three groups and novel analytical methods, but failed to reach a firm resolution.

Within the Euarchonta, most evidence supports a link between the Dermoptera and Scandentia relative to Primates. Under this hypothesis, the colugos and tree shrews are united under the clade name Sundatheria, which references their common geographic placement. An alternative scenario links the Dermoptera with Primates under the clade name Primatomorpha, to the exclusion of the Scandentia (e.g., Janecka et al. 2007). Herein we group the euarchontan orders into Primates + Sundatheria (Dermoptera and Scandentia), following the large-scale morphological analyses of O'Leary et al. (2013) and most, but not all, molecular studies.

Order PRIMATES

Primates as a group are quite diverse in body form, locomotory style, feeding mode, and social organization. Many of their morphological attributes are directly linked to specializations for feeding. Fruit eaters, such as the New World spider monkeys (Atelidae) or Old World baboons and mangabeys (Cercopithecidae: Cercopithecinae) have broad incisors, low rounded molar cusps, and long small intestines; leaf eaters, such as sportive lemurs (Lepilemuridae) or colobus monkeys (Cercopithecidae: Colobinae) have much smaller incisors, molar cusps with well-developed shearing crests, and either complex sacculated stomachs or an enlarged cecum for fermentation; gum eaters, such as many lemurs (Lemuridae) and the New World marmosets (Cebidae: Callitrichinae) have enlarged, procumbent, and stout lower incisors for cutting into bark, claws on the feet for clinging to trunks while feeding, and a very long cecum; and more insectivorous primates, such as the marmoset *Callimico* (Cebidae: Callitrichinae) or the potto (Lorisidae) have sharp molar cusps and short, simple guts.

Vision is the dominant sense among all primates, so it is not surprising that the eyes are large and the corresponding orbit of the skull is large as well (nocturnal primates have larger eyes and orbits than diurnal ones). Diurnal primates tend to have a smaller ratio of retinal cells to the number of nerves carrying information to the brain, but greater visual acuity, than do nocturnal species. All primates have color vision, but trichromatic color vision in this order is limited to catarrhine monkeys (a group that includes humans) and the howler monkeys of the New World (which are platyrrhine monkeys); some argue that trichromatic color vision evolved as an adaptation for foraging on young, red leaves.

Finally, locomotor adaptations can be found in many parts of the body, especially in the hands, limbs, and trunk. Arboreal quadrupedal primates tend to have a laterally placed and freely moving scapula, narrow thorax, long tail, grasping feet with long digits and opposable thumb (pollex) and big toe (hallux), and a deep ulna with a long olecranon process. Terrestrial quadrupeds also have a narrow thorax and an ulna with a posteriorly extended olecranon process, but have a more restricted shoulder joint, reduced tail, hands and feet with short digits with limited opposability of the hallux, and a robust radius. Arboreal leaping primates have a long lumbar region and long hind limbs with a

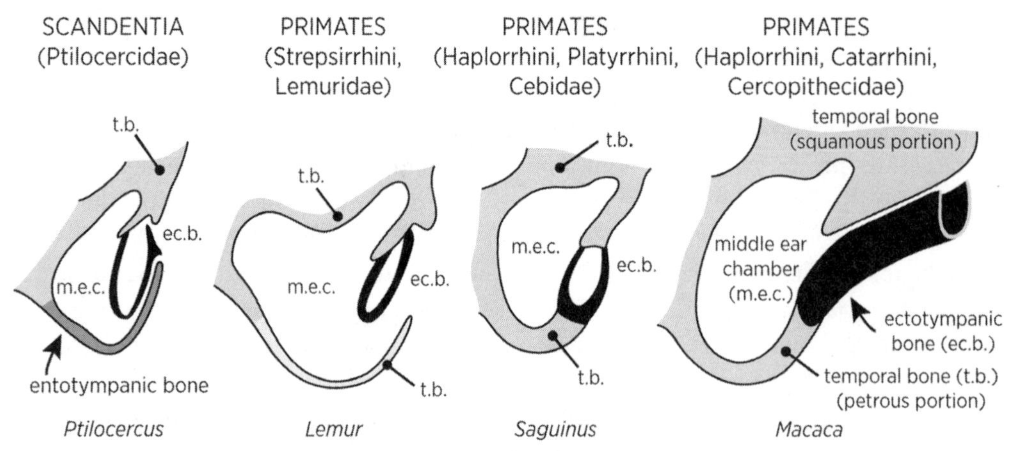

FIG. 286. Transverse view of the auditory bulla of a tree shrew (*Ptilocercus*: Ptilocercidae: Scandentia) compared with those of three primates (*from left to right*, common lemur [*Lemur*: Lemuridae: Strepsirrhini]; tamarin [*Saguinus*: Cebidae: Haplorrhini]; and rhesus [*Macaca*: Cercopithecidae: Haplorrhini]). Redrawn from Hershkovitz (1977). The temporal bone in primates is formed from the fusion of four bones (squamous, mastoid, petrous, and tympanic) that are typically separate in other mammals.

short-necked femur, deep femoral condyles (at the knee joint), and a narrow tibia. And suspensory primates (those that hang by hands, feet, and/or prehensile tail) have dorsally placed scapulae, a broad thorax, short lumbar region, often no tail, a very mobile hip joint, very long forelimbs, short olecranon process of the ulna, highly rotary wrist joint, and elongate and curved fingers, often with the pollex reduced or absent. The **intermembral index** is the ratio of forelimb length to hind limb length (hence, [humerus + radius]/[femur + tibia] × 100) and is useful in predicting locomotor patterns in primates. Values near 100 typify quadrupedal species, whereas values below and above 100 are found in animals preferentially using their hind limbs (leaping, bipedal) and forelimbs (brachiating), respectively.

Characters that unite this broadly diverse group include the following shared-derived attributes:

1. SKULL
   - **braincase relatively large, housing well-developed cerebral hemispheres**
2. STRUCTURES OF THE LOCOMOTORY APPARATUS
   - foot posture plantigrade, soles of feet naked, with enlarged pads
   - most with grasping hands and feet with **opposable pollex and/or hallux**
   - **hallux contains a nail**
   - **nails present on all or most digits** (may be modified secondarily into claws)
   - calcaneus elongate
   - hind limb dominance during locomotion (except in brachiators)
   - center of gravity shifted toward hind limbs
   - radius and ulna, tibia and fibula separate and subequal in length to humerus and femur, respectively
   - clavicle well developed
3. STRUCTURES ASSOCIATED WITH STEREOSCOPIC VISION
   - forward rotation of the orbits and narrowing of the interorbital distance
   - enlargement of the orbital cavity
   - exposure of the ethmoid bone on the inner orbital wall
   - stereoscopic vision in which approximately half of the retinal axons project to the ipsilateral side of the brain
   - complete postorbital bar or plate (the postorbital plate of haplorrhine primates is unique among mammals)
4. STRUCTURES OF THE AUDITORY BULLA
   - floor is formed by the petrosal bone, with tympanic ring or tube present (fig. 286)
5. STRUCTURES OF THE DENTITION
   - two incisors and three premolars (e.g., loss of one incisor and one premolar from primitive eutherian condition)
   - molars tritubercular or quadritubercular, bunodont and brachydont

PHYLOGENY AND CLASSIFICATION OF PRIMATES

Primates are traditionally divided into two suborders, the composition of which varies depending on the placement of the tarsiers (family Tarsiidae) (see table 5).

Table 5. Contrasting classifications and taxon diversity (genera/species) of primates given in three recent compendia.

	Groves (2005b)	*Mittermeier et al. (2013)*	*Fleagle (2013)*
Suborder Strepsirrhini			
Infraorder Lemuriformes			
Superfamily Cheirogaleoidea			
Family Cheirogaleidae[1]–dwarf lemurs	5/21	5/31	5/30
Superfamily Lemuroidea[1]			
Family Indriidae–woolly lemur and sifaka	3/11	3/19	3/19
Family Lemuridae–lemurs	5/19	5/21	5/22
Family Lepilemuridae–sportive lemurs	1/8	1/26	1/26
Infraorder Chiromyiformes			
Family Daubentoniidae–aye-aye	1/1	1/1	1/1
Infraorder Lorisiformes			
Family Galagidae (= Galagonidae)–galagos	3/19	4/12	5/18
Family Lorisidae (= Loridae)–lorises, pottos	5/9	5/18	4/12
Suborder Haplorrhini			
Infraorder Tarsiiformes			
Family Tarsiidae–tarsiers	1/7	3/11	3/11
Infraorder Simiiformes[1, 2]			
Parvorder Platyrrhini			
Family Atelidae–spider monkeys	5/24	5/25	4/26
Family Callitrichidae[3, 4]		4/47	
Family Cebidae–New World monkeys	6/56	3/29	11/87
Family Aotidae[4]–night monkeys	1/8	1/11	
Family Pitheciidae–uacari and saki monkeys	4/40	4/44	4/43
Parvorder Catarrhini			
Superfamily Cercopithecoidea			
Family Cercopithecidae–Old World monkeys	21/132	23/159	28/151
Superfamily Hominoidea			
Family Hylobatidae–gibbons	4/14	4/19	4/19
Family Hominidae–orangutan, gorilla, chimpanzee, human	4/7	4/7	4/7
totals	69/376	75/480	81/472

Sources: Groves (2005b), Mittermeier et al. (2013), and Fleagle (2013); note that the numbers of recognized genera and species have increased by 11%–12% (69 to 75–81) and 26%–28% (376 to 472–480), respectively, in the intervening 8 years between the publication dates. The classifications of Groves and Mittermeier et al. differ primarily from that of Fleagle with regard to the family or subfamily status of the platyrrhine night monkeys, tamarins + marmosets, and capuchin + squirrel monkeys. We distinguish Callitrichidae and Aotidae from Cebidae in this table but treat them as subfamilies in text.

1 Superfamilies Cheirogaleoidea and Lemuroidea and Infraorder Simiiformes not recognized by Mittermeier et al. (2013).

2 Equivalent to Anthropoidea of some authors (e.g., Fleagle 2013).

3 Callitrichidae treated as part of Cebidae by Groves (2005b).

4 Aotidae and Callitrichidae treated as subfamilies of Cebidae by Fleagle (2013) and herein.

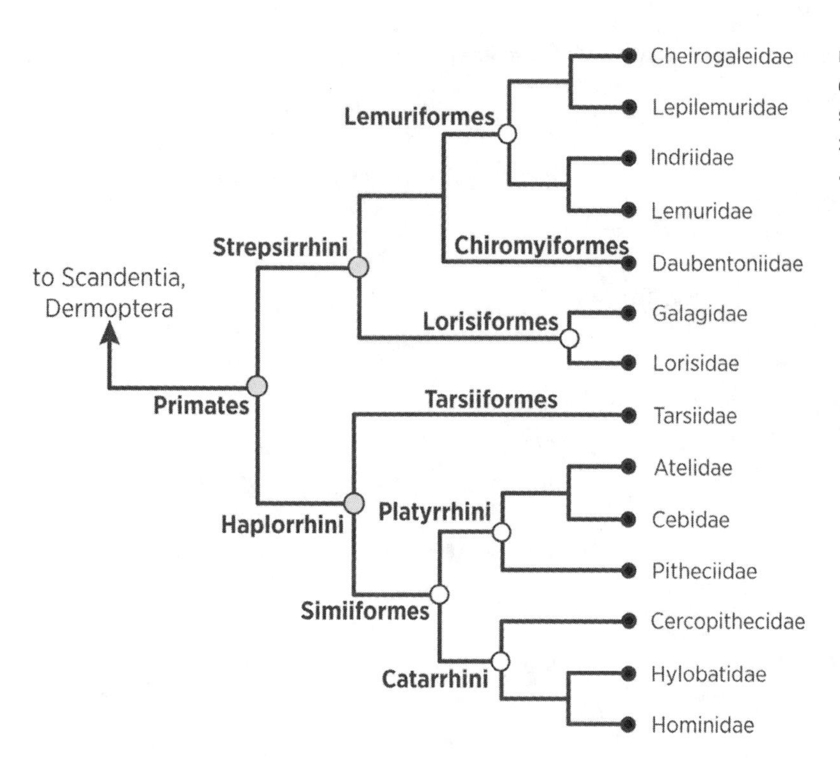

FIG. 287. Phylogenetic relationships and resulting classification of living primates derived from DNA sequence analyses (Perelman et al. 2011; Fleagle 2013; and the 10ktrees website [http://10ktrees.fas.harvard.edu/Primates]).

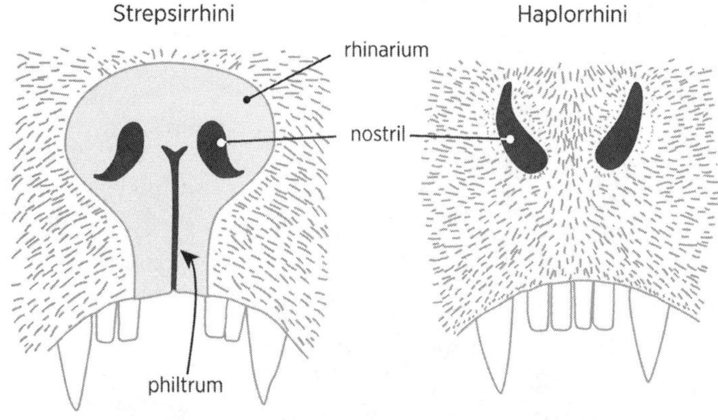

FIG. 288. Comparisons in the structure of the nose of strepsirrhine and haplorrhine primates. Redrawn from Martin (1990).

Morphological and molecular analyses have largely converged on a unified topology of relationships among primates, identifying the same major and secondary lineages. The only remaining disagreement in the resulting classifications involves the family or subfamily status of the New World platyrrhine lineages. The phylogeny depicted in fig. 287 represents one of two recent consensus treatments.

Older classifications placed tarsiers either with lemurs and lorises in the suborder Prosimii or with monkeys and apes in the suborder Anthropoidea. Cladistic analysis of both morphological and molecular data, however, unambiguously link tarsiers to monkeys and apes and thus define two suborders, Strepsirrhini and Haplorrhini, based on characteristics of the nose and the enclosed bony orbit. Strepsirrhine primates (the so-called lower primates) include the lemurs, aye-aye, lorises, and galagos. As a group, they possess a naked rhinarium with a median cleft (**philtrum**; fig. 288) and slit-like nostrils, and the eye socket, while bounded by a bony postorbital bar, is openly confluent with the temporal fossa. The strepsirrhines are divided into three lineages: the Chiromyiformes (which includes only the aye-aye of Madagascar), the Lemuriformes (the lemurs, also from Madagascar), and the Lorisiformes (the lorises, pottos, and galagos of Africa and Southeast Asia). The Haplorrhini ("higher" primates) include two major lineages (infraorders): the Tarsiiformes, the sole member of which is the Tarsiidae, and the Simiiformes, which is further subdivided into two parvorders, the Platyrrhini (the New World monkeys) and the Catarrhini (the Old World monkeys and great apes). All haplorrhine primates have a simple nose, lack a naked rhinarium, have a vestigial philtrum (or none at all), and have an orbit nearly to completely separated from the temporal fossa by a bony postorbital plate.

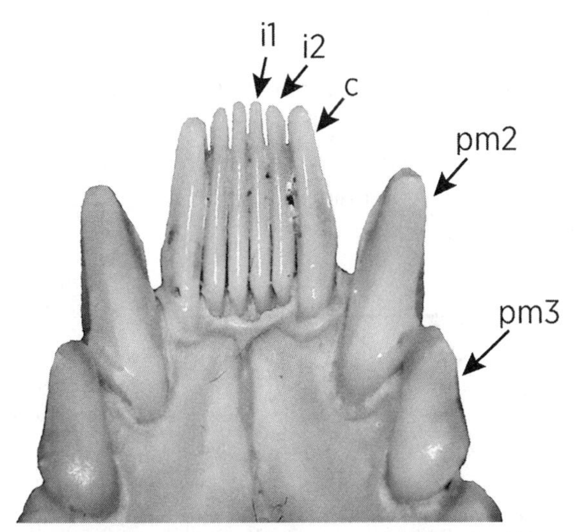

FIG. 289. Planar view of the anterior part of a lemur mandible illustrating the tooth comb, which is derived from the 1st and 2nd lower incisors (i1 and i2) plus the canine (c); the lower premolars (pm2 to pm3), not part of the tooth comb, are also identified.

Suborder STREPSIRRHINI (lemurs, galagos, and lorises)

DIAGNOSTIC CHARACTERS:

- **naked, glandular rhinarium with philtrum**; ventral extension of rhinarium fuses with gum, preventing great mobility of upper lip; **slit-like nostrils (diverticulum nasi) directed laterally**; this is the primitive condition for mammals in general (cf. description of nose in suborder Haplorrhini)
- **presence of a tooth comb**, composed of lower incisors and canine teeth (secondarily lost in Daubentoniidae) projecting forward from lower jaw (fig. 289), used in grooming; this is associated with reduced upper incisors (Tupaiidae [Scandentia] have a similar tooth comb, but it does not include the canines)
- **postorbital bar complete but orbital and temporal openings broadly confluent**
- **nasolacrimal duct long and oriented horizontally** (where it directs moisture to the rhinarium)
- cranial blood supply provided either by the stapedial artery (a branch of the internal carotid artery) or by the ascending pharyngeal artery (a branch of the external carotid artery), never by the promontory branch of the internal carotid
- all digits bear a nail, except for **grooming claw** on the 2nd digit of the hind foot
- eyes have a light-reflecting **tapetum lucidum** behind the retina to improve night vision (the tapetum lucidum reflects light back toward the retina; in so doing it also produces eyeshine); **lack a retinal fovea on posterior surface of eyeball**

Infraorder LEMURIFORMES

DIAGNOSTIC CHARACTERS:

- **bulla overgrows tympanic ring (the ectotympanic bone, which surrounds the eardrum), remaining separated from the petrosal part of the bulla (separation can be viewed through the auditory meatus)** (see fig. 286)
- **upper molars with three primary cusps (labial paracone and metacone, lingual protocone), but with secondary lingual cusplets providing generalized quadrate shape**

Superfamily CHEIROGALEOIDEA
Family CHEIROGALEIDAE (dwarf and mouse lemurs)

Small to medium-sized (head and body length 125–275 mm; mass 30–450 g); tail slightly shorter than to 50% longer than body (125–350 mm); hind limbs longer than forelimbs (intermembral index 68–72); head rounded, rostrum short, eyes large and close set; ears thin and membranous; coarsely ridged tactile pads on palms and soles of feet; 3rd and 4th digits subequal; possess a tapetum lucidum. Omnivorous, with bulk of food including insects and fruit; arboreal; nocturnal; mostly solitary but may sleep communally. Easily climb vertical tree trunks, aided by keeled, pointed nails for purchase; jump horizontal distances up to 5 m. *Cheirogaleus* can store fat in tail and become dormant during part of year in hollows in trees and even underground. Inhabit forests throughout Madagascar. Females with three pairs of nipples; give birth to twins.

RECOGNITION CHARACTERS:

1. fur long and soft and relatively unicolored gray-brown to reddish above, paler below
2. pollex not widely separated from digits 2–5
3. hind limbs slightly longer than trunk length (108%–117%)
4. **upper incisors long, separated by short midline diastema**
5. palatal-lacrimal contact in some, but not all, genera

6. tympanic ring usually free within auditory bulla (not connected to petrosal)
7. cranial blood supply via ascending pharyngeal artery; stapedial artery and stapedial canal very small
8. upper molars tritubercular or quadritubercular
9. baculum present in males

DENTAL FORMULA: $\frac{2}{2}\ \frac{1}{1}\ \frac{3}{3}\ \frac{3}{3} = 36$

TAXONOMIC DIVERSITY: About 31 species in five genera: *Allocebus* (hairy-eared dwarf lemur), *Cheirogaleus* (dwarf lemurs), *Microcebus* (mouse lemurs), *Mirza* (giant mouse lemurs), and *Phaner* (fork-crowned lemurs).

FIG. 290. Cheirogaleidae; gray mouse lemur, *Microcebus murinus*.

FIG. 291. Geographic range of family Cheirogaleidae: Madagascar.

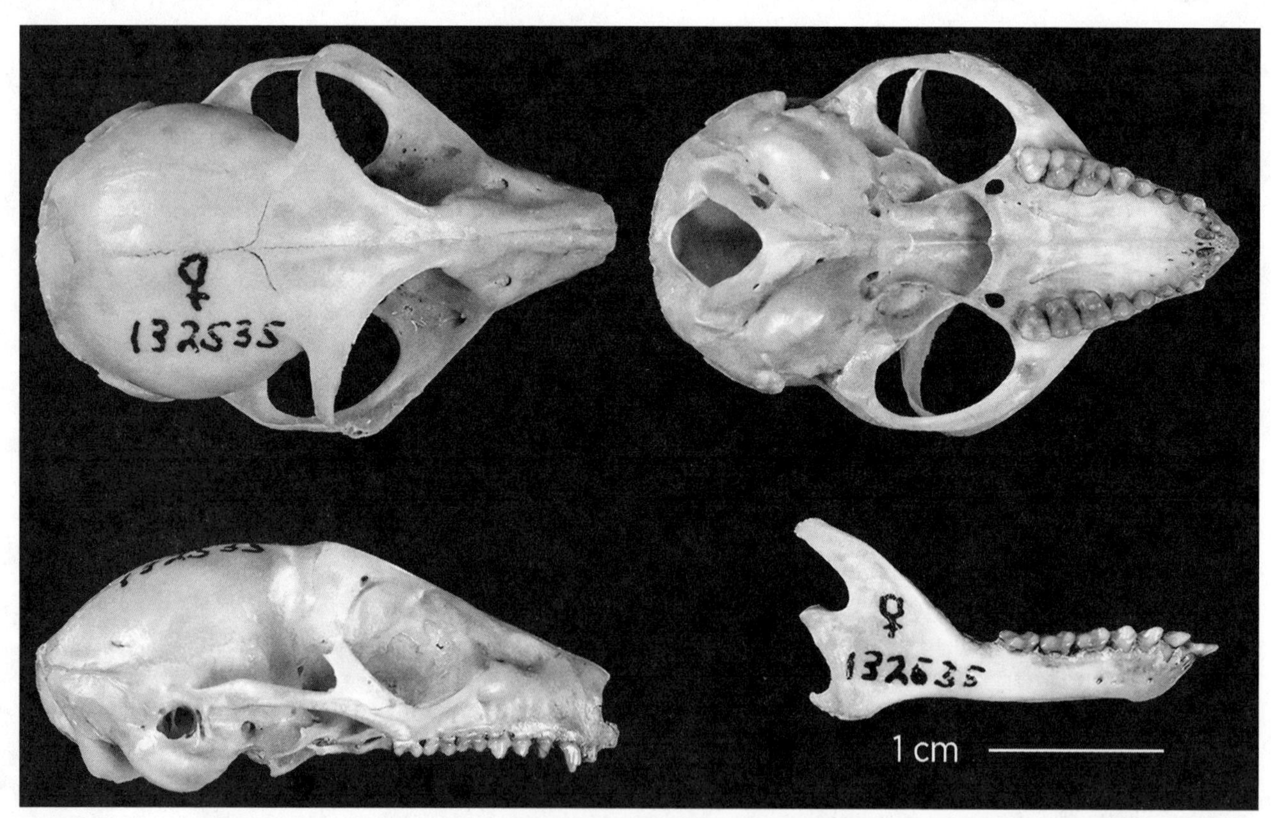

FIG. 292. Cheirogaleidae; skull of the gray mouse lemur, *Microcebus murinus*.

Superfamily LEMUROIDEA

Family INDRIIDAE (avahis, indri, and sifakas)

Medium-sized (body 300 mm; mass 1 kg) to large (body 900 mm; mass 10 kg); head round, muzzle short, eyes large, ears small; all with long tails except *Indri*, which has only a stump; **hind limbs about 1/3 longer than forelimbs** (intermembral index 58–64); hallux long, about 55% of foot length; **last four digits of hind feet joined by flaps of skin** and act as single unit opposing the hallux; **all digits with nails**. Fur silky to woolly, color ranges from striking black and white to browns and yellows in varied patterns; faces are somewhat shorter than in true lemurs. Herbivorous with a large cecum presumably for microbial fermentation; highly arboreal vertical clingers and leapers, but sifakas will descend to ground and hop from tree to tree; nocturnal (woolly lemurs) to diurnal (indri and sifakas); indri have powerful voices amplified by special laryngeal sacs. Habitat includes wet and dry forests. (The family name is often, but incorrectly, spelled Indridae.)

FIG. 293. Indriidae; indri, *Indri indri*.

RECOGNITION CHARACTERS:

1. rostrum relatively short; braincase rounded
2. **palatine and lacrimal bones separated by frontal-maxillary contact in orbit**
3. tympanic ring free within auditory bulla (not connected to petrosal)
4. cranial blood supply via stapedial artery; stapedial canal large
5. **upper incisors enlarged, unequal in size, separated by median diastema**
6. **two premolars in each quadrat**
7. lower canines absent, so tooth comb consists of four rather than six procumbent teeth
8. molars quadritubercular
9. baculum present in males

DENTAL FORMULA: $\frac{2}{2}\ \frac{1}{0}\ \frac{2}{2}\ \frac{3}{3} = 30$

TAXONOMIC DIVERSITY: 19 species in three genera: *Avahi* (avahis or woolly lemurs), *Indri* (indri), and *Propithecus* (sifakas).

FIG. 294. Geographic range of family Indriidae: Madagascar.

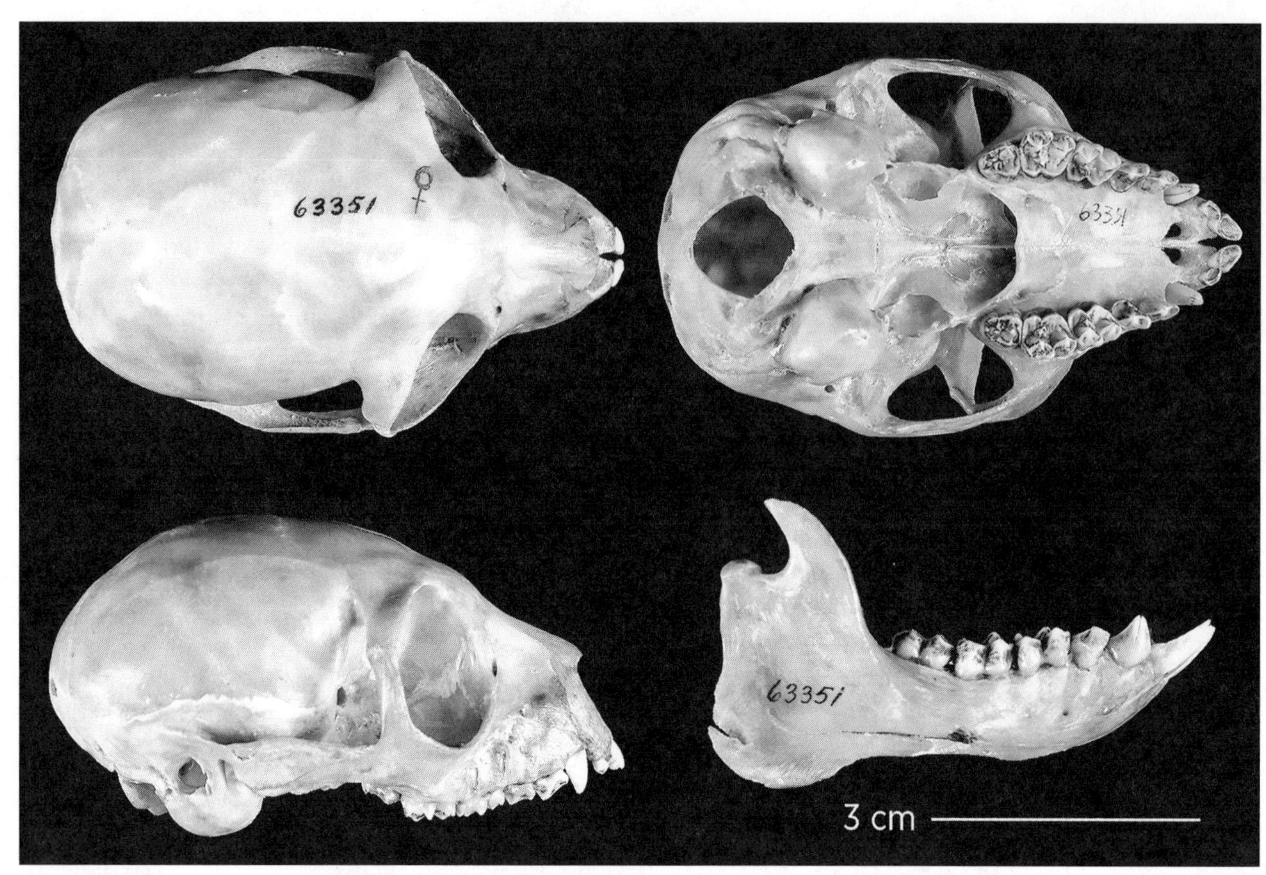

FIG. 295. Indriidae; skull and mandible of Milne-Edwards' sifaka, *Propithecus edwardsi*.

Family LEMURIDAE (lemurs)

Medium-sized (body length 320–560 mm; mass 700 g–3.6 kg); long, bushy tail (length 280–600 mm), equal to or slightly longer than body; hind limbs longer than forelimbs (intermembral index 64–72), slightly longer than trunk (~105%); locomotion generally fully quadrupedal; all but one species (*Lemur catta*, the ring-tailed lemur) lack a tapetum lucidum; fur soft and woolly, varying in colors and color patterns; some species sexually dichromatic. All partially arboreal but some spend considerable time on the ground; move in trees by running along branches and by leaping from vertical branches or trunks. Herbivorous, eating fruit, leaves, and nectar. Generally social, some living in large groups with dominance hierarchies. Habitat includes moist forest to dry woodlands, reeds in marshes, and bamboo thickets.

RECOGNITION CHARACTERS:

1. both pollex and hallux opposable
2. rostrum and braincase elongate
3. **upper incisors uniformly small, peg-like, with large median diastema**
4. **frontal and palatine bones in contact in orbit**
5. tympanic ring usually free within auditory bulla (not connected to petrosal)
6. cranial blood supply via stapedial artery; stapedial canal large
7. molars tritubercular or quadritubercular
8. baculum present in males

DENTAL FORMULA: $\frac{2}{2}\ \frac{1}{1}\ \frac{3}{3}\ \frac{3}{3} = 36$

TAXONOMIC DIVERSITY: About 20 species in five genera divided into two subfamilies:

Hapalemurinae—*Hapalemur* (bamboo lemurs), *Lemur* (ring-tailed lemur), and *Prolemur* (greater bamboo lemur)

Lemurinae—*Eulemur* ("true" lemurs) and *Varecia* (ruffed lemurs)

FIG. 296. Lemuridae: Lemurinae; black and white ruffed lemur, *Varecia variegata*.

FIG. 297. Geographic range of family Lemuridae: Madagascar and Comoro Islands.

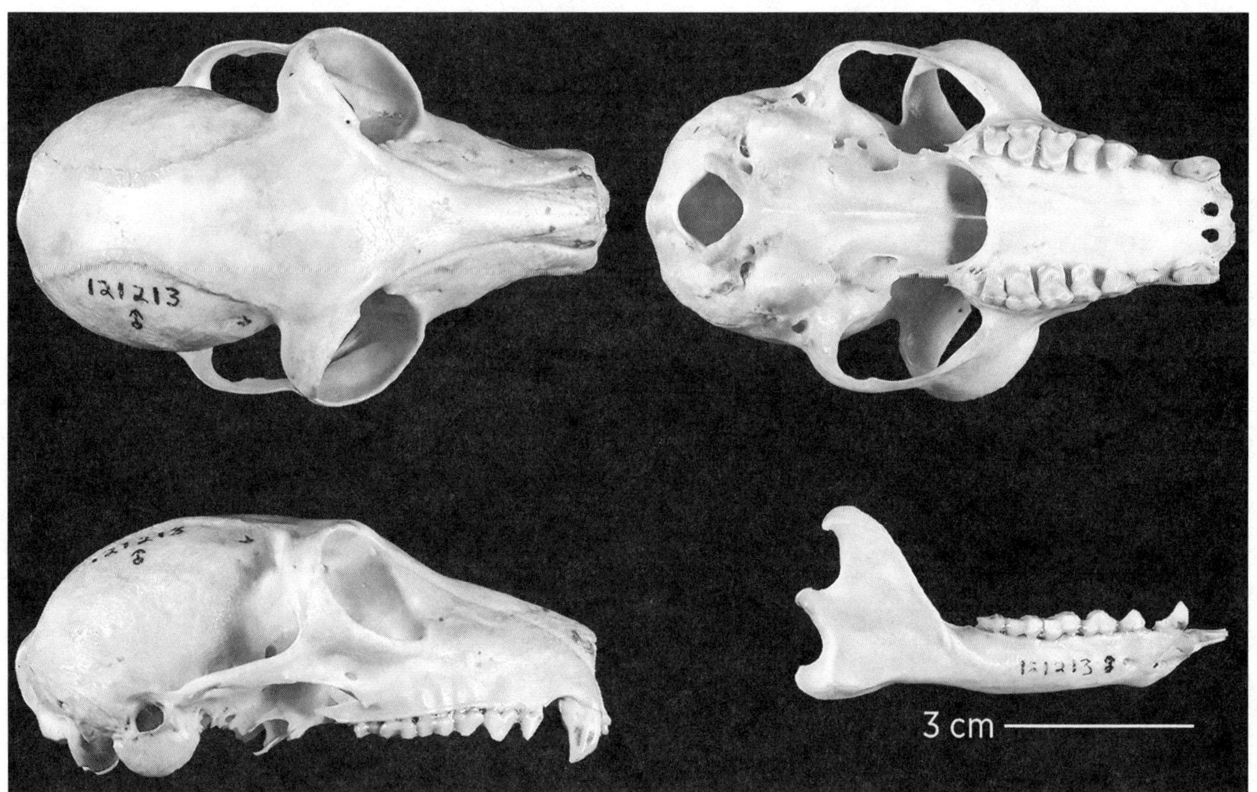

FIG. 298. Lemuridae: Lemurinae; skull and mandible of the brown lemur, *Eulemur fulvus*.

Family LEPILEMURIDAE (= Megaladapidae) (sportive or weasel lemurs)

Medium-sized (body length 300–350 mm; mass 500–900 g); head relatively short and pointed, ears rounded; tail equal in length to body; hind limbs much longer than forelimbs (intermembral index 60–65); feet slightly elongate; fur woolly and thick, generally brown to gray dorsally and whitish ventrally; possess a tapetum lucidum. Nocturnal and mainly arboreal; adept clingers and leapers, moving through trees in long jumps powered by elongate hind limbs; on ground they hop like kangaroos. Solitary and herbivorous, feeding mostly on leaves; large cecum presumably for microbial fermentation. Inhabit forested regions of Madagascar.

RECOGNITION CHARACTERS:

1. lack the bold markings often characterizing true lemurs
2. **large digital pads on hands and feet**
3. hind limbs long in relation to trunk length (140% versus ~105% for Lemuridae)
4. **lack upper incisors**
5. **condyloid process of mandible with dorsoventrally expanded articulation surface**
6. cranial blood supply via stapedial artery; stapedial canal large
7. upper molars quadritubercular
8. baculum presumably present in males

DENTAL FORMULA: $\frac{0}{2}\ \frac{1}{1}\ \frac{3}{3}\ \frac{3}{3} = 32$

TAXONOMIC DIVERSITY: The family is monotypic, with about 26 species in the single genus *Lepilemur*.

FIG. 300. Geographic range of family Lepilemuridae: Madagascar.

FIG. 299. Lepilemuridae; weasel sportive lemur, *Lepilemur mustelinus*.

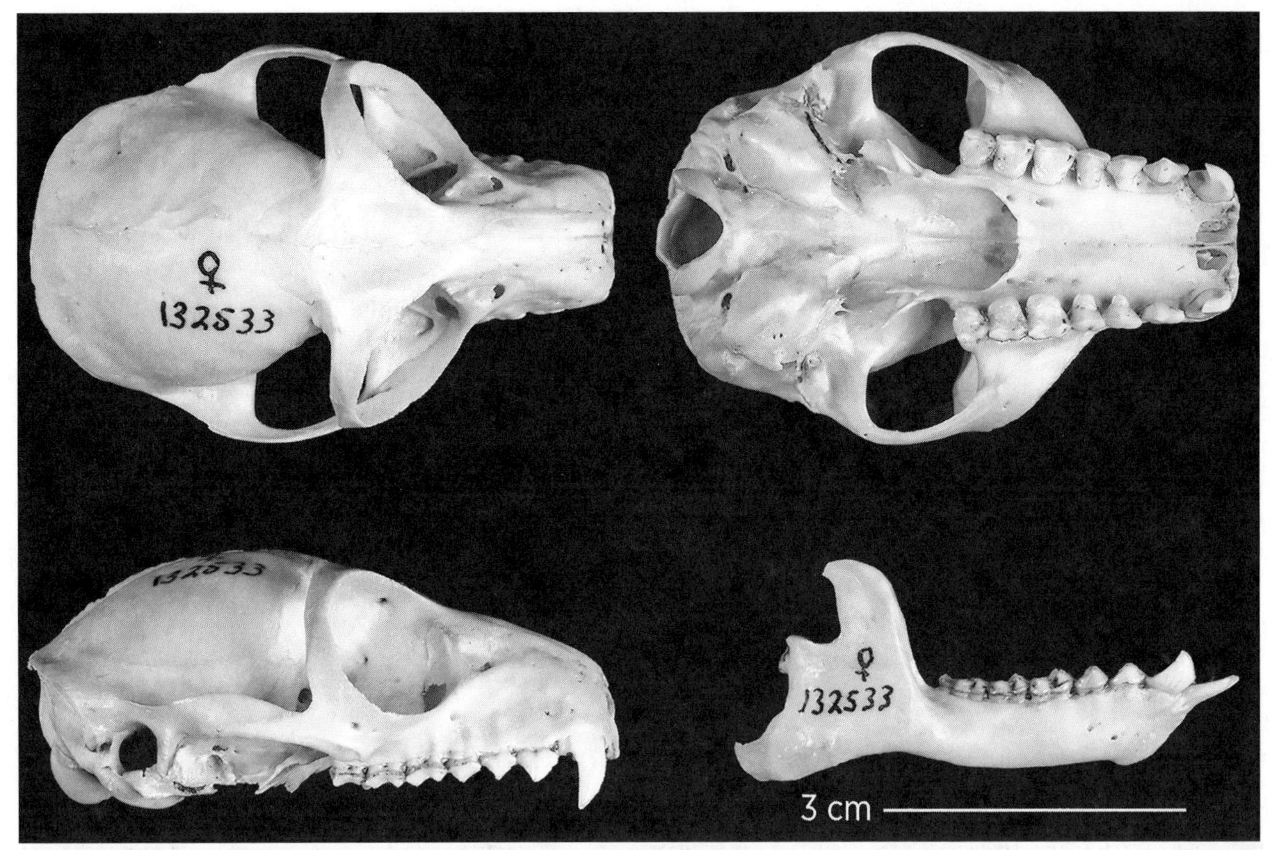

FIG. 301. Lepilemuridae; skull and mandible of the weasel sportive lemur, *Lepilemur mustelinus*.

Infraorder CHIROMYIFORMES

DIAGNOSTIC CHARACTERS:

- **incisors 1/1, chisel-shaped**
- **3rd digit on forefoot slender, much longer than other digits**

Family DAUBENTONIIDAE (aye-aye)

Medium-sized (body length 360–440 mm; mass 2.5 kg); head rounded, short; **eyes large; ears conspicuous**, membranous, somewhat pointed, movable; tail longer than body (560–600 mm), very bushy; hind legs longer than forelegs (intermembral index 71); all digits bear a claw except the hallux, which has a flat nail; rhinarium and lips flesh-colored; ears, hands, and feet black; pelage short, woolly underfur covered by **coarse, shaggy guard hairs**. Nocturnal; omnivorous, feeding mostly on fruit and insect larvae; uses elongate, very thin 3rd finger of manus to probe for insects and to groom; arboreal, good leaper. Habitat includes dense forest and bamboo thickets.

RECOGNITION CHARACTERS:

1. rostrum very short
2. braincase relatively large and globular
3. orbits medium-sized, without prominent orbital ridges
4. **frontal-maxillary contact in orbit**
5. cranial blood supply via stapedial artery; stapedial canal large
6. **dentition greatly reduced**
7. **single pair of greatly enlarged, chisel-shaped upper and lower incisors** with enamel only on anterior surfaces (rodent-like)
8. **molars basically quadritubercular with flat crowns and indistinct cusps**
9. baculum present in males

DENTAL FORMULA: $\frac{1}{1}\ \frac{0}{0}\ \frac{1}{0}\ \frac{3}{3} = 18$

TAXONOMIC DIVERSITY: The family is monotypic, with a single species, *Daubentonia madagascariensis* (aye-aye).

FIG. 302. Daubentoniidae; aye-aye, *Daubentonia madagascariensis*.

FIG. 303. Geographic range of family Daubentoniidae: Madagascar.

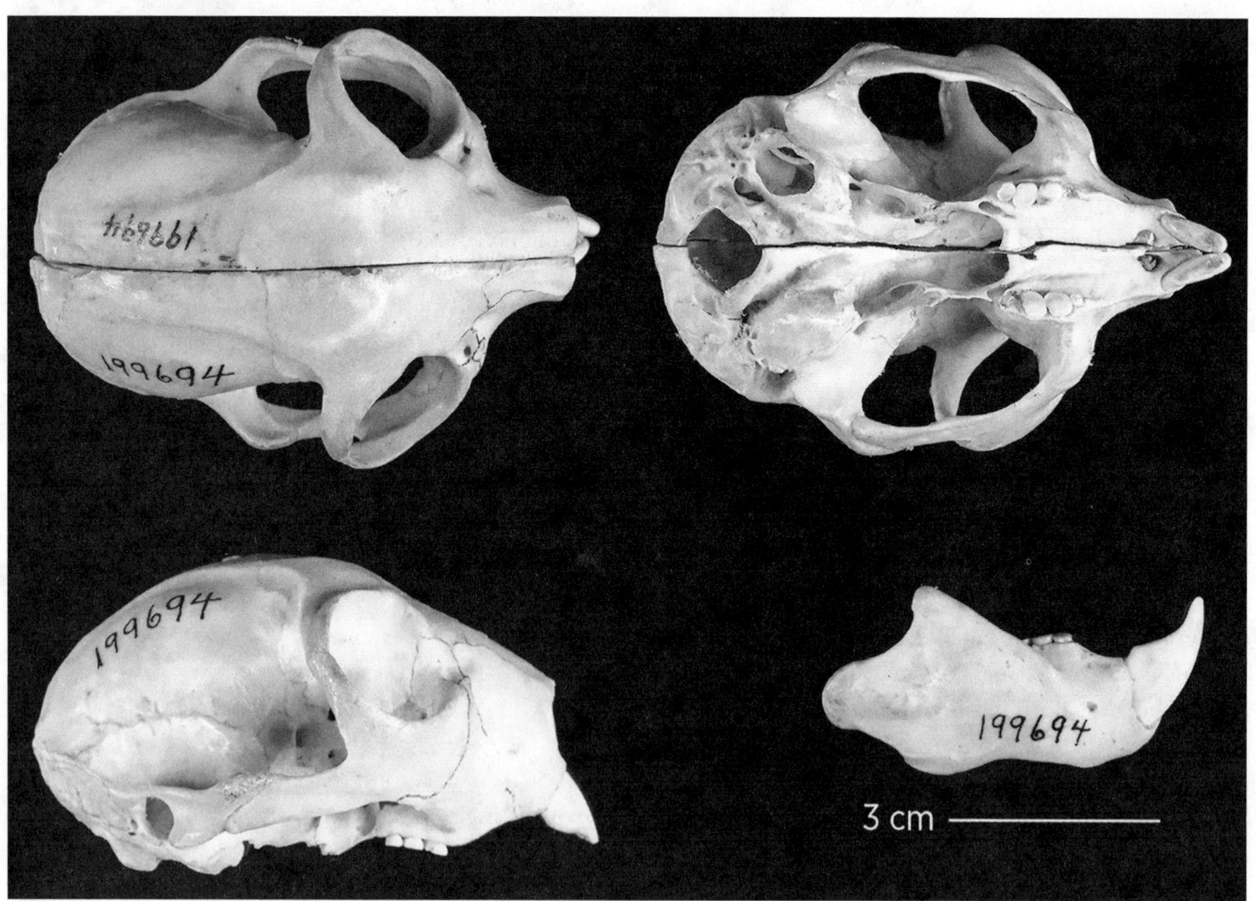

FIG. 304. Daubentoniidae; skull and mandible of the aye-aye, *Daubentonia madagascariensis*.

Infraorder LORISIFORMES

The Lorisiformes are the strepsirrhines of continental Africa and Asia. They share features with Malagasy strepsirrhines (tooth comb, grooming claw), but possess cranial features that distinguish them from most of that group. The tympanic ring is fused to the petrosal (in contrast to that of indriids, most lemurids, and most cheirogaleids), and the cranial blood supply is through the ascending pharyngeal artery (a feature shared with cheirogaleids). Lorisiformes have repeatedly been shown to be a separate radiation from Malagasy strepsirrhines, suggesting that their similarities to select Malagasy taxa are convergent. The Lorisiformes includes two families, both of which are nocturnal and arboreal, but which differ greatly in postcranial morphology and in patterns of movement: galagos are well known for leaping, whereas lorises are slow and methodical climbers.

DIAGNOSTIC CHARACTERS:

- **ectotympanic ring fused to petrosal part of bulla**
- **upper molars quadritubercular, with enlarged hypocone separated from trigon on distinct shelf** (fig. 305)
- **2nd digit of forefoot reduced**

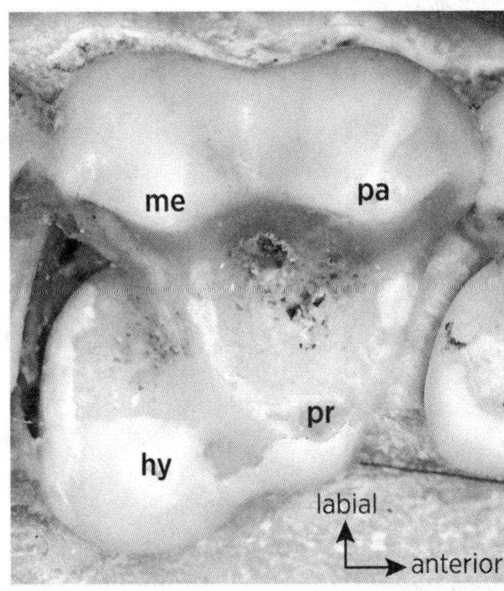

FIG. 305. Right upper first molar (M1) of the greater galago, *Otolemur crassicaudatus* (arrows indicate orientation, labial and anterior; pa = paracone, me = metacone, pr = protocone, hy = hypocone).

Family GALAGIDAE (galagos or bush babies)

Small to medium-sized (body length 75–465 mm; mass 50 g–2 kg); body slender; facial region reduced, eyes large, ears large and naked; tail equal to or longer than body (110–550 mm); limbs long, **hind limbs much longer than forelimbs** (intermembral index 52–70); forelimb 68%–77% and hind limb 104%–128% of trunk length); hind foot elongate, especially tarsus, for springing locomotion; **digits more slender than in lorisids; tail long and well furred**; fur generally woolly,

FIG. 306. Galagidae; lesser bush baby, *Galago senegalensis*.

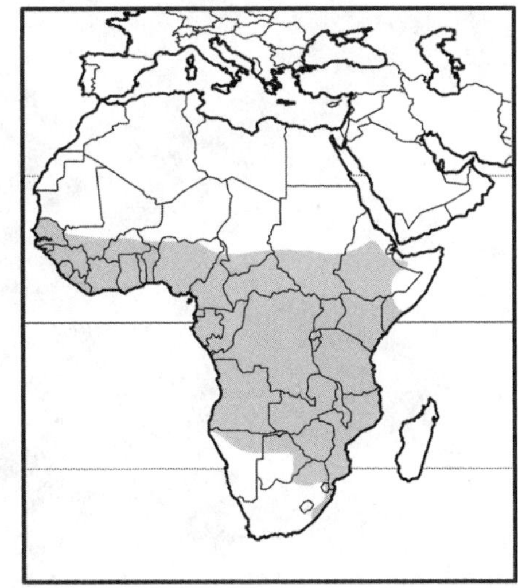

FIG. 307. Geographic range of family Galagidae: Africa, south of the Sahara.

gray to brown to black. Active and quick, known for leaping ability; some highly saltatorial; range from insectivorous to frugivorous to feeding exclusively on gum; some are canopy specialists, others live in undergrowth or are less specialized; nocturnal. Inhabit wet and dry forest, some in savannas.

The family name spelling Galagonidae has historical precedence, but the ICZN conserved the more commonly used Galagidae in 2002 (see Groves 2005b).

RECOGNITION CHARACTERS:

1. **skull with long rostrum**
2. **braincase globular, without a strong temporal ridge**
3. **postorbital bar not especially broad**
4. **external auditory meatus separate from zygomatic branch of squamosal**
5. cranial blood supply from ascending pharyngeal artery; stapedial artery and canal small
6. molars quadritubercular
7. baculum present in males

DENTAL FORMULA: $\frac{2}{2}\ \frac{1}{1}\ \frac{3}{3}\ \frac{3}{3} = 36$

TAXONOMIC DIVERSITY: About 18 species in five genera: *Euoticus* (needle-clawed bush babies), *Galago* (bush babies, lesser galagos), *Galagoides* (dwarf galagos), *Otolemur* (greater galagos), and *Sciurocheirus* (squirrel galagos).

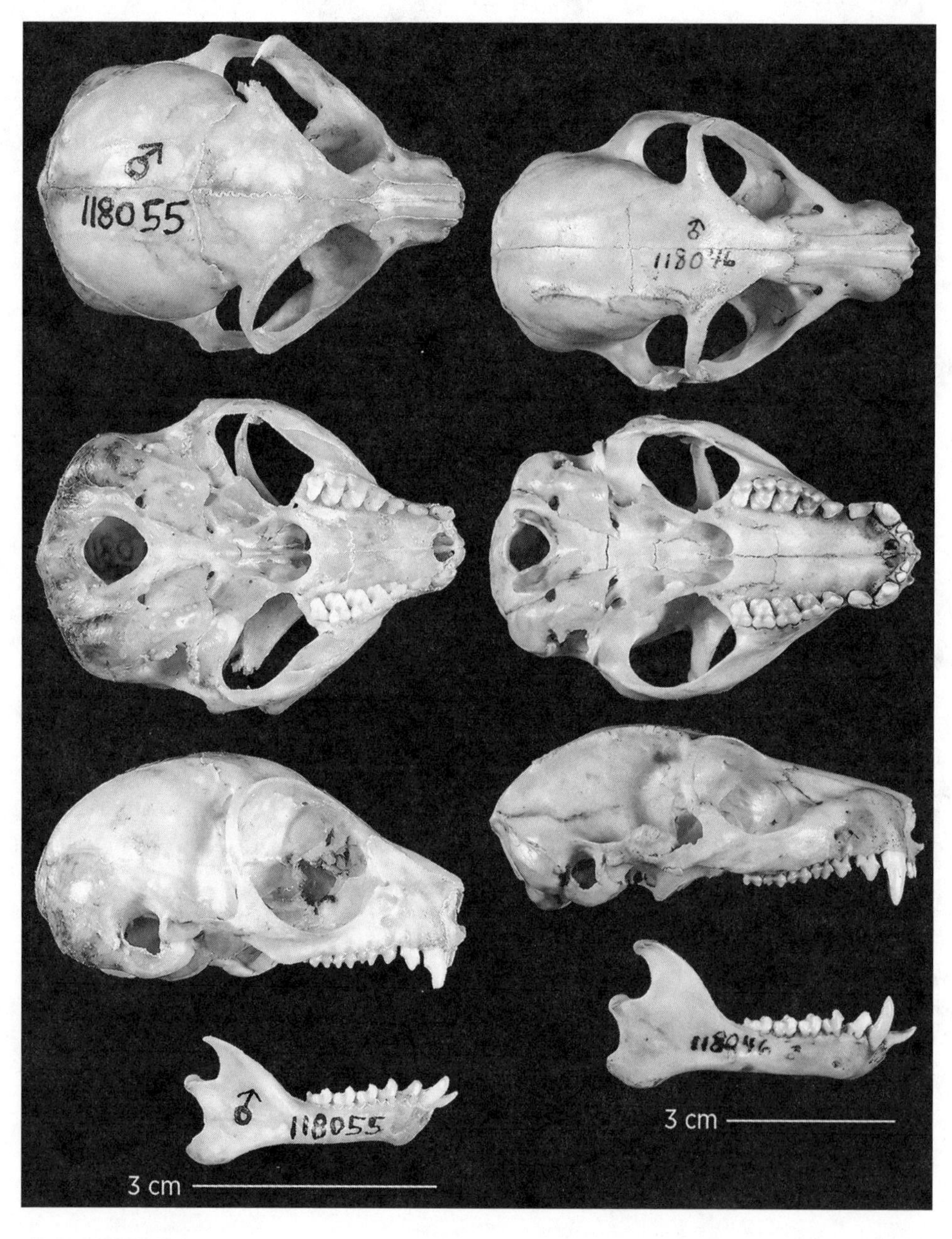

FIG. 308. Galagidae; skull and mandible of the lesser bush baby, *Galago senegalensis* (*left*), and brown greater galago, *Otolemur crassicaudatus* (*right*).

Family LORISIDAE (pottos, lorises)

Small to medium-sized (body length 190–370 mm; mass 200 g–1.5 kg); body chunky; tail short (about 15% of body length) or absent; **forelimbs and hind limbs subequal in length** (intermembral index 80–91), 77%–100% of trunk length; head rounded, facial region flattened, ears small; hands and feet robust, **digits stout and strong**, hallux and pollex strongly opposing other digits; **tail short** (perodictines) **or absent** (lorisines); fur generally short and woolly, gray to brown in color, with dorsal stripe and circumocular rings present in some. Insectivorous to carnivorous, but also feed on vegetable matter, such as fruits and gum; mostly nocturnal; almost exclusively arboreal; includes the only known venomous primate (*Nycticebus*), which produces toxin in specialized underarm glands, which is activated by mixing with saliva and delivered by biting. Climb with slow stealth; never jump or leap. Habitat is restricted to wet rainforest.

The family name spelling Loridae has historical precedence, but the ICZN conserved the more commonly used Lorisidae in 2002 (see Groves 2005b).

FIG. 309. Lorisidae: Lorisinae; red slender loris, *Loris tardigradus*.

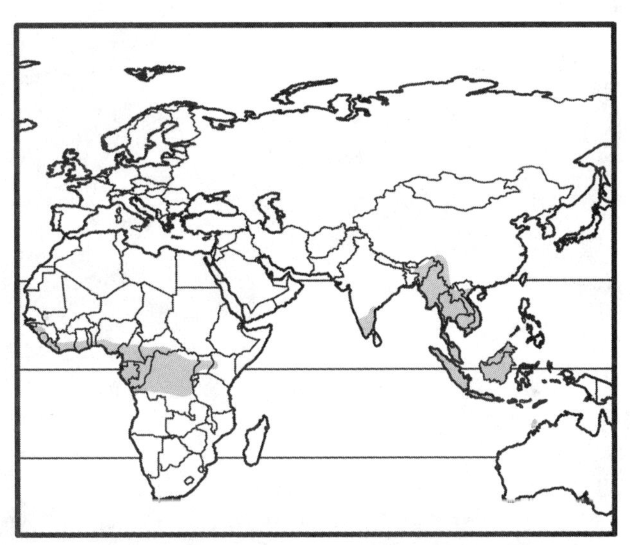

FIG. 310. Geographic range of family Lorisidae: tropical sub-Saharan Africa (Perodictinae); India and Sri Lanka; Southeast Asia; Indonesia, Philippines (Lorisinae).

RECOGNITION CHARACTERS:

1. **rostrum short**
2. **braincase rounded, with well-defined temporal ridges**
3. **postorbital bar very broad**
4. **external auditory meatus contacts zygomatic part of squamosal**
5. cranial blood supply from ascending pharyngeal artery; stapedial artery and canal small
6. molars quadritubercular
7. baculum present in males

DENTAL FORMULA: $\frac{2}{2}\ \frac{1}{1}\ \frac{3}{3}\ \frac{3}{3} = 36$

TAXONOMIC DIVERSITY: 12–13 species in four to five genera allocated to two subfamilies:

Lorisinae—*Loris* (slender lorises) and *Nycticebus* (slow lorises)

Perodicticinae—*Arctocebus* (pottos or angwantibos), *Perodicticus* (pottos), and *Pseudopotto* (false potto, included in *Perodicticus* by some)

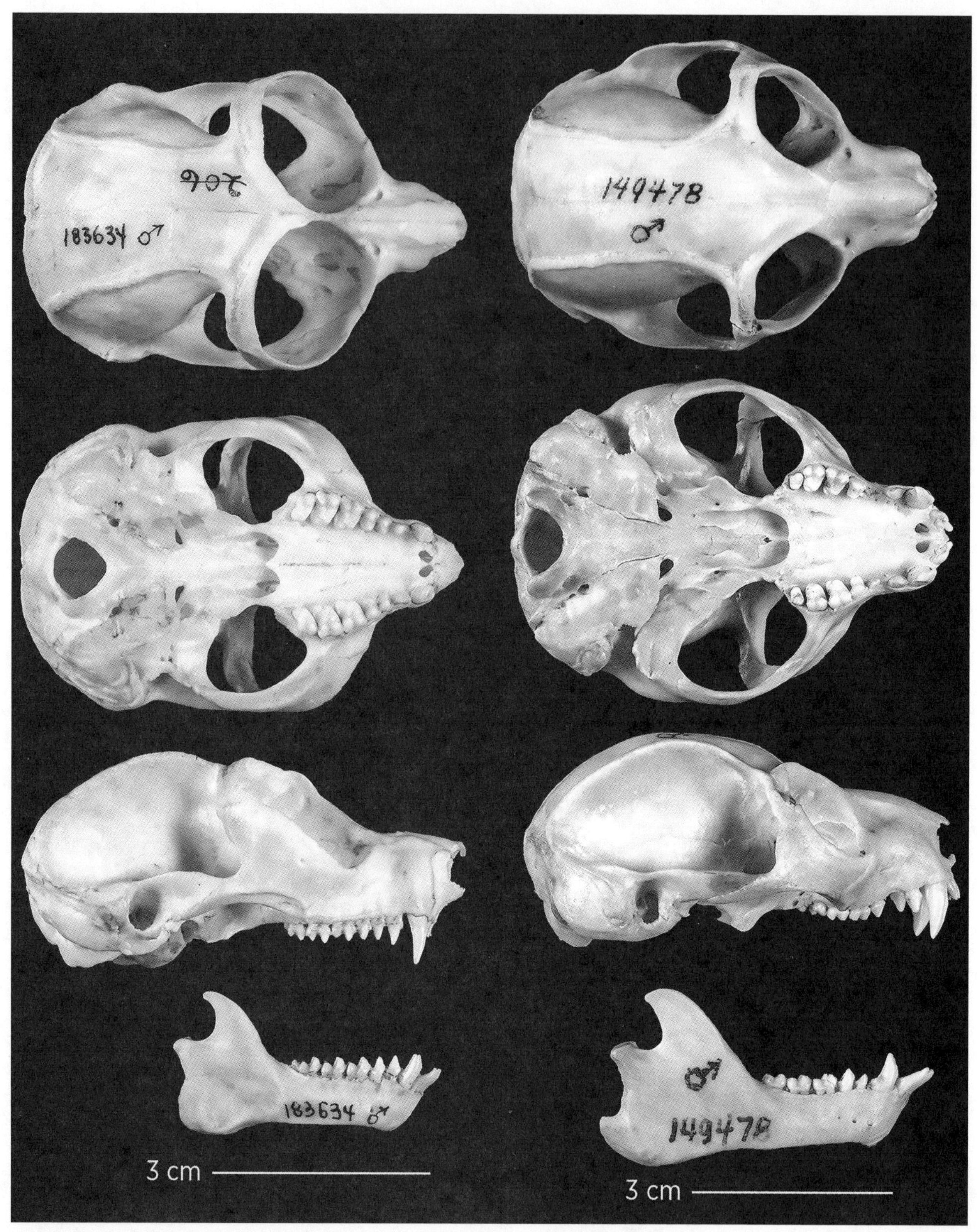

FIG. 311. Lorisidae; skull and mandible of a red slender loris, *Loris tardigradus* (Lorisinae; *left*) and a potto, *Perodicticus potto* (Perodictinae; *right*).

Suborder HAPLORRHINI (tarsiers, New World monkeys and tamarins, Old World monkeys, gibbons, and great apes)

DIAGNOSTIC CHARACTERS:

- **nose with simple nostrils, no rhinarium, philtrum vestigial or absent**, hence upper lip entirely free of upper gum; internarial breadth wide in *Tarsius* and all platyrrhines (with nostrils directed somewhat laterally) and comparatively narrow in all catarrhines (with nostrils directed forward) (comparison to rhinarium of suborder Strepsirrhini is given in that account)
- **orbital and temporal openings separated by postorbital plate**
- **braincase relatively large and rounded, not elongate**
- **nasolacrimal duct short and oriented vertically**
- cranial blood supply provided by the promontory branch of the internal carotid artery (in Strepsirrhini, brain receives blood through either the stapedial branch of the internal carotid or from the ascending pharyngeal branch of the external carotid)
- **foramen magnum directed more or less ventrally**
- **eyes face forward, good stereoscopic vision, lack a tapetum lucidum** (hence, no eyeshine); have **retinal fovea on posterior surface of eyeball**

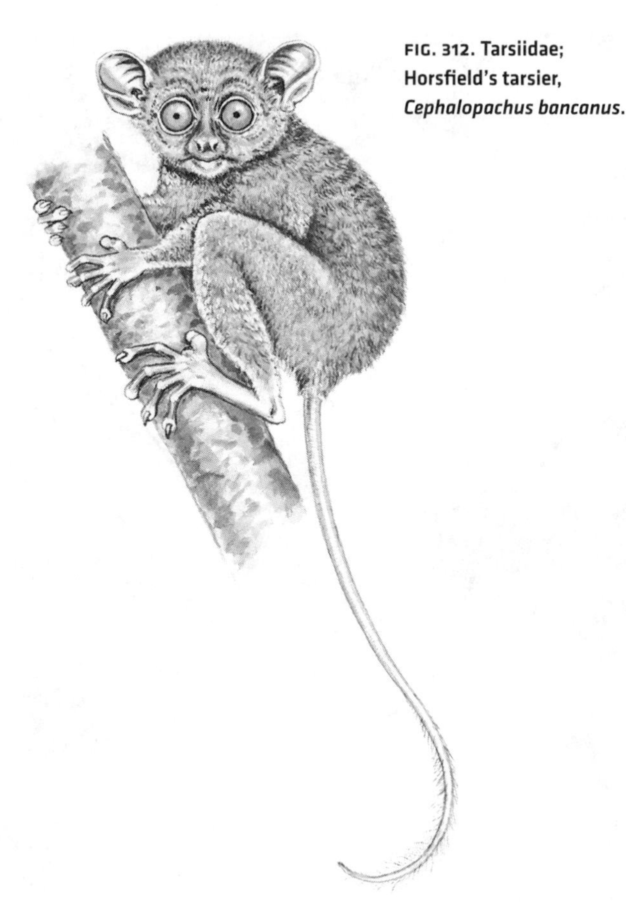

FIG. 312. Tarsiidae; Horsfield's tarsier, *Cephalopachus bancanus*.

Infraorder TARSIIFORMES

Family TARSIIDAE (tarsiers)

Small (body length 85–160 mm; mass 85–165 g); head globular, eyes enormous, ears conspicuous, membranous, mobile, and capable of folding; face flattened, nose covered by short hairs; neck short; hind limbs much longer than forelimbs (intermembral index 52–58; hind limb 158%–192% of trunk, forelimb 92%–105% of trunk); digits elongate, attenuate, and terminating in large pads; **all digits bear flat nails except 2nd and 3rd pedal digits, which have claws; tail longer than body** (135–275 mm), non-prehensile, naked to sparsely haired but **distally tufted** with **friction pad near base; fur silky and wavy.** Primarily arboreal, with vertical clinging and leaping common means of traveling, but do descend to ground. Insectivorous to carnivorous; crepuscular or nocturnal; social system from solitary to family groups. Habitat includes primary and secondary lowland and montane tropical forests.

FIG. 313. Geographic range of family Tarsiidae: Indonesian and Philippine islands.

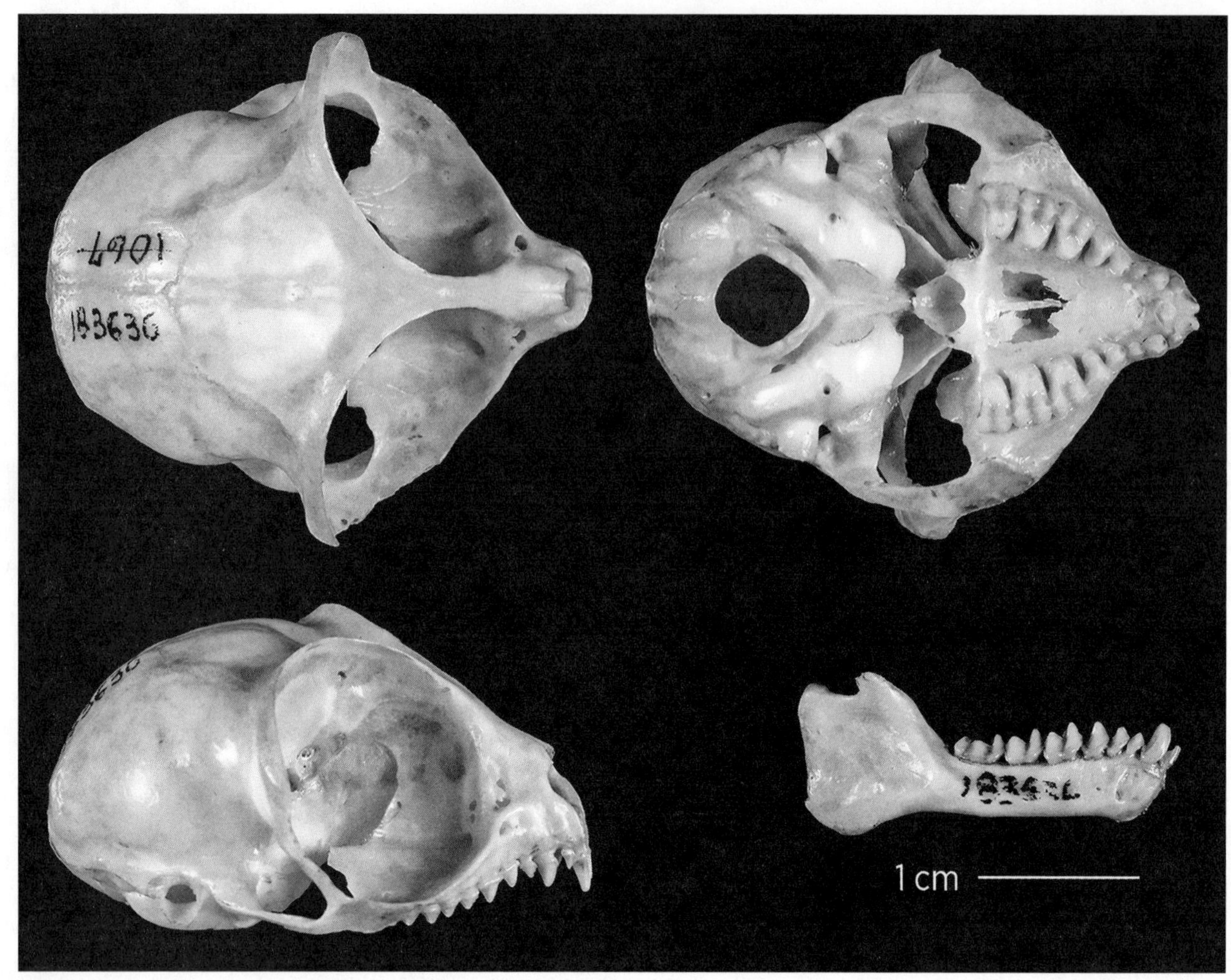

FIG. 314. Tarsiidae; skull and mandible of a tarsier, *Tarsius* sp.

RECOGNITION CHARACTERS:

1. rostrum short
2. **orbits enormous, with prominent ridge around each**
3. orbit incompletely walled off from temporal fossa
4. **auditory bulla extending laterally as a bony tube**
5. **upper and lower incisors well developed, not equal in size; no median diastema in upper incisors; lower incisors vertical, not procumbent**
6. **molars tritubercular**
7. no baculum in males

DENTAL FORMULA: $\frac{2}{1}\ \frac{1}{1}\ \frac{3}{3}\ \frac{3}{3} = 34$

TAXONOMIC DIVERSITY: About 11 species in three genera: *Cephalopachus* (western tarsier; Borneo and Sumatra, and nearby small islands), *Tarsius* (tarsiers; restricted to Indonesian island of Sulawesi and adjacent islands), and *Carlito* (Philippine tarsier; the Philippines).

Infraorder SIMIIFORMES (= Anthropoidea)

Simian, or anthropoid, primates—the so-called higher primates—include the majority of primate species. Historically one of two primary lineages of primates (along with Prosimii, which included Strepsirrhini + Tarsiidae), Simiiformes is now recognized as a well-supported lineage within the Haplorrhini.

RECOGNITION CHARACTERS: Simian primates are readily distinguished from strepsirrhine and tarsiiform primates by a suite of characters, the principal of which are the following:

1. **fused frontal bones**
2. **fused mandibular symphysis**

3. **complete postorbital plate (frontal, zygomatic [= jugal or malar], sphenoid bones)**
4. **lacrimal bone housed within the orbit**
5. upper molars with well-developed hypocones, yielding quadrate tooth structure
6. **nails on all digits**; no claws (but tegulae [= clawlike nails] present in Callitrichinae)

Parvorder PLATYRRHINI

Platyrrhine primates are exclusive to the New World. They were traditionally divided into the families Callitrichidae (tamarins and marmosets) and Cebidae (howler, spider, capuchin, woolly, squirrel, owl, titi, saki, and other monkeys), with Goeldi's marmoset (*Callimico*) sometimes placed in its own family, the Callimiconidae. Cladistic analyses of both morphological and molecular data have falsified these earlier hypotheses, but the two data types have not as yet converged on a common topology depicting platyrrhine relationships and thus a common family-level classification. Both data types support elevation of the titi, saki, and uacari to the family Pitheciidae and the howler, spider, and woolly monkeys to the family Atelidae. They disagree, however, on relationships among platyrrhine groups, and thus the categorical level at which to recognize the night monkey (*Aotus*) and tamarins + marmosets with respect to the capuchins (*Cebus*) and squirrel monkeys (*Saimiri*). Both Groves (2005b) and Mittermeier et al. (2013) place *Aotus* in its own family (Aotidae), but they differ in their treatment of the tamarins + marmosets versus capuchins + squirrel monkeys (see Phylogeny and Classification of Primates, above). High-resolution DNA sequence analyses, however, support only three family-level taxa within the platyrrhines, concatenating *Aotus* + tamarins/marmosets + capuchins/squirrel monkeys as subfamilies within the family Cebidae (fig. 315).

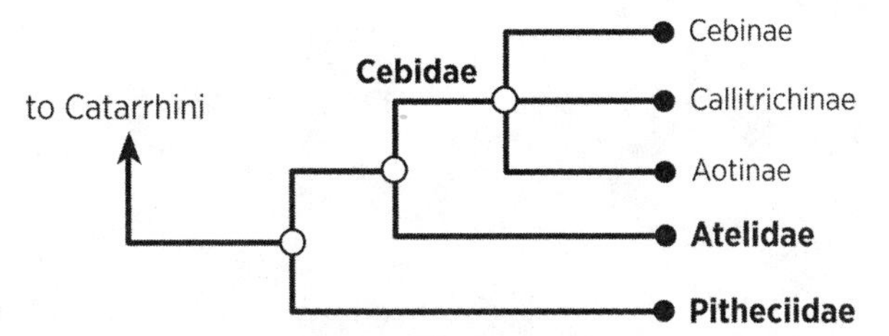

FIG. 315. Phylogenetic relationships and supported classification of the New World platyrrhine primates. After Wildman et al. (2009); Perelman et al. (2011); and Fleagle (2013).

RECOGNITION CHARACTERS: Platyrrhines can be differentiated from the Old World catarrhine monkeys and apes by the following characters:

1. **three premolars above and below**, instead of two
2. **nostrils directed laterally**, as opposed to forward (giving nose a flattened appearance, hence the name (*platy-* = [Greek] "broad, flat")
3. **prehensile tail present** in some genera (unique among primates)
4. no ischial callosities (also absent from the catarrhine family Hominidae)
5. **dichromatic color vision**, rather than trichromatic (except howler monkey, *Alouatta*, which has trichromatic color vision, and night monkeys, *Aotus*, which have monochromatic cone pigments and therefore lack color vision)
6. **parietal in contact with zygomatic (= jugal or malar) (and frontal not contacting sphenoid)**
7. **ectotympanic ring-shaped, not extending laterally as a bony tube**

We follow the three-family classification supported by molecular analyses, but emphasize the unresolved questions regarding the phylogenetic relationships among members of the Cebidae and the classificatory implications of alternative topologies.

Family ATELIDAE (*howler, spider, and woolly monkeys*)

Size large (body length 400–900 mm; mass 6–12 kg); all with **long, prehensile tail with friction ridges along distal part of ventral surface**; hind limbs subequal to or shorter than forelimbs (intermembral index 97–109); pollex reduced to absent, grasp with 2nd and 3rd elongate manal digits (schizodactyly); arboreal but quadrupedal "reachers and pullers," rarely leap; all are group living. Dental and cranial anatomy as well as diet and social organization very diverse. Two subfamilies: Howler monkeys (Alouattinae, *Alouatta*) highly sexually dimorphic; subequal forelimbs and hind limbs (intermembral index 97–98); folivorous; unique among platyrrhines with trichromatic color vision; inhabit lowland rainforests, secondary forests, dry deciduous forests, and montane forests. Woolly monkeys (Atelinae, *Lagothrix* and *Oreonax*) sexually dimorphic; subequal forelimbs and hind limbs (intermembral index 98); fur long, soft, and dense; primarily frugivorous; inhabit high rainforest, gallery forests, and montane cloud forests. Spider monkeys (Atelinae, *Ateles*) have very long forelimbs (intermembral index 103–109); arboreal, locomotion is quadrupedal and suspensory, including brachiation (forefeet lack pollex) and climbing, leaping on occasion; feed in suspensory mode, hanging by a combination of four limbs and tail; feed primarily on fruits or new leaves; habitat primary rainforests, with preference to upper canopy. The muriqui (Atelinae, *Brachyteles*) is the largest non-human primate in the New World; resembles *Ateles* in body and limb proportions (intermembral index 104) and lack of a pollex; small incisors and occlusal surface of molars more like that of *Alouatta* than *Ateles*; primarily folivorous, secondarily frugivorous; inhabit primary rainforest of Brazil's Atlantic coast.

FIG. 316. **Atelidae; Yucatan black howler monkey, *Alouatta pigra* (Alouattinae; *top*), and Geoffroy's spider monkey, *Ateles geoffroyi* (Atelinae; *bottom*).**

RECOGNITION CHARACTERS:

Subfamily Alouattinae:

1. skull with small braincase; limited cranial flexion
2. incisors small; canines large, strongly sexually dimorphic (males 26–43% larger)
3. **molars with labial shearing crests**
4. **mandible very large and deep**
5. **hyoid expanded into large, hollow resonating chamber**

Subfamily Atelinae:

1. braincase larger, globular
2. incisors large; canines moderate to small, sexual dimorphism limited (males 7% [*Oreonax*] to 22% [*Lagothrix*] larger)
3. **molars bunodont and brachydont** for crushing (spider and woolly monkeys) **or with lingual shearing crests** (muriqui)
4. mandible relatively shallow
5. baculum present only in male *Brachyteles*

DENTAL FORMULA: $\frac{2}{2}\ \frac{1}{1}\ \frac{3}{3}\ \frac{3}{3} = 36$

TAXONOMIC DIVERSITY: 25 species in four to five genera allocated to two subfamilies:

Alouattinae—*Alouatta* (howler monkeys)

Atelinae—*Ateles* (spider monkeys), *Brachyteles* (muriqui or woolly spider monkey), *Lagothrix* (woolly monkeys), and *Oreonax* (yellow-tailed woolly monkey; included in *Lagothrix* by some)

FIG. 317. Geographic range of family Atelidae: Neotropics from eastern and southern Mexico south to eastern Paraguay, northeastern Argentina, and southern Brazil.

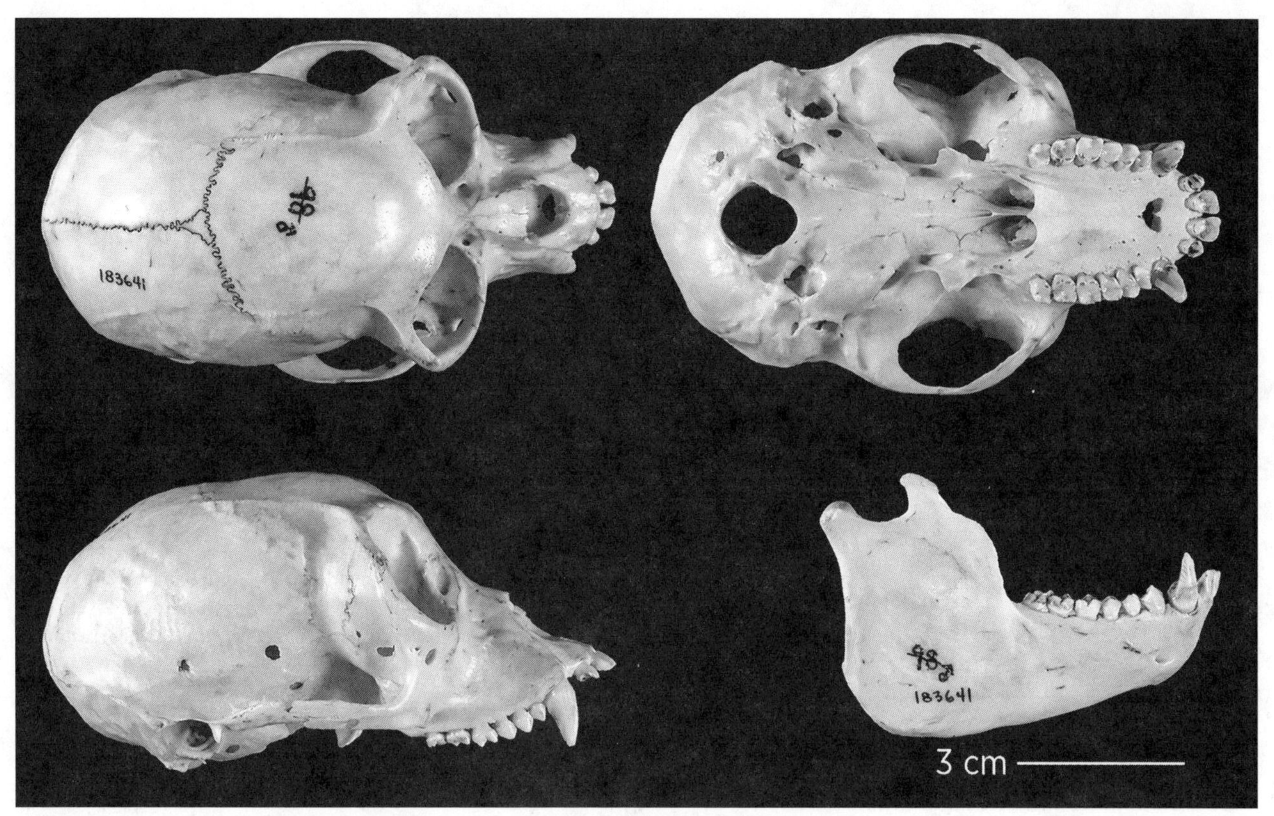

FIG. 318. Atelidae: Atelinae; skull and mandible of Geoffroy's spider monkey, *Ateles geoffroyi*.

Family CEBIDAE (*New World monkeys*)

The Cebidae includes Aotidae and Callitrichidae of other classifications. Members are highly varied in size, external attributes, craniodental characteristics, and ecology. Pollex not or only slightly opposable (hallux opposable). Aotinae (night or owl monkeys) are medium-sized (body length 240–370 mm; mass 700 g–1.3 kg) with tail longer than body (315–400 mm); rounded head and short face, with small round ears often hidden in fur; these are **the only nocturnal simiiform primates**, but lack tapetum lucidum; lack color vision; arboreal, predominantly quadrupedal but do leap; primarily frugivorous; monogamous, not sexually dimorphic; habitat lowland rainforest. Cebinae (capuchins and squirrel monkeys) are medium-sized (body length 260–570 mm; mass 650 g–3.7 kg); tail relatively short (300–560 mm), well furred, prehensile (except in adult *Saimiri*); hind limbs longer than forelimbs (intermembral index 80–83); fingers of forefeet short, pollex opposable (*Cebus* and *Sapajus*) or non-opposable (*Saimiri*); digits bear flattened or narrow, keeled nail; sexually dimorphic, group living; all species arboreal quadrupeds, capuchins use prehensile tail mainly while feeding; diet omnivorous, including hard nuts, fruits, and animal matter; highly social; inhabit virtually all Neotropical forest types. Callitrichinae (marmosets and tamarins) are small (body length 130–370 mm; mass 100–750 g), with long trunk, tail, and legs; hind limbs longer than forelimbs (intermembral index 69–89); all digits except hallux with clawlike rather than flattened nails (tegulae), an adaptation enabling trunk clinging while feeding on gum, sap, and insects; exploit marginal and disturbed forests; locomotor adaptations for quadrupedal walking, running, and leaping; social system varied, including monogamy, polygyny, polyandry, and polygynandry; females of most genera give birth to dizygotic twins.

RECOGNITION CHARACTERS:

Subfamily Aotinae:

1. large digital pads on manus and pes
2. pollex slightly opposable
3. **compressed, clawlike grooming nail on 2nd hind digit**
4. **orbits very large** (largest proportionally of any simiiform)

5. baculum present in males

Subfamily Callitrichinae:

1. canine subequal to incisors; larger in *Saguinus*
2. **molars tritubercular, lacking hypocone**
3. **only two molars above and below** in all genera except *Callimico*, with three (*Callimico* also has a small hypocone)
4. baculum present in males

Subfamily Cebinae:

1. **large premolars and quadrate molars; occlusal surface with low cusps** (*Cebus*) **or with sharp cusps** (*Saimiri*)
2. canines sexually dimorphic, large in males
3. baculum present in males

FIG. 319. Cebidae; gray-bellied night monkey, *Aotus lemurinus* (Aotinae; *top left*), Geoffroy's tamarin, *Saguinus geoffroyi* (Callitrichinae; *top right*), and white-headed capuchin, *Cebus capucinus* (Cebinae; *bottom*).

DENTAL FORMULA:

Aotinae and Cebinae: $\frac{2}{2}\ \frac{1}{1}\ \frac{3}{3}\ \frac{3}{3} = 36$

Callitrichinae: $\frac{2}{2}\ \frac{1}{1}\ \frac{3}{3}\ \frac{2\text{–}3}{2\text{–}3} = 32\text{–}36$

FIG. 320. Geographic range of family Cebidae: Neotropics from southern Mexico to northern Argentina.

TAXONOMIC DIVERSITY: About 90 species in 11 genera placed in three subfamilies, one with two tribes (treated as families and subfamilies, respectively, by some authors):

Aotinae—about 10 species in the genus *Aotus* (night monkeys)

Callitrichinae—*Callibella* (black-crowned dwarf marmoset), *Callimico* (Goeldi's marmoset), *Callithrix* (marmosets), *Cebuella* (pygmy marmoset), *Leontopithecus* (lion tamarins), *Mico* (marmosets), and *Saguinus* (tamarins)

Cebinae:

Cebini—*Cebus* and *Sapajus* (capuchins)

Saimiriini—about seven species in the genus *Saimiri* (squirrel monkeys)

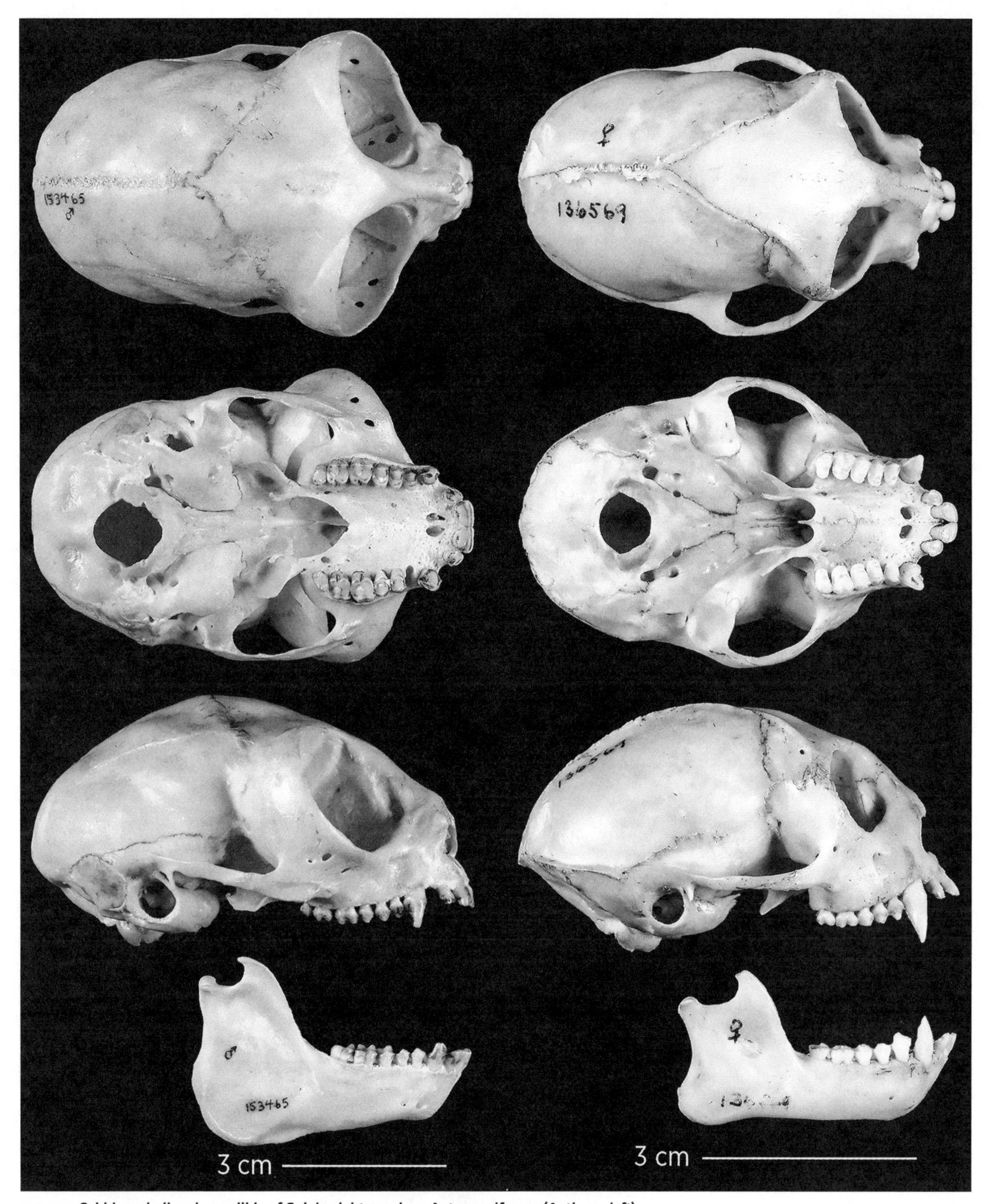

FIG. 321. Cebidae; skull and mandible of Spix's night monkey, *Aotus vociferans* (Aotinae; *left*), and the emperor tamarin, *Saguinus imperator* (Callitrichinae; *right*).

Family PITHECIIDAE
(*titi, uacari, and saki monkeys*)

Variable in size and external morphology, but less so than Atelidae or Cebidae. Two subfamilies: Callicebinae (titi monkeys) are least specialized in body form and dental characteristics; medium-sized (body length 230–460 mm; mass 800 g–1.2 kg); face short; body fluffily hairy; relatively large ears generally hidden by fur; tail long (260–550 mm), non-prehensile, and fluffy; legs long, with hind limbs longer than forelimbs (intermembral index 73–79); arboreal quadrupeds and leapers; mainly frugivorous; live in monogamous family groups; inhabit mature rainforests. Pitheciinae (saki and uacari monkeys) are larger (body length 300–705 mm; mass 1.6–4.5 kg); tail length subequal to that of body (255–545 mm) or considerably shorter (125–210 mm); hind legs longer than forelimbs, but less so than in Callicebinae (intermembral index 75–83); some quadrupedal, others among most saltatorial of New World primates, moving by spectacular leaps; both bearded sakis (*Chiropotes*) and uacaris (*Cacajao*) often feed using hind-limb suspension; most specialize on fruits with hard outer shells; social system extremely diverse, from monogamy to large groups; inhabit rainforests, including montane savanna forests, flooded lowland forests, dry forests, and liana forests.

FIG. 322. Pitheciidae: Pitheciinae; white-faced saki, *Pithecia pithecia* (*left*), and bald uacari, *Cacajao calvus* (*right*).

FIG. 323. Geographic range of family Pitheciidae: tropical South America east of the Andes.

RECOGNITION CHARACTERS:

Subfamily Callicebinae:

1. short muzzles, long skulls
2. **canines short, not sexually dimorphic**
3. molar teeth with simplified occlusal surface
4. baculum present in males

Subfamily Pitheciinae:

1. **prognathous snout**
2. **palate narrow, U-shaped**
3. **nasal bones enlarged**
4. **large, procumbent incisors**
5. **canines robust**
6. **premolars and molars relatively small, square, with low cusps**
7. baculum present only in male *Pithecia*

DENTAL FORMULA: $\frac{2}{2}\ \frac{1}{1}\ \frac{3}{3}\ \frac{3}{3} = 36$

TAXONOMIC DIVERSITY: About 44 species in four genera allocated to two subfamilies:

Callicebinae—about 31 species in the genus *Callicebus* (titi monkeys)

Pitheciinae—*Cacajao* (uacaris), *Chiropotes* (bearded sakis), and *Pithecia* (sakis)

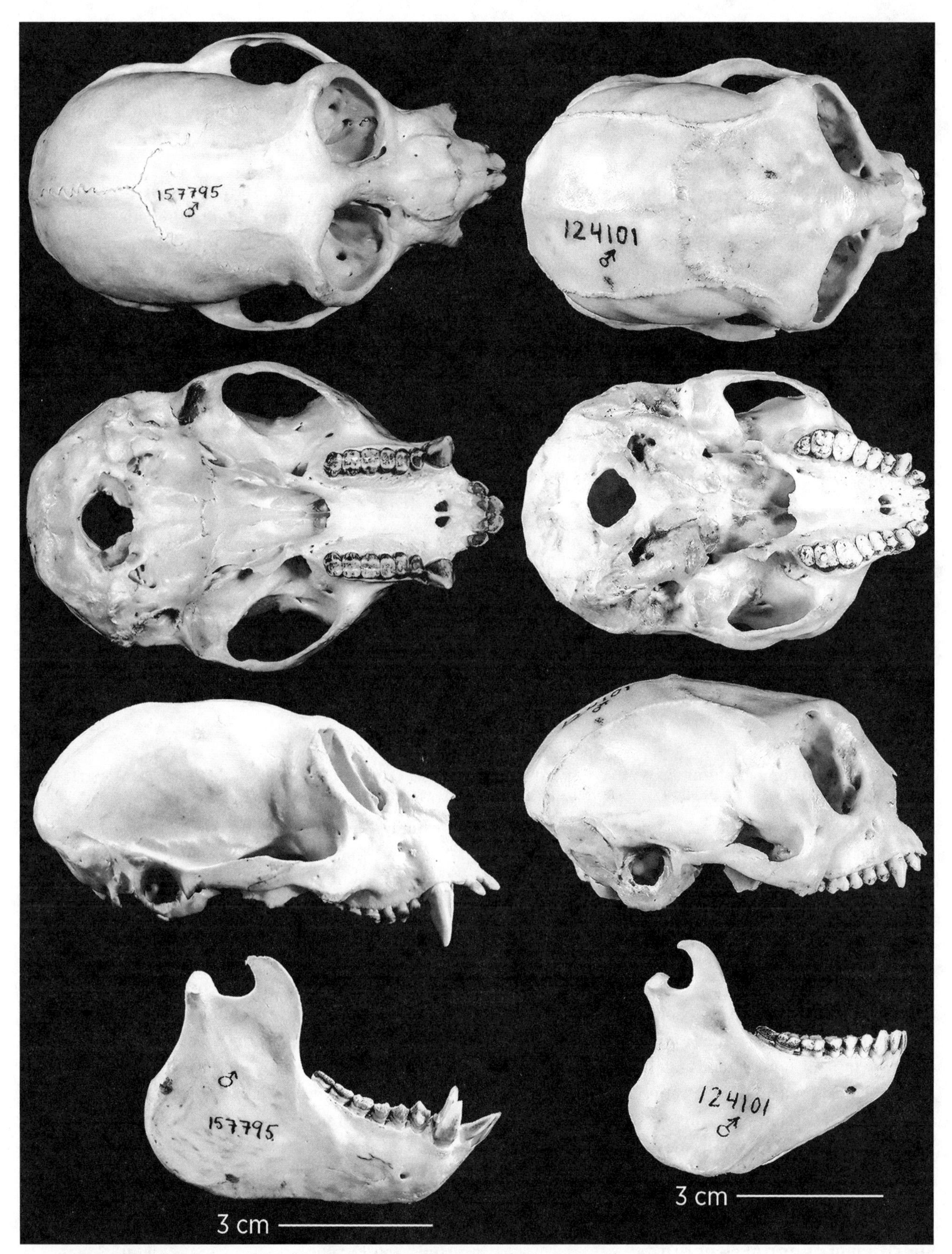

FIG. 324. Pitheciidae; skull and mandible of the Napo saki, *Pithecia napensis* (Pitheciinae; *left*), and the red-bellied titi, *Callicebus moloch* (Callicebinae; *right*).

Parvorder CATARRHINI

The catarrhines are exclusively Old World primates (except humans, now cosmopolitan). They include three families: Cercopithecidae (Old World monkeys, such as macaques, baboons, mandrills, langurs, and leaf-eating monkeys), Hylobatidae (gibbons), and Hominidae (great apes, including humans). Older classifications include the gibbons and great apes together in the family Pongidae and place humans (*Homo*) in the separate family Hominidae. However, both morphological and molecular data support the very close affinity of humans with chimpanzees (*Pan*) and gorillas (*Gorilla*), with the orangutan (*Pongo*) more distantly related to these three genera. Some recent classifications thus place *Gorilla*, *Homo*, and *Pan* in the family Hominidae and *Pongo* in a monotypic family, Pongidae; we follow Williamson et al. (2013) and present these two groups as subfamilies.

RECOGNITION CHARACTERS: Catarrhines can be distinguished from New World platyrrhines by the following characters:

1. **two premolars above and below**, instead of three
2. **nostrils close together and directed forward and downward, not laterally** (*cata-* = [Greek] "down, downward")
3. prehensile tail absent in all taxa, even those with long tails
4. lack an entepicondylar foramen on the humerus (present in some, but not all, Platyrrhini; see Garbino and de Aquino 2018).
5. **ischial callosities present** (except in most Hominidae)
6. trichromatic color vision (dichromatic in most platyrrhines)
7. **skull typically robust and heavily ridged, often with a sagittal crest**
8. **frontal in contact with sphenoid (and parietal not in contact with zygomatic [= jugal or malar])**
9. **ectotympanic bone extends laterally to form a tubular external auditory meatus** (see fig. 286)
10. **bony shelf (simian shelf) present at posterior border of jaw symphysis in most taxa** (notably absent in humans)

Superfamily CERCOPITHECOIDEA

The superfamily Cercopithecoidea is the anatomically most diverse and numerically successful lineage of living catarrhine primates. Characters distinguishing both fossil and living members of the superfamily Cercopithecoidea (family Cercopithecidae) from those of the superfamily Hominoidea (living families Hylobatidae and Hominidae) include the following:

DIAGNOSTIC CHARACTER:

- **upper and lower molars quadritubercular, square to rectangular, and bilophodont, with anterior (paracone-protocone/protoconid-metaconid) and posterior (metacone-hypocone/hypoconid-entoconid) pairs of cusps aligned to form transverse ridges** (fig. 325)

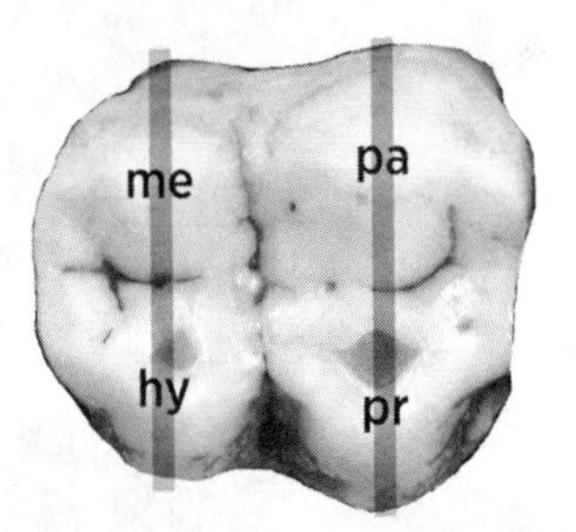

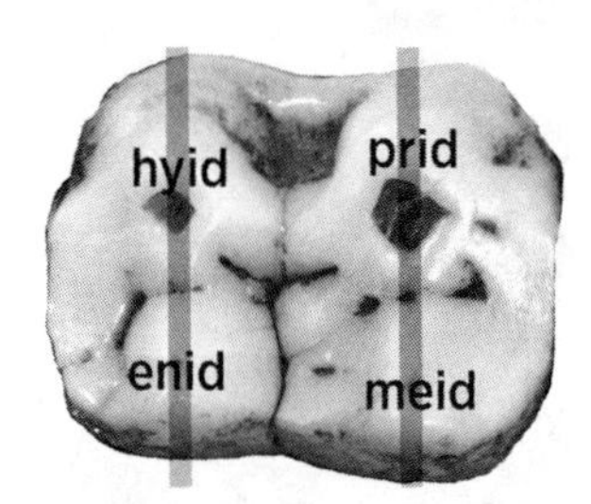

Fig. 325. Right M2 (*left*; pr = protocone, pa = paracone, me = metacone, hy = hypocone) and left m2 (*right*; prid = protoconid, meid = metaconid, hyid = hypoconid, and enid = entoconid) of the olive baboon (*Papio anubis*). Vertical lines indicate orientation of cusps that form transverse ridges.

RECOGNITION CHARACTERS (CERCOPITHECOIDEA VERSUS HOMINOIDEA):

1. **long versus short trunk**
2. **usually long tail versus no tail**
3. **narrow versus wide nasal aperture**
4. **narrow versus wide palate**

Family CERCOPITHECIDAE (macaques, baboons, langurs, leaf monkeys, and others)

Old World monkeys are exceedingly diverse in body size and form. Body size varies greatly, from about 1 kg (talapoin monkey, *Miopithecus*) to nearly 50 kg (mandrill, *Mandrillus*), and many aspects of form and other anatomical attributes, as well as behaviors and ecology, are related to dietary and habitat specializations. Terrestrial forms (e.g., baboons, *Papio*, and mandrills, *Mandrillus*) have stocky builds, long faces (and skulls with an elongate rostrum), a short to vestigial tail, forelimbs subequal to or slightly longer than hind limbs (intermembral index mostly above 90), short hands relative to the feet, 3rd manal digit longest, opposable pollex, and are quadrupedal and plantigrade. Arboreal monkeys are more gracile, have shorter faces (and short rostra), longer hind limbs relative to forelimbs (intermembral index mostly below 85), long tails, a short and less opposable pollex on the manus, and are quadrupedal but also capable of very long leaps through tree branches.

The family Cercopithecidae comprises two subfamilies that have notable differences in anatomy, many of which relate to dietary adaptations; differences between these groups are sufficient that some authors argue they should be elevated to family status. The Cercopithecinae comprises mostly frugivorous species (although their diets include invertebrates and vertebrates as well), whereas the Colobinae includes mostly folivores and granivores. Cercopithecines have cheek pouches; some colobines (e.g., *Colobus*, *Presbytis*) have specialized sacculated stomachs for fermentation. Cercopithecines have forelimbs and hind limbs that are generally similar in length, and they may have short or long tails; colobines have longer hind limbs and nearly all have long tails (the single exception is the pig-tailed langur, *Simias concolor*). Cranially, cercopithecines have a narrowed interorbital region, broader nasal openings, a shallower mandible, broader incisors, and cheek teeth with high crowns but relatively low cusps. The skull of colobines has a broader interorbital region, narrower nasal openings, a deeper mandible, narrower incisors, and cheek teeth with especially high and sharp cusps.

Most cercopithecids are arboreal, but some species of both subfamilies are extensively terrestrial; even the more terrestrial ones seek shelter in trees, especially at night. Social organization varies from small to quite large groups; males are commonly much larger than females; and all communicate by vocalizations and a variety of visual signals. Habitat is as varied as is size and body form, ranging from dense rainforest to mangrove swamps, open areas, and rocky areas, and from the tropics to temperate regions with strong seasonality.

FIG. 326. Cercopithecidae: Cercopithecinae: Cercopithecini; vervet monkey, *Chlorocebus pygerythrus*.

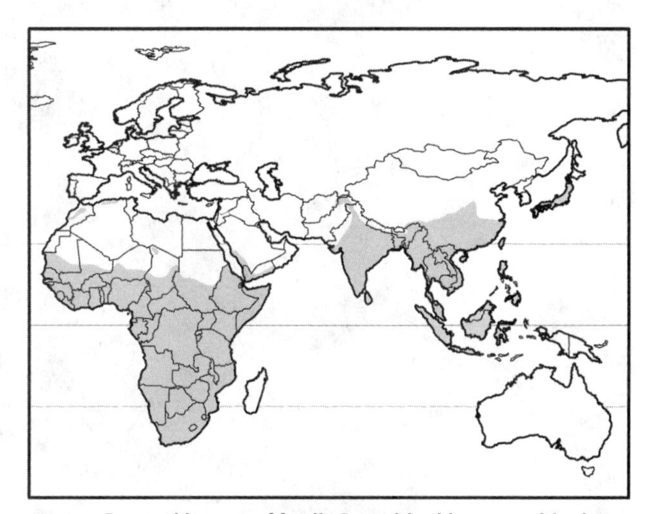

FIG. 327. Geographic range of family Cercopithecidae: cercopithecines are primarily African, with a single genus (*Macaca*) extending to Asia and the Iberian Peninsula; colobines occur in both Africa and Asia.

RECOGNITION CHARACTERS:

1. **ischial callosities present**
2. **auditory bulla extending laterally as a bony tube**
3. baculum present in males

DENTAL FORMULA: $\frac{2}{2}\ \frac{1}{1}\ \frac{2}{2}\ \frac{3}{3} = 32$

TAXONOMIC DIVERSITY: Nearly 160 species in about 23 genera in two subfamilies and four tribes. This group has been subject to substantial revision in recent years, with several genera being cleaved and a number of new species recognized.

Cercopithecinae:

Papionini—*Cercocebus* (capped mangabeys), *Lophocebus* (crested mangabeys), *Macaca* (macaques), *Mandrillus* (drill, mandrill), *Papio* (baboons), *Rungwecebus* (kipunji), *Theropithecus* (gelada baboon)

Cercopithecini—*Allochrocebus* (terrestrial guenons), *Allenopithecus* (Allen's swamp monkey), *Cercopithecus* (arboreal guenons), *Chlorocebus* (vervet monkeys), *Erythrocebus* (Patas monkey), and *Miopithecus* (talapoins)

Colobinae:

Colobini—*Colobus* (black and white colobus monkeys), *Piliocolobus* (red colobus monkeys), and *Procolobus* (olive colobus)

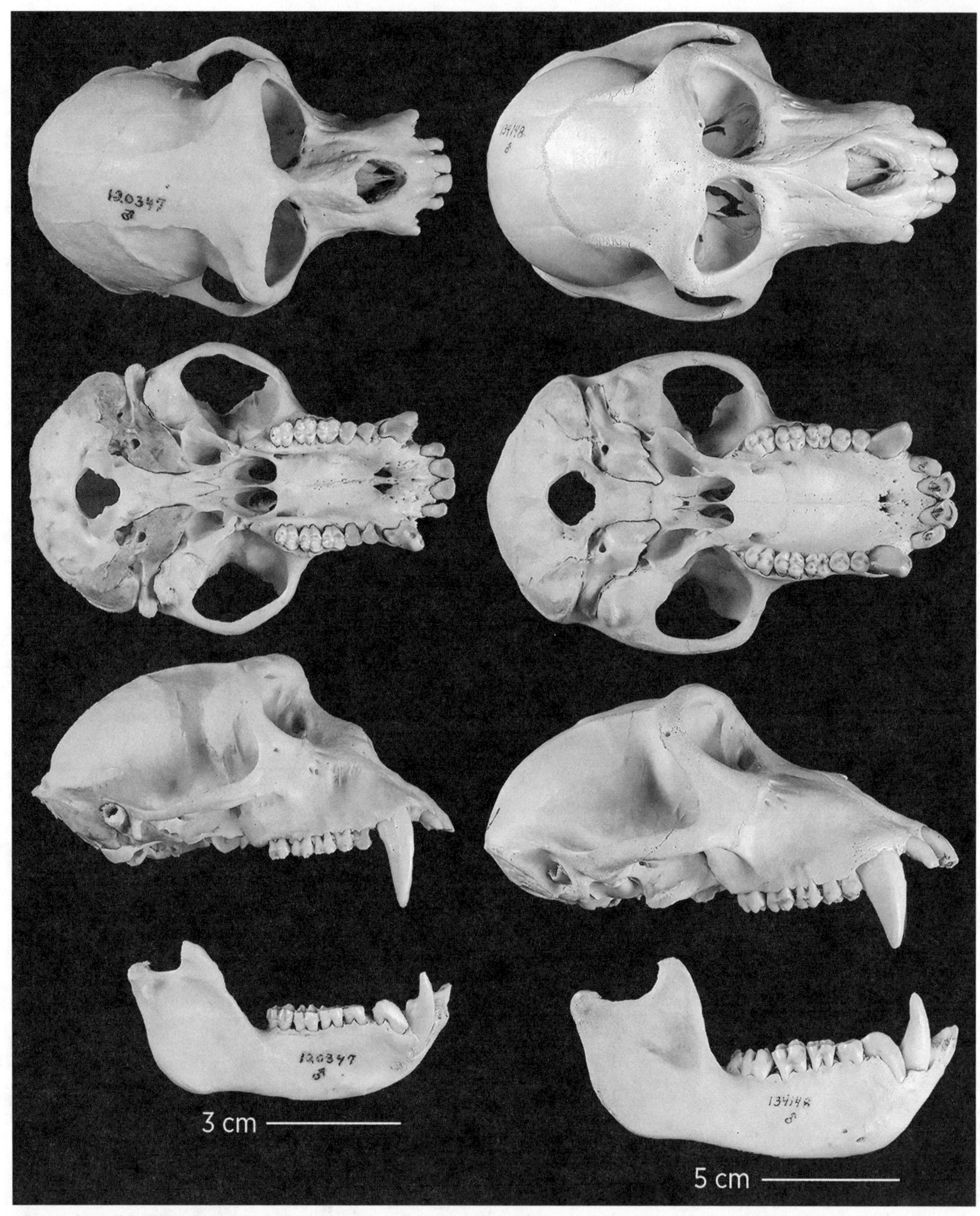

FIG. 328. Cercopithecidae: Cercopithecinae; skull and mandible of De Brazza's monkey, *Cercopithecus neglectus* (Cercopithecini; *left*), and the pig-tailed macaque, *Macaca nemestrina* (Papionini; *right*).

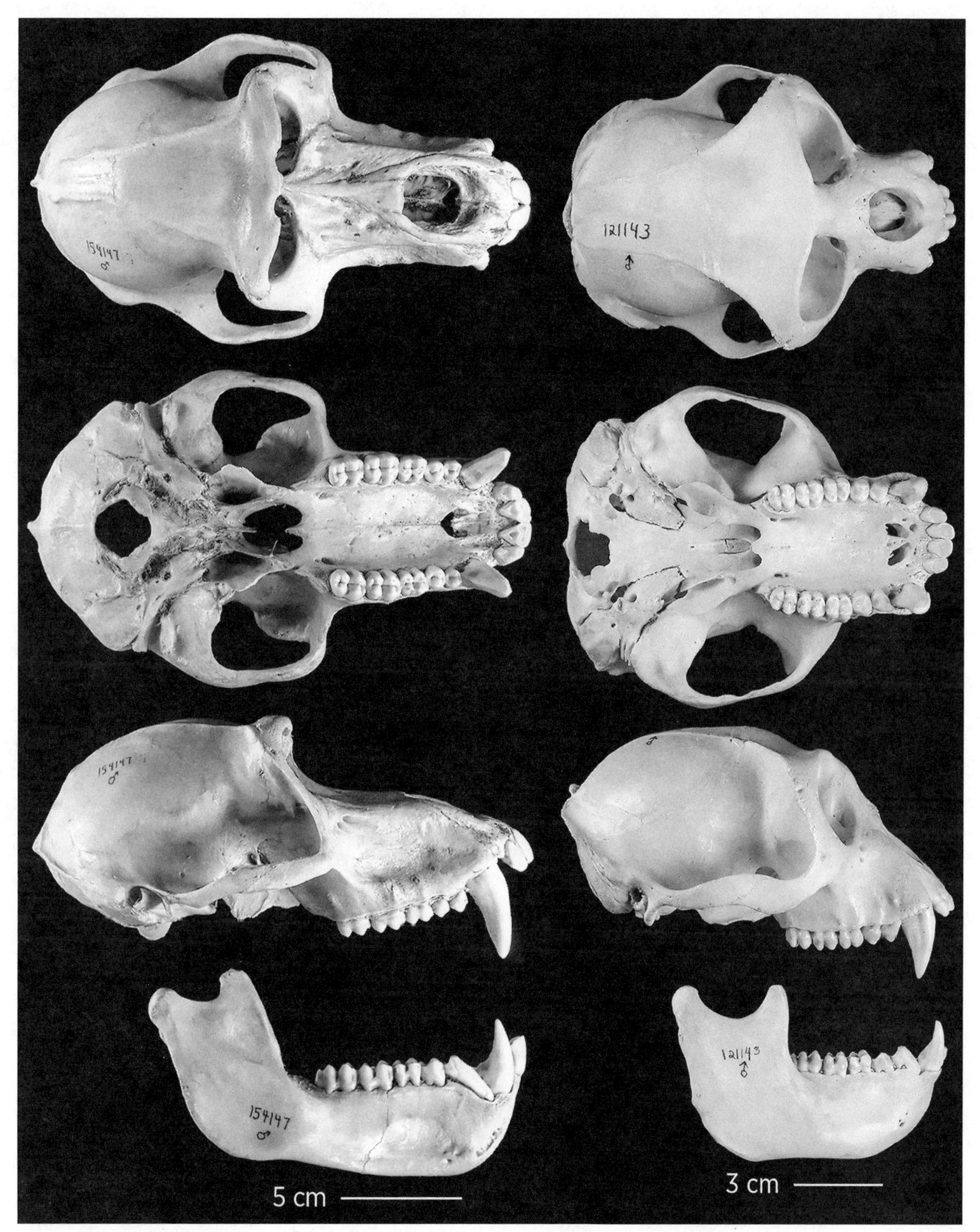

FIG. 329. Cercopithecidae; skull and mandible of the hamadryas baboon, *Papio hamadryas* (Cercopithecinae: Papionini; *left*), and the king colobus, *Colobus polykomos* (Colobinae: Colobini; *right*).

Presbytini—*Nasalis* (proboscis monkey), *Presbytis* (langurs or leaf-eating monkeys), *Pygathrix* (douc langurs), *Rhinopithecus* (snub-nosed monkeys), *Semnopithecus* (Indian langurs), *Simias* (simakobou or pig-tailed langur), and *Trachypithecus* (lutungs, langurs)

Superfamily HOMINOIDEA

Among catarrhine primates, the Hominoidea is much less diverse than the Cercopithecoidea, and with the exception of humans, this group is quite restricted in range, being found only in tropical forests of Africa and Southeast Asia.

DIAGNOSTIC CHARACTERS:

- **upper molars quadritubercular and bunodont with obliquely arranged cusps; a shearing crest (crista obliqua) connects metacone and protocone** (fig. 330)
- **lower molars with five primary cusps arranged to form a characteristic "Y5 pattern"** (fig. 331)

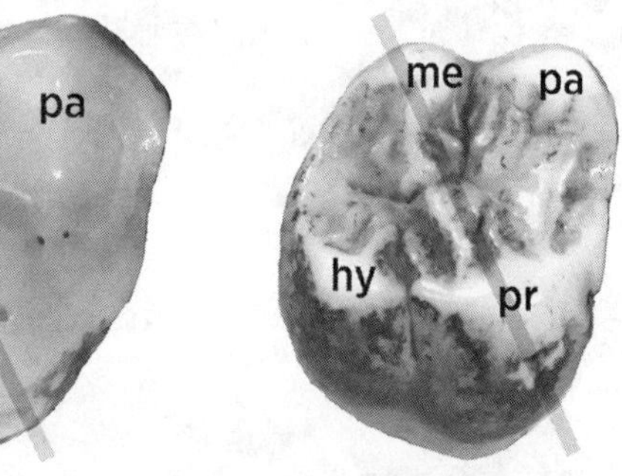

FIG. 330. Right M2 of the lowland gorilla (*Gorilla gorilla*; *left*) and of the chimpanzee (*Pan paniscus*; *right*); pr = protocone, pa = paracone, me = metacone, hy = hypocone; diagonal lines indicate the crista obliqua.

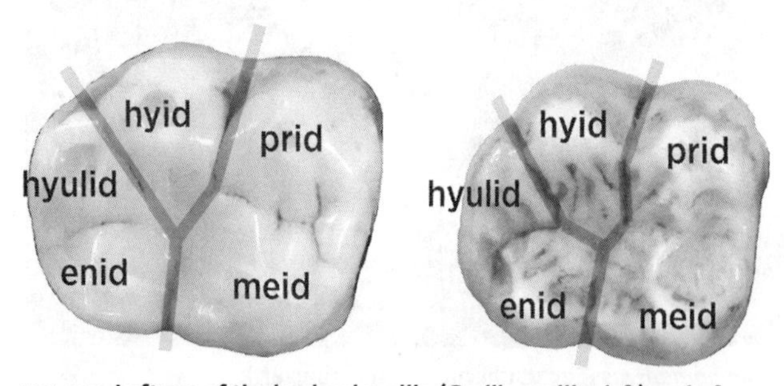

FIG. 331. Left m2 of the lowland gorilla (*Gorilla gorilla*; *left*) and of the chimpanzee (*Pan paniscus*; right); prid = protoconid, meid = metaconid, hyid = hypoconid, hyulid = hypoconulid, enid = entoconid; "Y5 pattern" highlighted by lines separating the trigonid from talonid and hypoconid from hypoconulid.

Family HYLOBATIDAE (gibbons and siamang)

Medium-sized (body length 400–630 mm; mass 3.9–12.7 kg) arboreal catarrhines with short muzzles and shallow faces, large eyes, and rounded heads; **no external tail; forelimbs very long**, proportionally longest of any living primate (intermembral index 126–147; forelimb 213%–274% of trunk length; hind limb 122%–166% of trunk length); both manus and pes with long, curved digits; manus proportionally longer than hind foot; **pollex reduced but opposable**, separated from 2nd digit by deep cleft; **digits have flattened or slightly keeled nails (except pollex and hallux, both of which lack either claws or nails)**; ischial callosities present, but reduced in size relative to cercopithecids; most show little sexual size dimorphism, but some are highly sexually dichromatic (both crested and hoolock gibbons); males of some species with well-developed throat sac. Active, highly arboreal brachiators; feed extensively on figs and other fruit, young leaves and shoots, flowers, and the occasional insect. Monogamous, live in small family groups, and territorial; very loud vocalizations are given in duets by both sexes. Restricted to lowland to lower montane tropical rainforest.

FIG. 332. Hylobatidae; lar or white-handed gibbon, *Hylobates lar*.

RECOGNITION CHARACTERS:

1. **skull with large orbits with protruding rims, wide interorbital distance, and no sagittal crest**
2. **auditory bulla extends laterally as bony tube**
3. canines prominent but not sexually dimorphic
4. baculum present in males

DENTAL FORMULA: $\frac{2}{2}\ \frac{1}{1}\ \frac{2}{2}\ \frac{3}{3} = 32$

TAXONOMIC DIVERSITY: About 19 species allocated to four genera: *Hoolock* (= *Bunopithecus*; hoolock gibbons), *Hylobates* (gibbons), *Nomascus* (crested gibbons), and *Symphalangus* (siamang).

FIG. 333. Geographic range of family Hylobatidae: Southeast Asia, including Indonesia.

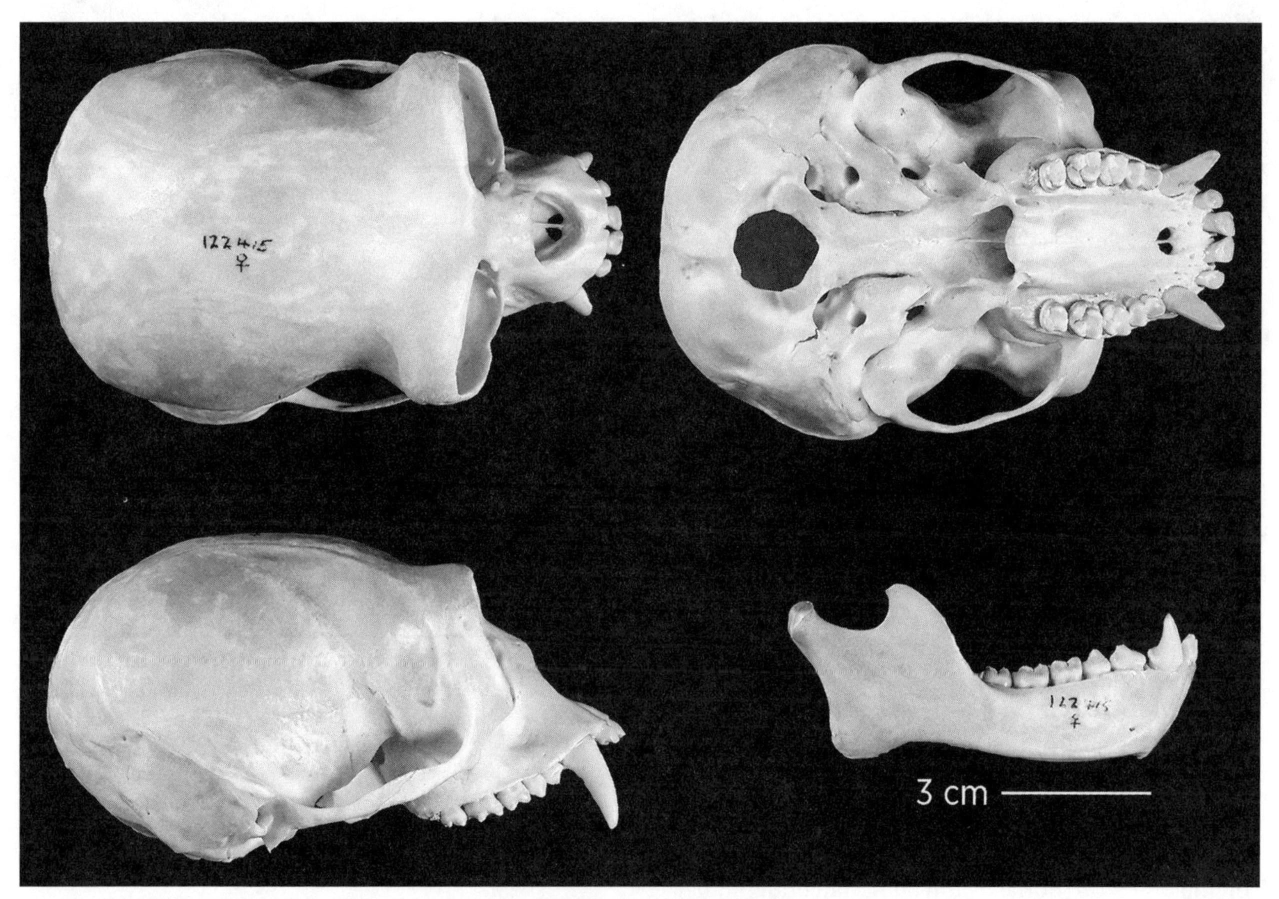

FIG. 334. Hylobatidae; skull and mandible of the silvery Javan gibbon, *Hylobates moloch*.

Family HOMINIDAE (great apes [orangutan, gorilla, chimpanzee, human])

Large (body length 700–1,050 mm; mass 40–180 kg) arboreal and terrestrial catarrhines; thorax broad, lumbar region short; all lack external tails; forelimbs typically longer than hind limbs (intermembral index 102 [in chimpanzee] to 139 [in orangutan]), but not in humans (intermembral index 72); face prominent and prognathous in great apes, flat in humans; hands of all with **well-developed opposable pollex; opposable hallux in great apes but not humans**; and **all digits have flattened nails.** Foot posture plantigrade; orangutans brachiate slowly, chimpanzees and gorillas are knuckle-walking quadrupeds, and humans are bipedal; ischial callosities usually absent. Strongly sexual dimorphic. Orangutans are mostly solitary; chimpanzees, gorillas, and humans are highly social; chimpanzees and humans are active hunters of vertebrate prey. Habitat of orangutan and gorilla largely restricted to dense tropical forests; that of chimpanzee more varied, from forest to forest-savanna mosaic to deciduous woodlands. Humans occupy nearly every terrestrial habitat. Remarkably, a new species of orangutan was discovered and named in 2017 (Nater et al. 2017).

FIG. 335. Hominidae: Homininae; lowland gorilla, *Gorilla gorilla*.

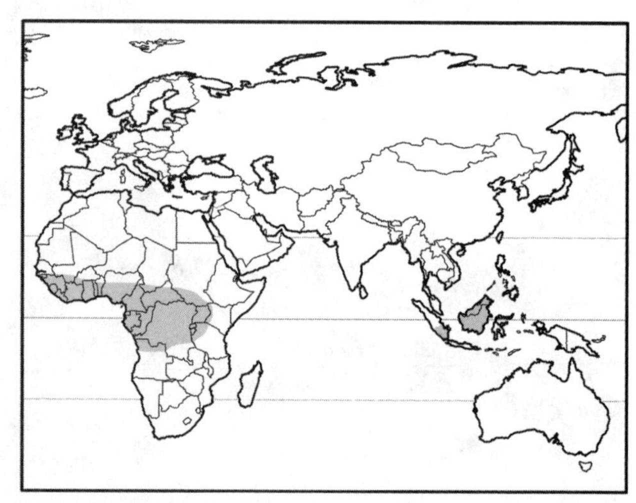

FIG. 336. Geographic range of family Hominidae: orangutan (*Pongo*)—Sumatra and Borneo; gorilla (*Gorilla*)—equatorial Africa; chimpanzee and bonobo (*Pan*)—equatorial Africa; human (*Homo*)—cosmopolitan.

RECOGNITION CHARACTERS:

1. **braincase large**
2. **sagittal crest often prominent**, except in humans
3. **auditory bulla extending laterally as a tube**
4. canines well developed, except in humans
5. baculum present in males of all genera except *Homo*

DENTAL FORMULA: $\frac{2}{2}\ \frac{1}{1}\ \frac{2}{2}\ \frac{3}{3} = 32$

TAXONOMIC DIVERSITY: About eight species in four genera often separated into two subfamilies:

Homininae—*Gorilla* (mountain [eastern] and lowland [western] gorillas), *Homo* (humans), and *Pan* (pygmy [= bonobo] and common chimpanzee)

Ponginae—*Pongo* (Bornean and Sumatran orangutans)

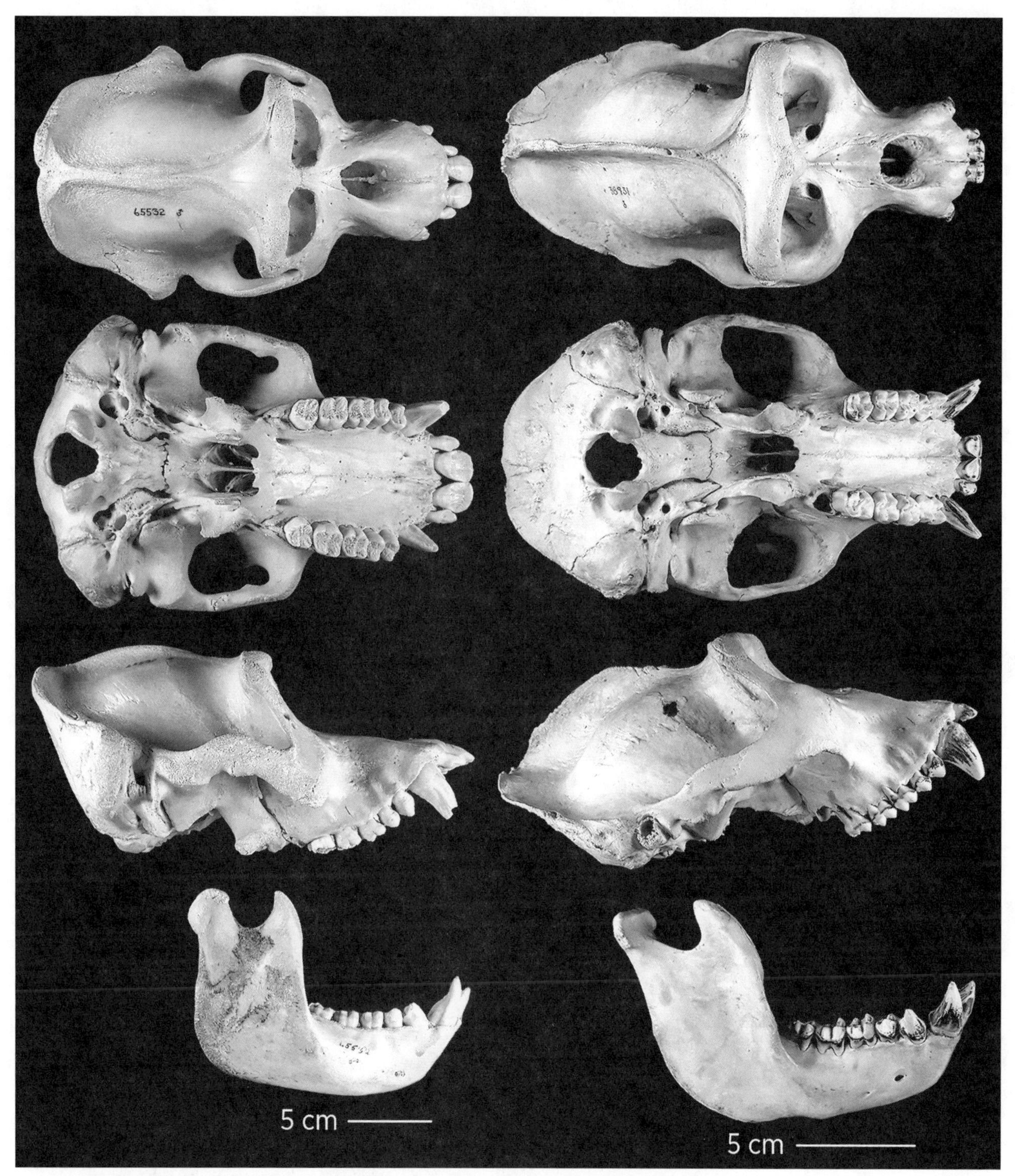

FIG. 337. Hominidae; skull and mandible of the Sumatran orangutan, *Pongo pygmaeus* (Ponginae; *left*) and lowland gorilla, *Gorilla gorilla* (Homininae; *right*).

Clade SUNDATHERIA

The clade Sundatheria contains Dermoptera and Scandentia and is sister to the Primates. This hypothesis is supported both by morphological analysis (O'Leary et al. 2013) and by some, though not all, molecular studies (see Gatesy et al. 2017; Esselstyn et al. 2017).

Order DERMOPTERA

Family CYNOCEPHALIDAE

(colugos or "flying lemurs")

Medium-sized (body length 340–420 mm; tail length 175–270 mm; mass 1–2 kg); large head with face resembling that of a lemuriform primate; ears short and nearly naked; forelimbs and hind limbs subequal in length; feet flat, retaining five laterally compressed digits terminating in sharp, recurved claws; tail long and broad; gliding membrane stretching from the neck to the manal digits, then to those of the pes, and finally to the tip of the tail. Generally nocturnal; herbivorous; live in hollows of trees sufficiently tall to allow escape glides; can glide up to 136 m with altitudinal loss of 12 m; exceptional climbers but slow and virtually helpless on the ground. Restricted to lowland tropical forests.

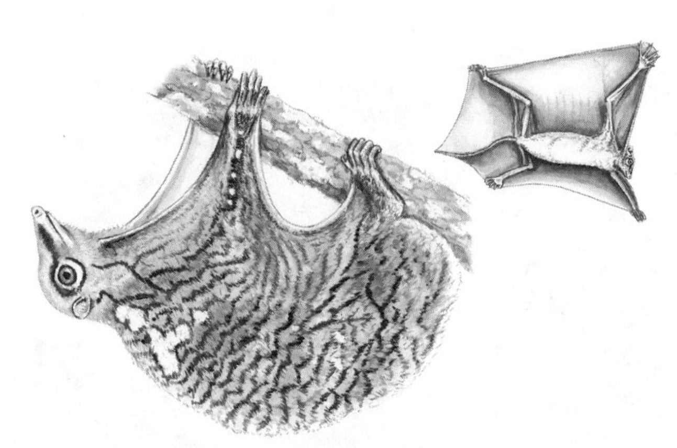

FIG. 338. Cynocephalidae; Sunda colugo, *Galeopterus variegatus*.

DIAGNOSTIC CHARACTERS:

- **membrane extending from neck to digits of forefeet, from forelimbs to digits of hind feet, and from hind limbs to tip of tail**
- **two inner lower incisors on each side wide, comblike (= pectinate) and procumbent; outer lower incisors multicuspidate** (fig. 339)

RECOGNITION CHARACTERS:

1. **cranium broad, flattened**
2. postorbital process well developed
3. temporal ridges prominent
4. upper incisors of left and right sides separated by diastema
5. canines indistinct, premolar-like
6. baculum not present in males

DENTAL FORMULA: $\frac{2}{3}\ \frac{1}{1}\ \frac{2}{2}\ \frac{3}{3} = 34$

TAXONOMIC DIVERSITY: Two monotypic genera, *Cynocephalus* (Philippine flying lemur) and *Galeopterus* (Sunda flying lemur).

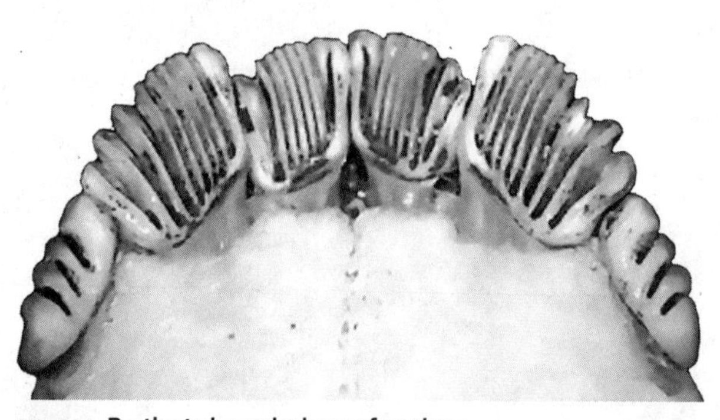

FIG. 339. Pectinate lower incisors of a colugo.

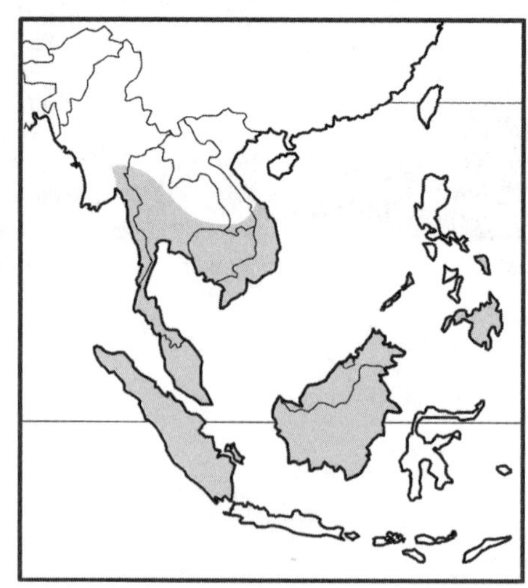

FIG. 340. Geographic range of family Cynocephalidae: Southeast Asia, including Indonesia and the Philippines.

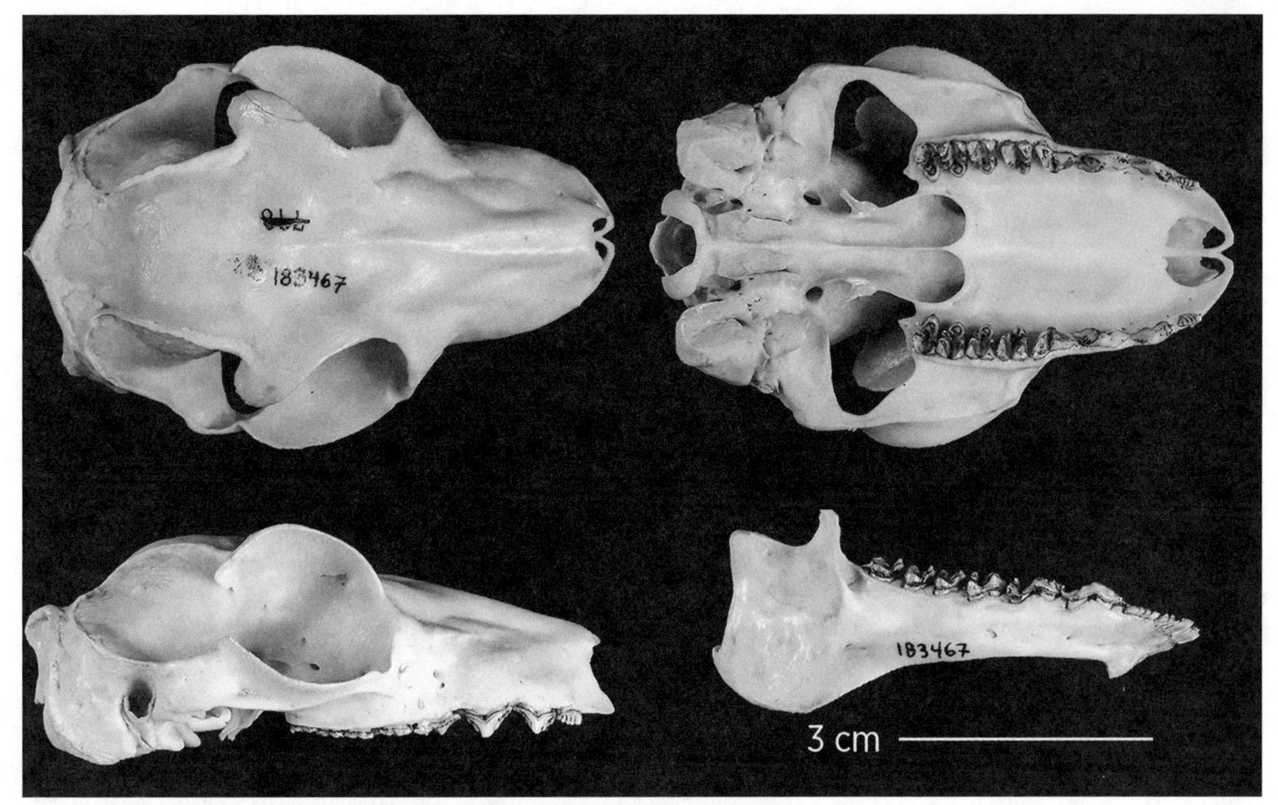

FIG. 341. Cynocephalidae; skull and mandible of the Sunda colugo, *Galeopterus variegatus*.

Order SCANDENTIA

Tupaiids and ptilocercids have been treated both as subfamilies of the family Tupaiidae and as separate families (see Helgen 2005b). These two groups diverged from a common ancestor approximately 63 million years ago, near the Cretaceous-Paleocene boundary (Janecka et al. 2007). While we recognize the families Tupaiidae and Ptilocercidae, we treat them together here and list their distinguishing features in table 6.

Families PTILOCERCIDAE and TUPAIIDAE (tree shrews)

Small (body length 100–230 mm; mass usually less than 350 g) and squirrel-like; head with an elongate muzzle; facial region without long vibrissae; ears prominent; tail length equal to, slightly shorter than, or longer than body length, lightly to heavily furred, and with lateral rows of elongate hairs, or tufts, on each side of the terminal portion of the tail (termed distichously tufted in table 6) in *Ptilocercus*; hind limbs only slightly longer than forelimbs; plantar surfaces of feet naked, with tubercle-like pads; digits 5/5, each equipped with curved claws. Terrestrial, semiarboreal, and arboreal; crepuscular to nocturnal; omnivorous, preferred foods are fruit and animal matter, particularly insects. Solitary or occur in pairs; some species live in larger social groups. Restricted to tropical deciduous forests.

RECOGNITION CHARACTERS:

1. auditory bulla not enlarged
2. no palatal perforations
3. **postorbital process large, contacting zygomatic arch to form complete postorbital bar**
4. zygomatic arch perforate
5. molars with three principal cusps, 4th cusp is very small
6. occlusal surface of upper molars with weakly developed W-shaped ectoloph
7. baculum not present in males

DENTAL FORMULA: $\frac{2}{3}\ \frac{1}{1}\ \frac{3}{3}\ \frac{3}{3} = 38$

TAXONOMIC DIVERSITY: About 20 species in five genera in two families or subfamilies:

Ptilocercidae—monotypic, with a single species, *Ptilocercus lowii* (pen-tailed tree shrew)

Tupaiidae—*Anathana* (Madras tree shrew), *Dendrogale* (smooth-tailed tree shrews), *Tupaia* (tree shrews), and *Urogale* (Philippine tree shrew)

Table 6. Comparison of behavioral and morphological character states between members of the families Ptilocercidae and Tupaiidae.

Character	*Ptilocercidae*	*Tupaiidae*
activity	nocturnal	diurnal
tail	distichously tufted; otherwise naked basally, scaly ventrum	lacks pencil; otherwise well haired or bushy
ears	large, membranous	small, cartilaginous
foot pads	relatively large and soft	modestly developed
supraorbital foramen	absent	well developed
temporal fossa	about equal to orbit in size	notably smaller than orbit
temporal ridges	roughly parallel	diverge anteriorly
orbits	larger, located anterior to those of tupaiids	smaller, more posteriorly located
dorsal aspect of skull	broader	slimmer
zygomatic arches	spread more broadly	not spreading
foramen rotundum	confluent with sphenoidal fissure	separate from sphenoidal fissure
upper incisors	I1 noticeably larger than I2	I1 and I2 of similar size
diastema	no conspicuous diastema between I2 and C	diastema between I2 and C, sometimes only modestly developed
2nd upper molar	bicuspid, with distinct posterior cusp	unicuspid
upper canine	double rooted; premolariform	usually single rooted; caniniform
upper molars	lack mesostyle; distinct cingulum encircles each molar	mesostyles well developed, bifurcated; distinctive cingulum lacking
lower molars	cingulum present on outer surface	cingulum lacking

FIG. 342. Tupaiidae; pygmy tree shrew, *Tupaia minor*.

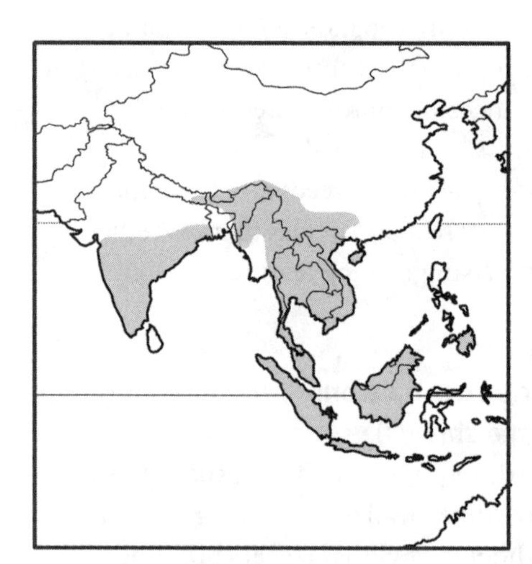

FIG. 343. Combined geographic range of families Ptilocercidae and Tupaiidae: India and Southeast Asia, including Indonesia and the Philippines.

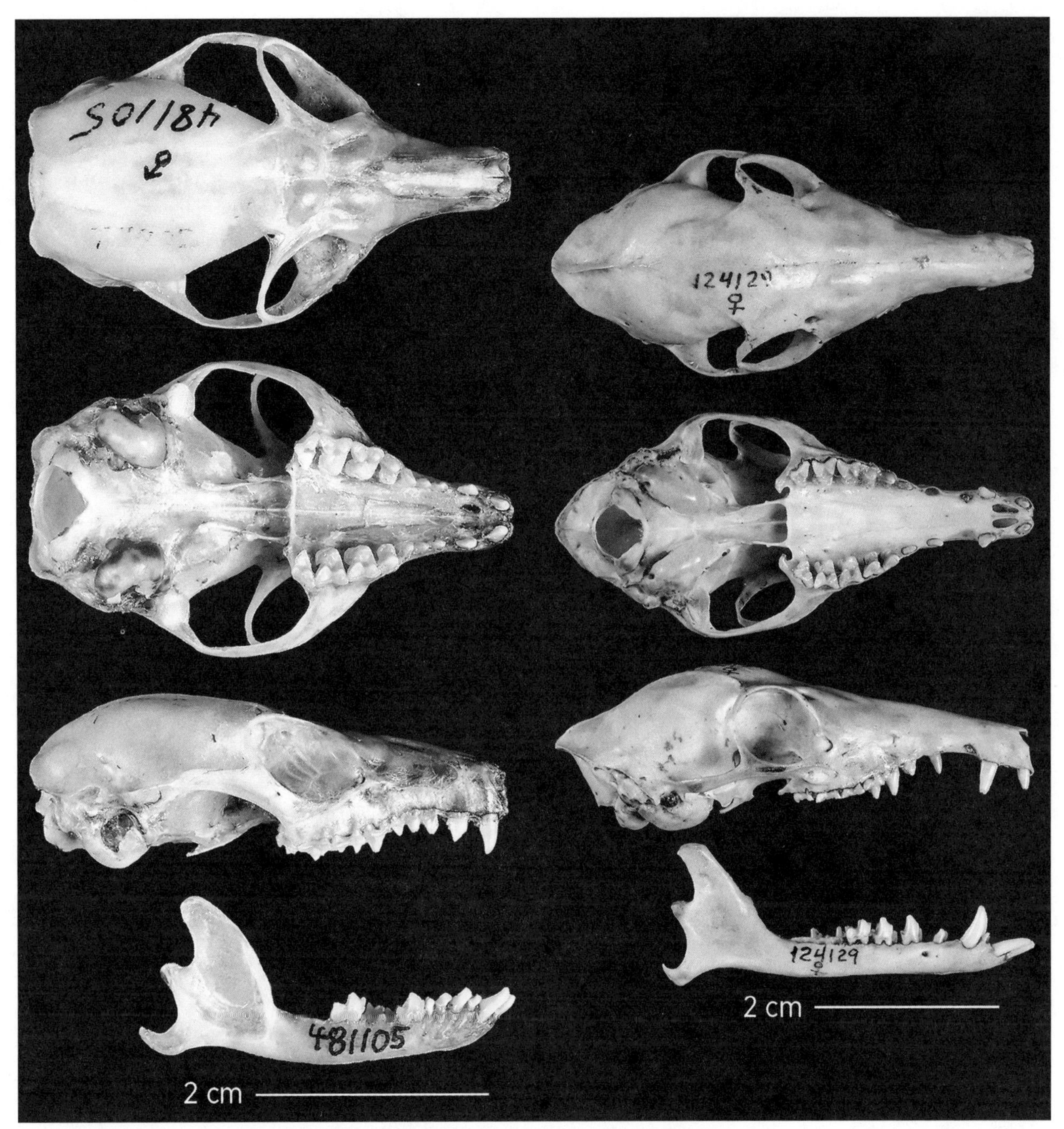

FIG. 344. Scandentia; skull and mandible of the pen-tailed tree shrew, *Ptilocercus lowii* (Ptilocercidae; *left*), and the Mindanao tree shrew, *Urogale everetti* (Tupaiidae; *right*).

Clade LAURASIATHERIA

The Laurasiatheria is the clade that unites the Lipotyphla (the solenodonts, hedgehogs, shrews, and moles) and Scrotifera (the pangolins, carnivorans, bats, all ungulates, and cetaceans) (see fig. 94 and 95).

Clade LIPOTYPHLA (= Eulipotyphla)

The clade uniting hedgehogs, shrews, moles, and solenodonts is varyingly called the Lipotyphla or Eulipotyphla; the latter term is used to avoid confusion with the earlier and taxonomically more expansive concept of the suborder Lipotyphla within the "waste-basket" order Insectivora (see the introduction to infraclass Eutheria or Placentalia). Relationships among these four lipotyphlan lineages remain unresolved, but the group is currently divided into the order Erinaceomorpha (which includes the single family Erinaceidae) and the order Soricomorpha (which includes the Nesophontidae, Solenodontidae, Soricidae, and Talpidae; see, for example, Hutterer 2005a, b), a scheme that we follow. The Nesophontidae occurred in the Greater Antilles and survived the late Pleistocene extinctions, but has probably been extinct for 200–500 years; it will not be considered here.

LIPOTYPHLAN RECOGNITION CHARACTERS:

1. very small to medium-sized
2. foot posture usually plantigrade
3. snout generally elongate
4. pelage usually consisting of only one kind of hair of uniform length
5. cerebral hemispheres smooth, lacking complex folding
6. cheek teeth relatively simple, zalambdodont, dilambdodont, or quadritubercular in some groups, with sharp edges to cusps

Order ERINACEOMORPHA

Family ERINACEIDAE (hedgehogs, moonrats or gymnures)

Mouse-sized (body length 90–125 mm; mass 45–80 g; lesser gymnure, *Hylomys*) to rabbit-sized (body length 240–460 mm; mass 0.5–2.0 kg; moonrat, *Echinosorex*); ears and eyes relatively large; rostrum usually long; tail length less than 70% of body length, usually less than 20% in erinaceines; forelimbs and hind limbs subequal in length; foot posture plantigrade, most with 5/5 digits, a few with 5/4; barbless spines covering body as highly effective spiny armor in members of the subfamily Erinaceinae; spines are absent in all members of the subfamily Galericinae, but the hair is coarse; a urogenital opening is well separated from the anus (no cloaca). Hedgehogs are nocturnal; gymnures may be either diurnal or nocturnal. Most erinaceomorphs are omnivorous, but animal matter, especially invertebrates, is their preferred food. Generally solitary; some hibernate; most species can swim and climb well, but all are generally terrestrial. Habitats range broadly, from tropical and deciduous forests to open desert scrub landscapes.

DIAGNOSTIC CHARACTER:

- **upper molars quadritubercular with short, rounded cusps, not sharp vertical shear faces (crowns lack W-shaped ectoloph)**

RECOGNITION CHARACTERS:

1. lacrimal and palatine separated by maxilla
2. pinna present, well developed
3. zygomatic arch present, complete, robust
4. auditory bulla composed of basisphenoid wing medially
5. **1st upper incisor large, canine-like**
6. pubic bones united in short symphysis
7. baculum absent in males

DENTAL FORMULA: $\frac{2\text{–}3}{3}\ \frac{1}{1}\ \frac{3\text{–}4}{2\text{–}4}\ \frac{3}{3} = 36\text{–}44$

TAXONOMIC DIVERSITY: About 24 species in 11 genera allocated to two subfamilies:

Erinaceinae—*Atelerix* (African hedgehogs), *Erinaceus* (Eurasian hedgehogs), *Hemiechinus* (long-eared hedgehog), *Mesechinus* (Chinese and Russian hedgehogs), and *Paraechinus* (desert and Brandt's hedgehogs)

Galericinae—*Echinosorex* (moonrat), *Hylomys* (Indonesian gymnures), *Neohylomys* (Hainan gymnure), *Neotetracus* (shrew gymnure), and *Podogymnura* (Philippine gymnures)

FIG. 345. Erinaceidae; short-tailed gymnure, *Hylomys suillus* (Galericinae; *left*), and southern African hedgehog, *Atelerix frontalis* (Erinaceinae; *right*).

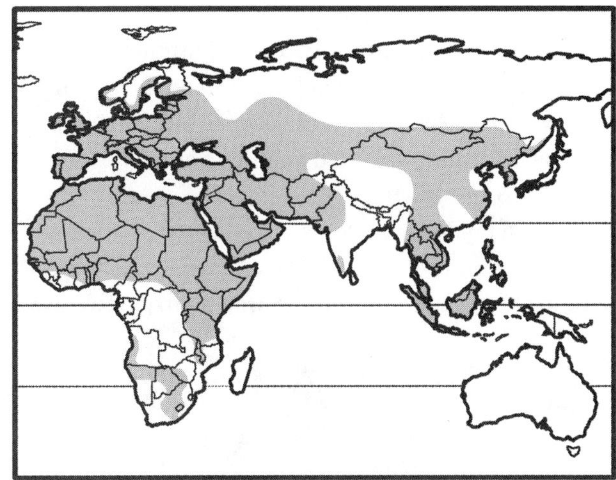

FIG. 346. Geographic range of family Erinaceidae: Old World—Eurasia and Africa (absent from New World, Australasia); gymnures occur only in Southeast Asia and Indonesia.

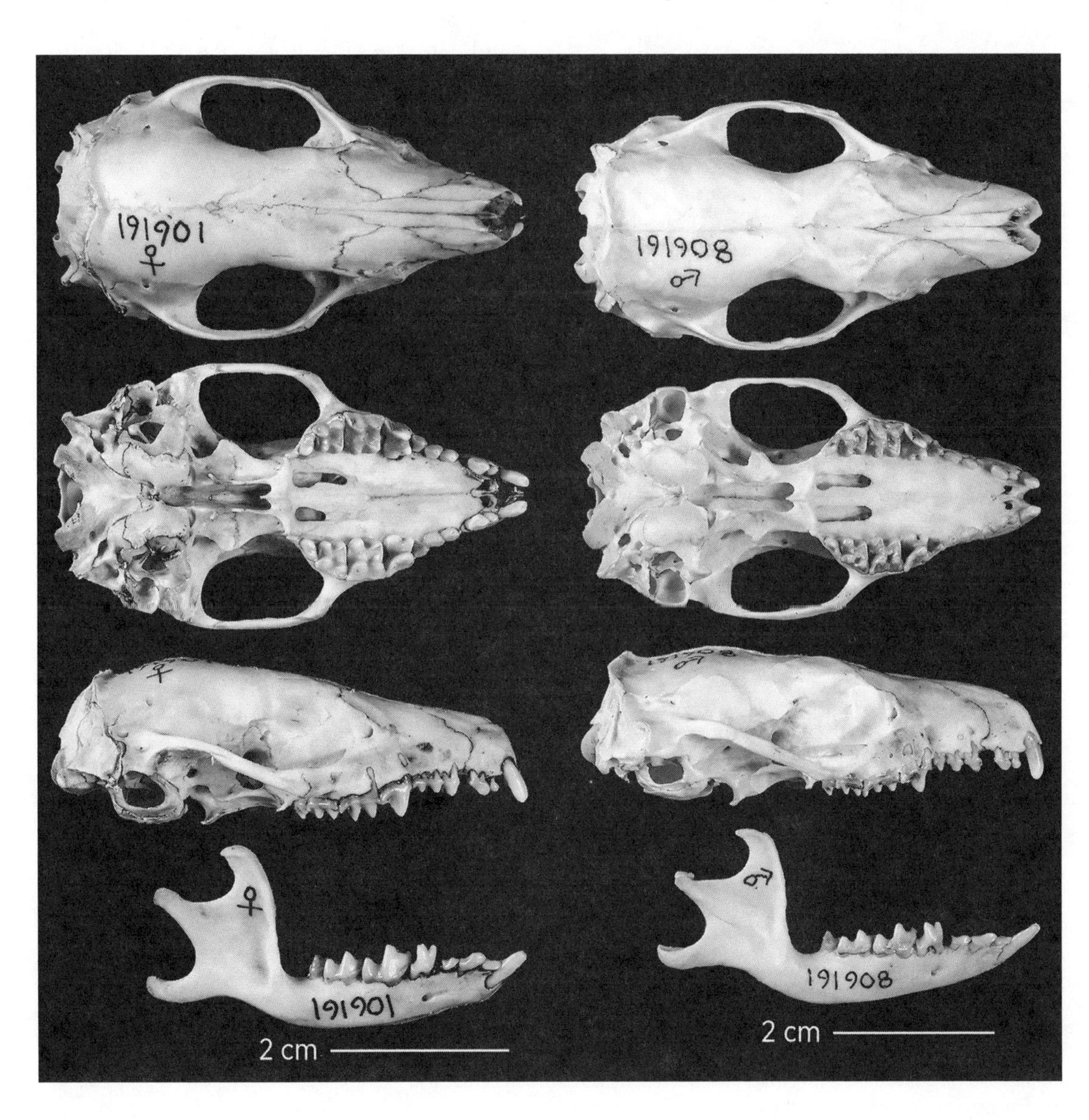

FIG. 347. Erinaceidae: Erinaceinae; skull and mandible of the long-eared hedgehog, *Hemiechinus auritus* (*left*), and Brandt's hedgehog, *Paraechinus hypomelas* (*right*), both members of the subfamily Erinaceinae.

Order SORICOMORPHA

Family SOLENODONTIDAE (solenodons)

Among the largest of the "insectivores" (body length 280–390 mm; mass 1 kg); nostrils open laterally; muzzle long, flexible, and supported at tip by a small sesamoid bone (the os proboscides); eyes small; ears well developed; tail long, but less than body length, nearly naked; forelimbs and hind limbs subequal in length; both manus and pes have five clawed digits; move with slow shuffle on entire sole (plantigrade) of forefeet but on toes (digitigrade) of hind feet; pelage long, coarse, and relatively sparse. Nocturnal; omnivorous, but feed mainly on invertebrates. Urogenital openings and anus well separated (cloaca absent). Produce toxic saliva from submaxillary glands located at base of deeply grooved lower incisors. Inhabit forest and shrublands, taking refuge in rock crevices, burrows, and hollow trees.

FIG. 348. Solenodontidae; Hispaniolan solenodon, *Solenodon paradoxus*.

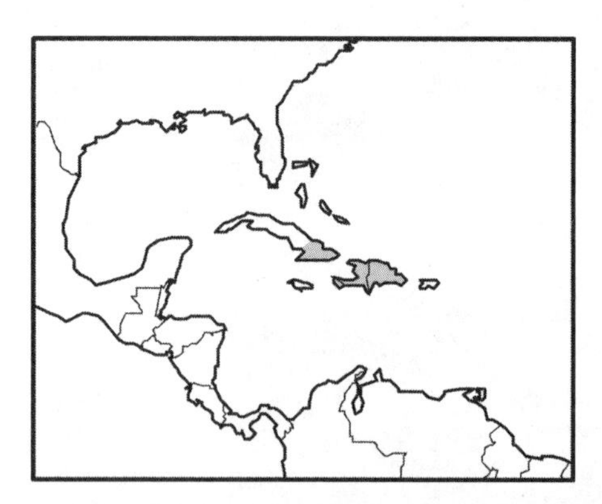

FIG. 349. Geographic range of family Solenodontidae: eastern Cuba and Hispaniola.

DIAGNOSTIC CHARACTER:

- **alisphenoid canal present** (absent in other lipotyphlans)

RECOGNITION CHARACTERS:

1. **zygomatic arch present, incomplete**; only maxillary and squamosal roots present, jugal absent
2. large lambdoidal crest projecting over occiput
3. no auditory bulla
4. **1st upper incisor greatly enlarged, directed slightly backward**; short diastema separating 1st and 2nd upper incisors
5. 2nd lower incisor greatly enlarged
6. **upper molars tritubercular; crowns with V-shaped ectoloph** (zalambdodont)
7. pubic bones united in short symphysis
8. baculum absent in males

DENTAL FORMULA: $\frac{3}{3}\ \frac{1}{1}\ \frac{3}{3}\ \frac{3}{3} = 40$

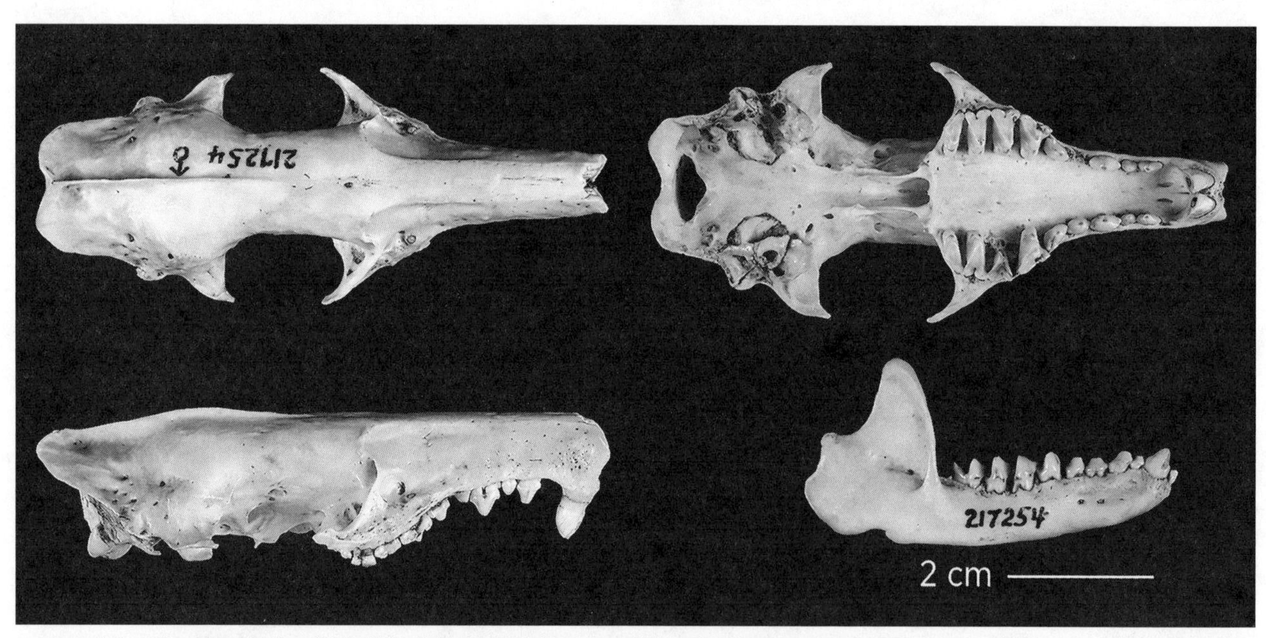

FIG. 350. Solenodontidae; skull and mandible of the Hispaniolan solenodon, *Solenodon paradoxus*. Note the incomplete zygomatic arches (jugal absent) that characterize this family.

TAXONOMIC DIVERSITY: The family is monotypic, with three living species—one recently rediscovered in southeastern Cuba—in the single genus *Solenodon*.

Family SORICIDAE (shrews)

Small (body length 60–300 mm; mass 2–180 g); snout elongate, pointed, and mobile; eyes minute, often partly concealed by fur; pinna present, but often reduced; tail length variable, but usually shorter than body length; feet unspecialized (except in aquatic forms, which have stiffened hairs along edge of soles and lateral toes); five digits on both forefeet and hind feet, with claws; soles naked; pelage soft, thick; shallow cloaca present in some. Most active at any time of day or night; high metabolic rate requires nearly constant consumption of food; diet mostly insects or soil invertebrates; aquatic species eat aquatic insect larvae and some take small fish. Some species produce toxic saliva, others echolocate. Occur in a wide range of habitats, mostly in moist forests, meadows, and marshes beneath leaf litter, but a few extend into arid environments.

DIAGNOSTIC CHARACTER:

- **condyloid process of lower jaw with two condyles (only a single condyle in all other mammals)**

RECOGNITION CHARACTERS:

1. **no zygomatic arch**
2. **no ossified auditory bulla**
3. **1st upper incisor large, hooked, and with cusp at proximal base of tooth**
4. **upper molars tritubercular to quadritubercular, with sharp cusps; crowns with W-shaped ectoloph** (dilambdodont)
5. pubic bones separate (no symphysis)
6. baculum absent in males of most taxa

DENTAL FORMULA: $\frac{3}{3}\ \frac{1}{1}\ \frac{3}{3}\ \frac{3}{3} = 40$

TAXONOMIC DIVERSITY: This large family includes over 370 species in about 26 genera distributed among three subfamilies, one of which has six tribes:

Crocidurinae (white-toothed shrews): *Crocidura* (white-toothed shrews), *Diplomesodon* (piebald shrew), *Feroculus* (Kelaart's long-clawed shrew), *Paracrocidura* (large-headed shrews), *Ruwenzorisorex* (Ruwenzori shrew), *Scutisorex* (armored shrew), *Solisorex* (Pearson's long-clawed shrew), *Suncus* (musk shrews), and *Sylvisorex* (forest shrews)

Myosoricinae (African mouse shrews): *Congosorex* (Congo shrews), *Myosorex* (mouse shrews), and *Surdisorex* (Aberdare and Mt. Kenya mole shrews)

Soricinae (red-toothed shrews):

Anourosoricini—*Anourosorex* (mole shrews)

Blarinellini—*Blarinella* (Asiatic short-tailed shrews)

FIG. 351. Soricidae: Soricinae; northern short-tailed shrew, *Blarina brevicauda* (Blarinini; *left*), and cinereus shrew, *Sorex cinereus* (Soricini; *right*).

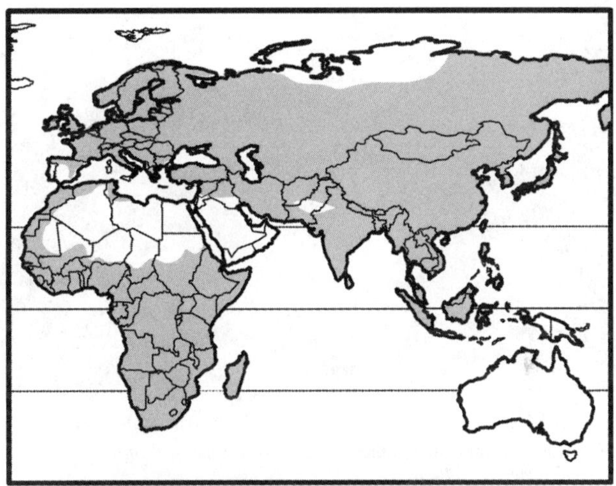

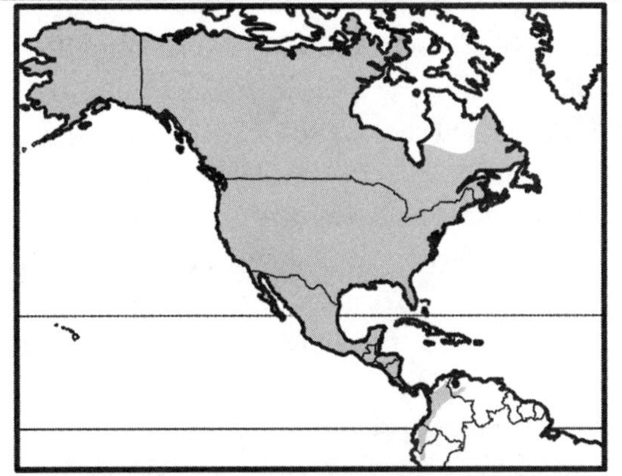

FIG. 352. Geographic range of family Soricidae: *Old World*—Eurasia and Africa (absent from Australasia); *New World*—North America south to northwestern South America (members of the Soricinae tribes Blarinini, Notiosoricini, and Soricini).

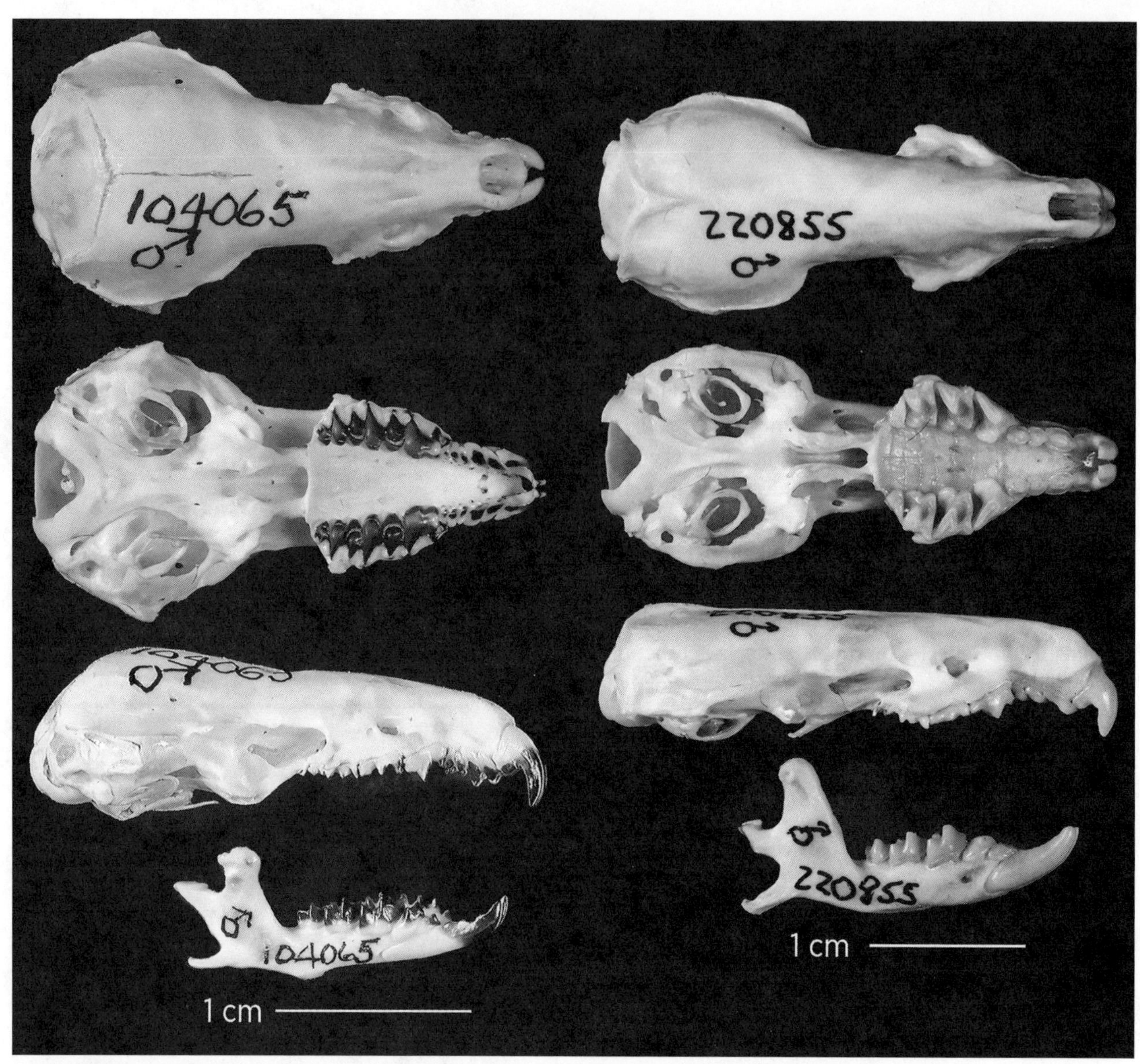

FIG. 353. Soricidae; skull and mandible of the northern short-tailed shrew, *Blarina brevicauda* (Soricinae: Blarinini; *left*), and white-toothed shrew, *Crocidura* sp. (Crocidurinae; *right*).

Blarinini—*Blarina* (North American short-tailed shrews) and *Cryptotis* (American small-eared and long-clawed shrews)

Nectogalini—*Chimarrogale* (Asiatic water shrews), *Chodsigoa* (East Asian shrews), *Episoriculus* (Asian brown-toothed shrews), *Nectogale* (elegant water shrew), *Neomys* (European water shrews), *Nesiotites* (Balearic and Sardinian shrews), and *Soriculus* (Himalayan shrew)

Notiosoricini—*Megasorex* (Mexican shrew) and *Notiosorex* (North American desert shrews)

Soricini—*Sorex* (long-tailed shrews)

Family TALPIDAE (moles)

Small to medium-sized (body length 95 mm in shrew moles, 450 mm in desmans); body fusiform; rostrum long and narrow, nose mobile; eyes minute and sometimes covered with skin; ears usually without external pinnae; tail short in most, longer and laterally compressed in *Desmana*; external appendages short and situated close to the body; pectoral girdle, forelimbs, and **manus highly modified for lateral-stroke burrowing** (rotation-thrust); manus turned permanently outward, usually broad and paddle-shaped, unspecialized or webbed in some; fur very soft, short, without spines; no cloaca. Subterranean or fossorial, except desmans, which are aquatic, and the Asian shrew mole, which is more scansorial than fossorial. Diet primarily earthworms, insects, and other invertebrates. Tunnel systems consist of near-surface feeding systems and deep, more permanent burrows. Most are solitary. Habitat ranges from lowland marshes and moist grasslands to subalpine forests and meadows and, in the case of desmans, montane streams.

FIG. 354. Talpidae; broad-footed mole, *Scapanus latimanus* (Scalopinae: Scalopini; *upper left*), Russian desman, *Desmana moschata* (Talpinae: Desmanini; *right*), and star-nosed mole, *Condylura cristata* (Scalopinae: Condylurini; *lower left*).

DIAGNOSTIC CHARACTERS:

- **humerus usually blocky, often nearly as wide as long, articulating with scapula and clavicle**
- **forefoot projecting outward and backward in most taxa; elbow rotated upward**

RECOGNITION CHARACTERS:

1. **zygomatic arch complete, slender**
2. **auditory bulla present, incomplete**
3. incisors simple; **1st upper incisor directed downward and backward**
4. **upper molars tritubercular to quadritubercular, with sharp cusps; crowns with W-shaped ectoloph** (dilambdodont)
5. pubic bones separate
6. baculum present in males

DENTAL FORMULA: $\frac{2\text{–}3}{1\text{–}3}\ \frac{1}{0\text{–}1}\ \frac{3\text{–}4}{3\text{–}4}\ \frac{3}{3} = 34\text{–}44$

TAXONOMIC DIVERSITY: Almost 40 species in about 17 genera in three subfamilies and seven tribes:

Scalopinae:

Condylurini—*Condylura* (star-nosed mole)

Scalopini—*Parascalops* (hairy-tailed mole), *Scalopus* (eastern mole), *Scapanulus* (Gansu mole), and *Scapanus* (western North American moles)

Talpinae:

Desmanini—*Desmana* (desman) and *Galemys* (Pyrenean desman)

Neurotrichini—*Neurotrichus* (shrew-mole)

Scaptonychini—*Scaptonyx* (long-tailed mole)

Talpini—*Euroscaptor* (Asian moles), *Mogera* (Japanese moles), *Parascaptor* (white-tailed mole), *Scaptochirus* (short-faced mole), and *Talpa* (common moles)

Urotrichini—*Dymecodon* (True's shrew mole) and *Urotrichus* (Japanese shrew mole)

Uropsilinae: *Uropsilus* (Chinese shrew moles)

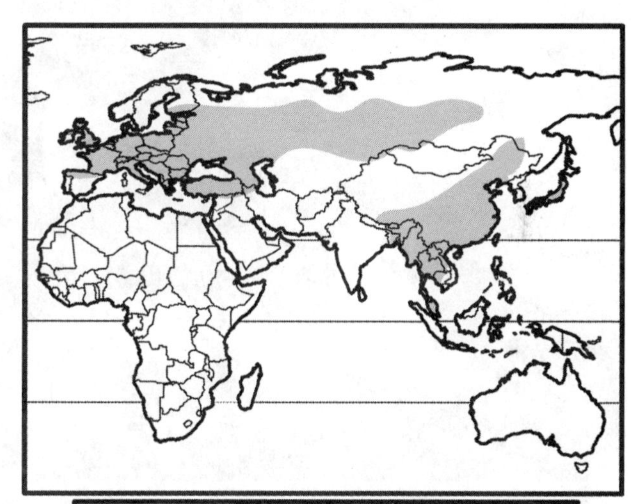

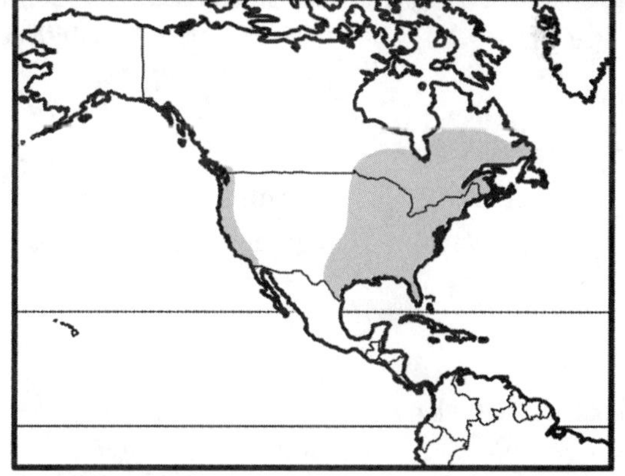

FIG. 355. Geographic range of family Talpidae: North America and Eurasia (absent from South America, Africa, and Australasia).

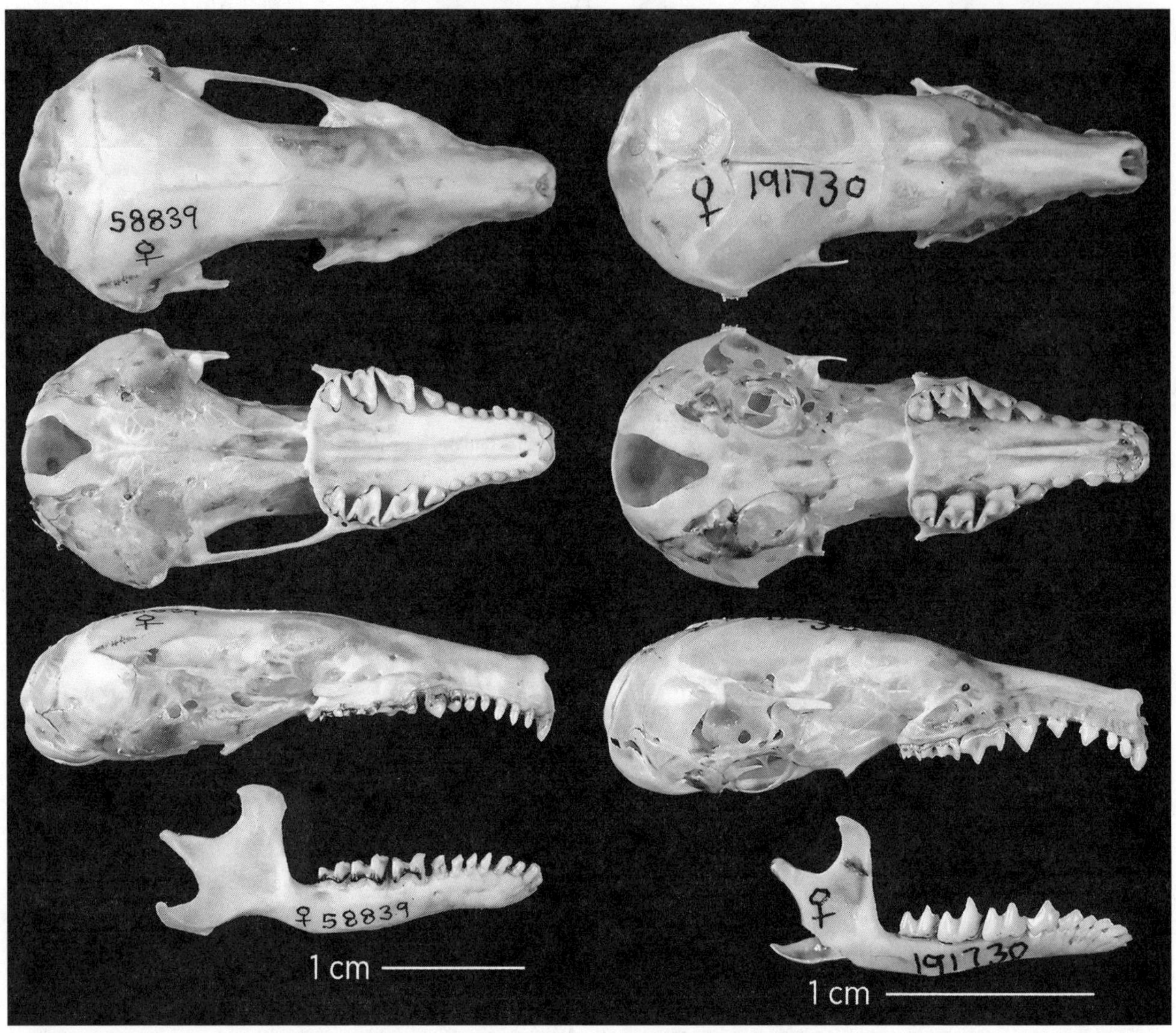

FIG. 356. Talpidae; skull and mandible of Townsend's mole, *Scapanus townsendii* (Scalopinae: Scalopini; *left*; right zygomatic arch broken), and the shrew mole, *Neurotrichus gibbsii* (Talpinae: Neurotrichini, *right*; zygomatic arch broken on both sides).

Clade SCROTIFERA

The Scrotifera is the clade that unites pangolins + carnivorans, bats, and all ungulates + cetaceans (see fig. 94).

Order CHIROPTERA

UNIQUELY DERIVED CHARACTER (FOR A LIST OF 33 MORPHOLOGICAL SYNAPOMORPHIES THAT DIAGNOSE THE CHIROPTERA, SEE SIMMONS AND GEISLER 1998: table 7):

- **forelimb modified for flight, with digits elongate and joined together by a membrane extending to side of body and hind limb (figs. 357, 358)**

RECOGNITION CHARACTERS:

1. ulna reduced and nonfunctional; radius relatively large
2. **sternum usually keeled**
3. clavicle present
4. **glenoid fossa of scapula directed dorsally**
5. trochiter (= greater tuberosity) frequently extends to proximal edge of head of humerus, where it provides a second articulation with the scapula (characters 139 and 100, respectively, in Simmons and Geisler 1998)
6. bones light, tubular
7. cervical and thoracic vertebrae without neural spines
8. number of bony phalanges variable: pollex with two,

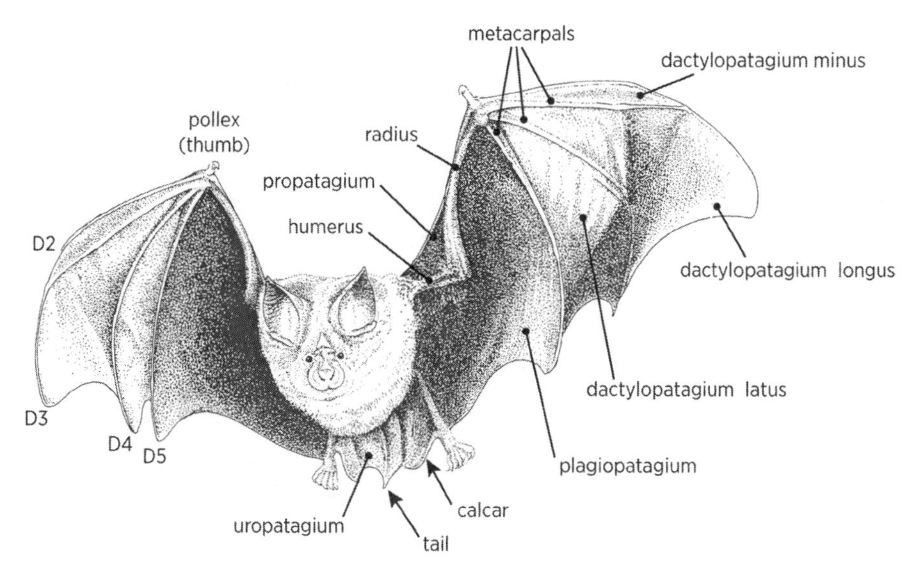

FIG. 357. Major external features of a generalized bat. D = digit; note that the forearm is dominated by the radius (not the ulna) and that the metacarpals are greatly lengthened to support wings. Modified from Altringham (2011).

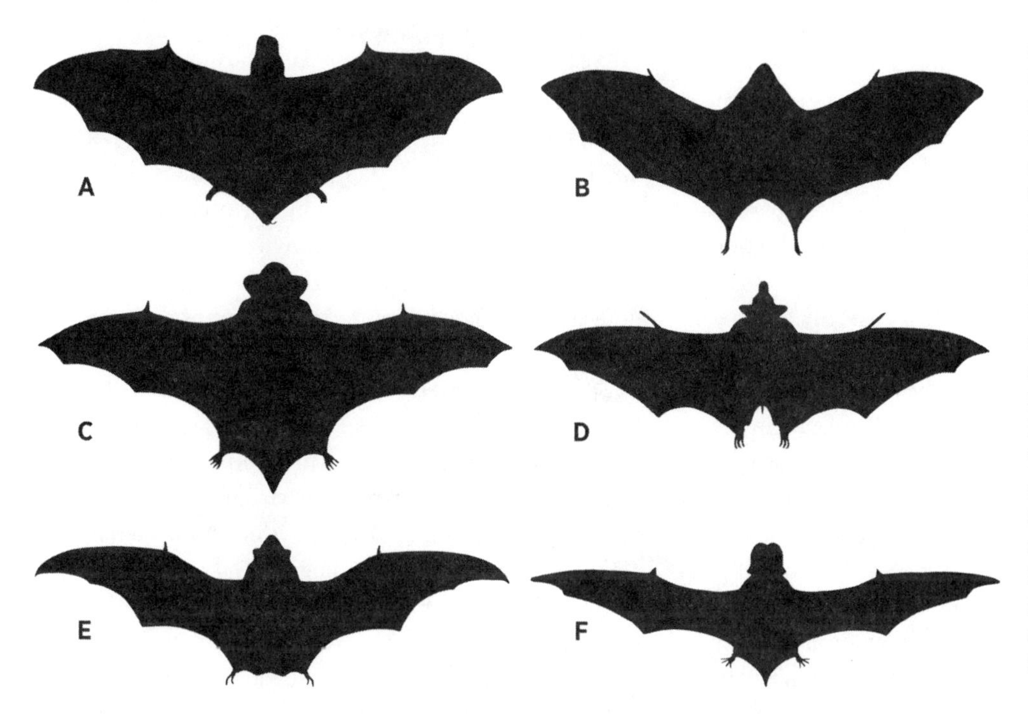

FIG. 358. Variation in wing aspect ratio (wingspan squared divided by wing area, approximated by length divided by depth): Short, deep wings (low aspect ratio; e.g., *A*, *B*, and *C*) occur in species with slow, highly maneuverable flight that glean insects in obstructed habitats (have low stalling speeds). Long, narrow wings (high aspect ratio; e.g., *D*, but especially *E* and *F*) occur in species that are rapid, but less maneuverable, fliers (have high stalling speeds) and that fly in open areas, often well above the canopy. *A*, *Nycteris* (Nycteridae); *B*, *Cardioderma* (Megadermatidae); *C*, *Mimetillus* (Vespertilionidae); *D*, *Eidolon* (Pteropodidae), *E*, *Saccolaimus* (Emballonuridae); *F*, *Mops* (Molossidae). Redrawn from Feldhamer et al. (2015).

2nd digit with zero or one except Rhinopomatidae (two) and Pteropodidae (three); 3rd digit with two or three; 4th and 5th digits with two bony phalanges and usually one cartilaginous one; phalangeal formulae include cartilaginous phalanges

9. cranium domed, often inflated in region of braincase and concave in frontal region
10. premaxillary bones variable; in many taxa, both nasal and palatal branches present, fused to each other and to the maxillary bone; in others, only one branch present, with varying degrees of fusion; one family (Megadermatidae) lacks premaxillae entirely
11. cheek teeth variable, but usually tritubercular
12. incisors relatively small
13. **knee directed posteriorly owing to rotation of hind limb for support of wing and tail membrane**
14. rhinarium and lips often with fleshy appendages (nose leaves, ridges, and/or chin leaves; see fig. 359)
15. tragus of ear usually well developed (see fig. 359)
16. cartilaginous rod (calcar) arising from inner side of ankle joint; present in most bats; absent in some Phyllostomidae and Pteropodidae (character 171 in Simmons and Geisler 1998, who incorrectly listed Rhinopomatidae as without calcar)
17. uropatagium (interfemoral membrane) supporting tail usually present

FIG. 359. Nasal conditions and appendages and ear structures of selected bats: Pteropodidae (*far left*): tube nose, tragus and antitragus absent; Vespertilionidae (*middle left*): simple muzzle, large tragus; Phyllostomidae (*middle right*): nose leaf, small tragus; Rhinolophidae (*far right*): nose leaf, antitragus.

PHYLOGENY AND CLASSIFICATION OF BATS

Relationships among major lineages of bats remain largely unresolved, mainly because disagreements among studies are generally confounded by different taxonomic samples, data sources, and analytical methods. Extensive cladistic analyses of morphological characters (e.g., Simmons and Geisler 1998) generally support the traditional division of the bats that places the flying foxes (family Pteropodidae)—those bats that are visual navigators and lack laryngeal echolocation—in the suborder Megachiroptera and all other bats in the suborder Microchiroptera. This basic division, however, has not been supported by molecular analyses, which uniformly link the pteropodids with the microchiropteran families traditionally placed in the superfamilies Rhinolophoidea and Rhinopomatoidea (see Teeling et al. 2012 for review of molecular analyses).

In figure 360, we compare the major competing phylogenetic hypotheses, and the classifications that result, derived from a supermatrix analysis of both morphological data and the molecular data then available (*left*) (Jones et al. 2002), with the consensus based on multi-gene DNA sequence analyses (*right*) (Teeling et al. 2012). Note in particular that there are only three clades in common across the two trees (the family pairs Rhinolophidae + Hipposideridae, Mormoopidae + Phyllostomidae, and Molossidae + Vespertilionidae). Both trees provide the hierarchical classification for each hypothesis via the names used for internal nodes. The supertree (*left*) supports the traditional division of bats into Megachiroptera and Microchiroptera, and it recognizes seven superfamilies within the latter. The molecular tree (*right*) divides bats into two major clades, the Yinpterochiroptera (also called Pteropodiformes), which is further subdivided into two lineages, and the Yangochiroptera (= Vespertilioniformes), with three superfamilial lineages.

In the most recent synthesis of bat classification, Simmons (2005) opted to avoid use of any categorical ranks above that of the family. We recognize the rationale underlying her arguments (most notably that understanding of arrangements was sufficiently poor and under such flux that any suprafamilial associations were provisional at that time), but we organize the accounts herein using the hierarchical arrangement developed from the increasingly sensitive molecular data sets, which is illustrated by the right-hand tree in figure 360.

Some have argued that bats are diphyletic, with the Megachiroptera more closely related to Primates and Dermoptera than to the Microchiroptera, based primarily on the structure of the visual senses in the brain. This suggestion, however, is not supported by cladistic analyses of morphological and molecular characters (see the introduction to infraclass Eutheria or Placentalia), and has now been discarded.

Although family-level classification of bats has been largely stable since Miller's (1907) landmark review, contemporary molecular analyses question some arrangements. For example, initial morphological and molecular data sets (Simmons 1998; Simmons and Geisler 1998) suggested that the pallid bats (*Antrozous* and *Bauerus*) were not members of the Vespertilionidae, but rather should be elevated to their own family (Antrozoidae) because of a sister relationship to the Molossidae. This hypothesis has not been supported by more recent studies (see Roehrs et al. 2010). Additionally, some authors have argued for family status for the vespertilionid subfamily Miniopterinae, in part because of the temporal depth of its relationship to other groups traditionally placed within the family

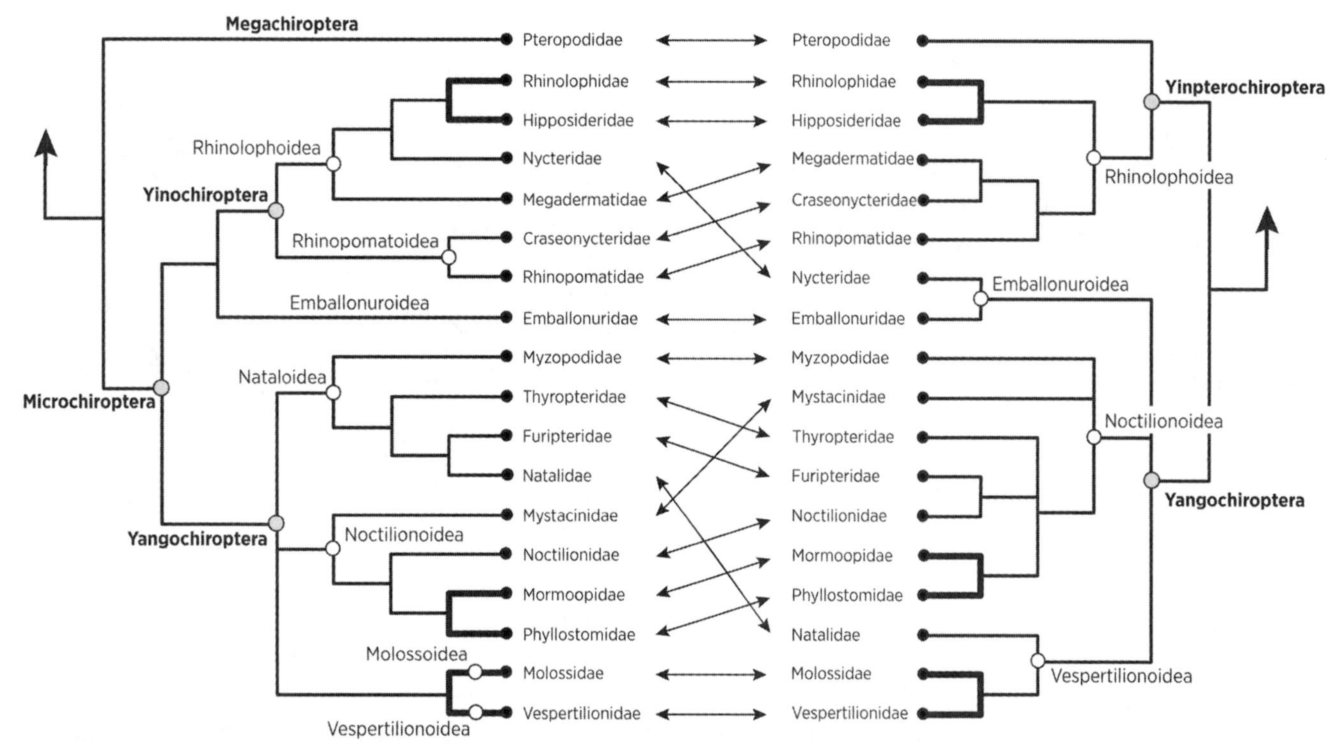

FIG. 360. Phylogenetic hypotheses and classifications linking the 18 families of living bats. The tree on the left is a supertree analysis of multiple independent morphological as well as molecular studies (Jones et al. 2002); that on the right is a general consensus of more recent molecular analyses (Teeling et al. 2012) and is the arrangement we follow herein. Nodes that correspond to major hierarchical groups are identified for both trees (*left*: the traditional groups organized by Simmons and Geisler 1998, modified based on Kirsch et al. 1998; see also Pierson et al. 1986; *right*: nodal groups used by Teeling et al. 2012). Polytomies indicate unresolved nodes; the arrows reflect differences in the placement of families in the two phylogenies; and the heavy lines connecting pairs of families are those clades shared between the two hypotheses. Note that Myzopodidae, part of a trichotomy within the Noctilionoidea in the tree on the right, is weakly placed as basal to the Vespertilionoidea families by Gatesy et al. (2017).

Vespertilionidae (e.g., Miller-Butterworth et al. 2007). Others have argued that the genus *Cistugo* (traditionally placed within *Myotis*) merits full family status (e.g., Cistugidae; see Lack et al. 2010). Pending further resolution of these relationships, we retain the traditional placement of both the miniopterines and *Cistugo* within the Vespertilionidae. However, these various efforts underscore the extent to which our understanding of relationships even among historically stable chiropteran groups remains subject to reassessment in the face of increasingly refined molecular analyses.

Clade YINPTEROCHIROPTERA (= Pteropodiformes)

The Yinpterochiroptera is the clade that joins the Old World flying foxes, Pteropodidae, as the sister group to the five families in the Rhinolophoidea (see the right-hand tree in figure 360).

Family PTEROPODIDAE (flying foxes, Old World fruit bats)

Size ranges from large (*Pteropus*, the largest bat, forearm length 220; mass 1.5 kg) to very small (*Macroglossus*, forearm length 37 mm; mass 15 g); wingspan up to 2.2 m; 2nd digit of wing with three bony elements, 3rd digit with two; pollex complete, with nail; tail usually short or absent, never enclosed in a uropatagium, if present; rostrum long, muzzle doglike (hence the common name "flying fox"). *Rousettus* echolocates by tongue clicking, all other members non-echolocating, navigate by vision and locate food by smell. Some migrate seasonally; none hibernate. Many form large roosting groups; some form large leks of calling males; others are solitary, roosting under leaves. Habitat includes tropical and subtropical regions that are at least partially wooded; extends over western Pacific and Indian Ocean islands.

DIAGNOSTIC CHARACTERS:

- **pinna simple, completely surrounding ear opening**
- **2nd digit of forelimb relatively free of 3rd digit and usually clawed**
- **postorbital processes large, forming a complete postorbital bar in some *Pteralopex* and *Pteropus***
- **angular process of lower jaw absent or low and broad**

RECOGNITION CHARACTERS:

1. no tragus in ear
2. eyes well developed
3. phalangeal formula of wing 2–3–2–3–3
4. **greater tuberosity (trochiter) small, not extending to level of proximal edge of humerus head**
5. dorsal articular facet on scapula absent (trochiter never articulates with scapula)
6. premaxillae with palatal and nasal branches, fused to each other and to maxillae
7. teeth highly modified for frugivorous diet
8. incisors small, styliform, somewhat flattened
9. canines always well developed
10. cheek teeth simple, crowns low, cusps indistinct, not obviously tuberculate

DENTAL FORMULA: $\frac{1\text{–}2}{0\text{–}2}\ \frac{1}{1}\ \frac{3}{3}\ \frac{1\text{–}2}{2\text{–}3} = 24\text{–}34$

TAXONOMIC DIVERSITY: About 46 genera and over 190 species in eight subfamilies and nine tribes.

Cynopterinae: Short-nosed fruit bats, chiefly Indo-Malayan in distribution; 14 genera and over 30 species

Balionycterini—11 genera and 17 species, including *Aethalops* (pygmy fruit bats), *Alionycteris* (Mindanao pygmy fruit bats), *Balionycteris* (Spotted-wing fruit bats), *Chironax* (black-capped fruit bats), *Dyacopterus* (Dayak fruit bats), and *Thoopterus* (swift fruit bats)

Cynopterini—three genera and 14 species, including *Cynopterus* (short-nosed fruit bats), *Megaerops* (tailless fruit bats), and *Ptenochirus* (musky fruit bats)

Eidolinae: straw-colored fruit bats; monotypic, with two species in the genus *Eidolon*

Harpyionycterinae: Harpy and bare-backed fruit bats; 17 species in four genera, including *Dobsonia* (naked-backed fruit bats) and *Harpyionycteris* (harpy fruit bats)

FIG. 361. Pteropodidae; masked flying fox, *Pteropus capistratus* (*left*), and broad-striped tube-nosed fruit bat, *Nyctimene aello* (*right*).

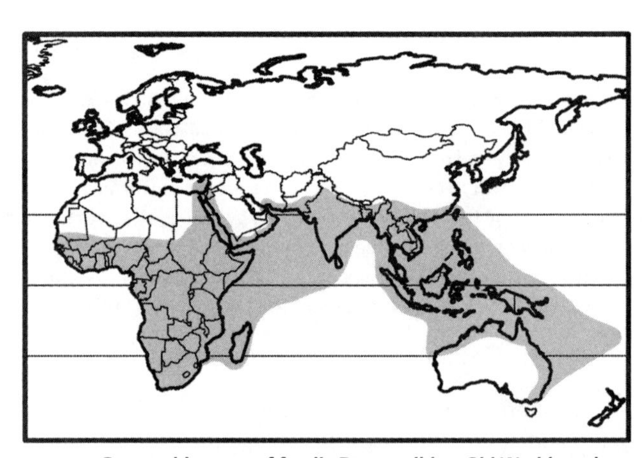

FIG. 362. Geographic range of family Pteropodidae: Old World tropics, including Africa, India and Southeast Asia, Australasia, and many Indian and Pacific Ocean islands.

Macroglossinae: five species in two genera, *Macroglossus* (long-tongued blossom bats) and *Syconycteris* (blossom bats)

Notopterinae: long-tailed fruit bats; monotypic, with two species in the genus *Notopterus*

Nyctimeninae: tube-nosed fruit bats; 17 species in at least two genera, *Nyctimene* and *Paranyctimene*

Pteropodinae: the largest subfamily of fruit bats, with 76 species in nine genera, including *Acerodon* (flying foxes), *Desmalopex* (Philippines flying foxes), *Nesonycteris* (blossom bats), *Pteralopex* (monkey-faced fruit bats), *Pteropus* (flying foxes), and *Styloctenium* (stripe-faced fruit bats)

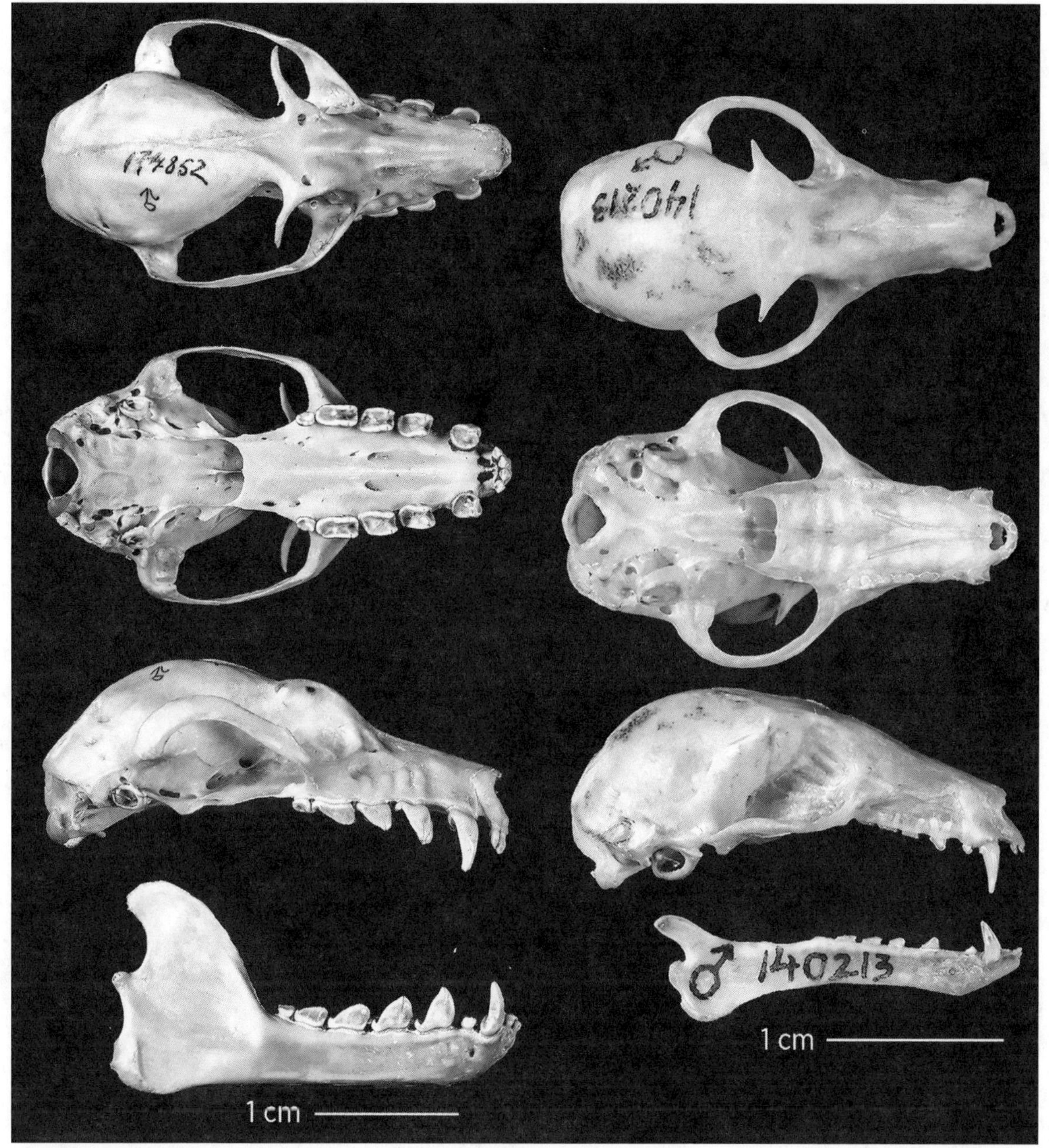

FIG. 363. Pteropodidae; skull and mandible of the Ryukyu flying fox, *Pteropus dasymallus* (*left*), and the dagger-toothed long-nosed fruit bat, *Macroglossus minimus* (*right*).

Rousettinae: diverse subfamily, including 13 genera and over 40 species in seven tribes.

- Eonycterini—dawn bats; monotypic, with three species in the genus *Eonycteris*
- Epomophorini—15 species in four genera, including *Epomophorus* (epauletted fruit bats), *Epomops* (epauletted fruit bats), and *Hypsignathus* (hammer-headed bat)
- Myonycterini—seven species in three genera, including *Myonycteris* (collared fruit bats)
- Plerotini—monotypic, with a single species in the genus *Plerotes* (broad-faced fruit bat)
- Rousettini—rousette bats; monotypic, with eight species in the single genus *Rousettus*
- Scotonycterini—African rainforest fruit bats; six species in two genera, *Casinycteris* and *Scotonycteris*
- Stenonycterini—long-haired fruit bat; monotypic, with a single species in the genus *Syconycteris*

Superfamily RHINOLOPHOIDEA
Family CRASEONYCTERIDAE
(Kitti's hog-nosed bat)

Smallest of all bats (forearm length 22–26 mm; mass 1.5 g); **muzzle swollen laterally, terminates in vertical pad surrounded by a ridgelike outgrowth**; pinnae very large, rounded, separate, not united at midline; tragus present; 2nd digit of wing with one bony phalanx only, 3rd digit with two; pollex relatively small; hind limb and hind foot slender, with relatively large claws; **no calcar**; **no tail** but large uropatagium and broad wings for slow, hovering flight. Insectivorous, foraging over trees and bamboo clumps. Associated with limestone caves in which they roost; known from only 43 caves at two localities in Thailand and Myanmar; not discovered until 1973.

DIAGNOSTIC CHARACTER:

- **palatal branches of premaxillae fused anteriorly and posteriorly (no sign of suture) enclosing a large vacuity; nasal branches fused posteriorly, extending upward to lie on surface of nasals and maxillae**

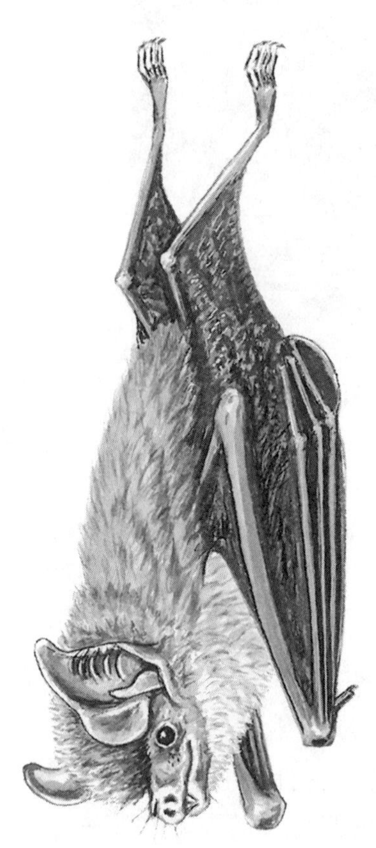

FIG. 364. Craseonycteridae; Kitti's hog-nosed bat, *Craseonycteris thonglongyai*.

RECOGNITION CHARACTERS:

1. phalangeal formula of wing 2–1–2–3–3
2. trochiter extends proximally well beyond level of humerus head
3. dorsal articular facet on scapula present (trochiter articulates with scapula)
4. no postorbital process
5. cheek teeth tritubercular

DENTAL FORMULA: $\frac{1}{2}\ \frac{1}{1}\ \frac{1}{2}\ \frac{3}{3} = 28$

TAXONOMIC DIVERSITY: The family is monotypic, with a single species, *Craseonycteris thonglongyai*.

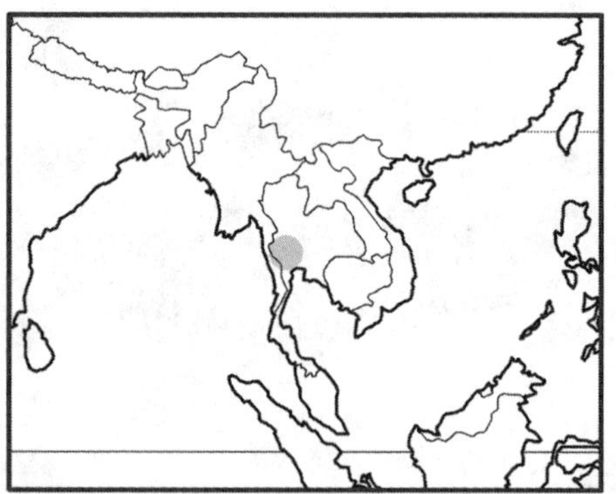

FIG. 365. Geographic range of family Craseonycteridae: west-central Thailand and Myanmar.

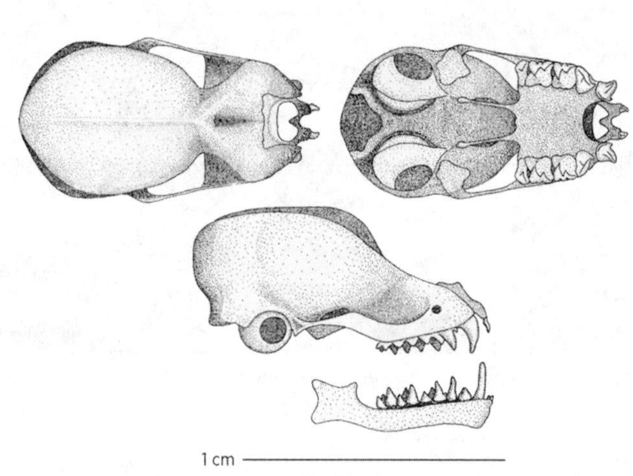

FIG. 366. Craseonycteridae; skull and mandible of Kitti's hog-nosed bat, *Craseonycteris thonglongyai*. Redrawn from Hill (1974).

Families HIPPOSIDERIDAE and RHINOLOPHIDAE (horseshoe-nosed bats, leaf-nosed bats)

Small to medium-sized (forearm length 32–102 mm; mass 5–60 g); **muzzle with prominent outgrowths (often horseshoe-shaped) and depressions**; pinnae small to large, pointed or rounded, united in some, without ventral extension under eye; **no tragus**; 2nd digit of wing absent (well-developed metacarpal remaining), 3rd digit with two phalanges; pollex relatively small; hind limb and hind foot slender, with only two phalanges per digit in some (Hipposideridae; table 7), and with relatively large claws; calcar present, well developed; tail enclosed within and extending to distal margin of uropatagium. Insectivorous. Roost in caves, rock crevices, hollow trees, and buildings; emit ultrasonic pulses through nostrils, directing sound with aid of nose leaf. Hibernate in northern latitudes; females store sperm. Inhabit forests, savannas, and deserts in tropical, subtropical, and temperate regions.

DIAGNOSTIC CHARACTERS:

- **no tragus** (antitragus present)
- **premaxillae represented by palatal branches only, free of maxillae and extending anteriorly from palate as a spatula-like process**

RECOGNITION CHARACTERS:

1. phalangeal formula of wing 2–0–2–3–3
2. trochiter extends proximally well beyond level of humerus head
3. dorsal articular facet on scapula present (trochiter articulates with scapula, although limitedly in some Hipposideridae)
4. no postorbital process
5. cheek teeth tritubercular

DENTAL FORMULA: $\frac{1}{2}\ \frac{1}{1}\ \frac{1\text{–}2}{2\text{–}3}\ \frac{3}{3} = 28\text{–}32$

TAXONOMIC DIVERSITY: 11 genera in two families (or subfamilies):

Rhinolophidae: *Rhinolophus* (horseshoe-nosed bats), with nearly 80 species

Hipposideridae: 10 genera containing over 90 species: *Anthops* (flower-faced bat), *Asellia* (trident leaf-nosed bats), *Ascelliscus* (trident bats), *Cleotis* (short-eared trident bat), *Coelops* (tailless leaf-nosed bats), *Hipposideros* (Old World leaf-nosed bats), *Paracoelops* (Vietnamese leaf-nosed bat), *Paratriaenops* (Malagasy trident bats), *Rhinonicteris* (orange leaf-nosed bat), and *Triaenops* (trident bats)

Rhinolophids and hipposiderids are each monophyletic assemblages that are phylogenetic sisters; the two lineages have been treated either as subfamilies of the Rhinolophidae (e.g., Simmons and Geisler 1998) or as distinct families (Simmons 2005). Recent phylogenetic work (Foley et al. 2014) argues that the Rhinonycterina, a subtribe within the Hipposideridae, merits family status (e.g., Rhinonycteridae, containing the genera *Cleotis*, *Rhinonicteris*, *Paratriaenops*, and *Triaenops*).

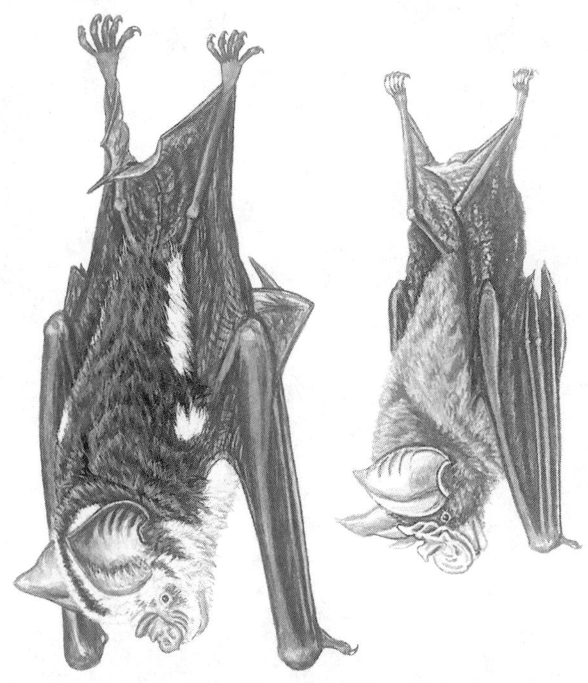

FIG. 367. Diadem leaf-nosed bat, *Hipposideros diadema* (Hipposideridae; *left*), and smaller horseshoe bat, *Rhinolophus megaphyllus* (Rhinolophidae; *right*).

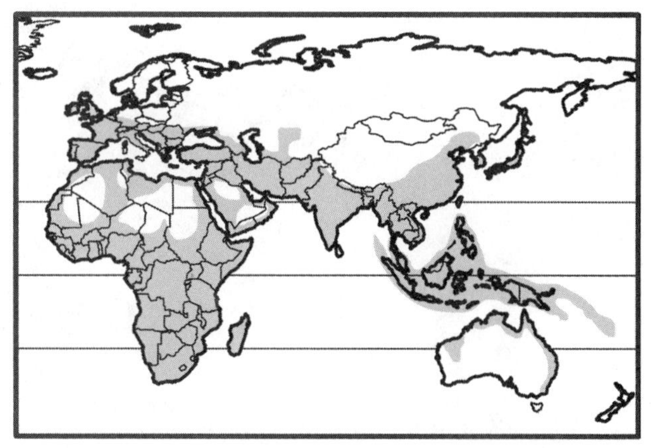

FIG. 368. Combined geographic range of families Hipposideridae and Rhinolophidae: Old World, from Europe and Africa to Japan, Philippines, and Australia.

Table 7. Characters that distinguish Rhinolophidae from Hipposideridae + Rhinonycteridae.

Trait	*Rhinolophidae*	*Hipposideridae (including Rhinonycteridae)*
Nose leaf	usually spear-shaped, pointed; includes "sella" (flattened leaflet in middle of nose leaf)	usually more rounded; no "sella"
Posterior (pes) phalanges	3 (2 in hallux)	2
Lower premolars	2	3 (pm3 small)

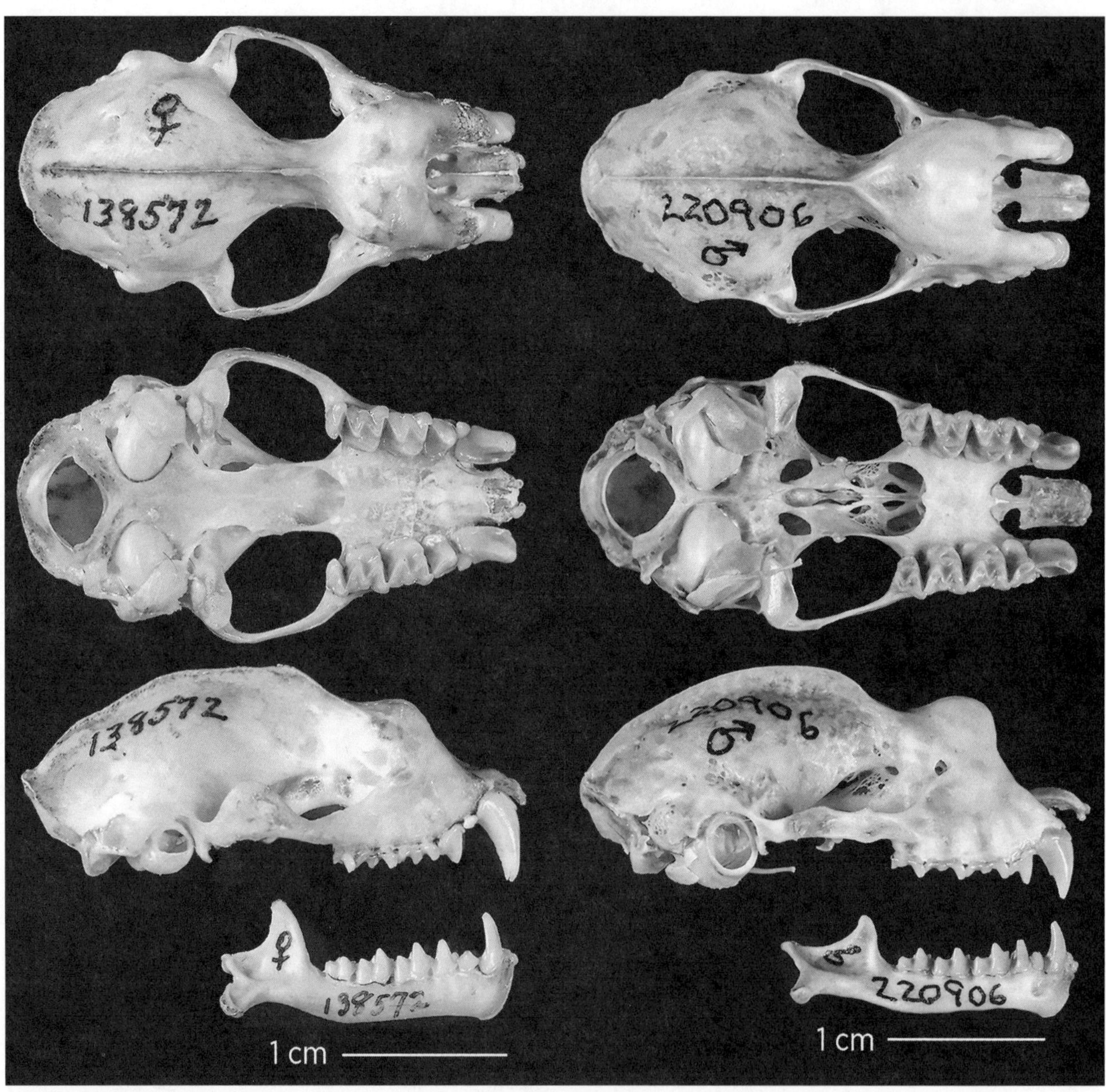

FIG. 369. Skull and mandible of the diadem leaf-nosed bat, *Hipposideros diadema* (Hipposideridae; *left*), and Hildebrandt's horseshoe bat, *Rhinolophus hildebrandtii* (Rhinolophidae; *right*).

Family MEGADERMATIDAE (false vampire bats, yellow-winged bat)

Medium-sized (forearm length 54–107 mm; mass 20–170 g); muzzle with conspicuous long, erect nose leaf; **pinnae large, rounded, connected across forehead by high ridge of skin**, without ventral extension under eye; **tragus present, large, forked**; 2nd digit of wing with one bony phalanx, 3rd digit with two, pollex relatively large; hind limb and hind foot slender, with relatively large claws; calcar present, well developed; **tail short or absent**; uropatagium large. Insectivorous or carnivorous. Most are colonial, roosting in large numbers in caves or in small numbers in hollow trees; nocturnal, but some can be diurnal. Inhabit tropical forests and savannas.

DIAGNOSTIC CHARACTER:

- **upper canine large, notably procumbent, with large secondary cusp**

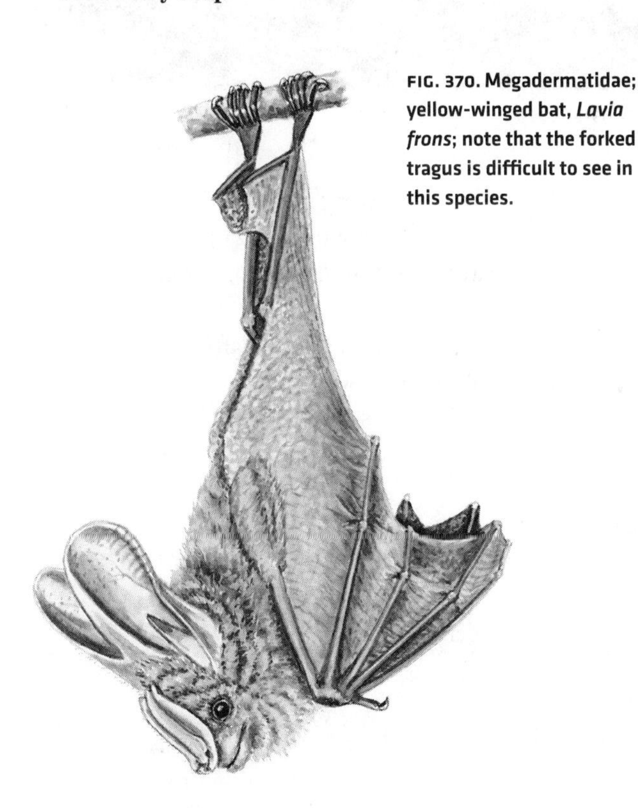

FIG. 370. Megadermatidae; yellow-winged bat, *Lavia frons*; note that the forked tragus is difficult to see in this species.

RECOGNITION CHARACTERS:

1. phalangeal formula of wing 2–1–2–3–3
2. trochiter extends only to level of proximal edge of humerus head
3. dorsal articular facet on scapula present (trochiter articulates with scapula)
4. **no premaxillae; hence, large gap between maxillae**
5. postorbital process present, small (often obscured by large supraorbital ridge)
6. cheek teeth tritubercular

DENTAL FORMULA: $\frac{0}{2}\ \frac{1}{1}\ \frac{1\text{–}2}{2}\ \frac{3}{3} = 26\text{–}28$

TAXONOMIC DIVERSITY: Five species in four genera: *Cardioderma* (heart-nosed bat), *Lavia* (yellow-winged bat), *Macroderma* (Australian false vampire bat), and *Megaderma* (greater and lesser false vampire bats).

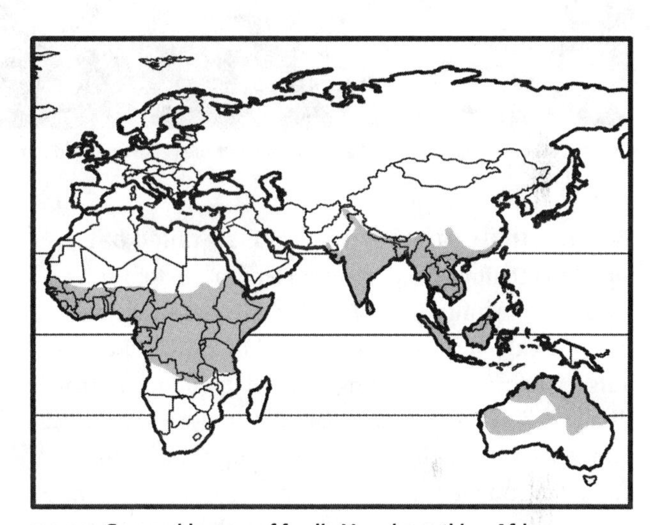

FIG. 371. Geographic range of family Megadermatidae: Africa, southern Asia, Indonesian and Philippine islands, Australia.

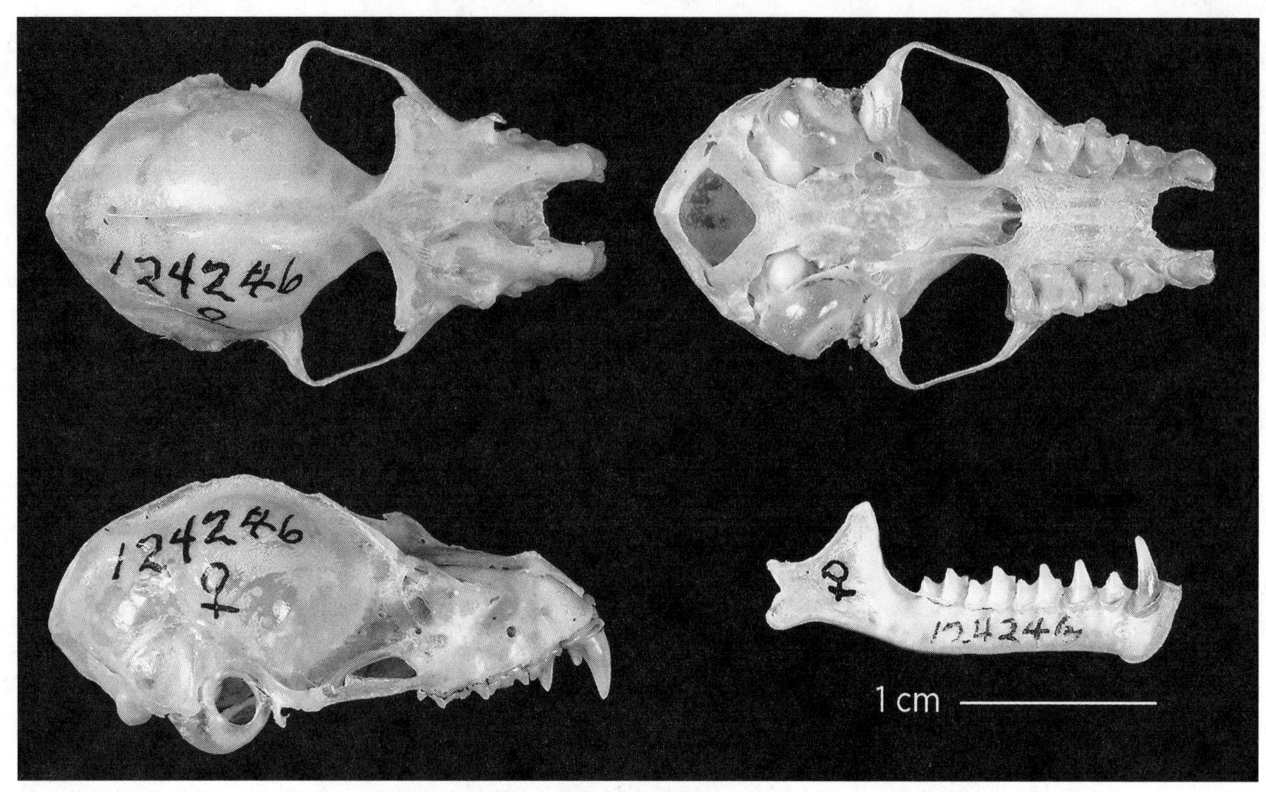

FIG. 372. Megadermatidae; skull and mandible of the yellow-winged bat, *Lavia frons*.

Family RHINOPOMATIDAE (mouse-tailed bats)

Small to medium-sized (forearm length 42–70 mm; mass 6–14 g); **muzzle swollen laterally, terminating in a vertical pad that is surrounded by a ridgelike outgrowth**; nostrils opening forward in oblique narrow slits; nasal chamber dilated laterally; pinnae very large, rounded, united at bases, without ventral extension under eye; tragus present, prominent; 2nd digit of wing with two bony phalanges, 3rd digit with two; hind limb and hind foot slender, with relatively large claws; calcar present (contra Simmons and Geisler 1998); pelage soft and dull brown; face, rump, and part of venter usually naked. Insectivorous. Roost in buildings, caves, and rock clefts, in large colonies or smaller groups. Inhabit chiefly desert and semidesert regions.

DIAGNOSTIC CHARACTERS:

- **nostrils slit-like, valvular**
- tail very long, roughly equal to length of body, extending freely beyond distal margin of wide uropatagium

RECOGNITION CHARACTERS:

1. phalangeal formula of wing 2–2–2–3–3
2. trochiter extends only to level of proximal edge of humerus head
3. **dorsal articular facet on scapula absent (trochiter never articulates with scapula)**
4. no postorbital process
5. cheek teeth tritubercular

DENTAL FORMULA: $\frac{1}{2}\ \frac{1}{1}\ \frac{1}{2}\ \frac{3}{3} = 28$

TAXONOMIC DIVERSITY: The family is monotypic, with four species in the single genus *Rhinopoma*.

FIG. 373. Rhinopomatidae; mouse-tailed bat, *Rhinopoma hardwickii*.

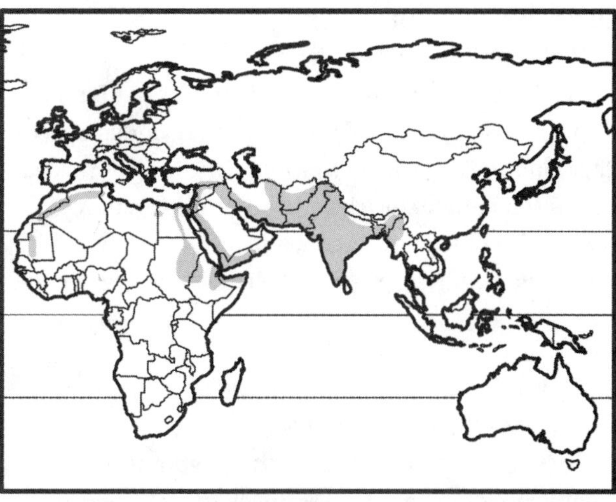

FIG. 374. Geographic range of family Rhinopomatidae: northern Africa and southern Asia.

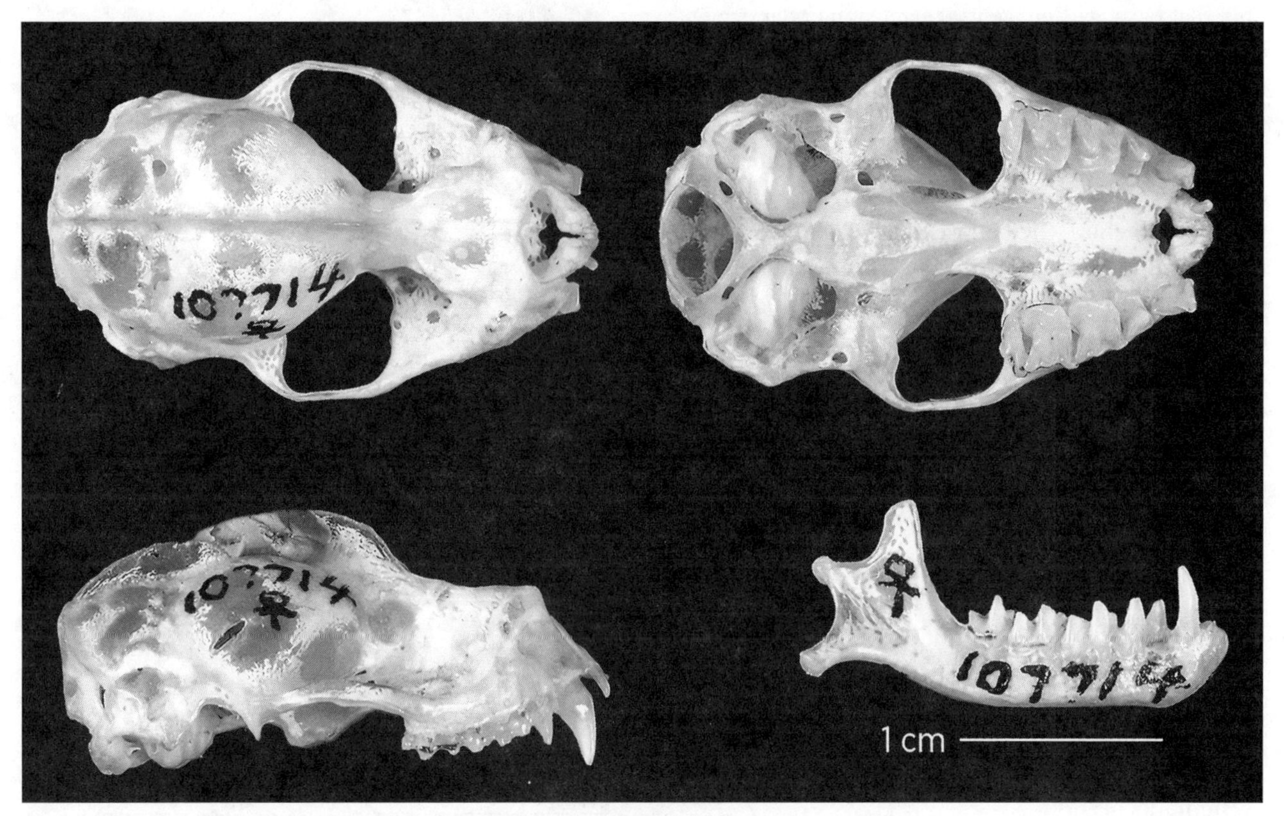

FIG. 375. Rhinopomatidae; skull and mandible of the small mouse-tailed bat, *Rhinopoma muscatellum*.

Clade YANGOCHIROPTERA (= Vespertilioniformes)

The clade Yangochiroptera includes three superfamilies with currently unresolved phylogenetic relationships: the Emballonuroidea, with two families; the Noctilionoidea, with seven families, of which relationships among several also remain uncertain; and the Vespertilionoidea, with three families (see the right-hand tree in fig. 360).

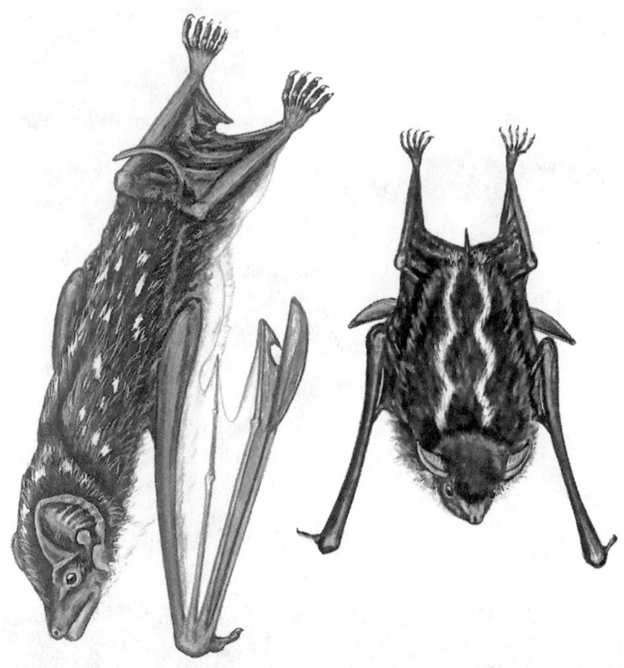

FIG. 376. Emballonuridae; naked-rumped pouched bat, *Saccolaimus saccolaimus* (Taphozoinae; *left*), and greater white-lined bat, *Saccopteryx bilineata* (Emballonurinae; *right*).

Superfamily EMBALLONUROIDEA

Family EMBALLONURIDAE

(sac-winged bats, sheath-tailed bats)

Small to medium-sized (forearm length 35–95 mm; mass 5–105 g); muzzle without skin outgrowths on face; pinnae moderately large, rounded, often united at base, without ventral extension under eye; tragus present, small; 2nd digit of wing absent (well-developed metacarpal remains), 3rd digit with two bony phalanges, pollex unremarkable; hind limb and hind foot slender, with relatively large claws; **distal portion of tail free, resting on top of but not extending beyond long uropatagium when fully extended; surface of propatagium near elbow with sac housing subdermal glands** that secrete a substance with a strong odor, especially developed in males. Roost solitarily or in colonies on vertical rock faces or tree trunks, in caves, or in tree hollows. Insectivorous. Inhabit forests to deserts in tropical and subtropical regions.

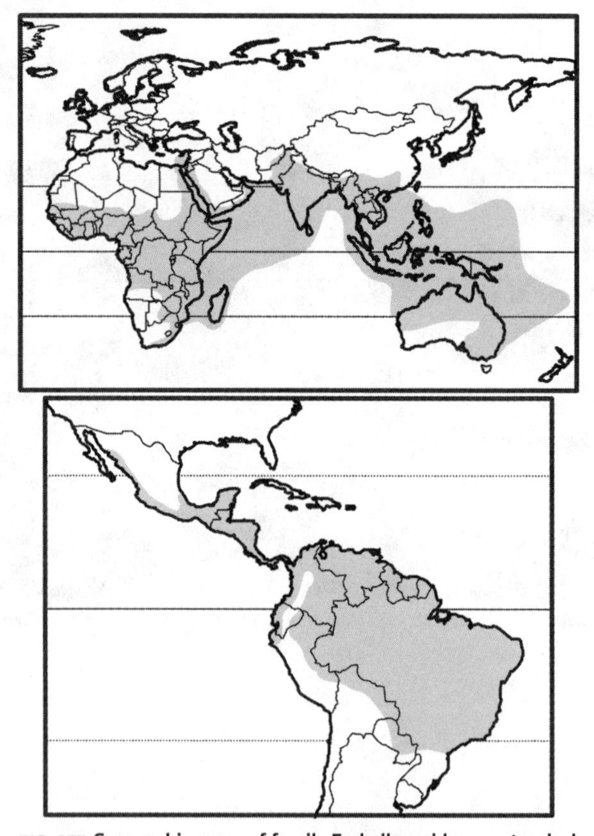

FIG. 377. Geographic range of family Emballonuridae: pantropical—*New World*, from Mexico to Argentina; *Old World*, from sub-Saharan Africa east through Southeast Asia to Australasia and some Pacific islands.

DIAGNOSTIC CHARACTER:

- **glandular sac present in propatagium in most genera** (larger in males; absent in all other microchiropterans)

RECOGNITION CHARACTERS:

1. phalangeal formula of wing 2–0–2–3–3
2. trochiter extends only to level of proximal edge of humerus head
3. dorsal articular facet on scapula present or absent (where present, trochiter articulates with scapula)
4. premaxillae represented by nasal branches only; small, separated, not fused to surrounding bones
5. **postorbital process present, prominent but long and thin**
6. cheek teeth tritubercular

DENTAL FORMULA: $\frac{1\text{–}2}{2\text{–}3}\ \frac{1}{1}\ \frac{2}{2}\ \frac{3}{3} = 30\text{–}34$

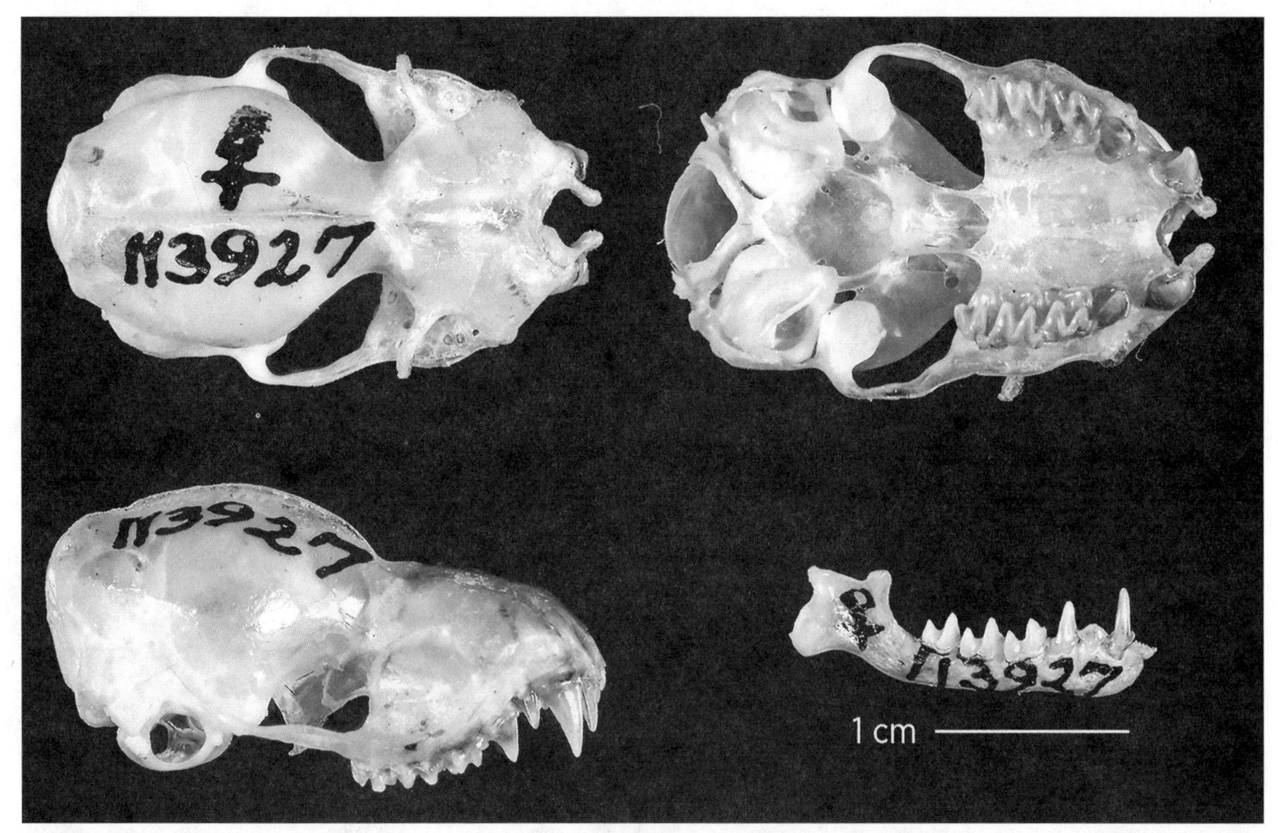

FIG. 378. Emballonuridae: Emballonurinae; skull and mandible of the greater doglike bat, *Peropteryx kappleri*.

TAXONOMIC DIVERSITY: Over 50 species in 13 genera allocated to two subfamilies:

Emballonurinae—11 genera, including *Balantiopteryx* (gray sac-winged bats), *Coleura* (African sheath-tailed bats), *Diclidurus* (white bats), *Emballonura* (Old World sheath-tailed bats), *Peropteryx* (doglike bats), *Rhynchonycteris* (proboscis bat), and *Saccopteryx* (white-lined bats)

Taphozoinae—*Saccolaimus* (pouched bats) and *Taphozous* (tomb bats)

Insectivorous; usually colonial, living in hollow trees, under vegetation, in caves, and in buildings. Habitat ranges from tropical forests to semideserts.

DIAGNOSTIC CHARACTERS:

- **tail long, fully enclosed in large uropatagium, and terminating in T-shaped cartilaginous tip**
- **cranium with large depression between orbits**
- **fibula absent**

Family NYCTERIDAE (slit- or hollow-faced bats)

Small to medium-sized (forearm length 35–65 mm; mass 10–45 g); **top of muzzle with complex outgrowths of skin alongside a deep concavity**, usually split vertically between the nares (hence common name); pinnae relatively large, rounded, separate, without ventral extension under eye; tragus present, small; 2nd digit of wing absent (well-developed metacarpal remaining), 3rd digit with two bony phalanges; pollex relatively small; hind limb and hind foot slender; claws not enlarged; calcar present, well developed.

FIG. 379. Nycteridae; hairy slit-faced bat, *Nycteris hispida*.

RECOGNITION CHARACTERS:

1. phalangeal formula of wing 2-0-2-3-3
2. trochiter extends only to level of proximal edge of humerus head
3. dorsal articular facet on scapula present (trochiter limitedly articulates with scapula)
4. premaxillae represented by palatal branches only, fused to maxillae and sometimes with each other
5. **postorbital process present, relatively small, extending from large supraorbital ridge**
6. cheek teeth tritubercular

DENTAL FORMULA: $\frac{2}{3}\ \frac{1}{1}\ \frac{1}{2}\ \frac{3}{3} = 32$

TAXONOMIC DIVERSITY: The family is monotypic, with about 16 species in the single genus *Nycteris*.

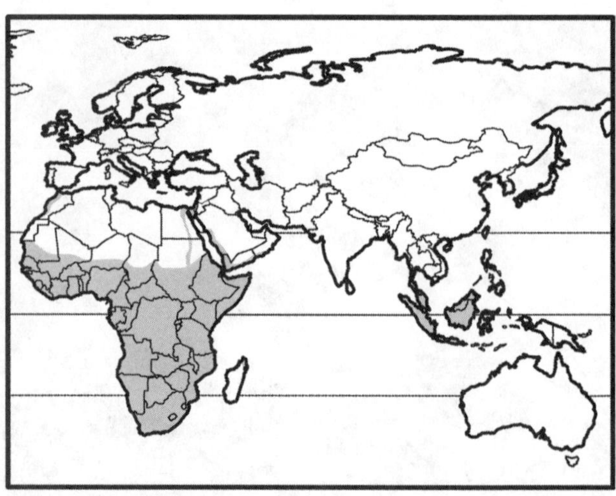

FIG. 380. Geographic range of family Nycteridae: Africa, Middle East, India, Southeast Asia.

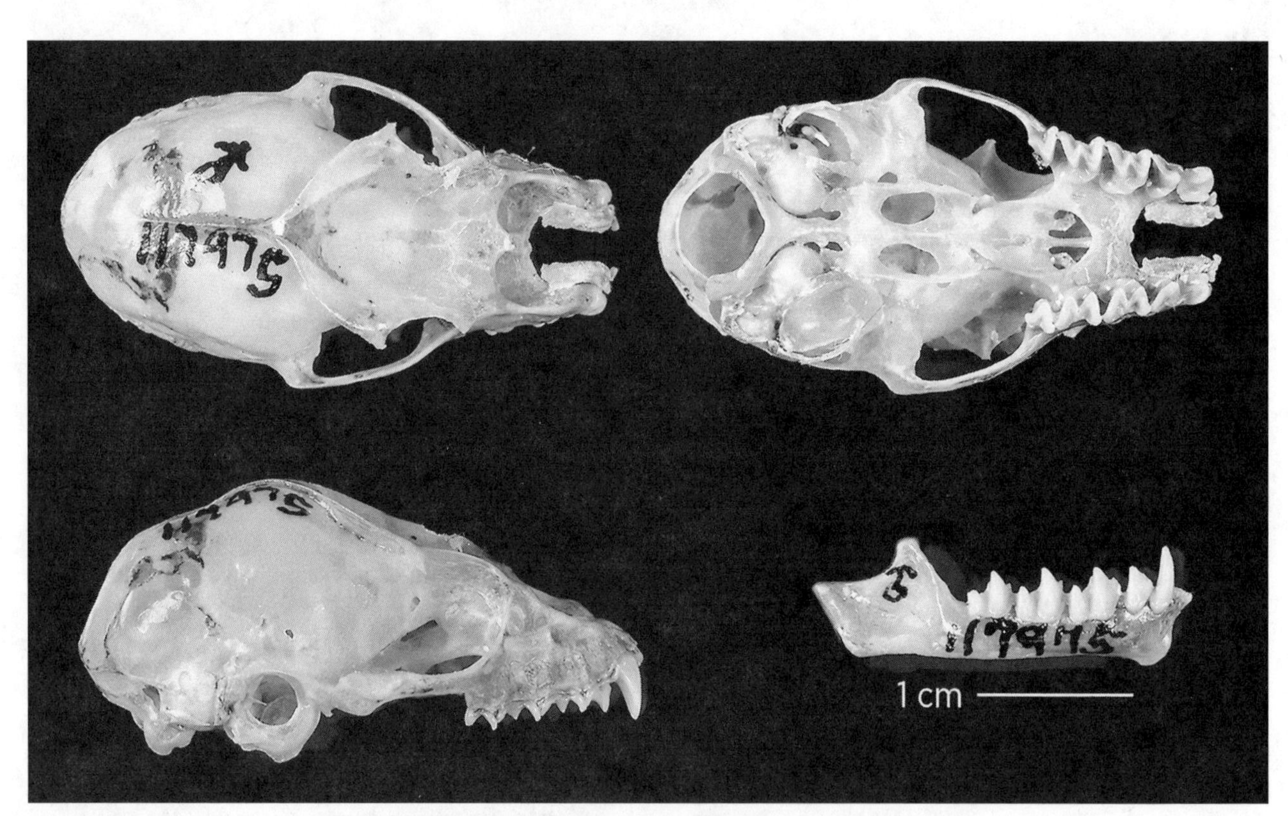

FIG. 381. Nycteridae; skull and mandible of the Egyptian slit-faced bat, *Nycteris thebaica*.

Superfamily NOCTILIONOIDEA

The Noctilionoidea includes a group of seven families with largely Southern Hemisphere distributions: five families live in the New World, mainly in the Neotropics (Furipteridae, Mormoopidae, Noctilionidae, Phyllostomidae, and Thyropteridae); one family is restricted to New Zealand (Mystacinidae), although fossils are known from Australia; and another family is restricted to Madagascar (Myzopodidae), although fossils have been recorded from northern Africa. Molecular phylogenetic analyses support the Myzopodidae as basal to the group, followed successively by the Mystacinidae and then the New World families. Therefore, the Noctilionoidea probably originated in eastern Gondwanaland and subsequently dispersed southward into Australia (mystacinids) and then westward into South America, probably across Antarctica (Gunnell et al. 2014).

FIG. 382. Furipteridae; thumbless bat, *Furipterus horrens*.

Family FURIPTERIDAE (smoky bats)

Size small (forearm 30–40 mm; mass 3–5 g); muzzle plain; lower lip with (*Amorphochilus*) or without (*Furipterus*) three triangular fleshy prominences; **pinnae large, funnel-shaped, separate**, without ventral extension under eye; **tragus present, small, triangular**; 2nd digit of wing absent (metacarpal remaining), 3rd digit with two bony phalanges, pollex much reduced; hind limb and hind foot slender, small, with relatively small claws; calcar present, well developed; tail long, fully enclosed in large uropatagium. Insectivorous, roost in caves; little is known about ecology or behavior. Inhabit very arid coast of Peru and Chile as well as tropical forests.

FIG. 383. Geographic range of family Furipteridae: New World tropics and subtropics from Central America to Brazil.

DIAGNOSTIC CHARACTER:

- **pollex much reduced, functionless, mostly enclosed in wing membrane**

RECOGNITION CHARACTERS:

1. phalangeal formula of wing 2–0–2–3–3
2. trochiter extends proximally well beyond level of humerus head
3. dorsal articular facet on scapula present (trochiter articulates with scapula)
4. premaxillae with reduced and unossified palatal branches only, fused to each other and to maxillae
5. no postorbital process
6. cheek teeth tritubercular

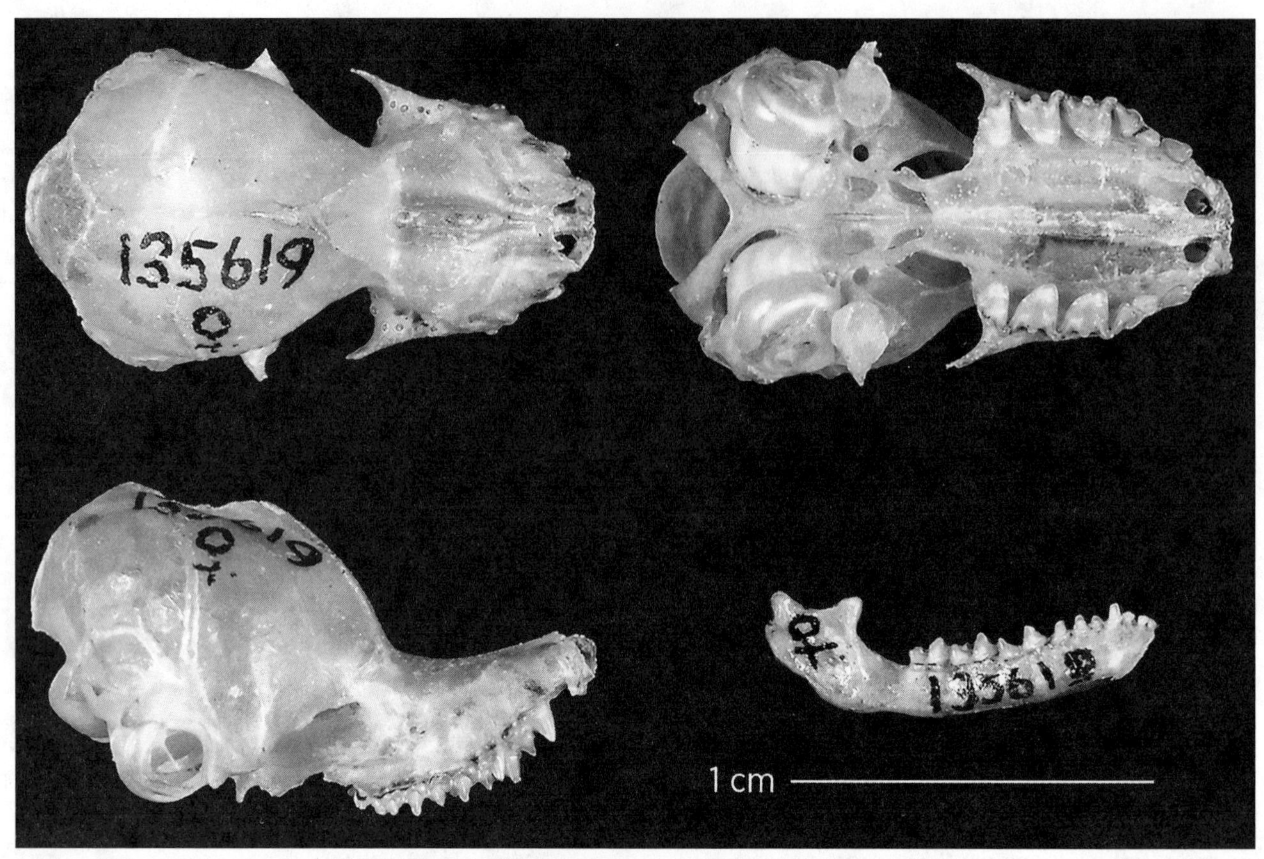

FIG. 384. Furipteridae; skull and mandible of the smoky bat, *Amorphochilus schnablii*.

DENTAL FORMULA: $\frac{2}{3}\ \frac{1}{1}\ \frac{2}{3}\ \frac{3}{3} = 36$

TAXONOMIC DIVERSITY: Two monotypic genera, *Amorphochilus* (smoky bat) and *Furipterus* (thumbless bat).

Family MORMOOPIDAE (leaf-chinned, mustached, naked-back bats)

Small to medium-sized (forearm 35–65 mm; mass 7–44 g); rudimentary nose leaf on muzzle; ears variable, either narrow (*Pteronotus*) or funnel-shaped (*Mormoops*); **tragus variable but always with secondary fold**; 2nd digit of wing with one small bony phalanx, 3rd digit with three, pollex relatively large; hind limb and hind foot slender, with relatively large claws; calcar present, well developed; **tail short, protruding from dorsal surface of large uropatagium**; wing membranes extend over back to meet at midline in some *Pteronotus* (hence common name "naked-back bats"). Highly gregarious, roosting in large colonies, generally in caves; insectivorous, usually foraging near water. Habitat ranges from semidesert to forest in subtropical and tropical regions.

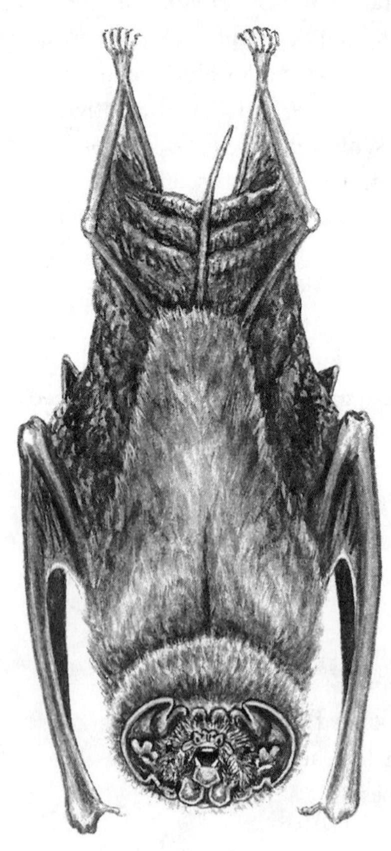

FIG. 385. Mormoopidae; ghost-faced bat, *Mormoops megalophylla*.

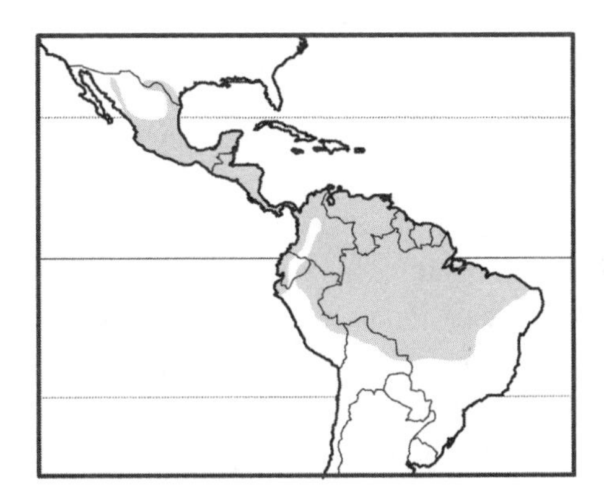

FIG. 386. Geographic range of family Mormoopidae: New World—northwestern Mexico and southern Texas south to Bolivia and Brazil.

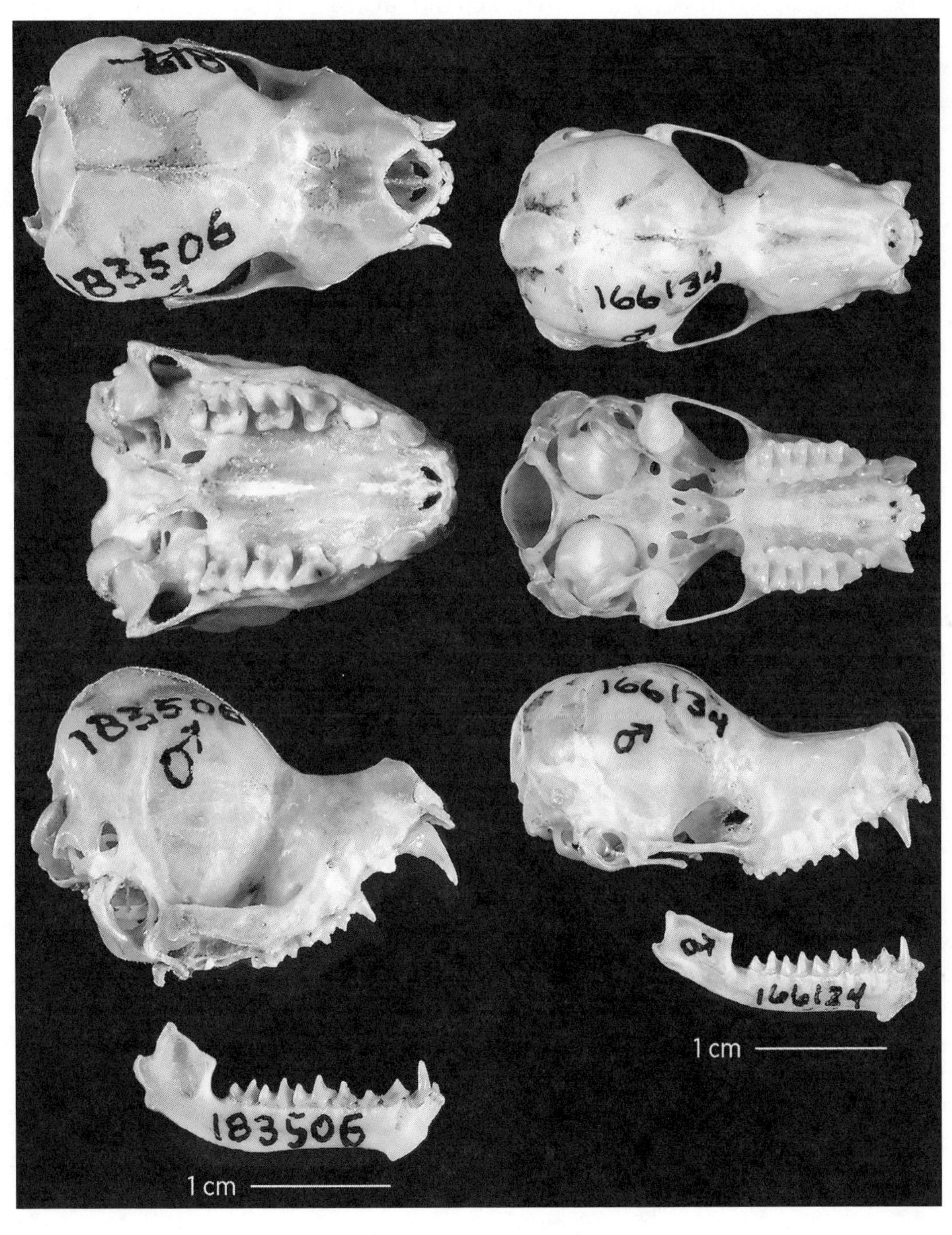

FIG. 387. Mormoopidae; skull and mandible of Peter's ghost-faced bat, *Mormoops megalophylla* (*left*), and common mustached bat, *Pteronotus parnellii* (*right*).

DIAGNOSTIC CHARACTERS:

- **lower lip with leaflike or platelike outgrowths**
- **pinnae large, often united at bases, with ventral extension under eye**

RECOGNITION CHARACTERS:

1. phalangeal formula of wing 2–1–3–3–3
2. trochiter extends proximally well beyond level of humerus head
3. dorsal articular facet on scapula present (trochiter articulates with scapula)
4. premaxillae with nasal and palatal branches, fused to each other and to maxillae
5. no postorbital process
6. cheek teeth tritubercular

DENTAL FORMULA: $\frac{2}{2}\ \frac{1}{1}\ \frac{2}{3}\ \frac{3}{3} = 34$

TAXONOMIC DIVERSITY: 10 species in two genera: *Mormoops* (ghost-faced bats) and *Pteronotus* (mustached bats).

Family MYSTACINIDAE (short-tailed bats)

Small (forearm 35–50 mm; mass 12–15 g); **muzzle with small pad that houses hairs bearing spoon-shaped tips**; pinnae large, pointed, separate, without ventral extension under eye; **tragus present, long**; 2nd digit of wing with one small bony phalanx, 3rd digit with three, pollex relatively large; **hind limb and hind foot short, broad, with relatively large, sharp claws with a basal talon, pes with grooved soles**; calcar present, well developed; **tail short, protruding from dorsal surface of narrow tail membrane.** Primarily insectivorous, but also eat fruit and nectar; roost in crevices and hollow trees or in burrows they excavate with their upper incisors (the only bats known to burrow); adept at moving quickly on solid surfaces with wings folded. Inhabit temperate forests.

DIAGNOSTIC CHARACTERS:

- **1st phalanx of each manal digit folding to outer (rather than the usual inner) side of metacarpal when wing is folded**
- **claws of pollex and feet with basal talon (talon absent in all other microchiropterans)**

RECOGNITION CHARACTERS:

1. phalangeal formula of wing 2–1–3–3–3
2. trochiter extends proximally well beyond level of humerus head
3. dorsal articular facet on scapula present (trochiter articulates with scapula)
4. premaxillae with nasal and palatal branches, fused to each other and to maxilla
5. no postorbital processes
6. cheek teeth tritubercular

DENTAL FORMULA: $\frac{1}{1}\ \frac{1}{1}\ \frac{2}{2}\ \frac{3}{3} = 28$

TAXONOMIC DIVERSITY: The family is monotypic, with two species in the single genus *Mystacina*.

FIG. 388. Mystacinidae; lesser short-tailed bat, *Mystacina tuberculata*.

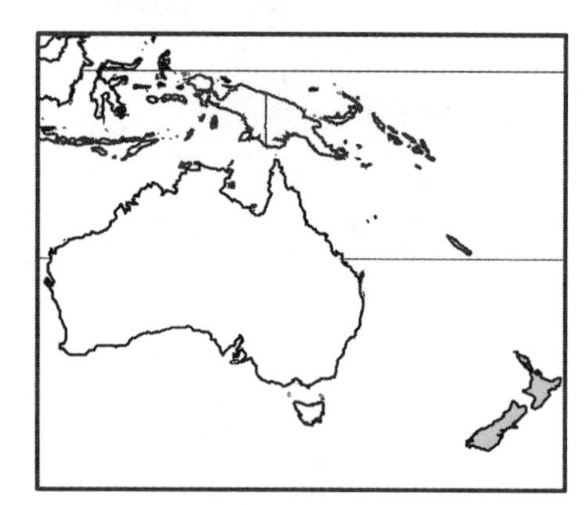

FIG. 389. Geographic range of family Mystacinidae: restricted to New Zealand.

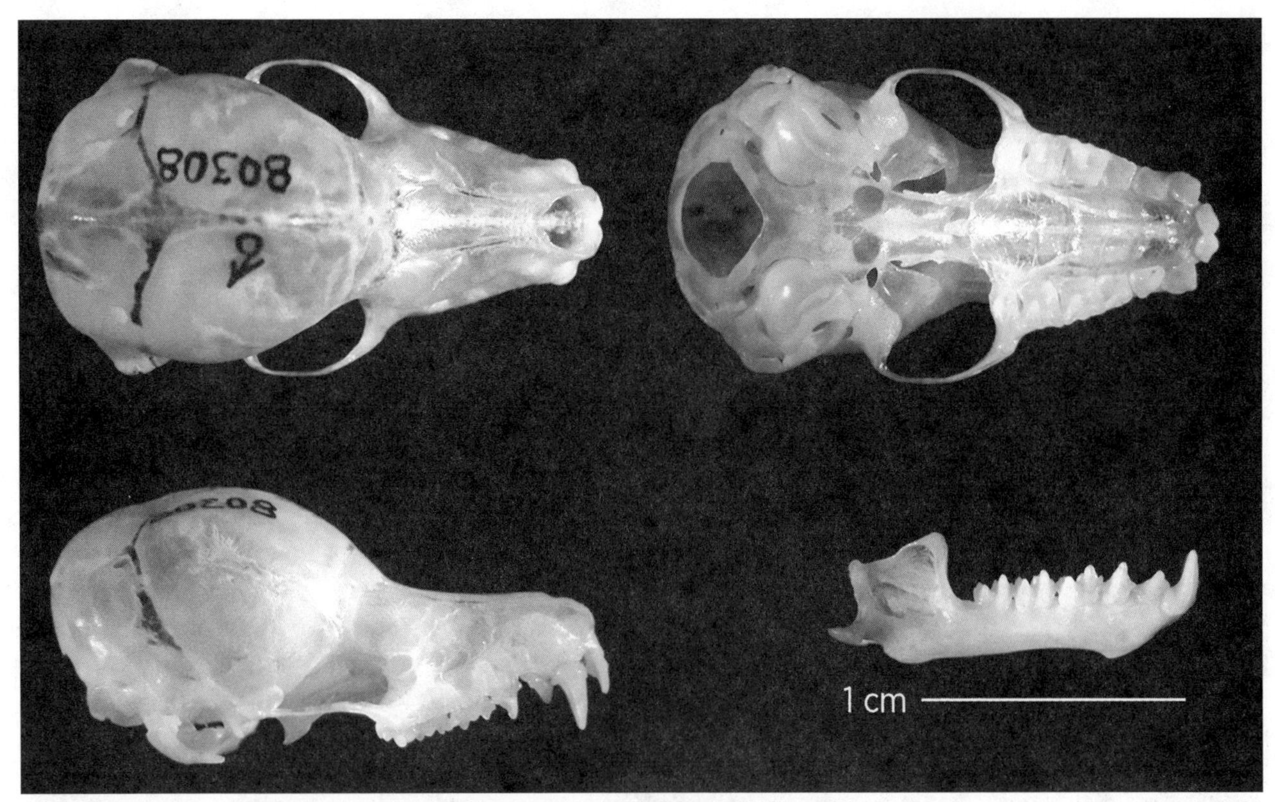

FIG. 390. Mystacinidae; skull and mandible of the lesser short-tailed bat, *Mystacina tuberculata* (image kindly provided by Judith Eger through the Royal Ontario Museum).

Family MYZOPODIDAE (sucker-footed bats)

Medium-sized (forearm 44–50 mm; mass 8–10 g); muzzle plain, no skin outgrowths; pinnae very large, rounded, without ventral extension under eye; **tragus present, bound to pinna; prominent unstalked adhesive disc on wrist and ankle** (cf. Thyropteridae); 2nd digit of wing with one cartilaginous phalanx only, 3rd digit with three bony phalanges, **pollex greatly reduced, functionless, with minute claw; hind limb slender; hind foot small, only two fused phalanges per digit, with small claws**; calcar present, well developed; tail completely enclosed in large membrane. Habits poorly known; habitat rainforest.

DIAGNOSTIC CHARACTERS:

- **all digits of hind foot (excluding claws) syndactylous**
- **auditory meatus partially closed by mushroom-shaped process**

RECOGNITION CHARACTERS:

1. phalangeal formula of wing 2-1-3-3-3
2. trochiter extends proximally well beyond level of humerus head
3. dorsal articular facet on scapula present (trochiter articulates with scapula)
4. premaxillae with nasal and palatal branches, fused to each other and to maxillae
5. no postorbital process
6. cheek teeth tritubercular

DENTAL FORMULA: $\frac{2}{3}\ \frac{1}{1}\ \frac{3}{3}\ \frac{3}{3} = 38$

TAXONOMIC DIVERSITY: The family is monotypic, with two species in the single genus *Myzopoda*.

FIG. 391. Myzopodidae; Madagascar sucker-footed bat, *Myzopoda aurita*.

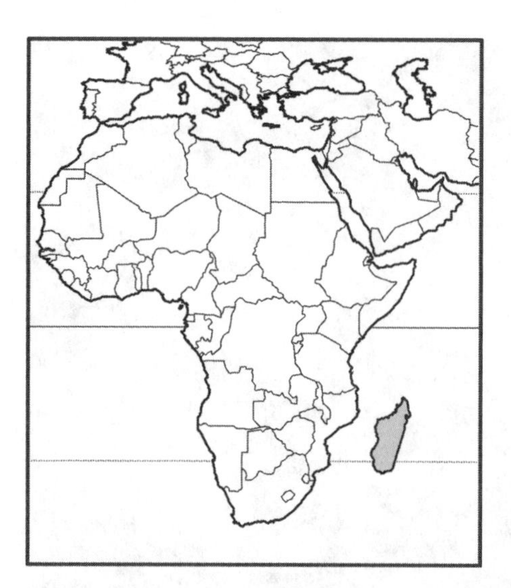

FIG. 392. Geographic range of family Myzopodidae: restricted to Madagascar.

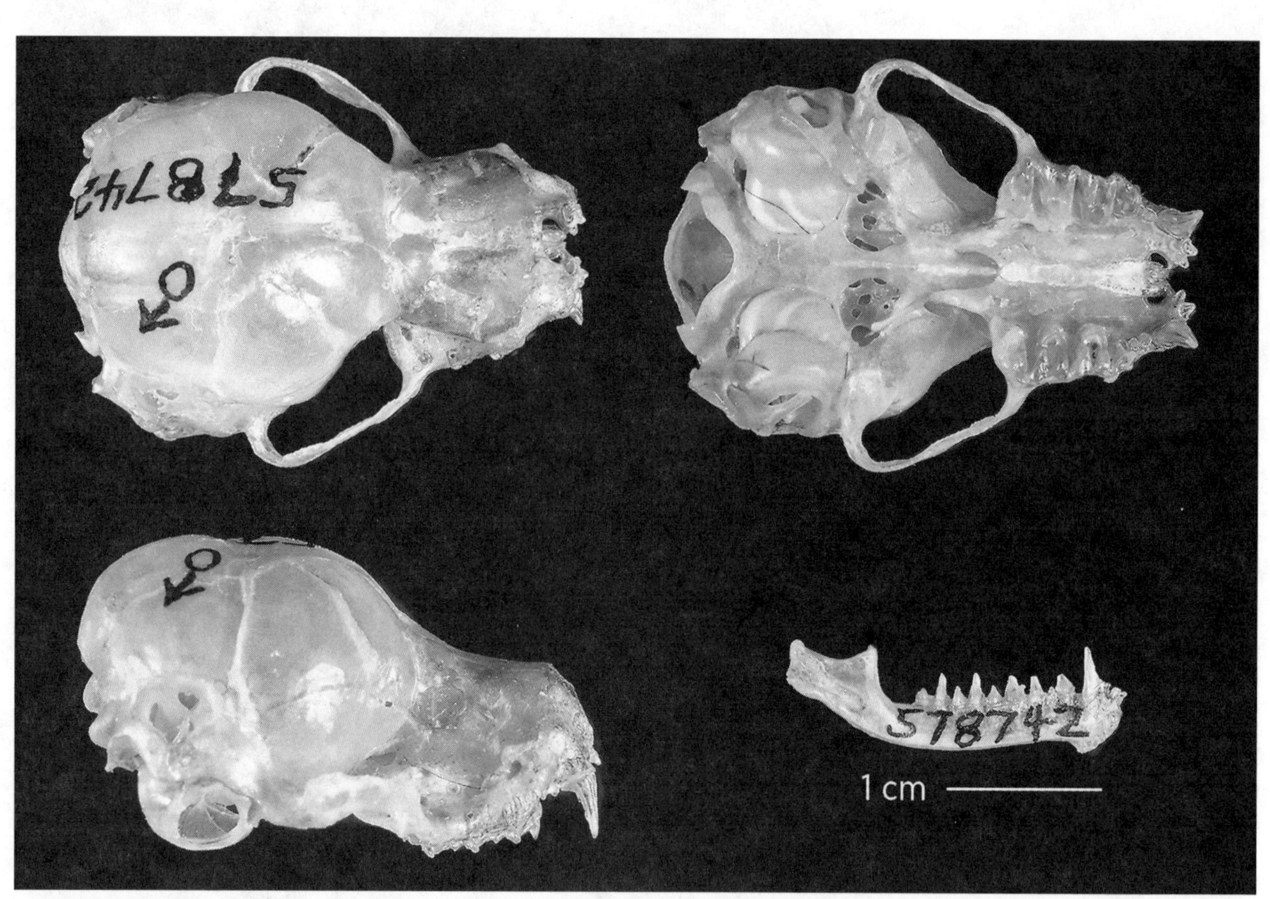

FIG. 393. Myzopodidae; skull and mandible of the Madagascar sucker-footed bat, *Myzopoda aurita*.

Family NOCTILIONIDAE (fishing bats or bulldog bats)

Medium-sized (forearm length 54–92 mm; mass 20–45 g); muzzle pointed, plain (no skin outgrowths); pinnae large, pointed, separate, without ventral extension under eye; tragus present, small; 2nd digit of forelimb with one vestigial bony phalanx, 3rd digit with two, pollex large; **hind limb large; hind foot with large, strongly curved claws**; calcar present, well developed; **tail extending to middle of large uropatagium.** Insectivorous to piscivorous, catching fish on the fly by using enlarged hind claws as gaffs. Roost in narrow fissures or hollow trees, usually near water; occasionally occupy caves or buildings. Slow fliers; can swim well. Inhabit subtropical and tropical lowland forests near water.

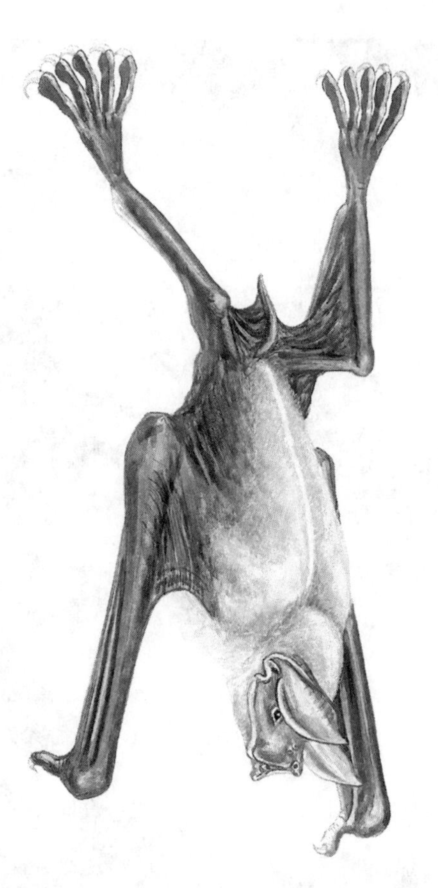

FIG. 394. Noctilionidae; greater bulldog bat, *Noctilio leporinus*.

DIAGNOSTIC CHARACTER:

- **lips thick, folded, with median cleft, and forming distinct cheek pouches**

RECOGNITION CHARACTERS:

1. phalangeal formula of wing 2–1–2–3–3
2. trochiter extends only to level of proximal edge of humerus head
3. **dorsal articular facet on scapula absent (trochiter never articulates with scapula)**
4. premaxillae with nasal and palatal branches, fused to each other and to maxillae
5. no incisive foramina
6. no postorbital process
7. prominent sagittal crest
8. cheek teeth tritubercular

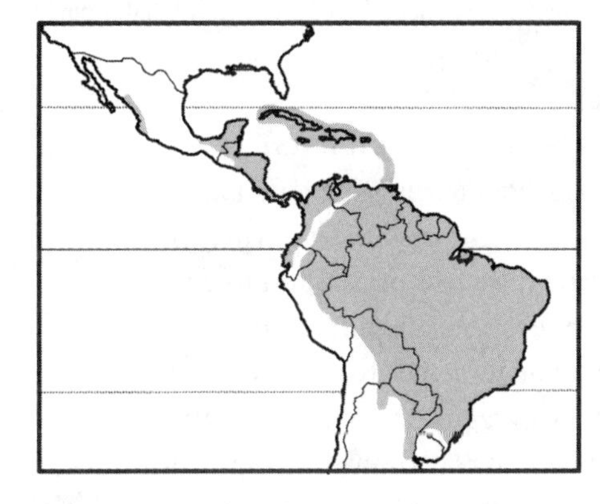

FIG. 395. Geographic range of family Noctilionidae: Neotropics, from Mexico south to Argentina and Uruguay, including Greater and Lesser Antilles.

DENTAL FORMULA: $\frac{2}{1}\ \frac{1}{1}\ \frac{1}{2}\ \frac{3}{3} = 28$

TAXONOMIC DIVERSITY: The family is monotypic, with two species in the single genus *Noctilio*.

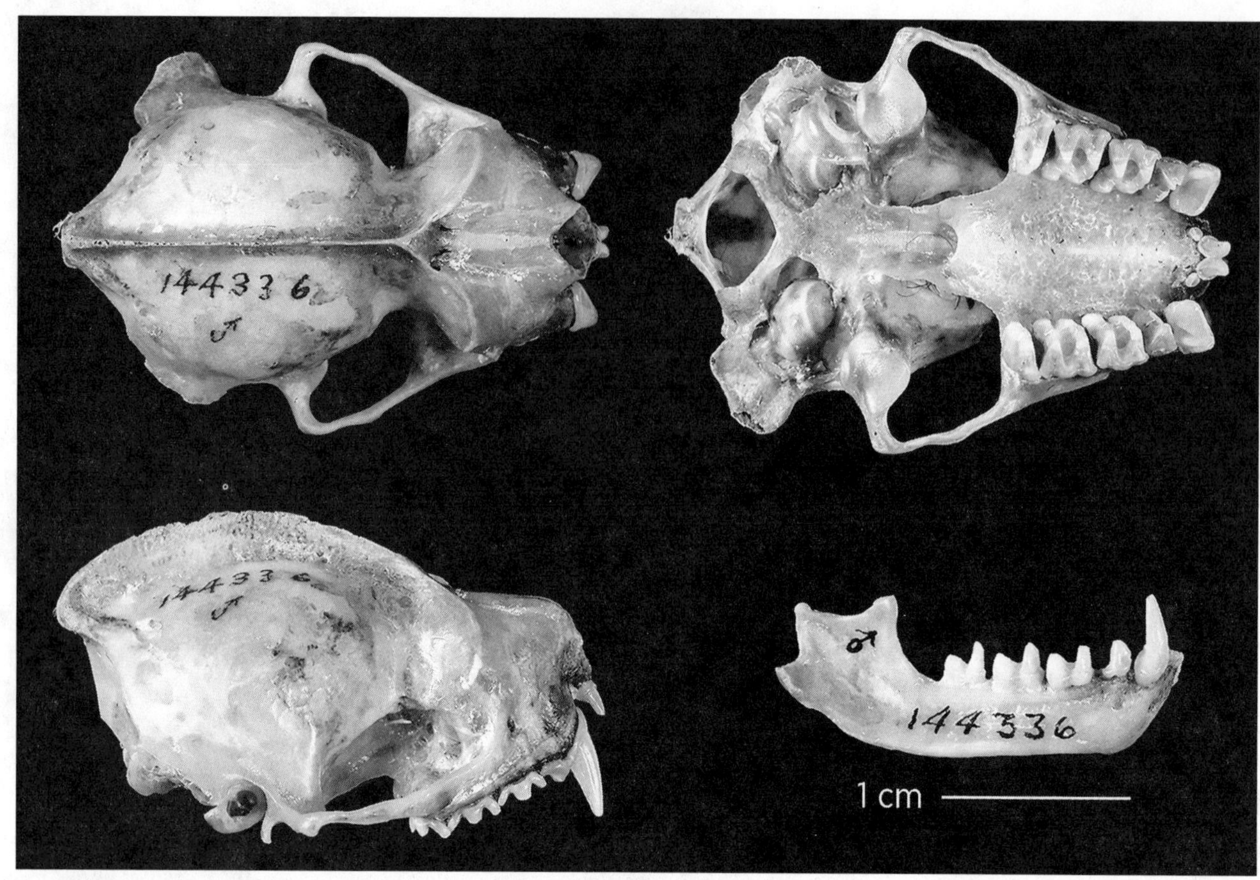

FIG. 396. Noctilionidae; skull and mandible of the greater bulldog bat, *Noctilio leporinus*.

Family PHYLLOSTOMIDAE (New World leaf-nosed or fruit bats)

Size small (forearm 25 mm, mass 12 g in *Ametrida*) to large (forearm 110 mm mass 190 g in *Vampyrum*); **conspicuous, erect nose leaf present on muzzle** (rudimentary in a few genera); pinnae small to large, variable in form, usually separate, without ventral extension under eye; tragus present, small; 2nd digit of wing with one small bony phalanx, 3rd digit with three, pollex relatively large; hind limb and hind foot slender, with relatively large claws; calcar absent, small, or well developed (in different subfamilies); tail variable; if present, may be shorter, longer, or equal in length to uropatagium. Food habits highly variable but generally consistent within subfamilies, including sanguivory (Desmodontinae), carnivory (Phyllostominae), frugivory (Stenodermatinae), and nectarivory (Glossophaginae). Most are gregarious, or roost singly or in small groups; occupy caves, buildings, trees, culverts; some make shelters of large leaves. Habitat ranges from deserts and savanna to both subtropical and tropical wet and dry forests.

RECOGNITION CHARACTERS:

1. phalangeal formula of wing 2–1–3–3–3
2. trochiter extends proximally well beyond level of humerus head
3. dorsal articular facet on scapula present (trochiter articulates with scapula)
4. premaxillae with nasal and palatal branches, fused to each other and to maxillae
5. no postorbital process
6. cheek teeth variable, simple crushing type to tritubercular

DENTAL FORMULA: $\frac{1\text{–}2}{0\text{–}2}\ \frac{1}{1}\ \frac{1\text{–}3}{2\text{–}3}\ \frac{1\text{–}3}{1\text{–}3} = 20\text{–}34$

TAXONOMIC DIVERSITY: Nearly 220 species in 60 genera allocated to 11 subfamilies, each generally characterized by different food habits. The taxonomy used herein generally reflects that of Wilson and Reeder (2005) although taxon numbers have been updated to reflect Solari et al. (2019), which appeared after this book went to press. The most diverse and widespread subfamilies of Phyllostomidae are the following:

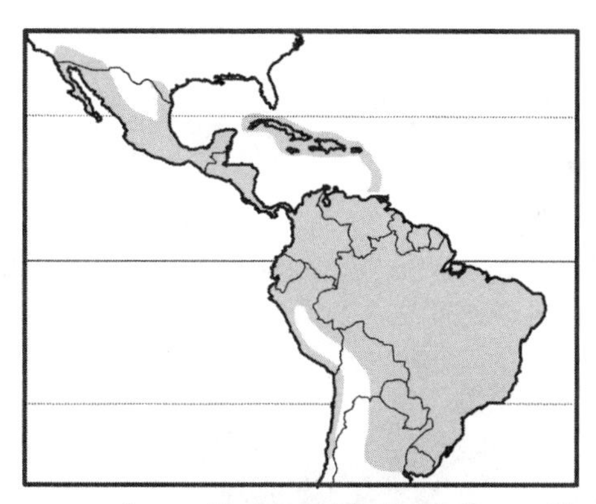

FIG. 397. Geographic range of family Phyllostomidae: New World, from southwestern United States to Argentina, including Greater and Lesser Antilles; most genera are tropical.

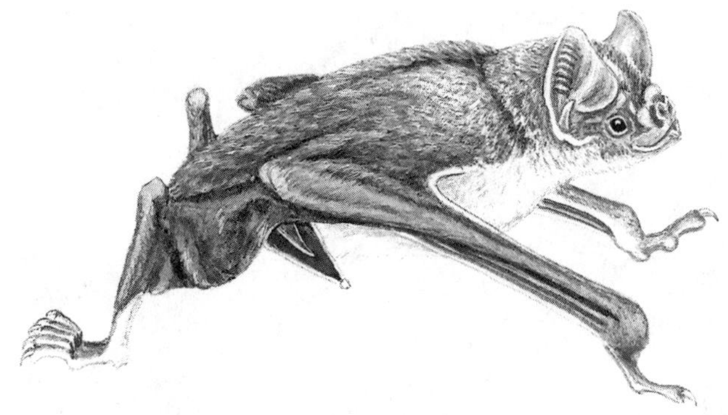

FIG. 398. Phyllostomidae: Desmodontinae; common vampire bat, *Desmodus rotundus*.

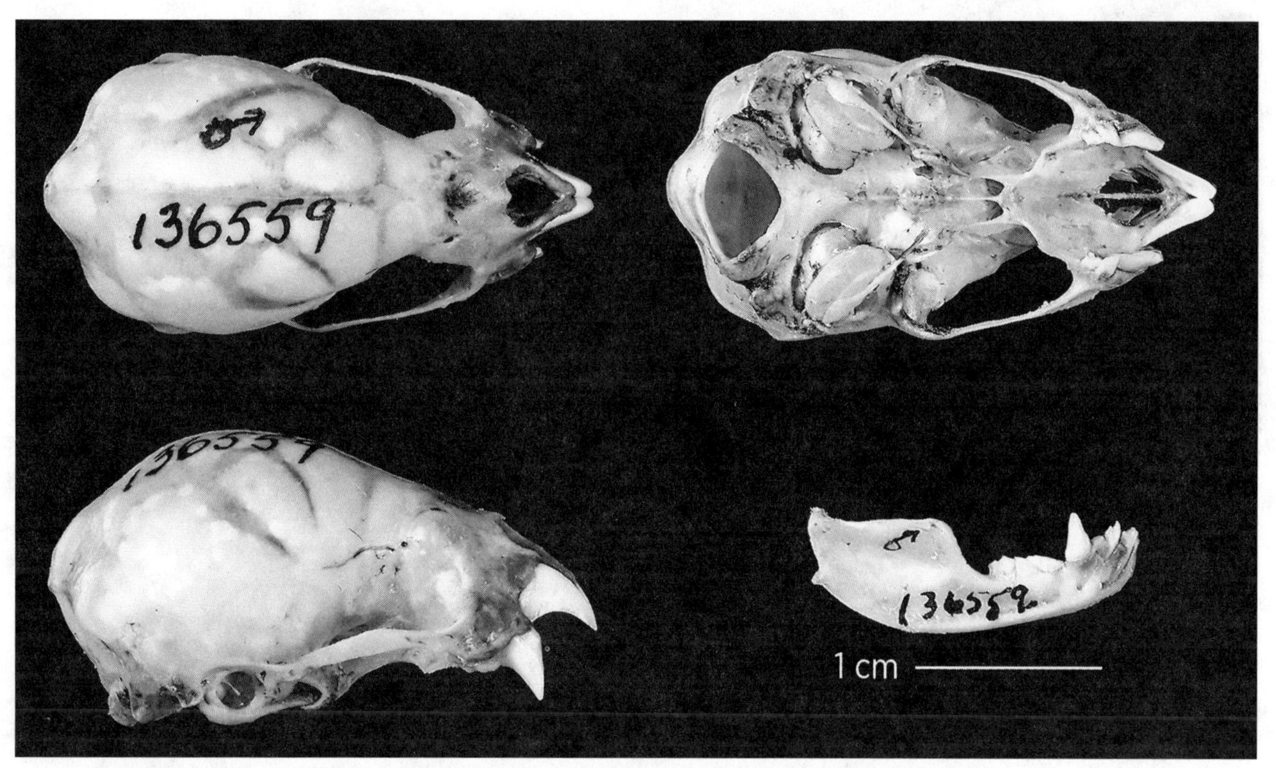

FIG. 399. Phyllostomidae: Desmodontinae; skull and mandible of the common vampire bat, *Desmodus rotundus*.

Desmodontinae: True vampire bats [obligate blood feeders, or sanguivores], with greatly enlarged, compressed, and bladelike upper incisors and canines, with cheek teeth greatly reduced in size and lacking shear surfaces; remarkable agility on ground, with quadrupedal locomotion, elongate pollex, and stout hind legs; digestive tract simple. Comprises 3 monotypic genera: *Desmodus* (common vampire), *Diaemus* (white-winged vampire), and *Diphylla* (hairy-legged vampire).

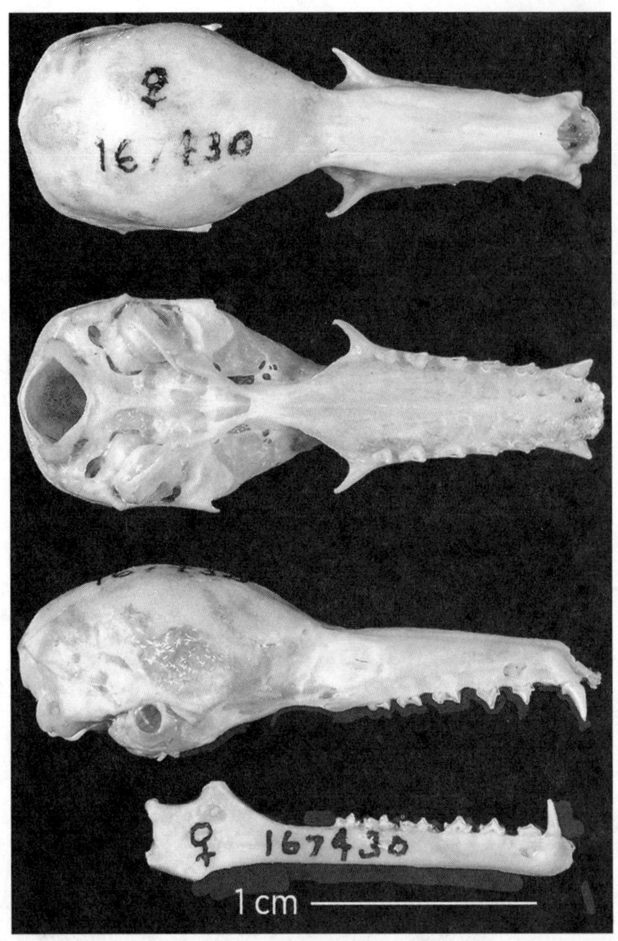

FIG. 400. Phyllostomidae: Glossophaginae: Glossophagini; skull and mandible of the Mexican long-tongued bat, *Choeronycteris mexicana*.

Glossophaginae: Pollen and nectar feeders, with elongate snouts, extremely long tongues, reduced teeth; called "hummingbird" bats. About 13 genera divided into two tribes:

- Glossophagini—16 genera, including *Anoura* (hairy-legged long-tongued bats), *Choeronycteris* (long-tongued bat), *Glossophaga* (common long-tongued bats), and *Leptonycteris* (long-nosed bats)
- Lonchophyllini—4 genera, the most widespread of which is *Lonchophylla* (spear-nosed long-tongued bats)

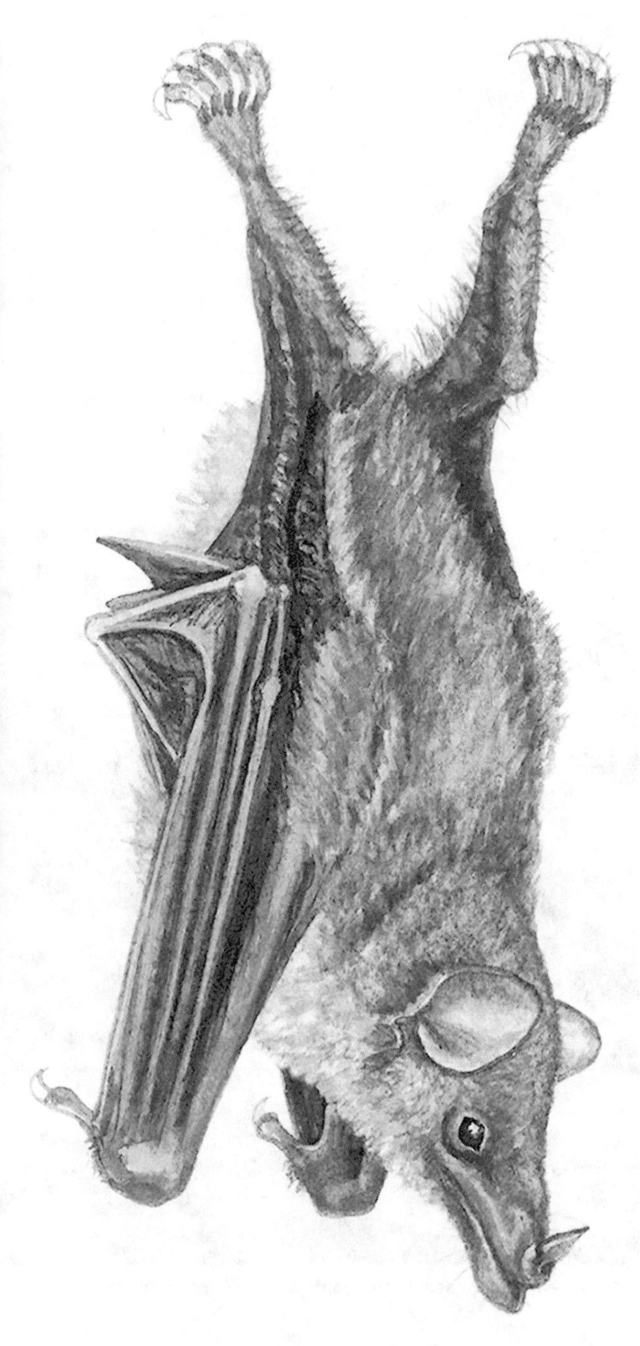

FIG. 401. Phyllostomidae: Glossophaginae: Glossophagini; Geoffroy's hairy-legged bat, *Anoura geoffroyi*.

Phyllostominae: Vertebrate and large insect predators with large dilambdodont cheek teeth, large uropatagium, large ears, and large body size. Seventeen genera, including *Chrotopterus* (woolly false vampire bat), *Lonchorhina* (sword-nosed bats), *Lophostoma* (round-eared bats), *Macrophyllum* (long-legged bat), *Macrotus* (large-eared leaf-nosed bats), *Micronycteris* (little big-eared bats), *Mimon* (spear-nosed or golden bats), *Phyllostomus* (spear-nosed bats), *Tonatia* (round-eared bats), *Trachops* (fringe-lipped bats), and *Vampyrum* (spectral bat)

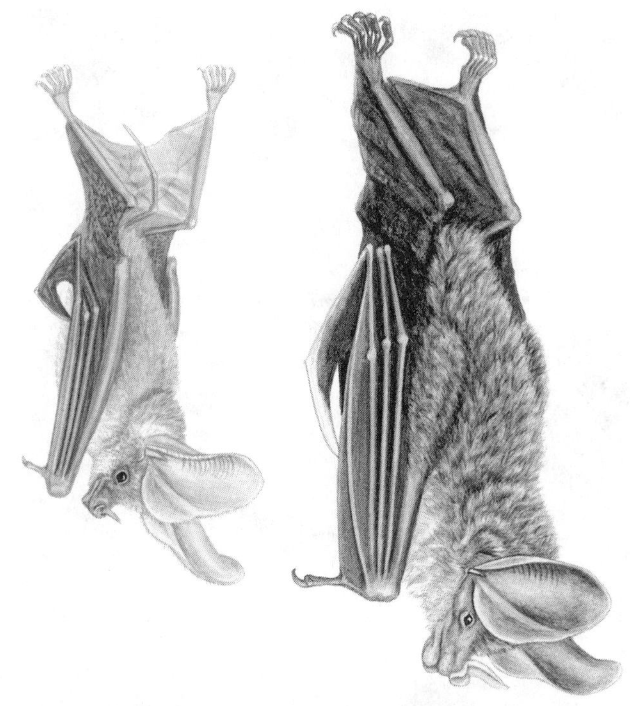

FIG. 402. Phyllostomidae: Phyllostominae; California leaf-nosed bat, *Macrotus californicus* (*left*), and woolly false vampire, *Chrotopterus auritus* (*right*).

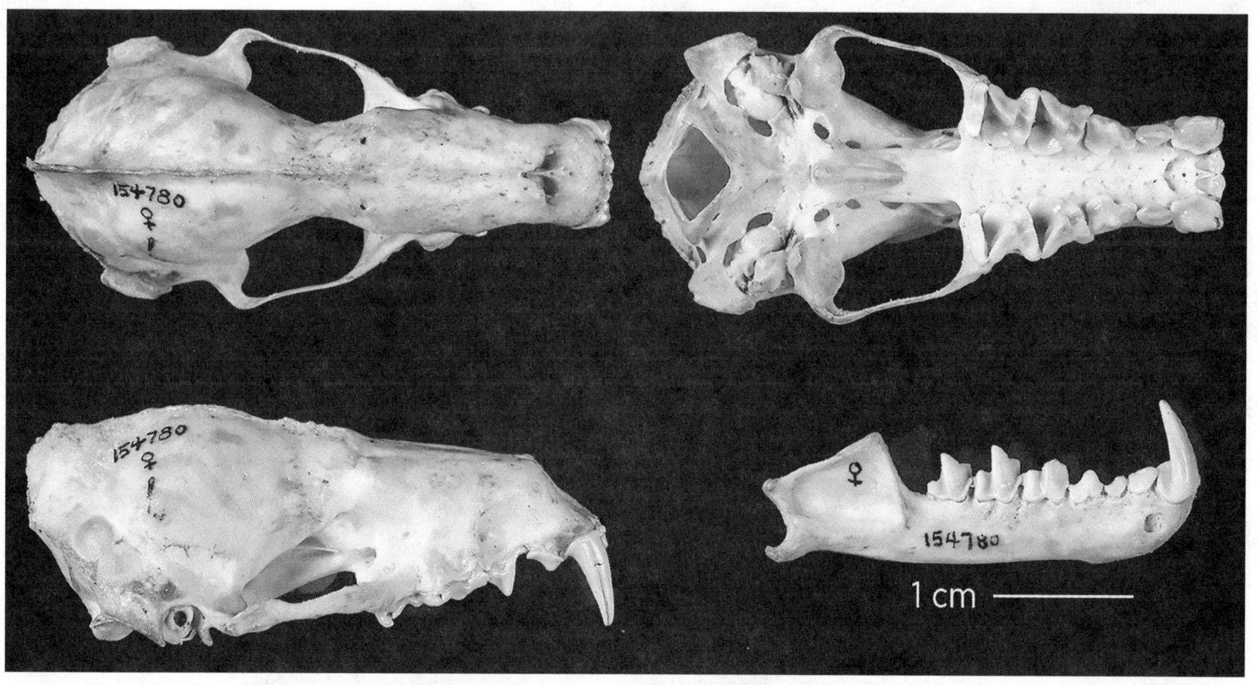

FIG. 403. Phyllostomidae: Phyllostominae; skull and mandible of the spectral bat, *Vampyrum spectrum*.

Stenodermatinae: Fruit eaters, with enlarged and blocky molars with reduced surface topology, short face with round dental arcade, short nose leaf, and small uropatagium with tail very short or absent. Twenty-one genera divided into three tribes (some authors subsume the Ectophyllini within the Stenodermatini):

- Ectophyllini—12 genera, including *Artibeus* (large fruit-eating bats), *Chiroderma* (big-eyed bats), *Dermanura* (dwarf fruit-eating bats—treated by some as a subgenus under *Artibeus*), *Ectophylla* (white bats), *Platyrrhinus* (white-lined fruit bats), *Uroderma* (tent-making bats), and *Vampyressa* (yellow-eared bats)
- Sturnirini—*Sturnira* (hairy-legged fruit bats)
- Stenodermatini—8 genera, including *Ametrida* (white-shouldered bats) and *Centurio* (wrinkle-faced bat)

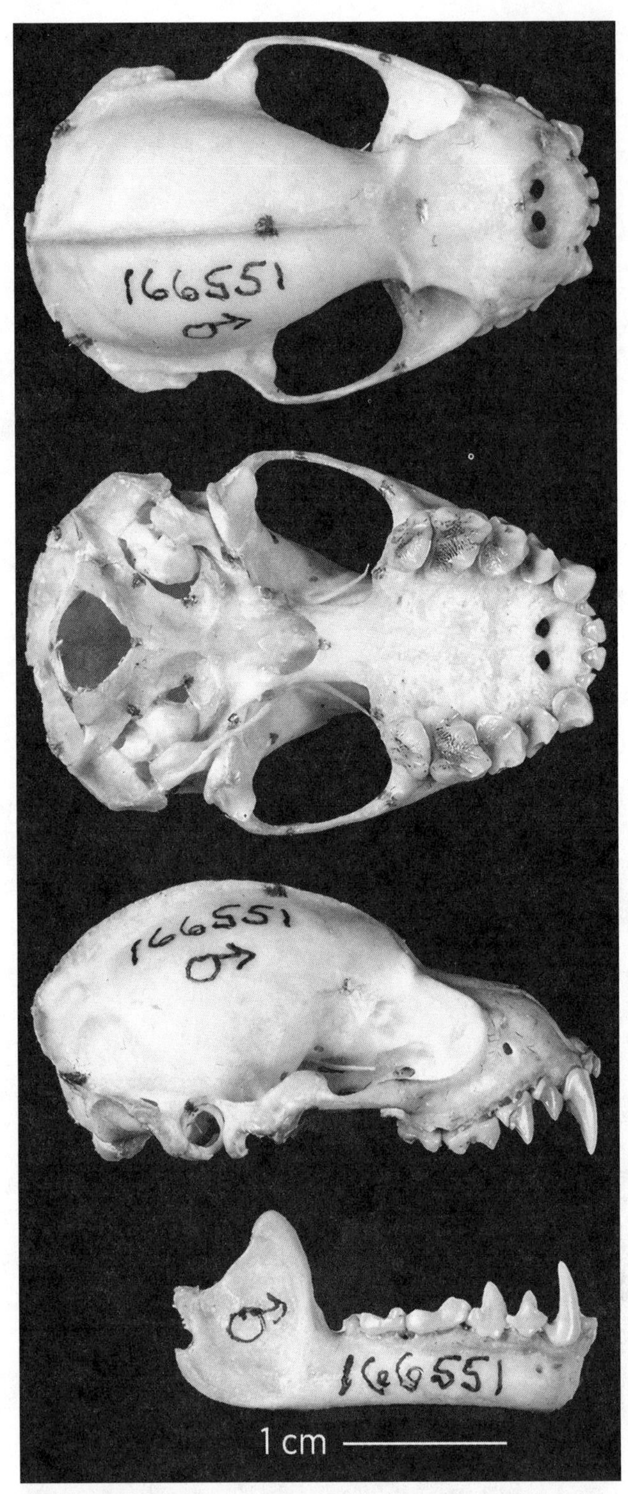

FIG. 404. Phyllostomidae: Stenodermatinae: Stenodermatini; skull and mandible of the Jamaican fruit-eating bat, *Artibeus jamaicensis*.

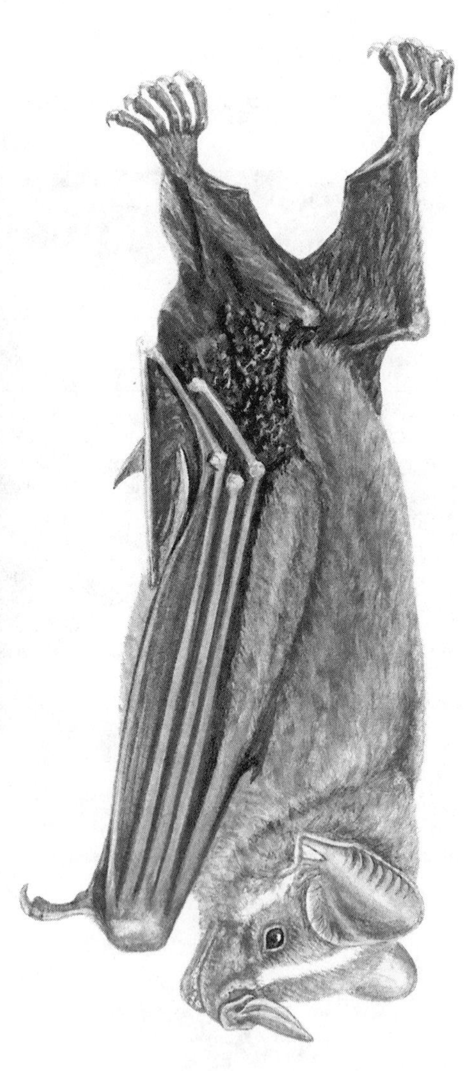

FIG. 405. Phyllostomidae: Stenodermatinae: Stenodermatini; great fruit-eating bat, *Artibeus lituratus*.

Family THYROPTERIDAE (disc-winged bats)

Size small (forearm 30–38 mm; mass 4–6 g); muzzle plain, no skin outgrowths; **pinnae large, funnel-shaped**, separate, without ventral extension under eye; **tragus present, small**; 2nd digit of wing absent (vestigial metacarpal remaining), 3rd digit with three bony phalanges, **pollex much reduced, but with well-developed claw; hind limb slender, hind foot small; only two unfused phalanges per digit**; claws reduced; calcar present, well developed; tail extending slightly beyond relatively large uropatagium. Insectivorous; roost individually or in small groups under large leaves or inside rolled-up leaf, clinging to surface with stalked adhesive discs (cf. Myzopodidae) on wrist and ankle. Inhabit tropical lowland forests, especially near water.

FIG. 406. Thyropteridae; Spix's disc-winged bat, *Thyroptera tricolor*.

DIAGNOSTIC CHARACTERS:

- **3rd and 4th digits of hind foot syndactylous (including claws)**
- **prominent stalked adhesive disc present at wrist and ankle**

RECOGNITION CHARACTERS:

1. phalangeal formula of wing 2-0-3-3-3
2. trochiter extends proximally well beyond level of humerus head
3. dorsal articular facet on scapula present (trochiter articulates with scapula)
4. premaxillae with nasal and palatal branches fused to each other and to maxillae
5. no postorbital process
6. cheek teeth tritubercular

DENTAL FORMULA: $\frac{2}{3}\ \frac{1}{1}\ \frac{3}{3}\ \frac{3}{3} = 38$

TAXONOMIC DIVERSITY: The family is monotypic, with four species in the single genus *Thyroptera*.

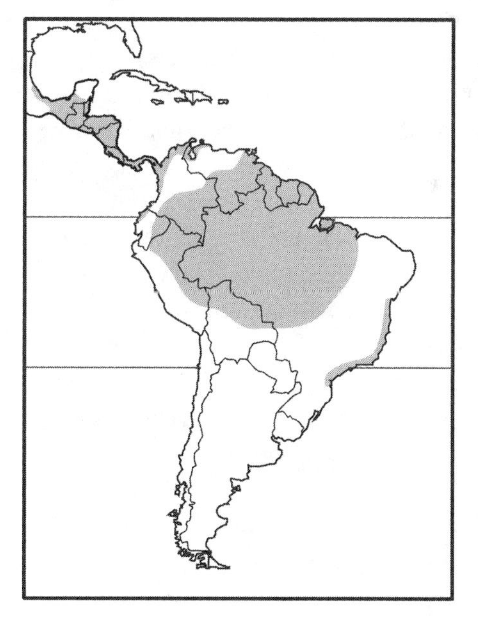

FIG. 407. Geographic range of family Thyropteridae: Neotropics, from southern Mexico to southern Brazil.

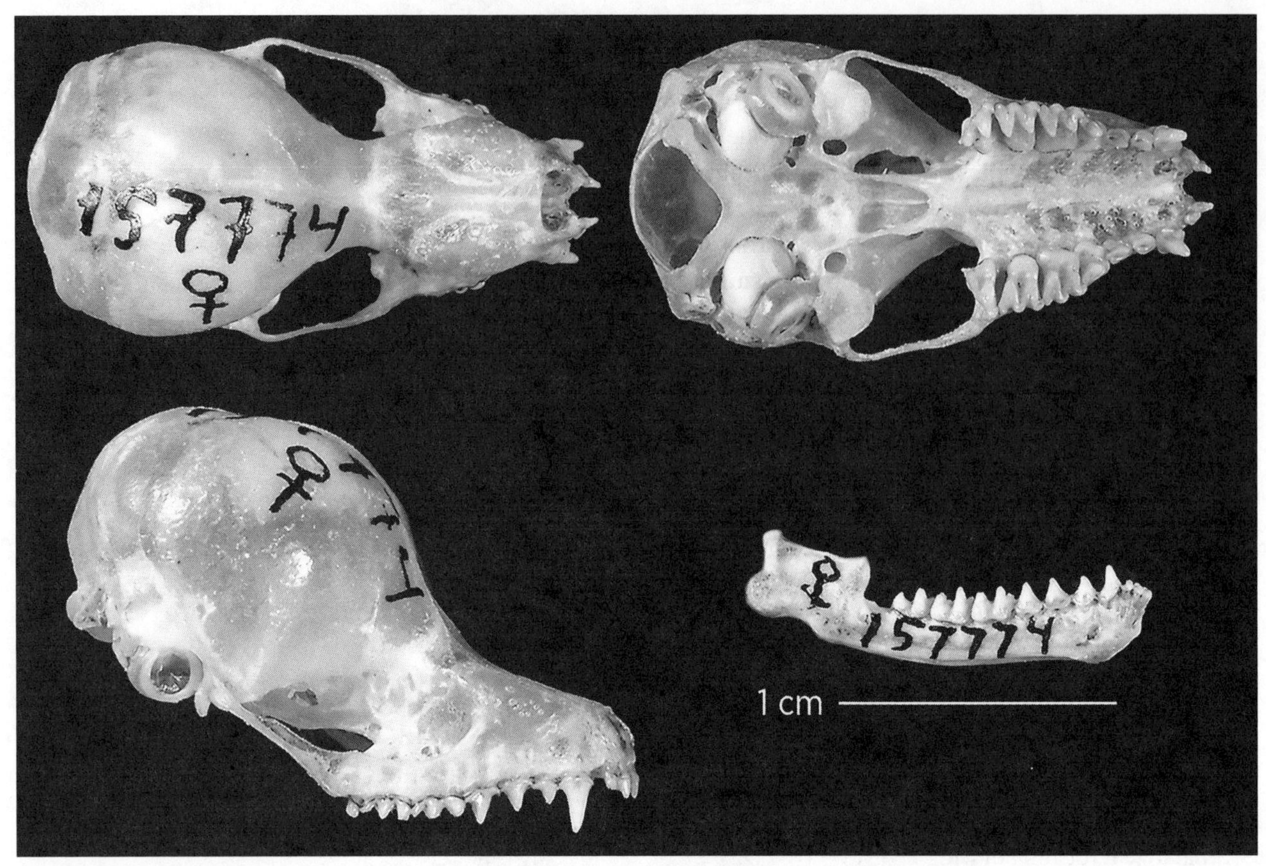

FIG. 408. Thyropteridae; skull and mandible of Spix's disc-winged bat, *Thyroptera tricolor.*

Superfamily VESPERTILIONOIDEA

Family MOLOSSIDAE (free-tailed bats)

Small to moderately large (forearm 25–90 mm; mass 5–100 g); **muzzle plain, broad, with short hairs bearing spoon-shaped tips; muzzle often with tiny bumps or vertical wrinkles on upper surface**; pinnae usually large, pointed or rounded, often united across forehead, and usually projecting forward rather than vertically above head; **tragus very small or absent**; 2nd digit of wing with one bony phalanx, 3rd digit with two plus one basally ossified cartilaginous phalanx, pollex relatively large; **hind limb and hind foot short, broad**; calcar present, well developed. Insectivorous. Frequent caves, where colonies may number in the millions, hollow trees, or crevices of rock cliffs. Some migrate seasonally over long distances, flying fast (up to 95 km/hr), often in groups. Most forage in the open, well above any tree canopy. Habitats include deserts, semideserts, savannas, and forests.

DIAGNOSTIC CHARACTERS:

- **1st phalanx of 3rd finger folding to outer (rather than the usual inner) side of metacarpal when wing is folded**
- **1st and 5th digits of hind foot with fringe of stiff bristles**
- **tail extending well beyond margin of relatively short uropatagium**

FIG. 409. Molossidae: Molossinae; greater bonneted bat, *Eumops perotis* (*bottom left*), and big free-tailed bat, *Nyctinomops macrotis* (*top right*).

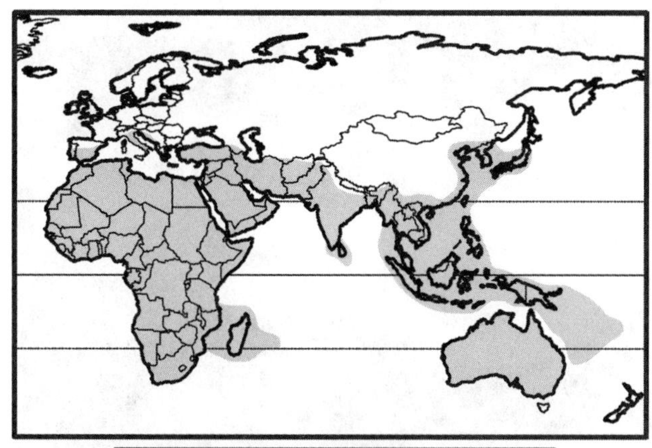

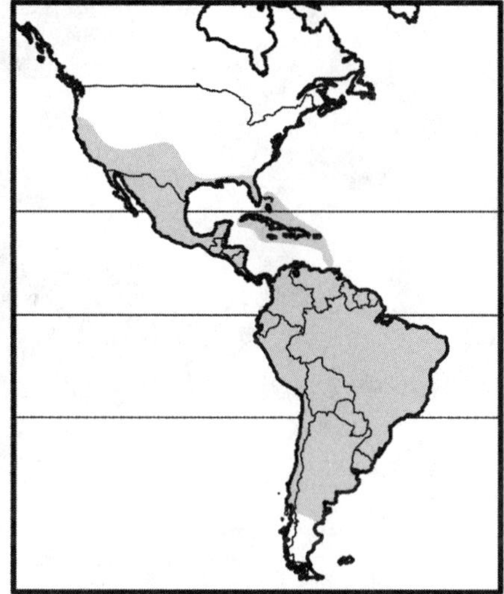

FIG. 410. Geographic range of family Molossidae: Cosmopolitan (North and South America, southern Eurasia, Africa, Australia).

RECOGNITION CHARACTERS:

1. phalangeal formula of wing 2-1-3-3-3
2. trochiter extends proximally well beyond level of humerus head
3. dorsal articular facet on scapula present (trochiter articulates with scapula)
4. premaxillae separated or united, fused to maxillae, with or without palatal branches
5. cheek teeth tritubercular, with well-developed W-shaped ectoloph (dilambdodont)

DENTAL FORMULA: $\frac{1}{1\text{–}3}\ \frac{1}{1}\ \frac{1\text{–}2}{2}\ \frac{3}{3} = 26\text{–}32$

TAXONOMIC DIVERSITY: About 100 species in 16 genera allocated to two subfamilies:

Molossinae—*Chaerephon* (free-tailed bats), *Cheiromeles* (naked bats), *Cynomops* (dog-faced bats), *Eumops* (bonneted bats), *Molossops* (dog-faced bats), *Molossus* (mastiff bats), *Mops* (free-tailed bats), *Mormopterus* (mastiff bats), *Myopterus* (winged-mouse bats), *Nyctinomops* (free-tailed bats), *Otomops* (giant mastiff bats), *Platymops* (Peter's flat-headed bat), *Promops* (crested mastiff bats), *Sauromys* (Roberts's flat-headed bat), and *Tadarida* (common free-tailed bats)

Tomopeatinae—*Tomopeas* (blunt-eared bat)

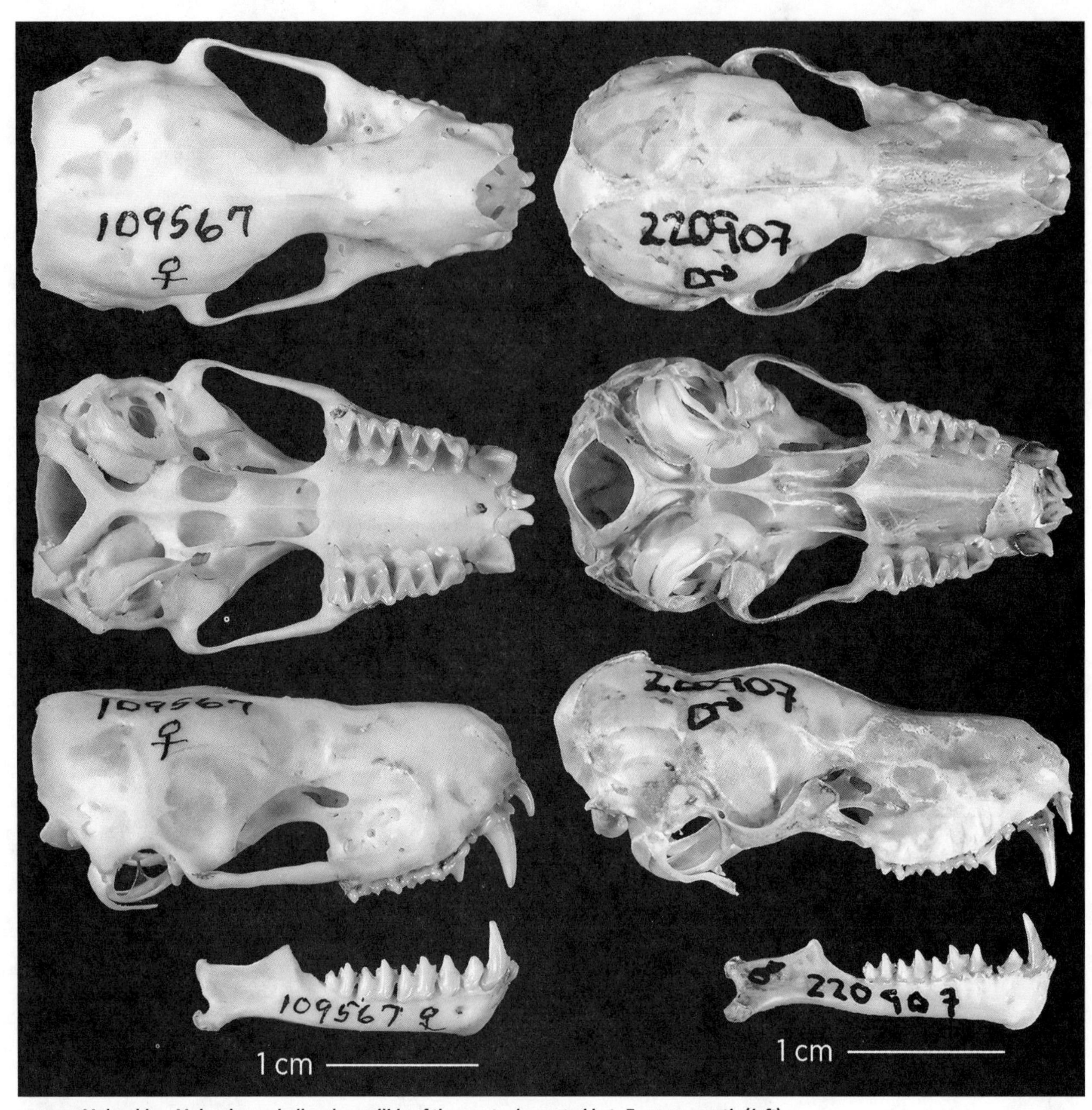

FIG. 411. Molossidae: Molossinae; skull and mandible of the greater bonneted bat, *Eumops perotis* (*left*) and large-eared giant mastiff bat, *Otomops martiensseni* (*right*).

Family NATALIDAE (funnel-eared bats)

Small (forearm 25–45 mm; mass 4–10 g); muzzle plain; no skin outgrowths; males possess a "**natalid organ**," a mass of glandular tissue on the dorsal surface of the muzzle (beneath the forehead), which "secretes a translucent greenish, viscous liquid that may function in communication" (Tejedor 2011, 11); **pinnae large, funnel-shaped**, separate, without ventral extension under eye; **tragus present, small, triangular**; 2nd digit of wing absent (metacarpal remaining); 3rd digit with two phalanges; pollex relatively small; **hind limb and hind foot very slender, small, with relatively small claws**; calcar present, well developed; **tail long, fully enclosed in large uropatagium**. Insectivorous. Usually roost in caves. Habitat ranges from dry cactus scrub to wet forest, but is typically deciduous to semideciduous subtropical and tropical forest.

RECOGNITION CHARACTERS:

1. phalangeal formula of wing 2-0-2-3-3
2. trochiter extends proximally well beyond level of humerus head

FIG. 412. Natalidae; Mexican greater funnel-eared bat, *Natalus mexicanus*.

3. dorsal articular facet on scapula present (trochiter articulates with scapula)
4. premaxillae with nasal and palatal branches, fused to each other and to maxilla
5. no postorbital process
6. cheek teeth tritubercular

DENTAL FORMULA: $\frac{2}{3}\ \frac{1}{1}\ \frac{3}{3}\ \frac{3}{3} = 38$

TAXONOMIC DIVERSITY: 12 species in three genera: *Chilonatalus* (funnel-eared bats), *Natalus* (greater funnel-eared bats), and *Nyctiellus* (Gervais's funnel-eared bat).

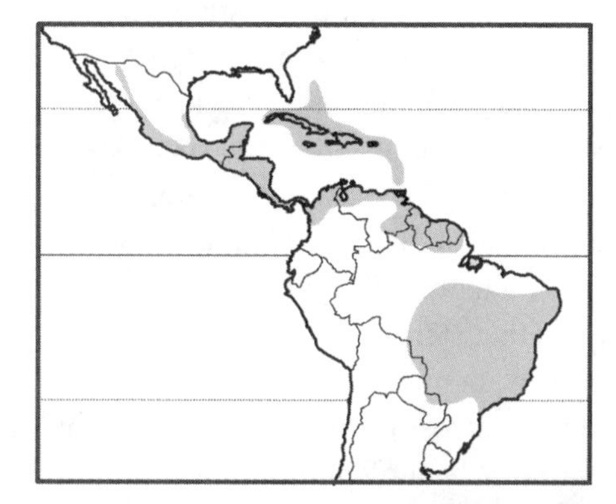

FIG. 413. Geographic range of family Natalidae: New World tropics and subtropics, from Mexico to southern Brazil, including Greater and Lesser Antilles.

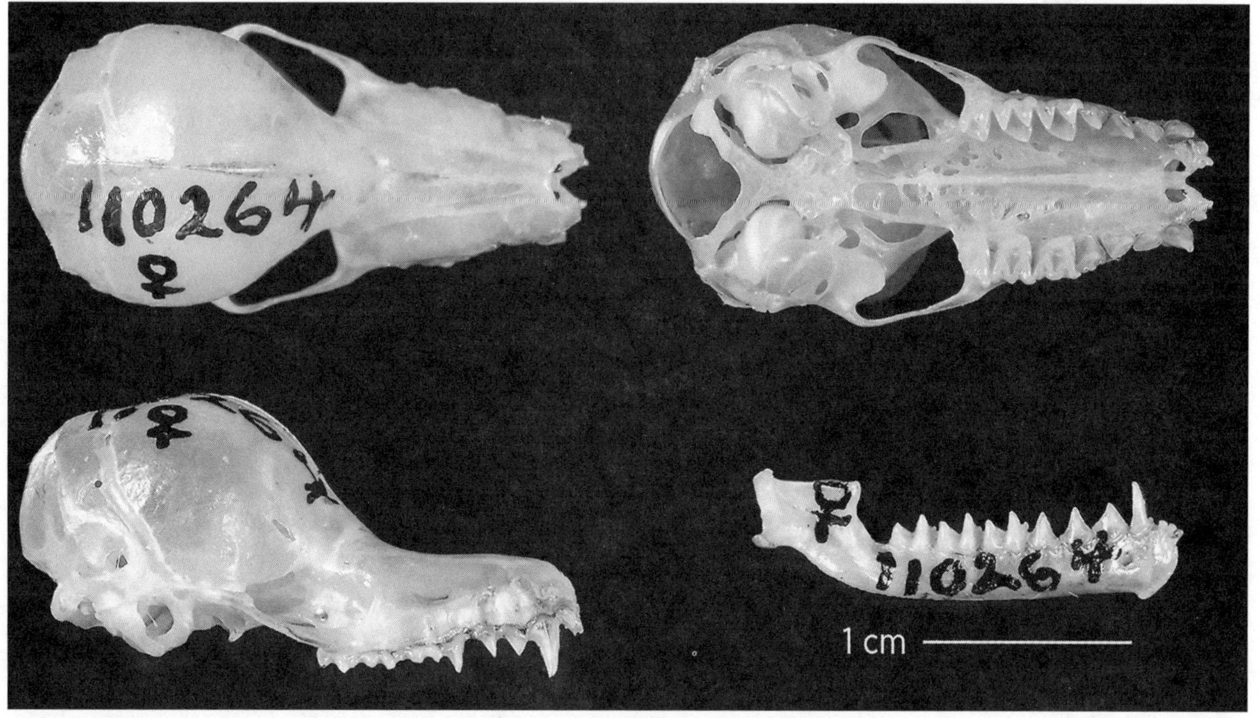

FIG. 414. Natalidae; skull and mandible of the Mexican greater funnel-eared bat, *Natalus mexicanus*.

Family VESPERTILIONIDAE (common bats)

Small to medium-sized (forearm 25–90 mm; mass 4–50 g); muzzle plain, no skin outgrowths; pinnae variable, small to very large, pointed or rounded, usually separate without ventral extension under eye; **tragus present, prominent**; small sucker-like pad on wrist or ankle (or both) in some (*Eudiscopus*, *Glischropus*, *Hesperoptenus*, *Tylonycteris*, some *Pipistrellus*); 2nd digit of forelimb with one small bony phalanx, 3rd digit with two plus one basally ossified cartilaginous phalanx, pollex relatively large; hind limb and hind foot slender, usually with moderately large claws (very large in fish-eating bat, *Myotis vivesi*); calcar present, well developed; tail long, enclosed within and reaching posterior edge of a large uropatagium. Most insectivorous, a few piscivorous; insects may be caught in mid-flight or taken from surfaces (ground, leaves). Roosting sites, which vary extensively, include caves, hollow trees, under tree bark, hanging from branches, rock crevices, and buildings. Many, but not all, temperate species hibernate; some migrate seasonally. Over-winter sperm storage and delayed implantation are common. Some are solitary, others commonly occur in pairs, still others are highly colonial; females of many colonial species segregate into nursing colonies to bear young. Habitat ranges broadly across all biomes, from tropical forests at the equator to the limit of tree growth in northern and southern latitudes. As noted above, the antrozoines, miniopterines, and the genus *Cistugo* have been argued to warrant family status (Antrozoidae, Miniopteridae, and Cistugidae, respectively); pending further phylogenetic resolution, we retain these groups within Vespertilionidae.

RECOGNITION CHARACTERS:

1. phalangeal formula of wing 2-1-3-3-3
2. trochiter extends proximally well beyond level of humerus head
3. dorsal articular facet on scapula present (trochiter articulates with scapula)
4. premaxillae separated, but fused to maxillae; only nasal branches present
5. no postorbital process
6. cheek teeth tritubercular

DENTAL FORMULA: $\frac{1\text{–}2}{2\text{–}3}\ \frac{1}{1}\ \frac{1\text{–}3}{2\text{–}3}\ \frac{3}{3} = 28\text{–}38$

TAXONOMIC DIVERSITY: About 48 genera and more than 400 species arranged in six subfamilies, one with seven tribes:

Antrozoinae: *Antrozous* (pallid bat) and *Bauerus* (Van Gelder's bat)

Kerivoulinae: *Kerivoula* (woolly bats) and *Phoniscus* (trumpet-eared bats)

Miniopterinae: *Miniopterus* (long-fingered bats)

Murininae: *Harpiocephalus* (hairy-winged bat) and *Murina* (tube-nosed bats)

Myotinae: *Cistugo* (wing-gland bats), *Lasionycteris* (silver-haired bat), and *Myotis* (myotis bats)

Vespertilioninae:

- Eptesicini—*Arielulus* (sprites), *Eptesicus* (serotines or brown bats), and *Hesperotenus* (serotines)
- Lasiurini—*Aeorestes* (hoary bats), *Dasypterus* (yellow bats), and *Lasiurus* (red bats)
- Nycticeiini—*Nycticeinops* (twilight bat), *Nycticeius* (mysterious and evening bats), *Rhogeessa* (yellow bats), *Scoteanax* (broad-nosed bat), *Scotoecus* (house bats), *Scotomanes* (harlequin bat), and *Scotophilus* (house bats)
- Nyctophilini—*Nyctophilus* (long-eared bats) and *Pharotis* (big-eared bat)
- Pipistrellini—*Glischropus* (thick-thumbed bats), *Nyctalus* (noctules), *Parastrellus* (canyon bat), *Pipistrellus* (pipistrelles), and *Scotozous* (pipistrelles)
- Plecotini—*Barbastella* (barbastelles), *Corynorhinus* (big-eared bats), *Euderma* (spotted bat), *Idionycteris* (big-eared bat), *Otonycteris* (desert bat), and *Plecotus* (long-eared bats)
- Vespertilionini—*Chalinolobus* (wattled bats), *Eudiscopus* (disk-footed bat), *Falsistrellus* (pipistrelles), *Glauconycteris* (butterfly bats), *Histiotus* (big-eared brown bats), *Hypsugo* (pipistrelles), *Ia* (evening bat), *Laephotis* (long-eared bats), *Mimetillus* (mimic bat), *Neoromicia* (pipistrelles and serotines), *Tylonycteris* (bamboo bats), *Vespadelus* (forest and cave bats), and *Vespertilio* (particolored bats)

FIG. 415. Vespertilionidae; hoary bat, *Aeorestes cinereus* (Vespertilioninae: Lasiurini; *left*), diminutive serotine, *Eptesicus diminutus* (Vespertilioninae: Eptesicini; *top right*), spotted bat, *Euderma maculata* (Vespertilioninae: Plecotini; *middle right*), and pallid bat, *Antrozous pallidus* (Antrozoinae; *bottom right*).

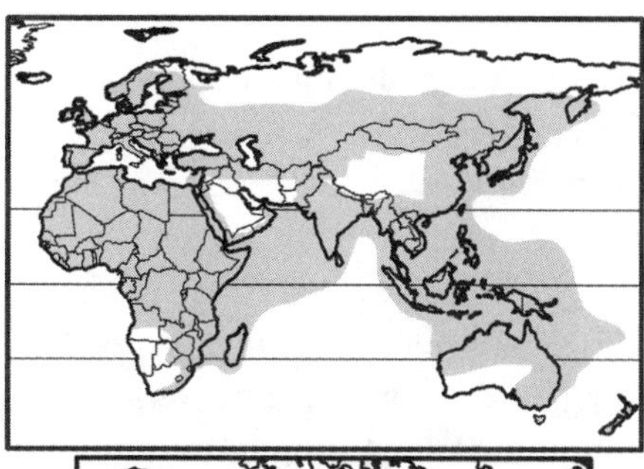

FIG. 416. Geographic range of family Vespertilionidae: cosmopolitan (North and South America, Eurasia, Africa, Australia).

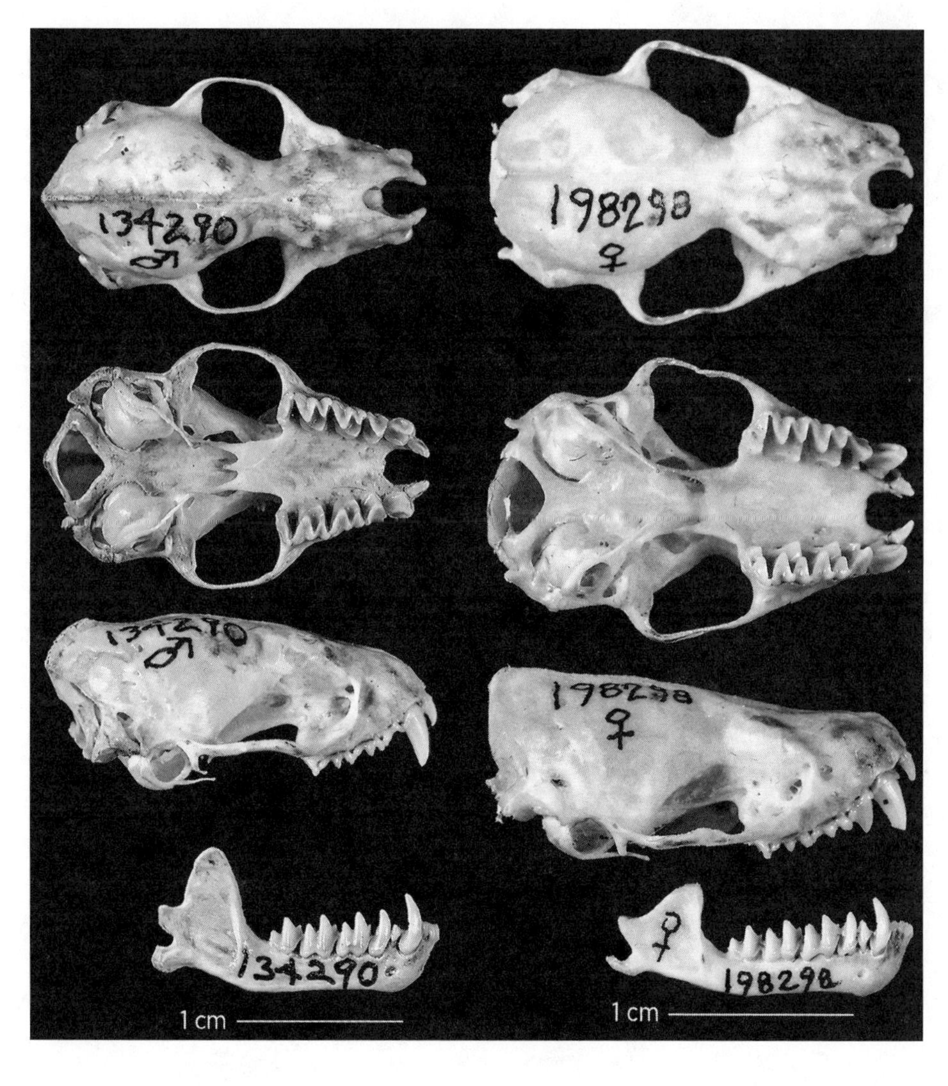

FIG. 417. Vespertilionidae; skull and mandible of the pallid bat, *Antrozous pallidus* (Antrozoinae; *left*), and big brown bat, *Eptesicus fuscus* (Vespertilioninae: Eptesicini; *right*).

Clade FERAE

The clade Ferae links the pangolins (order Pholidota) with the carnivorans (order Carnivora). These two orders are completely unexpected phyletic sisters, but their relationship is firmly supported by molecular data. Simpson (1945) had earlier placed the pangolins within a paraphyletic group (cohort Unguiculata) that also included the "insectivores," dermopterans, bats, primates, and xenarthrans. He placed the carnivorans in the monotypic superorder Ferae within the cohort Ferungulata, which otherwise included the tubulidentates, elephants, hyraxes, sirenians, perissodactyls, and artiodactyls.

Order PHOLIDOTA

DIAGNOSTIC CHARACTER:

- **body covered with keratinized, overlapping scale-like plates**

RECOGNITION CHARACTERS:

1. **no teeth**
2. skull conical, lacking crests and ridges
3. nasals large, broadly expanded at base
4. orbital and temporal fossae confluent
5. zygomatic arch incomplete
6. **no jugal or interparietal bone**
7. lacrimal absent in *Manis* and *Smutsia*, present in *Phataginus*
8. palate long and narrow
9. pterygoids separate, not forming part of palate
10. tympanic bone forming a bulla not closed dorsal to auditory meatus and not fused to other elements
11. no extra zygapophyses on vertebrae

DENTAL FORMULA: No teeth (edentulous)

Family MANIDAE (pangolins or scaly anteaters)

Medium-sized to large (total length 650–1,750 mm; mass 1.6–33 kg); body elongate, tapering anteriorly and posteriorly, **dorsal surface covered with imbricate, large, movable epidermal scales of varying shapes** (no scales on snout, belly, throat, chin, sides of face, or inner surfaces of limbs); eyes small; pinnae small or reduced to a thickened ridge, or absent; nostrils curved slits; mouth opening small and ventral; tongue elongate, protractile, strap-like, serves as major masticatory organ; tail variously prehensile, transversely convex above, flat below, completely scaled in most; **plantigrade, forefoot with three principal digits, hind foot with five; all digits with well-developed claws**; feed primarily on ants and termites; take refuge in self-dug burrows or those made by other animals; burrows large and can extend laterally 40 m and reach depths of 6 m. Terrestrial and arboreal; quadrupedal, walk on outer side of wrists, knuckles, or palms of forefeet with lumbering, shuffling gait. Timid, protecting themselves by rolling into a ball exposing only thickened scales. Inhabit lowland and upland areas in the tropics, including woodlands, savannas, sandy country, and rainforests.

FIG. 418. Manidae: Smutsiinae; Temminck's ground pangolin, *Smutsia temminckii*.

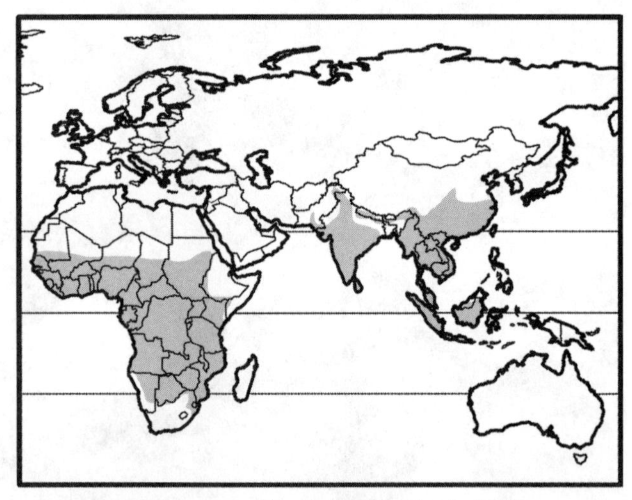

FIG. 419. Geographic range of family Manidae: tropical Africa (Smutsiinae), southern and Southeast Asia, including Indonesia (Maninae).

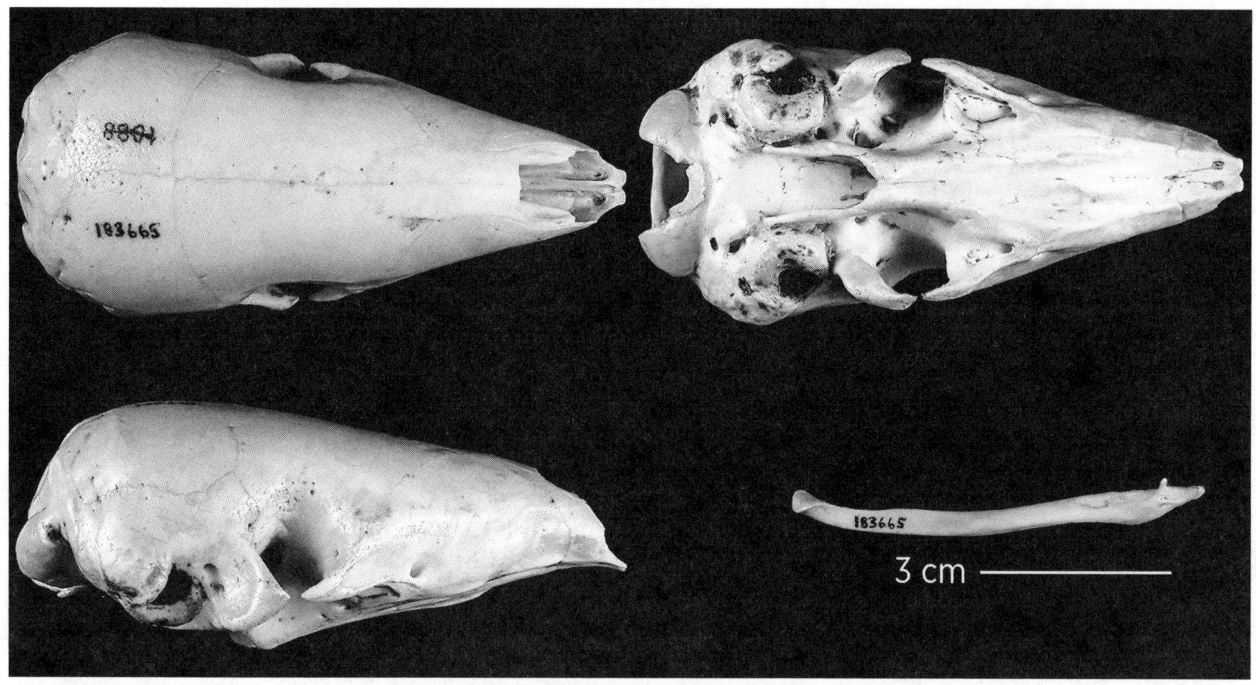

FIG. 420. Manidae: Smutsiinae; skull and mandible of the long-tailed pangolin, *Phataginus tetradactyla*.

TAXONOMIC DIVERSITY: Eight species in three genera divided into two subfamilies (Gaudin et al. 2009; Hassanin et al. 2015):

Maninae—*Manis* (four species—Asian pangolins)

Smutsiinae—*Smutsia* (two species—African ground pangolins) and *Phataginus* (two species—African tree pangolins)

Order CARNIVORA

The classification of carnivores has changed substantially since Simpson (1945), who separated carnivorans into the suborder Pinnipedia ("wing [= finned] foot") to contain the seals, sea lions, and walrus (based primarily on their extreme modifications of the skeleton and teeth for aquatic locomotion and feeding) and the suborder Fissipeda ("split [= toed] foot"), which grouped the largely terrestrial families (dogs, cats, hyenas, and so forth). While some authors elevated these groups to separate orders (e.g., Stains 1967; Corbet 1978; Hall 1981, but not Stains 1984; Corbet and Hill 1986), strong congruence in cladistic analyses of both extensive morphological character sets and DNA sequences now support the inclusion of the pinnipeds with other carnivorans in a single order. An emerging consensus of 16 families divided into two suborders is currently recognized (Wilson and Mittermeier 2009; Flynn et al. 2010; Nyakatura and Bininda-Emonds 2012; and Gatesy et al. 2017):

Suborder Feliformia
- Family Eupleridae (endemic Malagasy carnivores)
- Family Felidae (cats)
- Family Herpestidae (mongooses)
- Family Hyaenidae (aardwolf and hyenas)
- Family Nandiniidae (African palm civet, *Nandinia*)
- Family Prionodontidae (linsang, *Prionodon*)
- Family Viverridae (civets, genets, and relatives)

Suborder Caniformia
- Superfamily Canoidea
  - Family Canidae (dogs and relatives)
- Superfamily Arctoidea (all non-dog caniforms)
  - Family Ursidae (bears)
  - Clade Musteloidea
    - Family Ailuridae (red panda)
    - Family Mephitidae (skunks and stink badgers)
    - Family Mustelidae (weasels, badgers, otters)
    - Family Procyonidae (raccoons and relatives)
  - Clade Pinnipedia
    - Family Odobenidae (walrus)
    - Family Otariidae (eared seals, sea lions)
    - Family Phocidae (true seals)

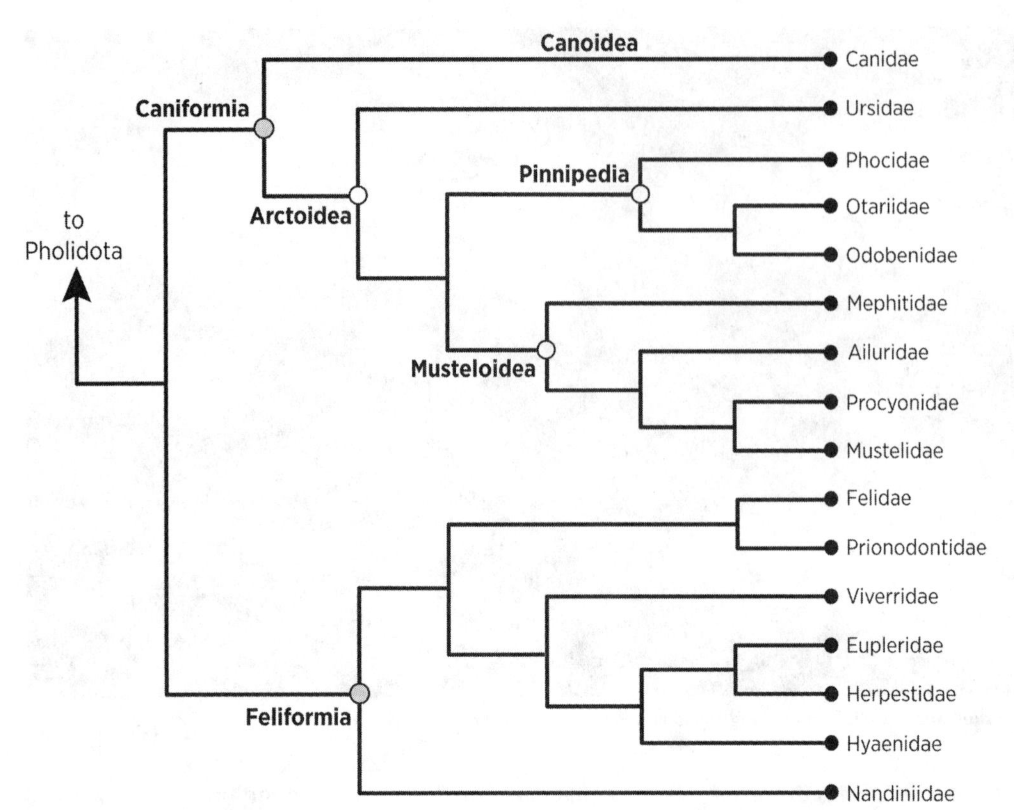

FIG. 421. Phylogenetic tree depicting relationships among the 16 carnivoran families (based on Gatesy et al. 2017).

Several nodes in the carnivoran phylogeny have been problematic, either because of conflicts between different data sets and analyses or because they remained ambiguous in most analyses (e.g., Flynn et al. 2010; Nyakatura and Bininda-Emonds 2012). The most recent study (Gatesy et al. 2017), however, provided a fully resolved tree, which we depict in figure 421.

Caniform and feliform carnivorans differ in the structure of the auditory bulla (fig. 422) as well as in carotid circulation. The further division of caniforms into Canoidea (Canidae) and Arctoidea (remaining Caniformia) also reflects (in part) bullar structure. Generally, the ectotympanic is the dominant bullar element in arctoids, whereas the entotympanic is larger in canoids and Feliformia. Moreover, most extant feliforms (except Nandiniidae) have a bony septum that divides the bulla into two chambers, an anterior chamber dominated by the ectotympanic, and a posterior entotympanic. When visible, the orientation of this septum (horizontal vs. diagonal) is useful in distinguishing some families. In contrast, the bulla of most caniforms has a single chamber and no septum (Canidae have a partial septum). In caniforms, the brain receives blood through the internal carotid artery, whereas this artery is reduced in feliforms, and flow to the brain occurs via the external carotid.

Most carnivores have anal sacs, which are paired vesicular cutaneous invaginations located just internal to the anus or on either side of it (= perianal sacs). These sacs are connected to modified sebaceous glands that produce variable secretions, and the walls of the sacs may be well muscled, allowing expulsion of accumulated secretions to some distance (although most carnivores merely "mark" vegetation or other structures as a means of chemical communication). In some carnivores (Herpestidae, Hyaenidae, some Eupleridae), the anal sacs lie within anal pouches, in which the skin around the anus is invaginated and often closed. Finally, some carnivores (Nandiniidae, Viverridae, some Eupleridae) have perineal glands (also called perfume glands), which are compact masses of glandular tissue located between the anus and the vulva or penis, and which open onto a naked or sparsely haired area, and which may be infolded to form a storage pouch. (See Macdonald 1985 for further detail on carnivore glands.)

Characters that diagnose carnivorans relative to other mammals include the following:

1. FISSIPED CARNIVORES:
    - canines large, pointed, curved
    - **cheek teeth secodont or bunodont**; specialized bladelike, slicing carnassial pair (PM4/m1)

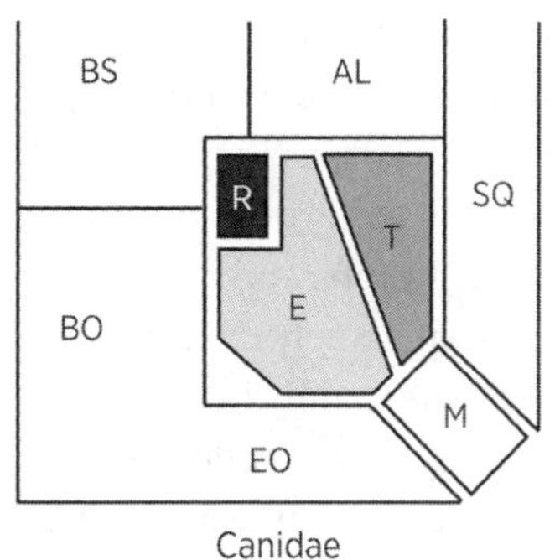

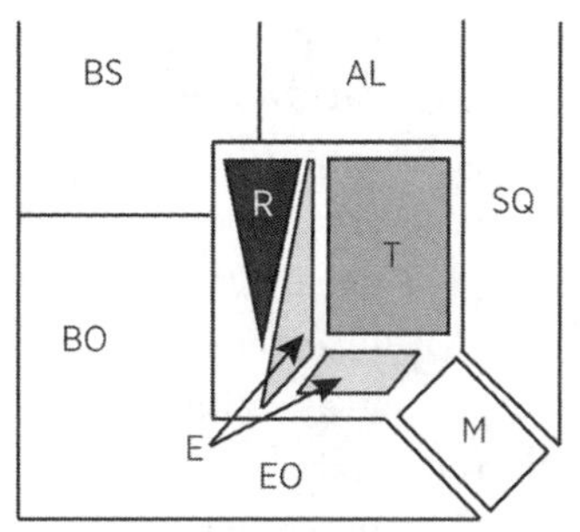

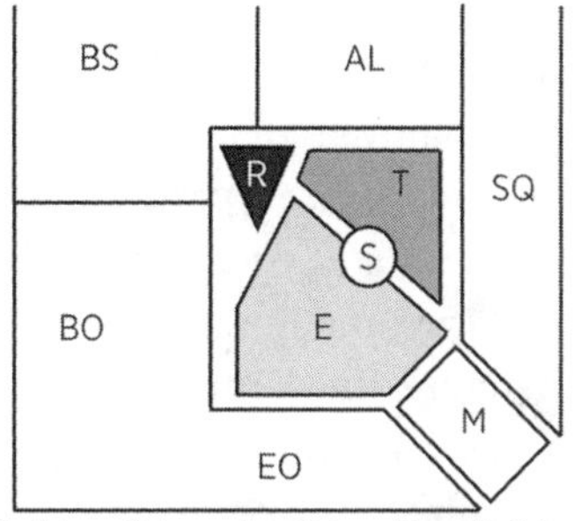

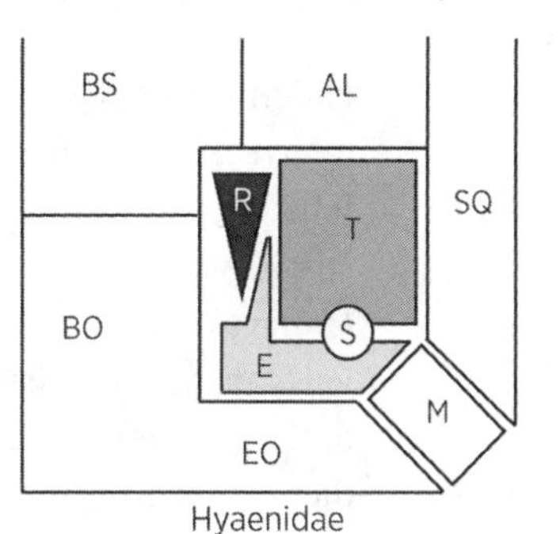

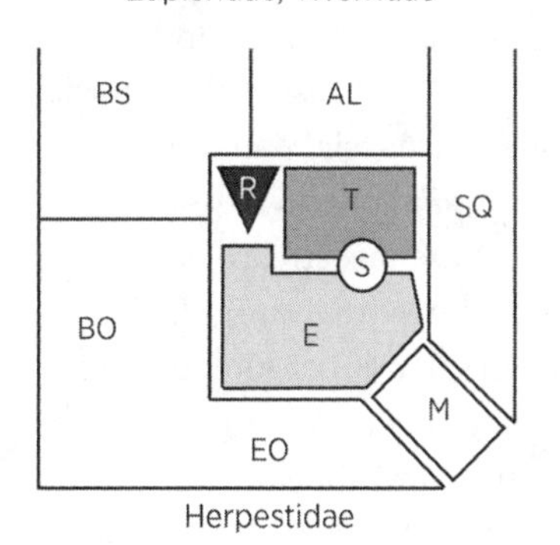

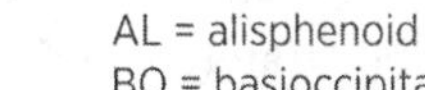

FIG. 422. Diagrammatic illustrations, in ventral view, of the left bullar structure and bony components in different carnivoran groups. *Top left*, Caniformia: Canoidea; *top right*, Caniformia: Arctoidea; *middle and bottom*, Feliformia. Redrawn from Hunt and Tedford (1993).

present (weakly developed to undeveloped in bears, most procyonids, a few mustelids); molars usually with crushing surfaces (except m1 in most)

- **incisors 3/3** (3/2 in sea otter, *Enhydra* [Mustelidae])
- limbs variable in length but always with claws
- feet either plantigrade or digitigrade
- **pinnae prominent**
- eyes variable in position, usually well separated
- **postorbital processes on frontal and zygomatic arch usually present**
- **lacrimal foramen present**

2. PINNIPED CARNIVORES:
   - canines large (upper canine enormous in walrus *Odobenus* [Odobenidae]), pointed, curved
   - **cheek teeth relatively simple,** conical or with secondary cusps anterior and posterior to main cusp, without crushing surfaces; no carnassial slicing pair
   - **incisors 1–3/0–2**
   - forelimbs and hind limbs modified into flippers
   - body streamlined, torpedo-shaped; hair very short
   - **pinnae present** (Otariidae) **or reduced to absent** (Odobenidae, Phocidae)
   - eyes positioned forward on face, close together
   - **prominent postorbital process present only in Otariidae**
   - **no lacrimal foramen**

Suborder FELIFORMIA

The number and membership of family-rank groups within the Feliformia has been confused and confusing for well over a century. Modern molecular methods, combined in some, but not all, cases with cladistic analyses of morphological characters, have resolved the diversity of distinct lineages within the suborder, but have also highlighted the often high level of convergence in morphological traits around which the historically confused taxonomy has revolved. Gaubert et al. (2005) provide an illuminating example of these convergences for the viverrid-like carnivorans, especially the mosaic nature of character combinations in the Eupleridae, Nandiniidae, and Prionodontidae. Readers are directed to this paper for a more thorough understanding. Here we describe the seven families of feliform carnivorans currently recognized, but, as will be evident in some accounts, identifying shared-derived characters diagnosing some is not possible at the moment.

In addition to the bullar and carotid characteristics outlined and illustrated above (see fig. 422 and accompanying description), feliform carnivores tend to differ from caniform carnivores in having shorter rostra, with fewer teeth and more specialized carnassials. They tend to be more carnivorous, often employing ambush hunting. Feliforms also tend to be digitigrade; many have retractile or semi-retractile claws; and many are arboreal or semiarboreal.

Family EUPLERIDAE (fossa, falanouc, Malagasy civet, Malagasy mongoose)

Small (*Mungotictis*, with body length 30 cm, mass 500 g) to medium-sized (*Cryptoprocta*, with body length 80 cm, mass 9 kg); body elongate, but otherwise external features vary extensively, with head ranging from elongate and angular to flat and rounded, legs from short to long, and tails always long but slightly shorter than body, with or without extensive brush; dorsal color pattern may be a uniform monotone or coupled with longitudinal strong (*Galidictis*) or weak (*Mungotictis*) stripes or with spots (*Fossa*); the tail may similarly be unicolored or with incomplete (*Fossa*) to complete bands (*Galidia*). Ecology and behavior are as varied as morphology: *Cryptoprocta* is diurnal or nocturnal, solitary, an agile climber, and carnivorous, specializing on small mammals. *Eupleres* is nocturnal, terrestrial, is solitary or lives in small groups, and is a specialized earthworm feeder with an elongate muzzle and conical cheek teeth. *Fossa* is nocturnal, terrestrial to arboreal, with a social system consisting of male-female pairs, and is carnivorous, feeding on both vertebrates and invertebrates. *Galidia* is largely diurnal, terrestrial to arboreal, lives in family groups, and is carnivorous. *Galidictis* is nocturnal, strictly terrestrial, solitary or paired, and carnivorous, specializing on small vertebrates. *Mungotictis* is largely diurnal, usually terrestrial, lives in family units, and is largely insectivorous. *Salanoia* is diurnal, largely terrestrial, lives solitarily or in pairs, and eats mainly insect larvae. Habitat is equally varied, ranging from dry deciduous forest to rainforest to montane forest, but extends into semidesert and above tree line.

The Eupleridae is a monophyletic assemblage endemic to Madagascar that collectively mimics (converges on) the combined adaptive radiation of felids (e.g., *Cryptoprocta*), viverrids (e.g., *Eupleres* and *Fossa*), and herpestids (e.g., *Galidia*, *Galidictis*, *Mungotictis*, and *Salanoia*) of Africa and Eurasia. Perhaps not surprisingly, therefore, the dentition of these genera varies greatly as a function of their dietary specializations: the carnivorous *Cryptoprocta* has a large bladelike carnassial pair and a single, very reduced upper molar, much like a cat; *Fossa* and the four galidiine genera have an all-purpose dentition with a moderately well-developed carnassial pair, but with a large protocone, and one (*Galidia*) or two (*Fossa*) large triangular upper molars; and the insectivorous *Eupleres* has an overall reduced dentition with a simplified carnassial pair and two triangular upper molars. While euplerids are strongly supported as a monophyletic lineage on molecular grounds, and appear as sister to the Herpestidae (see fig. 421), there are no unique morphological characters that unite all members of this family.

RECOGNITION CHARACTERS:

1. foot posture digitigrade (plantigrade in *Cryptoprocta*)
2. digits 5/5; interdigital webbing in some galidiines
3. claws variable across genera, nonretractile (*Fossa*, *Galidia*, *Galidictis*) or semi-retractile (*Cryptoprocta*)
4. perineal glands present in some genera (*Galidia*, *Galidictis*)
5. anal glands present; simple and opening to a small anal sac (*Eupleres*, *Fossa*, *Galidia*) or opening into an anal pouch (*Cryptoprocta*)
6. postorbital processes of frontal long (>125% of postorbital width) in *Cryptoprocta*, *Fossa*, and *Galidia*; absent in *Eupleres*
7. postorbital constriction immediately posterior to postorbital processes

FIG. 423. Eupleridae: Euplerinae; fossa, *Cryptoprocta ferox*.

FIG. 424. Geographic range of family Eupleridae: Madagascar.

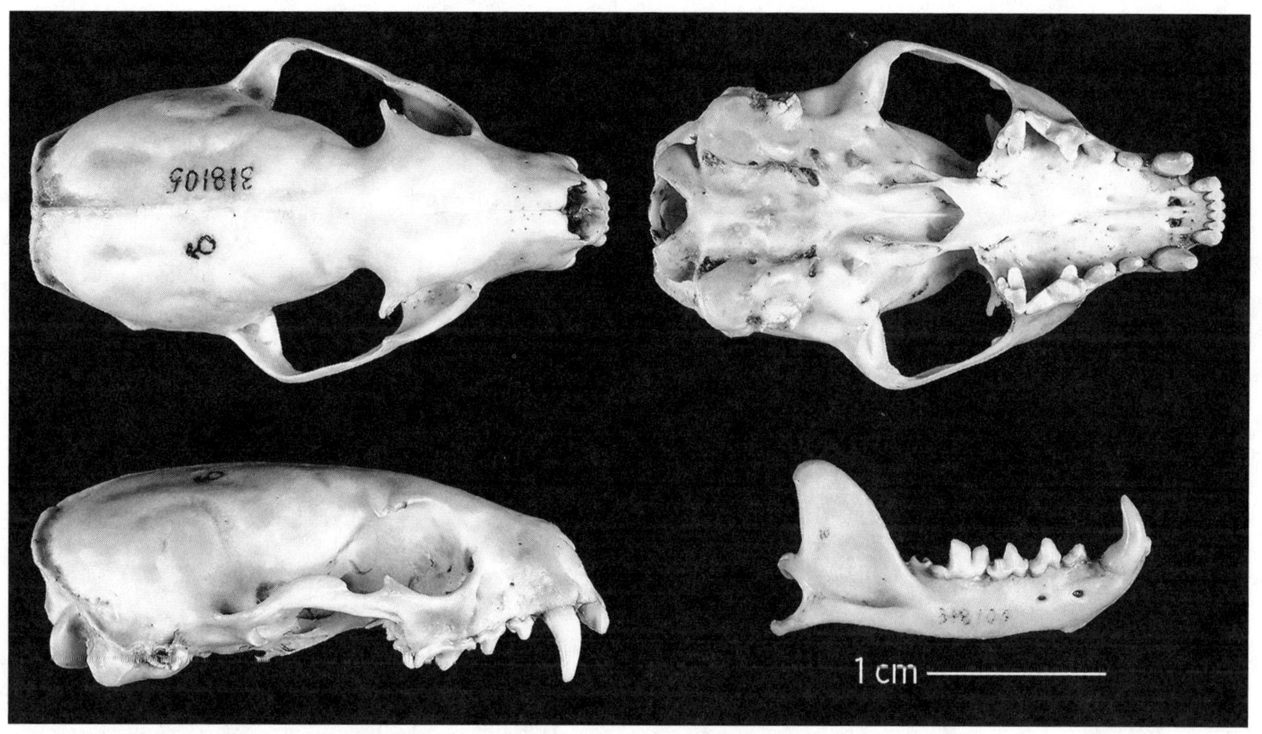

FIG. 425. Eupleridae: Galidiinae; skull and mandible of the ring-tailed mongoose, *Galidia elegans*.

8. single sagittal crest small (*Cryptoprocta*, *Fossa*, *Galidia*) or absent (*Eupleres*)
9. paroccipital process short, cupped around posterior edge of bulla
10. bulla with enlarged entotympanic and diagonal septum (*Cryptoprocta*, *Eupleres*, *Fossa*) or chambers separated by horizontal septum (*Galidia*)
11. no alisphenoid canal
12. baculum present (long in *Cryptoprocta*; short in *Eupleres*, *Fossa*, and *Galidia*)

DENTAL FORMULA: $\frac{3}{3}\ \frac{1}{1}\ \frac{3\text{–}4}{3\text{–}4}\ \frac{1\text{–}2}{1} = 32\text{–}38$

TAXONOMIC DIVERSITY: About eight species in seven genera divided into two subfamilies:

Euplerinae—*Cryptoprocta* (fossa), *Eupleres* (falanouc), and *Fossa* (spotted fanaloka)

Galidiinae—*Galidia* (ring-tailed mongoose), *Galidictis* (broad-striped Malagasy mongoose, Grandidier's mongoose), *Mungotictis* (narrow-striped mongoose), and *Salanoia* (brown-tailed mongoose)

Family FELIDAE (cats)

Medium-sized to relatively large (total length 75–370 mm; mass 2 kg [smaller *Felis*, *Leopardus*, and *Pronailurus*] to 275 kg [*Panthera tigris*]); face short and broad; tail short (*Lynx*) to long (10–110 cm), not bushy; legs short (*Puma yagouaroundi*) to long (*Acinonyx*); anal glands present, moderate in size; ears more or less triangular, some with long tufts forming acute points (*Lynx*, *Caracal*); eyes proportionally large; pelage color varied, from black to various shades of brown or russet, uniform or with spots, rosettes, or stripes (melanism, in particular, common in some species). All cats are carnivorous, feeding on vertebrates, but some also consume fruits or invertebrates; most are nocturnal and more or less arboreal, especially smaller species; social system varies from solitary to family groups or male coalitions; senses of smell, hearing, and touch highly acute. Inhabit virtually all terrestrial biomes, from grasslands, shrublands, and woodlands to both temperate and tropical forests and sea level to alpine habitats.

FIG. 426. Felidae; jaguar, *Panthera onca* (Pantherinae; *top*), and Canadian lynx, *Lynx canadensis* (Felinae; *bottom*).

RECOGNITION CHARACTERS:

1. foot posture digitigrade
2. **digits 5/4**, but pollex (dewclaw) is located on side of foot, not contacting ground
3. no metatarsal pads
4. **claws sharp, strongly curved, retractile (= fully sheathed**; only partly so in the cheetah, *Acinonyx*)
5. perineal glands absent
6. anal glands simple, opening to a small anal sac; anal pouches absent
7. **skull short, rounded dorsally; rostrum very short, blunt**
8. postorbital processes of frontal long (>125% of postorbital width), sometimes contacting jugal
9. postorbital constriction immediately posterior to postorbital processes
10. sagittal crest single, large or small
11. paroccipital process long or short, cupped around posterior edge of bulla
12. bulla with dominant entotympanic chamber and diagonal division between ectotympanic and entotympanic
13. no alisphenoid canal
14. **carnassials very well developed, bladelike**; PM4 with reduced protocone and expanded parastyle; m1 without protoconid or talonid
15. last upper molar tiny, round
16. baculum absent or vestigial

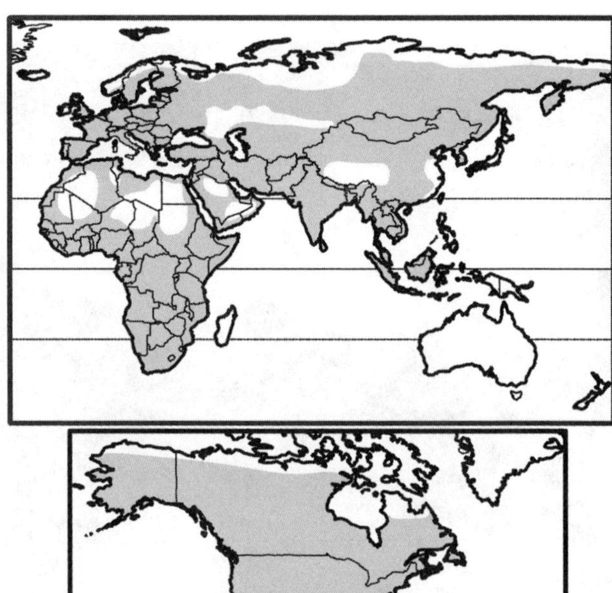

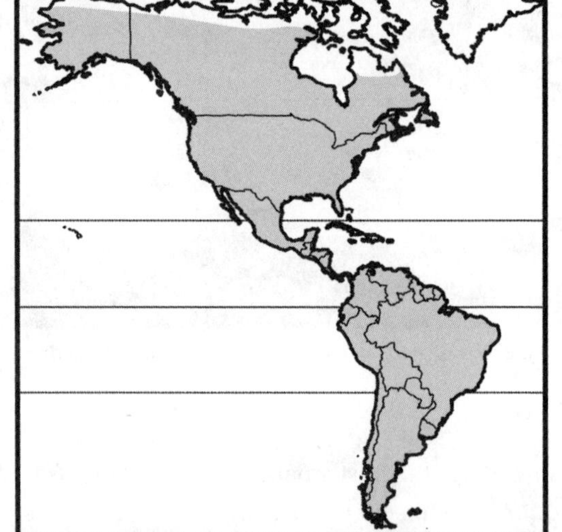

FIG. 427. Geographic range of family Felidae: native to every continent except Australia and Antarctica; absent from Greenland, Madagascar, New Guinea, and New Zealand; domesticated, feral cats (*Felis catus*) now widely introduced.

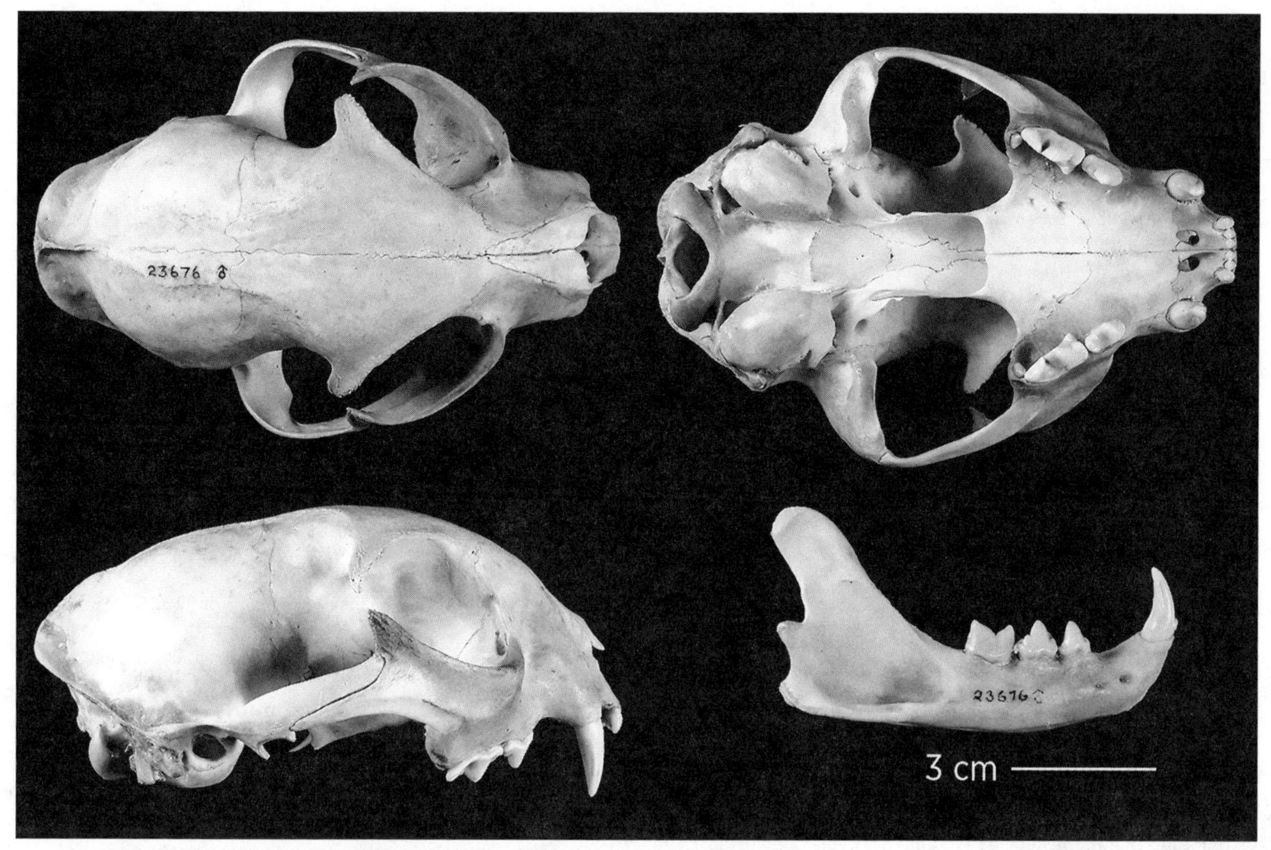

FIG. 428. Felidae: Felinae; skull and mandible of the bobcat, *Lynx rufus*.

DENTAL FORMULA: $\frac{3}{3}\ \frac{1}{1}\ \frac{2\text{–}3}{2}\ \frac{1}{1} = 28\text{–}30$

TAXONOMIC DIVERSITY: Alpha taxonomy within the Felidae remains a work in progress. The most recent syntheses (Wozencraft 2005; Sunquist and Sunquist 2009; Kitchener et al. 2017) recognize about 37–41 species in 14–15 genera divided into two subfamilies:

Felinae—*Acinonyx* (cheetah), *Caracal* (caracal), *Catopuma* (bay cat, Asian golden cat), *Felis* (wildcat, domesticated cat, black-footed cat, jungle cat, sand cat), *Leopardus* (ocelot, margay, tiger cat, colocolo, Geoffroy's cat, Andean mountain cat), *Leptailurus* (serval), *Lynx* (bobcat, lynx), *Otocolobus* (Pallas's cat; included in *Felis* by Wozencraft 2005), *Pardofelis* (marbled cat), *Prionailurus* (leopard cat, flat-headed cat, fishing cat), *Profelis* (African golden cat; included in *Caracal* by Kitchener et al. 2017), and *Puma* (puma, jaguarundi; the latter species is placed in the genus *Herpailurus* by Kitchener et al. 2017)

Pantherinae—*Neofelis* (clouded leopard) and *Panthera* (lion, tiger, jaguar, leopard, snow leopard [separated as genus *Uncia* by Wozencraft 2005])

Family HERPESTIDAE (mongooses)

Slender and small carnivores (body length 300–1,700 mm; mass 200 g [common dwarf mongoose, *Helogale*] to 5 kg [white-tailed mongoose, *Ichneumia*]); muzzle and body long; legs short; ears small and rounded, **without the bursae (= pockets) on lateral margins** typical of most other carnivores; tail shorter than head and body but tapering and bushy; feet semi-plantigrade to digitigrade; digits in most 5/5 but 4/4 in some, partially webbed in semiaquatic taxa (marsh [*Atilax*], short-tailed [*Herpestes brachyurus*], and long-nosed mongooses [*Xenogale*]); claws well developed, elongate (especially on forefeet), and nonretractile; mainly terrestrial, with poor climbing abilities; **pelage uniform in color** (light gray or yellow to brown or black), long and coarse, never spotted and rarely striped; feet, legs, tail, or tail tip often differently colored; transverse stripes present in Mungotinae. Diurnal or nocturnal; primarily carnivorous, with diet varying among taxa, in which emphasis may be on invertebrate or vertebrate prey; plant material rarely consumed. Herpestines tend to be solitary; mungotines are highly social, with many forming large and complex

FIG. 429. Herpestidae: Herpestinae; ruddy mongoose, *Herpestes smithii*.

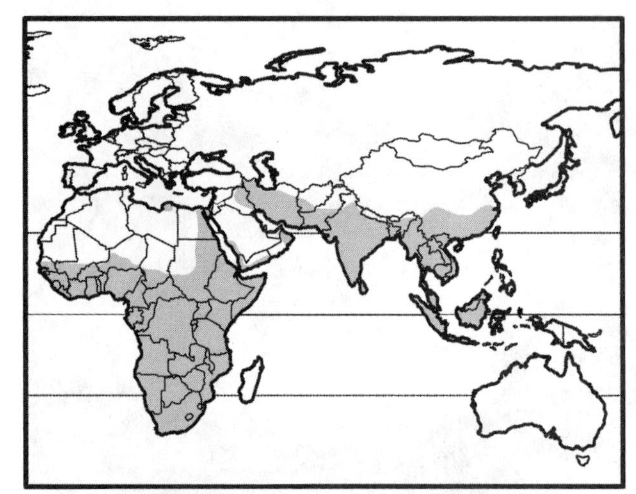

FIG. 430. Geographic range of family Herpestidae: throughout Africa (except the Sahara), the Middle East, and the Arabian Peninsula across southern Asia to Southeast Asia and Indonesia.

groups. Habitat ranges from open areas, including deserts, grasslands, and savannas, to closed forests across lowland to montane areas.

Earlier classifications included the herpestids as a subfamily of the Viverridae, but both morphological and molecular studies confirm the family status of the mongooses and a close relationship between them and the Hyaenidae and the Malagasy endemic Eupleridae (see fig. 421).

RECOGNITION CHARACTERS:

1. foot posture semi-digitigrade to digitigrade
2. **digits usually** 5/5 (4/4 or 5/4 in some), webbing reduced or absent
3. **claws well developed, long and protracted on front feet, nonretractile**
4. perineal glands absent
5. anal glands present in anal sac, in turn enclosed in anal pouch
6. **skull elongate**; rostrum moderately long
7. postorbital processes of frontal long (>125% of postorbital width), sometimes contacting jugal
8. postorbital constriction immediately posterior to postorbital processes
9. sagittal crest single, large or small
10. paroccipital process short, cupped around posterior edge of bulla
11. alisphenoid canal present
12. **median lacerate foramen present**
13. **bulla large, division between ectotympanic and entotympanic horizontal; auditory tube present**
14. **1st lower incisor slightly out of line of others in row**
15. carnassials weakly developed; protocone of PM4 broadly expanded; trigonid of m1 not bladelike and talonid expanded
16. **last upper molar large, triangular**
17. baculum present, small

DENTAL FORMULA: $\frac{3}{3}\ \frac{1}{1}\ \frac{3\text{–}4}{3\text{–}4}\ \frac{1\text{–}2}{1\text{–}2} = 32\text{–}40$

TAXONOMIC DIVERSITY: About 34 species in 15 genera divided into two subfamilies (Gilchrist et al. 2009):

Herpestinae—*Atilax* (marsh mongoose), *Bdeogale* (bushy-tailed, black-legged, Jackson's mongooses), *Cynictis* (yellow mongoose), *Galerella* (Cape gray mongoose, slender mongooses), *Herpestes* (Indian, Egyptian, Javan, and other mongooses), *Ichneumia* (white-tailed mongoose), *Paracynictis* (Selous's mongoose), *Rhynchogale* (Meller's mongoose), and *Xenogale* (long-nosed mongoose)

Mungotinae—*Crossarchus* (cusimanses), *Dologale* (Pousargues's mongoose), *Helogale* (dwarf mongooses), *Liberiictis* (Liberian mongoose), *Mungos* (Gambian and banded mongoose), and *Suricata* (meerkat)

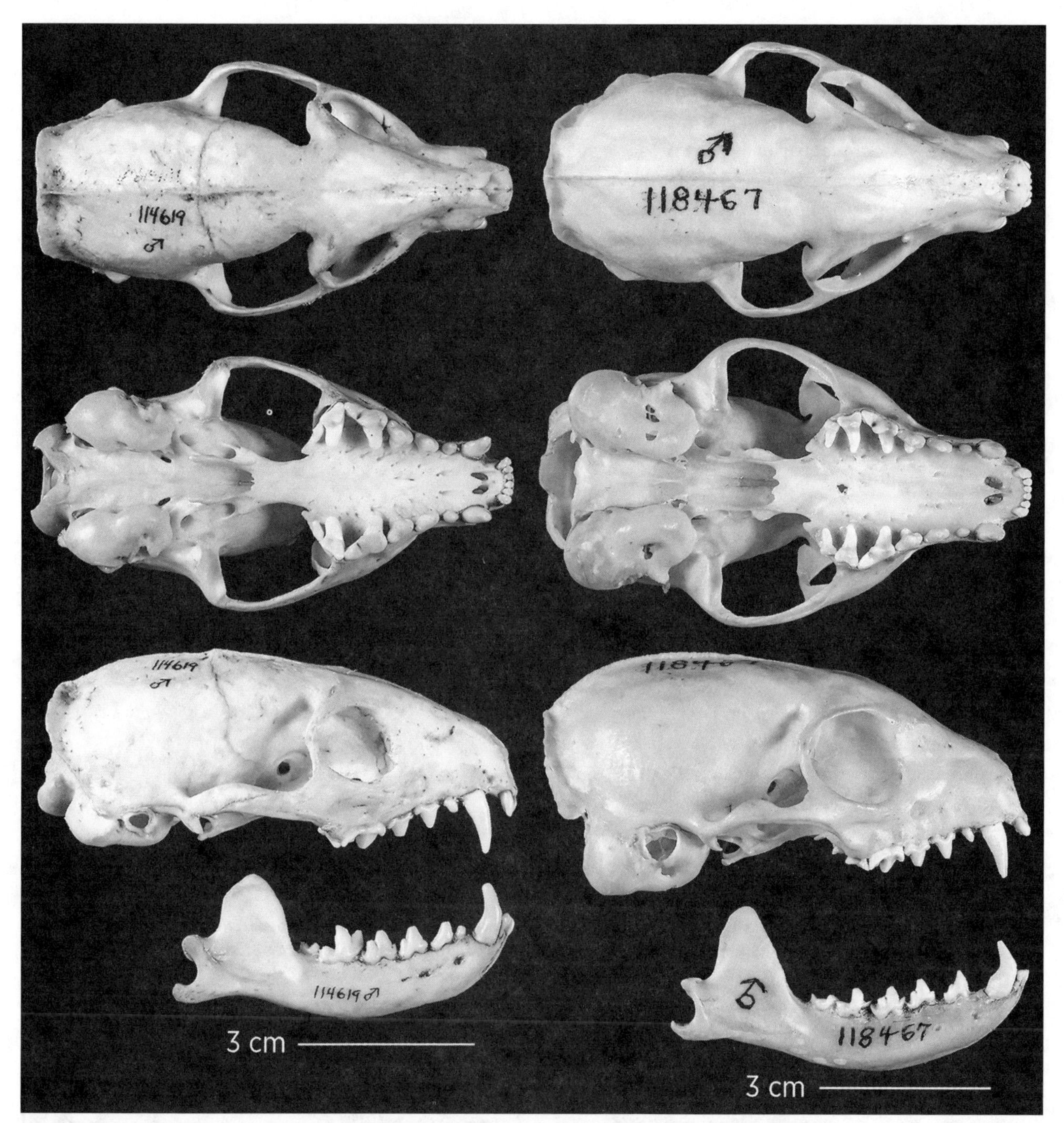

FIG. 431. Herpestidae; skull and mandible of the small Indian mongoose, *Herpestes javanicus* (Herpestinae; *left*), and the banded mongoose, *Mungos mungo* (Mungotinae; *right*).

Family HYAENIDAE (aardwolf, hyenas)

Medium-sized to relatively large (total length 75–195 cm; mass 10–82 kg); tail of medium length (20–35 cm), bushy; ears large and rounded; color brown with dark spots or stripes; short mane usually present. Aardwolf is insectivorous; other Hyaenidae specialize on terrestrial vertebrates, obtained by both active hunting and scavenging; noted for bone-crushing capabilities, evidenced by cranial and dental modifications; terrestrial; solitary to group living, with complex social hierarchy and female dominance (especially *Crocuta*). Female hyenas (especially *Crocuta*) are "masculinized," with enlarged and erectile phallus that also serves as birth canal. Olfaction is a more important sense than either sight or hearing. Inhabit primarily open-vegetation communities.

RECOGNITION CHARACTERS:

1. foot posture digitigrade; **hind limbs shorter than forelimbs**
2. **digits 4/5** (*Crocuta*, *Hyaena*) **or** 5/5 (*Proteles*)
3. claws blunt, nonretractile
4. perineal glands absent
5. anal glands present, open within anal pouch, located in anterior position
6. **skull massive** (*Crocuta*, *Hyaena*) **or delicate** (*Proteles*); rostrum relatively long
7. postorbital processes of frontal long (>125% of postorbital width)
8. postorbital constriction immediately posterior to postorbital processes
9. sagittal crest large and single (*Crocuta*, *Hyaena*) or small and double (*Proteles*)
10. paroccipital process long (*Proteles*) or short (*Crocuta*, *Hyaena*), cupped around posterior edge of bulla
11. bulla with reduced entotympanic and enlarged ectotympanic; septum horizontal (*Crocuta*, *Hyaena*) or vertical (*Proteles*)
12. no alisphenoid canal
13. carnassials well developed, similar to those of Felidae (*Crocuta*, *Hyaena*), or not developed at all (*Proteles*)
14. **upper molar, if present, very small, round or elongate**
15. baculum absent

DENTAL FORMULA:

hyenas (*Crocuta*, *Hyaena*): $\frac{3}{3}\ \frac{1}{1}\ \frac{4}{3}\ \frac{1}{1} = 34$

aardwolf (*Proteles*): $\frac{3}{3}\ \frac{1}{1}\ \frac{3}{1\text{–}2}\ \frac{1}{1\text{–}2}$ = 28–32, usually 30

TAXONOMIC DIVERSITY: Four species in three or four genera segregated in two subfamilies:

Hyaeninae—*Crocuta* (spotted hyena), *Hyaena* (striped hyena), *Parahyaena* (brown hyena; variously treated as a distinct genus or as a subgenus within *Hyaena*)

Protelinae—*Proteles* (aardwolf)

FIG. 432. Hyaenidae; aardwolf, *Proteles cristatus* (Protelinae; *upper right*), and spotted hyena, *Crocuta crocuta* (Hyaeninae; *lower left*).

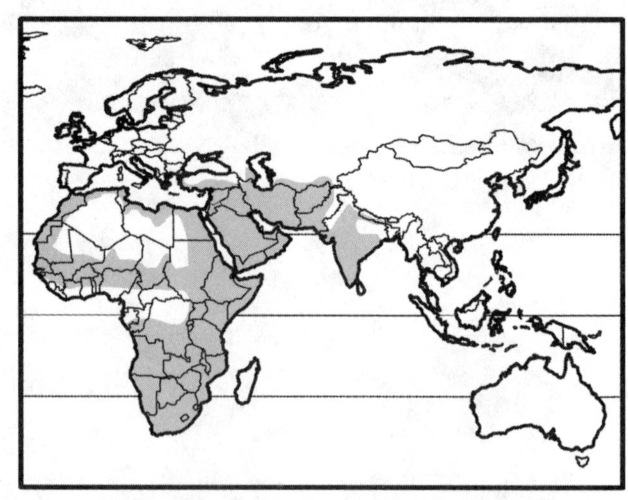

FIG. 433. Geographic range of family Hyaenidae: Africa, except Sahara and Congo basin; Middle East from Turkey through Arabian Peninsula to India.

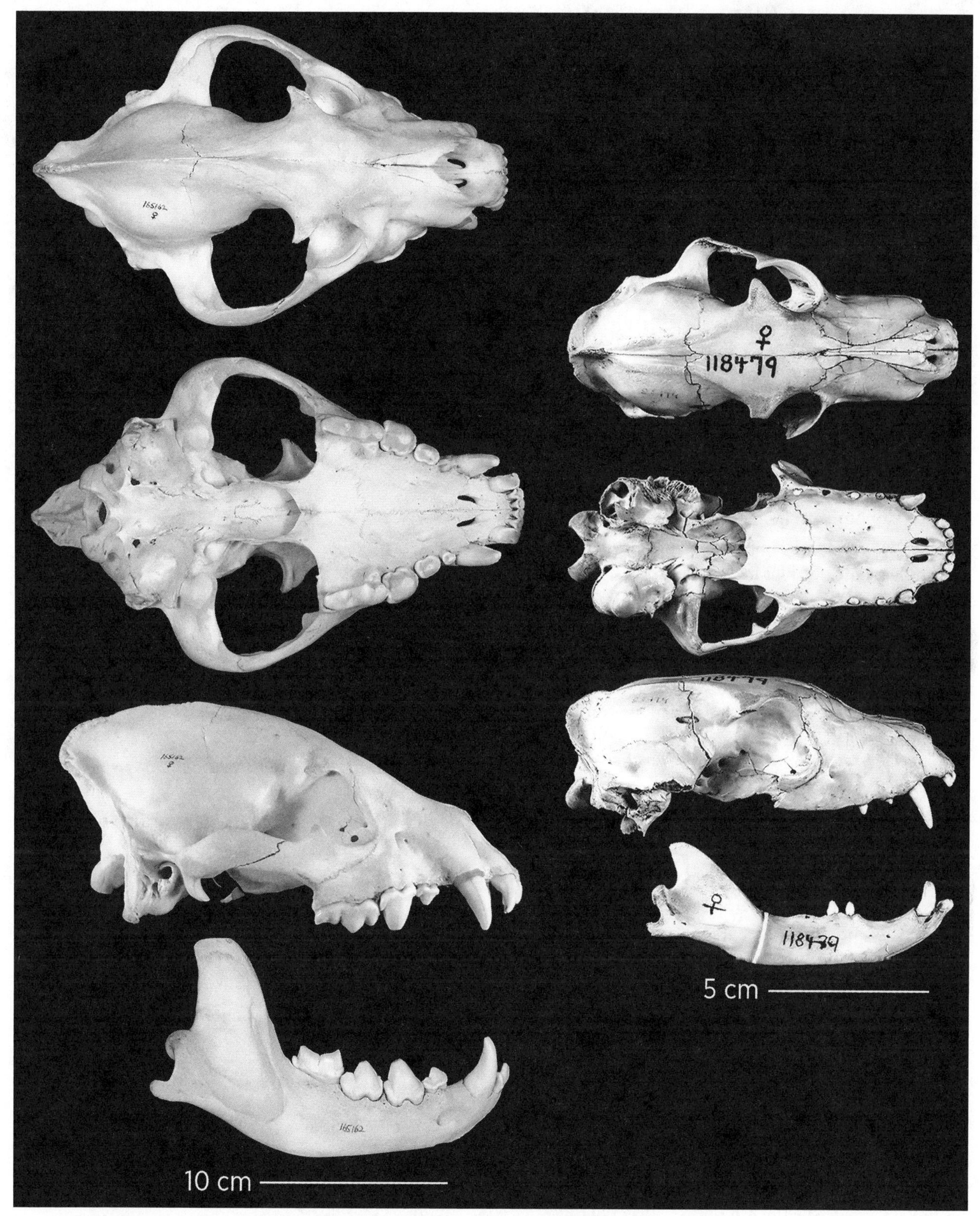

FIG. 434. Hyaenidae; skull and mandible of the spotted hyena, *Crocuta crocuta* (Hyaeninae; *left*), and the aardwolf, *Proteles cristatus* (Protelinae; *right*).

Family NANDINIIDAE (African palm civet)

Small (body length 35–65 cm; mass 1–3 kg); heavily built genet-like animal with a short muzzle, short legs, and tail 110%–120% longer than body (up to 75 cm in length); nose pad pale brown; ears short, broad at base and rounded; **color pattern of back with characteristic** pair of yellow spots on shoulders and irregularly distributed dark spots that may partly coalesce on the neck; the tail has narrow dark rings, generally incomplete ventrally; both forefeet and hind feet are plantigrade, each foot has five clawed digits; claws are partially retractile; foot pads are well developed, with **hairy area between plantar and digital pads**. Food habits primarily frugivorous, but will feed on invertebrate and vertebrate prey; nocturnal and somewhat secretive; arboreal; solitary, scent-mark territorial boundary. Habitat includes rainforest and deciduous forest from coastal to montane areas as well as savanna brushlands.

The taxonomic position of *Nandinia* has long been debated; it has been considered either a morphologically peculiar member of the Asiatic palm civets (Viverridae: Paradoxurinae) or the monotypic subfamily Nandiniinae of Viverridae. It is now placed in its own family as part of the basal polytomy of the Feliformia (see fig. 421), largely due to the unique location and structure of its perianal glands and its basicranial conformation, a position supported by molecular analyses.

FIG. 435. Nandiniidae; African palm civet, *Nandinia binotata*.

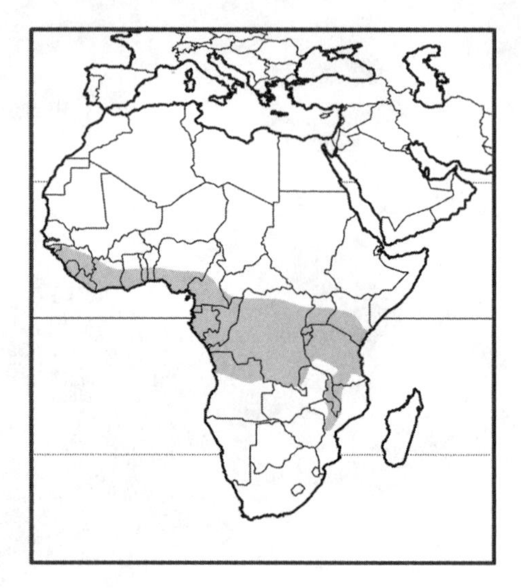

FIG. 436. Geographic range of family Nandiniidae: tropical Africa south of the Sahara.

DIAGNOSTIC CHARACTERS:

- **posterior chamber of bulla (caudal entotympanic bone) cartilaginous throughout life**
- **septum between bullar components indistinct, as caudal entotympanic is cartilaginous (present in most extant carnivorans)**
- **perineal glands in pouch anterior to genitals in both sexes**

RECOGNITION CHARACTERS:

1. foot posture plantigrade
2. digits 5/5
3. claws semi-retractile
4. **area between plantar and digital pads (except digit 1) hairy; pads of 3rd and 4th digits widely separated**
5. anal glands simple, open to small anal sac; anal pouches absent
6. lactating females present additional scent glands on foot sole, chin, and belly
7. skull with moderately elongate rostrum and wide zygomatic arches
8. postorbital processes of frontal long (>125% of postorbital width)
9. postorbital constriction immediately posterior to postorbital processes
10. sagittal crest small, single
11. paroccipital process very short, free from bulla
12. bulla with dominant entotympanic chamber and diagonal division between entotympanic and ectotympanic
13. alisphenoid canal present
14. carnassial pair well developed; PM4 with reduced protocone, but without enlarged parastyle; m1 with reduced talonid
15. **last upper molar, if present, small and round**
16. baculum present, small

DENTAL FORMULA: $\frac{3}{3}\ \frac{1}{1}\ \frac{4}{4}\ \frac{2}{2} = 40$

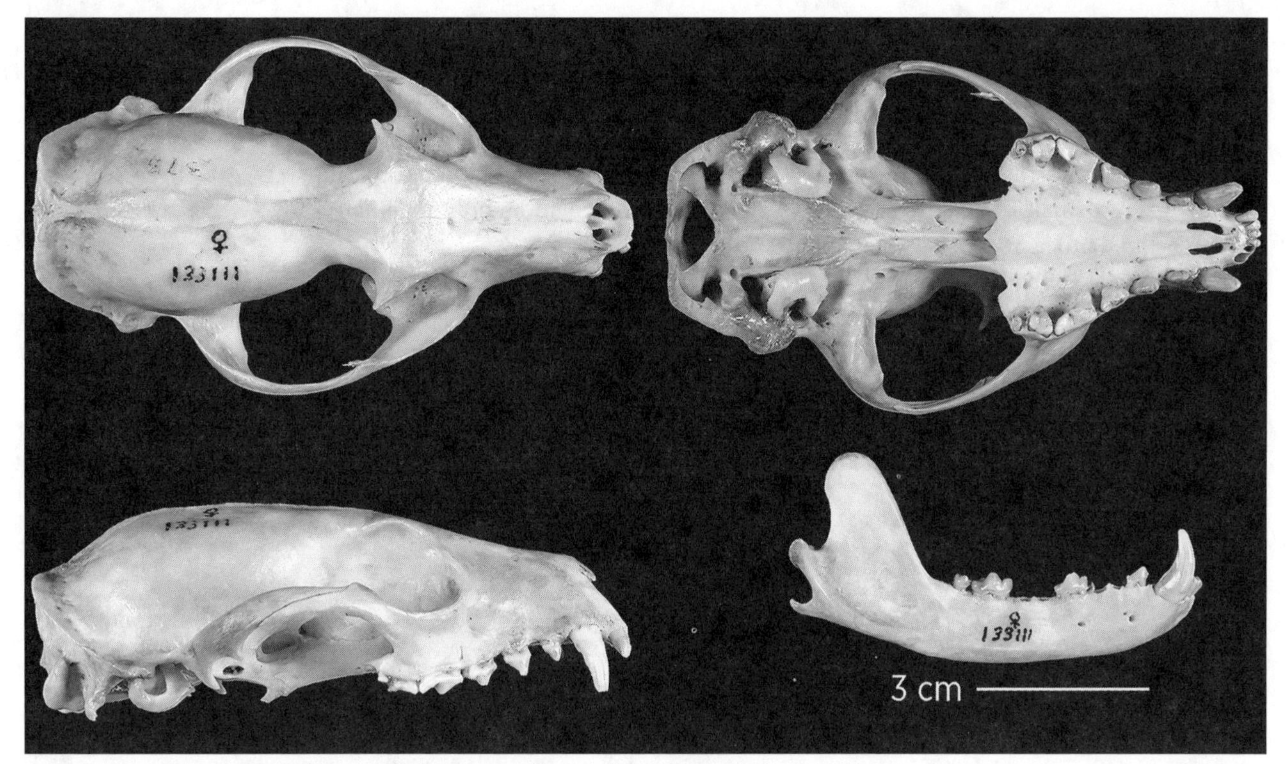

FIG. 437. Nandiniidae; skull and mandible of the African palm civet, *Nandinia binotata*.

TAXONOMIC DIVERSITY: The family is monotypic, with a single species, *Nandinia binotata.*

Family PRIONODONTIDAE (Asian linsangs)

Small (body length 30–45 cm; mass 550–1,200 g), slender; genet-like in appearance with pointed muzzle, elongate neck, tail almost as long as body (length 30–40 cm); brownish-pink nose pad; ears short, broad at base and rounded; background coat color relatively uniform, varying from pale gray to yellow rufous, but back, sides, and upper thighs spotted, and a pair of large stripes on the nape; tail with alternate light and dark bands; climbs well; forefeet plantigrade but hind feet digitigrade (Gaubert 2009). Primarily carnivorous, feeding on rodents, frogs, snakes, and small birds as well as large insects; terrestrial to arboreal; nocturnal; probably solitary, but social system unknown. Habitat includes primary and secondary moist evergreen and mixed deciduous forests.

Often placed as a subfamily of Viverridae (e.g., Wozencraft 2005), the Prionodontidae is now widely recognized as a clade separate from the civets, one perhaps sister to the Felidae (Gaubert et al. 2005; Flynn et al. 2010). Supertree analyses of a large body of molecular data, however, are conflicting in the supported relationships between the linsangs and cats. Nyakatura and Bininda-Emonds (2012) place the linsangs as part of a polytomy at the base of the Feliformia, whereas Gatesy et al. (2017) confirm a sister relationship between the linsangs and cats (see fig. 421). Many of the structural characters of linsangs are a mosaic of shared homoplasies (characters shared due to convergence and not derived from a common ancestor) with euplerids, nandiniids, and some African civets. Similarities between linsangs and felids (especially hypercarnivory and foot structure) are thus likely to be symplesiomorphies (shared ancestral traits not necessarily reflecting proximity of relationship) rather than supporting a close relationship.

RECOGNITION CHARACTERS:

1. foot posture digitigrade (hind feet) and plantigrade (forefeet)
2. digits 5/5
3. no metatarsal pads
4. **claws fully retractile** (= fully sheathed)
5. perineal glands absent
6. anal glands simple, opening to small anal sacs; anal pouches absent

7. skull elongate, rostrum narrow
8. distinct postorbital processes absent, although supraorbital region is inflated
9. postorbital constriction immediately posterior to postorbital inflation area
10. sagittal crest small, double
11. paroccipital process very short, not cupped around posterior edge of bulla
12. bulla with dominant entotympanic chamber and diagonal division between ectotympanic and entotympanic
13. alisphenoid canal present
14. dentition felid-like, with well-developed carnassials, pm4 with reduced talonid, and reduced crushing function of molars
15. upper molar reduced, transversely elongate
16. baculum present, small

DENTAL FORMULA: $\frac{3}{3}\ \frac{1}{1}\ \frac{4}{4}\ \frac{1}{2} = 38$

TAXONOMIC DIVERSITY: The family is monotypic, with two species in the single genus *Prionodon*.

FIG. 438. Prionodontidae; banded linsang, *Prionodon linsang*.

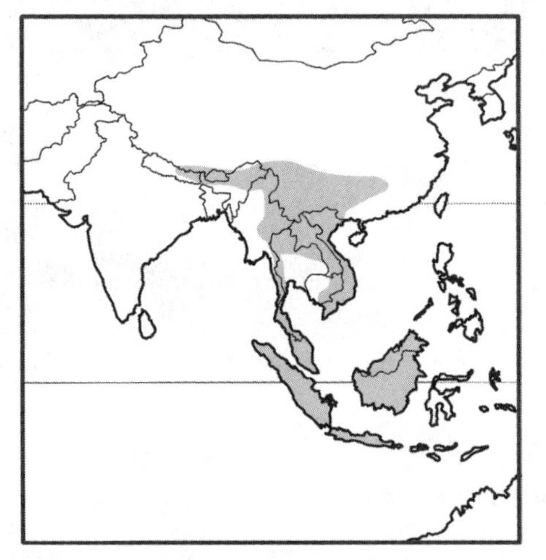

FIG. 439. Geographic range of family Prionodontidae: Southeast Asia from Nepal, Bhutan, and northeastern India to southern China and south through Malay Peninsula; Indonesian islands of Sumatra, Java, and Borneo.

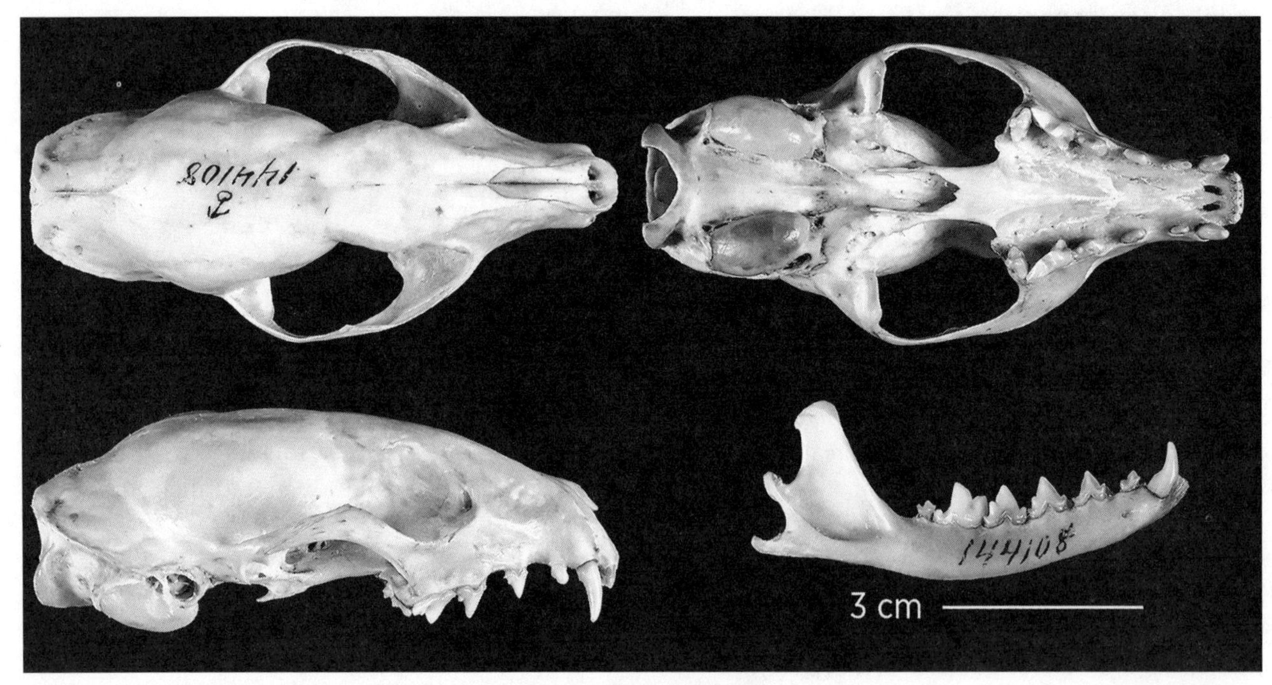

FIG. 440. Prionodontidae; skull and mandible of the banded linsang, *Prionodon linsang*.

Family VIVERRIDAE (civets, genets, and relatives)

Medium-sized carnivorans (body length 300–1,700 mm; mass <1–20 kg); body long and slender; head relatively small and short; ears erect, with rounded tips; muzzle relatively long; limbs short; tail long (about equal to body), well furred but tapering, typically without elongate brush; most species have **stripes, spots, or bands on body**, and **tails are often ringed** with contrasting colors; foot posture is digitigrade (terrestrial species) to semi-plantigrade (arboreal species); digits 5/5; claws are semi-retractile; most have perianal glands that produce a strong-smelling substance, sufficiently potent in some species to ward off predators (the secretion, called "civet," is used as a perfume base and in local medicine). Most are nocturnal, feeding on a variety of vertebrates and invertebrates, but some consume substantial quantities of fruits or other plant parts. Most are strong climbers; the binturong (*Arctictis*) has a prehensile tail. Often found in pairs, but never in large groups. Habitat includes most biomes in southern temperate and tropical regions of the Old World.

FIG. 441. Viverridae: Viverrinae; Cape genet, *Genetta tigrina*.

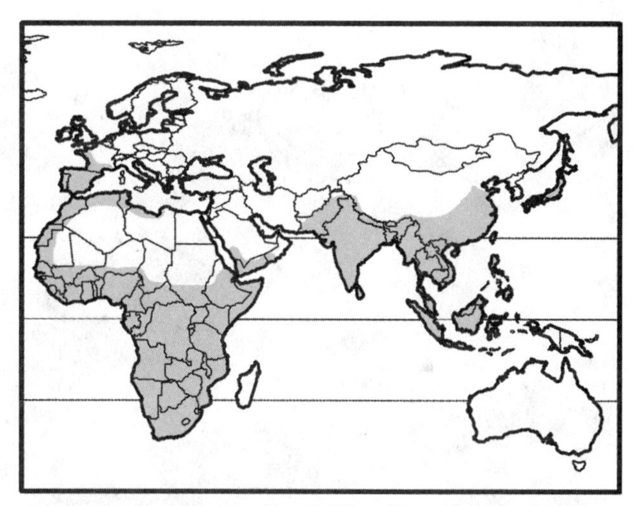

FIG. 442. Geographic range of family Viverridae: widely distributed from Africa through the southern Palearctic to much of the Oriental region; Paradoxurinae—Asia, from China to India and Indonesia; Hemigalinae—Southeast Asia and Indonesia; Viverrinae—southwestern Europe, Africa to the Middle East and Arabian Peninsula, and eastern Asia, from China through Southeast Asia.

RECOGNITION CHARACTERS:

1. foot posture plantigrade, semi-plantigrade, or digitigrade
2. digits 5/5, webbing between toes, hallux and pollex reduced in some species
3. claws well developed, semi- to fully retractile; some with protective skin sheath
4. perineal glands present, either centered around vulva (*Chrotogale*, *Diplogale*, *Hemigalus*, *Macrogalidia*, *Paradoxurus*), restricted to area between vulva and anus (*Arctictis*, *Cynogale*, *Genetta*, *Viverra*), or extending anterior to vulva (*Arctogalidia*); in males, absent (*Macrogalidia*), restricted to area between penis and anus (most genera), encircles base of penis like a collar (*Paguma*), or extends anterior to penis (*Arctogalidia*)
5. anal glands simple, opening to a small sac; anal pouches absent
6. skull elongate; rostrum moderately long
7. postorbital processes of frontal present, short (<125% of postorbital width) and blunt; rarely long and narrow (*Arctogalidia*)
8. postorbital constriction immediately posterior to postorbital processes
9. sagittal crest variable, usually small and single, rarely large and either single or double (e.g., *Arctictis*)
10. paroccipital process either long or short, cupped around posterior edge of bulla
11. median lacerate foramen absent
12. bulla with dominant entotympanic chamber and diagonal septum between ectotympanic and entotympanic
13. alisphenoid canal present

14. **2nd lower incisor slightly out of line of (raised above) remainder of incisors**
15. carnassials moderately well developed; PM4 with reduced protocone and usually without parastyle, but m1 with large talonid
16. **1st upper molar relatively large**, transversely elongate or squarish, without medial constriction
17. baculum present, small; possibly absent in *Paguma*, *Paradoxurus*

DENTAL FORMULA: $\frac{3}{3}\ \frac{1}{1}\ \frac{3\text{–}4}{3\text{–}4}\ \frac{1\text{–}2}{1\text{–}2} = 32\text{–}40$

TAXONOMIC DIVERSITY: About 34 species divided among 14 genera in four subfamilies (for a slightly altered, molecular-based arrangement that places *Macrogalidia* with hemigaline genera, see Veron et al. 2017):

Genettinae—*Genetta* (genets) and *Poiana* (African linsangs)

Hemigalinae—*Chrotogale* (Owston's palm civet), *Cynogale* (otter civet), *Diplogale* (Hose's palm civet), and *Hemigalus* (banded palm civet)

Paradoxurinae—*Arctictis* (binturong), *Arctogalidia* (small-toothed palm civet), *Macrogalidia* (Sulawesi palm civet), *Paguma* (masked palm civet), and *Paradoxurus* (common palm civets)

Viverrinae—*Civettictis* (African civet), *Viverra* (large Indian, large spotted, Malabar, and Malay civets), and *Viverricula* (small Indian civet)

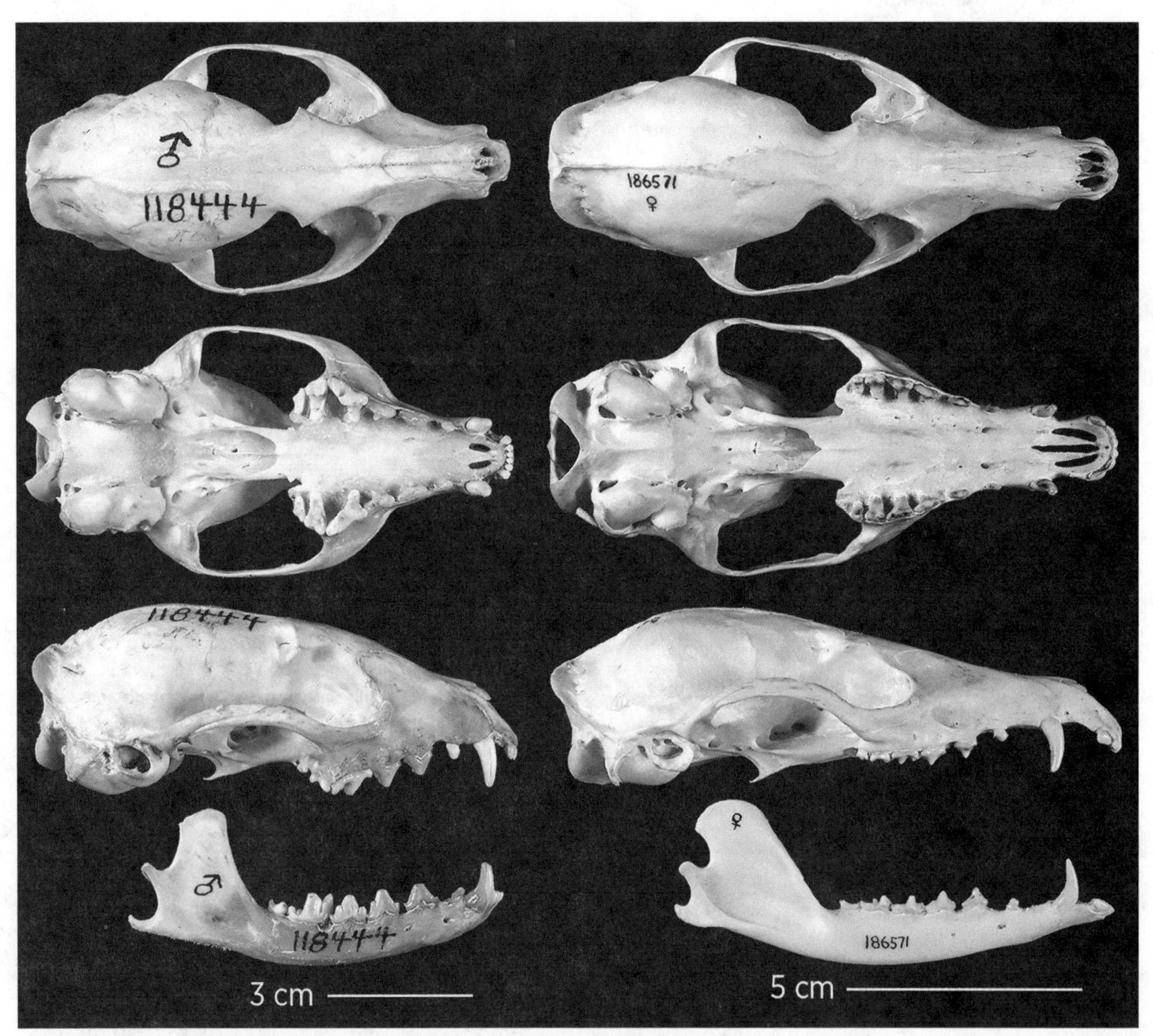

FIG. 443. Viverridae; skull of the common genet, *Genetta genetta* (Viverrinae; *left*), and Owston's palm civet, *Chrotogale owstoni* (Hemigalinae; *right*).

Suborder CANIFORMIA

Caniform carnivores are united by the bullar and carotid features outlined above (see fig. 422 and accompanying description). Additionally, caniform carnivores tend to have longer jaws, with more teeth, than do feliforms. Their carnassial teeth are generally less specialized, and caniforms tend more toward omnivory and opportunistic feeding. Most caniforms are terrestrial and tend toward plantigrade foot structure (the Canidae are a clear exception), with nonretractile claws.

FIG. 444. Canidae; gray fox, *Urocyon cinereoargenteus* (*top*), and timber wolf, *Canis lupus* (*bottom*).

Superfamily CANOIDEA

The superfamily Canoidea includes only one family and is diagnosed largely by bullar characteristics (see fig. 422).

Family CANIDAE (coyotes, dogs, foxes, wolves)

Medium-sized (total length 500 mm–1.9 m; mass 1.5 kg [*Vulpes*] to 80 kg [*Canis*]; **tail relatively long** (110–550 mm) **and distinctly bushy; muzzle typically elongate**; ears erect and pointed, quite large in some (*Otocyon, Vulpes*); legs proportionally long in most (extremely so in *Chrysocyon*; short in *Speothos*); foot posture digitigrade; digits 5/4 (pollex [dewclaw] elevated on side of foot and not in contact with ground) or 4/4 (*Lycaon*); claws nonretractile; scent gland ("violet gland") may be present on dorsal part of tail near base. Food habits typically versatile, ranging from carnivorous (including both vertebrate and invertebrate prey) to opportunistically omnivorous, including substantial vegetable matter in the diet; scavenging is widespread among canids. Most are terrestrial, but a few (*Urocyon*) climb readily; can be either diurnal or nocturnal. Social organization ranges from solitary to quasi-permanent male-female pairs to large and complex social groups. Habitat includes virtually every terrestrial biome, from the Arctic to desert to tropical rainforest.

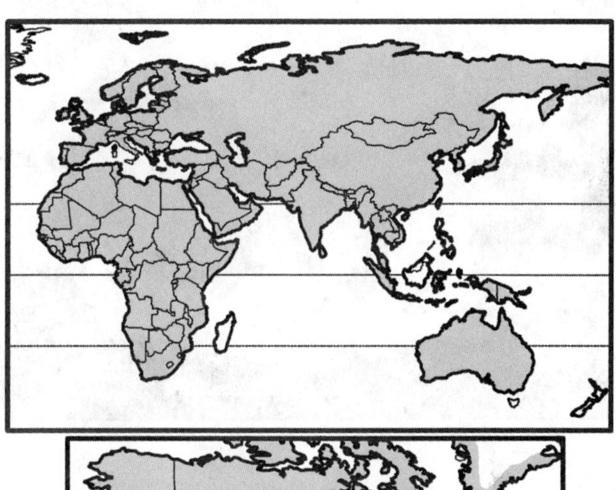

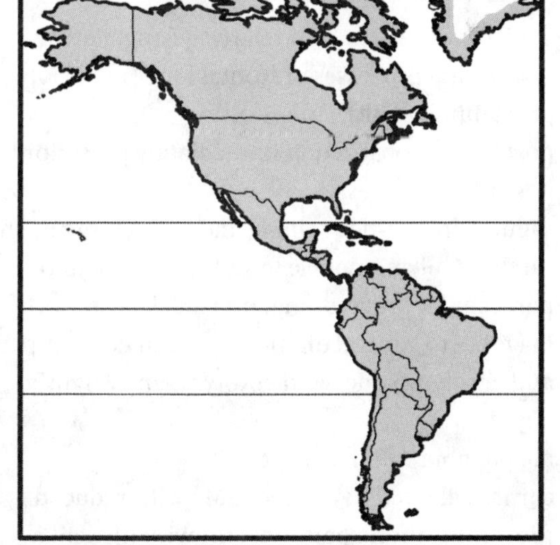

FIG. 445. Geographic range of family Canidae: cosmopolitan, except Madagascar, New Zealand, and Philippines (domestic dog, *Canis lupus familiaris*, introduced).

RECOGNITION CHARACTERS:

1. foot posture digitigrade
2. **digits 5/4 or 4/4** (pollex elevated on side of foot, not in contact with ground; pollex absent in *Lycaon*)
3. claws well developed, relatively straight, nonretractile
4. perineal glands absent
5. anal sacs present, which empty via excretory duct into rectum; anal pouches absent

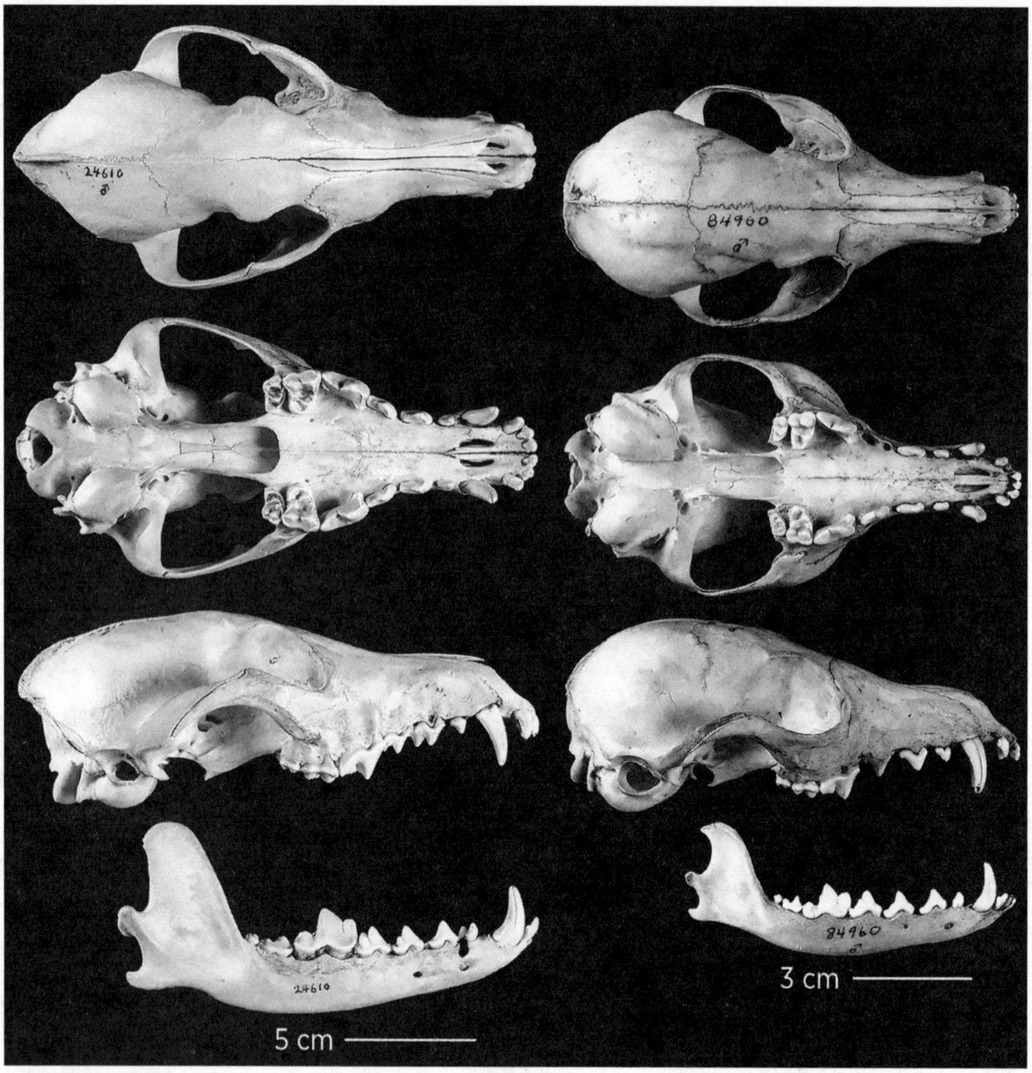

FIG. 446. Canidae; skull and mandible of the coyote, *Canis latrans* (*left*), and kit fox, *Vulpes macrotis* (*right*).

6. skull elongate; rostrum relatively long, narrow
7. postorbital processes of frontal long (>125% of postorbital width)
8. postorbital constriction immediately posterior to postorbital processes
9. sagittal crest usually single and either large or small; rarely small and double (e.g., *Vulpes*, *Urocyon*)
10. paroccipital process long, protrudes posteriorly
11. diagonal to vertical division between ectotympanic and entotympanic, with entotympanic slightly larger
12. alisphenoid canal present
13. carnassials well developed; PM4 with reduced protocone but no parastyle; m1 with bladelike trigonid but well-developed talonid
14. **last upper molar relatively large, transversely elongate**
15. baculum present, long

DENTAL FORMULA: $\frac{3}{3}\ \frac{1}{1}\ \frac{4}{4}\ \frac{1\text{–}4}{2\text{–}5} = 38\text{–}50$

Canid dentition varies little across taxa, varying only in number of molars.

TAXONOMIC DIVERSITY: About 35 species in 13 extant genera: *Alopex* (Arctic fox), *Atelocynus* (short-eared dog of South American tropical forests), *Canis* (dog, dingo, coyote, jackal, wolf), *Cerdocyon* (crab-eating fox of South America), *Chrysocyon* (maned wolf of South America), *Cuon* (dhole), †*Dusicyon* (Falklands Islands wolf), *Lycalopex* (= *Pseudalopex*; South American foxes), *Lycaon* (African wild dog), *Nyctereutes* (raccoon dog), *Otocyon* (African bat-eared fox), *Speothos* (bush dog of South America), *Urocyon* (gray fox), and *Vulpes* (foxes).

Superfamily ARCTOIDEA

The Arctoidea differ from the Canoidea in bullar structure (see fig. 422). This superfamily includes the family Ursidae, the musteloid carnivores in the clade Musteloidea, and the pinniped carnivores in the clade Pinnipedia.

FIG. 447. Ursidae: Ursinae; brown, or grizzly, bear, *Ursus arctos*.

Family URSIDAE (bears)

Medium-sized (*Helarctos*, mass 25 kg) **to very large** (*Ursus*, 760 kg); total length 1.2–3.0 m; **tail short** (7–13 cm); ears small; lips free from gums, mobile and protrusible; color typically uniform brownish, blackish, or white (polar bear, *U. maritimus*), but some species have white or orange on face (*Tremarctos*), neck, or chest. Omnivorous, but polar bear exclusively carnivorous; others feed on wide range of insects, plant matter, and various vertebrates; diurnal or nocturnal; terrestrial, but most climb readily; plantigrade foot posture, locomotion quadrupedal. Usually solitary; smell is the dominant sense, with eyesight and hearing generally poorly developed. Occur in terrestrial biomes from the open Arctic (including ice floes) to temperate and tropical forests.

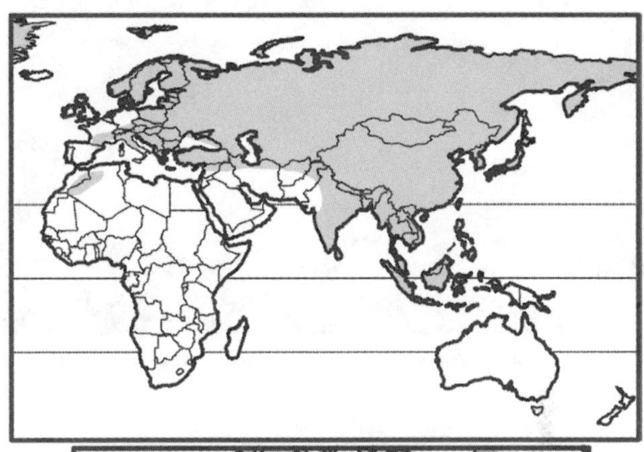

FIG. 448. Geographic range of family Ursidae: Old World—Eurasia, northwestern Africa; New World—North America, northern and central Andes of South America.

RECOGNITION CHARACTERS:

1. foot posture plantigrade
2. digits 5/5; hallux and pollex in contact with ground
3. **claws large, curved**, nonretractile
4. perineal glands absent
5. anal glands and anal sacs present in at least some species (*Ailuropoda*, *Ursus*); anal pouches absent
6. skull elongate; rostrum relatively long, narrow
7. postorbital processes of frontal long (>125% of postorbital width)
8. postorbital constriction considerably posterior to postorbital processes
9. sagittal crest large and single in males, smaller and tending toward double in females
10. paroccipital process large, separate, protrudes posteriorly
11. enlarged ectotympanic; very small, bipartite entotympanic; no true septum dividing components
12. alisphenoid canal present
13. **carnassial pair bunodont, specialized for crushing rather than slicing**
14. **last upper molar very large, elongate anteroposteriorly**
15. baculum present, well developed

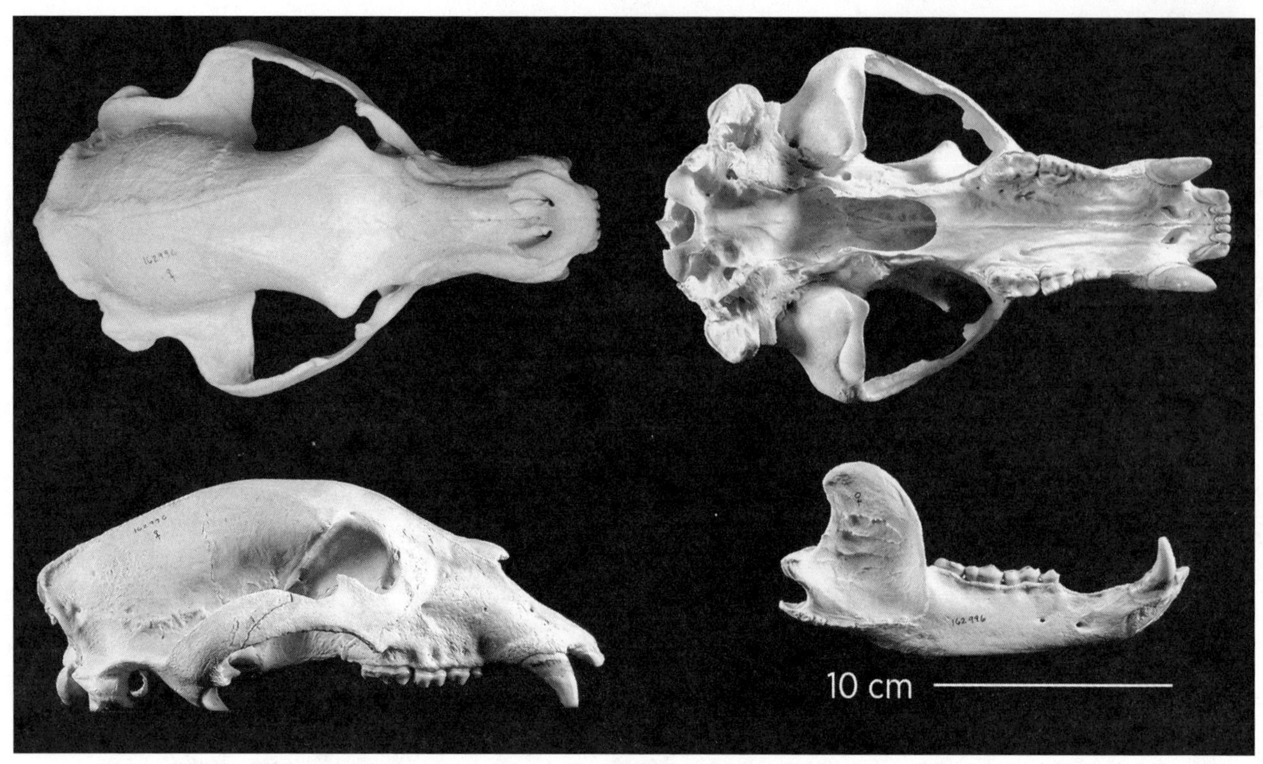

FIG. 449. Ursidae: Ursinae; skull and mandible of the North American black bear, *Ursus americanus*.

DENTAL FORMULA: $\frac{2\text{–}3}{3}\ \frac{1}{1}\ \frac{4}{4}\ \frac{2}{3} = 40\text{–}42$

Melursus has lost I1, leaving a gap through which it sucks termites; ursids in general exhibit high individual variation in the number of premolars, which in some taxa are very small and may either be lost or never develop.

TAXONOMIC DIVERSITY: Eight species in five genera separated into three subfamilies (Garshelis 2009):
Ailuropodinae—*Ailuropoda* (giant panda)
Tremarctinae—*Tremarctos* (spectacled bear)
Ursinae—*Helarctos* (sun bear), *Melursus* (sloth bear), *Ursus* (brown, black, and polar bears; the Asian black bear was previously separated in *Selenarctos* and the polar bear in *Thalarctos*)

Clade MUSTELOIDEA

The clade Musteloidea includes four families of small, terrestrial arctoid carnivores.

Family AILURIDAE (red, or lesser, panda)

Size similar to that of a raccoon (body length 550–625 mm; mass 3–6 kg); head round, with short muzzle; ears prominent, pointed; tail somewhat shorter than body (30–50 cm), bushy, non-prehensile, with alternating red and buff-colored rings; feet have hairy soles; five digits on each forefoot and hind foot; claws are semi-retractile; small white patches above eyes; muzzle, lips, and edges of ears also white; dorsum chestnut brown; venter and limbs blackish. Specialized bamboo feeders with associated very low metabolic rate; excellent climbers, scansorial, aided by very flexible wrist and ankle joints and pelvic and pectoral girdles; feet plantigrade, front legs angle inward, leading to waddling gait. Generally solitary, except during breeding season. Occur in montane deciduous or coniferous forests with bamboo-thicket understory.

The unique morphology of *Ailurus* involving traits associated with its peculiar specialization for herbivory has led to a confused taxonomic history. Some classifications placed *Ailurus* within the Ursidae, while others suggested a basal position within the Procyonidae. The most recent syntheses place *Ailurus* in its own family within the Musteloidea, as the sister to Mustelidae + Procyonidae (see Gatesy et al. 2017 and the cladogram in fig. 421).

RECOGNITION CHARACTERS:

1. **distinctive reddish-brown color pattern with white on face and ears**
2. **tail with alternating red and buff-colored bands**
3. foot posture plantigrade
4. digits 5/5; both hallux and pollex in contact with ground
5. claws semi-retractile
6. perineal glands absent
7. anal glands and anal sacs present; anal pouches absent
8. **skull robust, rostrum short**
9. postorbital processes of frontal small (<125% of postorbital width)
10. postorbital constriction immediately posterior to postorbital processes
11. sagittal crest large, single
12. postmandibular process large and anteriorly recurved
13. paroccipital process long, free from bulla
14. bullae small; horizontal and vertical divisions between equal-sized entotympanic and ectotympanic parts
15. alisphenoid canal present
16. **coronoid process long and robust, hooked posteriorly**
17. no bladelike carnassial pair—both PM3 and PM4 with well-developed paracone and hypocone
18. 1st and 2nd upper molars with especially well-developed hypocone
19. baculum present, well developed

DENTAL FORMULA: $\frac{3}{3}\ \frac{1}{1}\ \frac{3}{4}\ \frac{2}{2} = 38$

TAXONOMIC DIVERSITY: The family is monotypic, with a single species, *Ailurus fulgens*.

FIG. 450. Ailuridae; red panda, *Ailurus fulgens*.

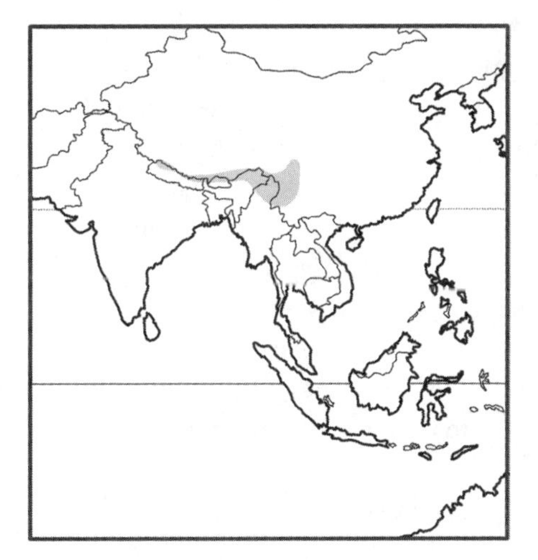

FIG. 451. Geographic range of family Ailuridae: Nepal, Bhutan, northeastern India, northern Myanmar, and southern China.

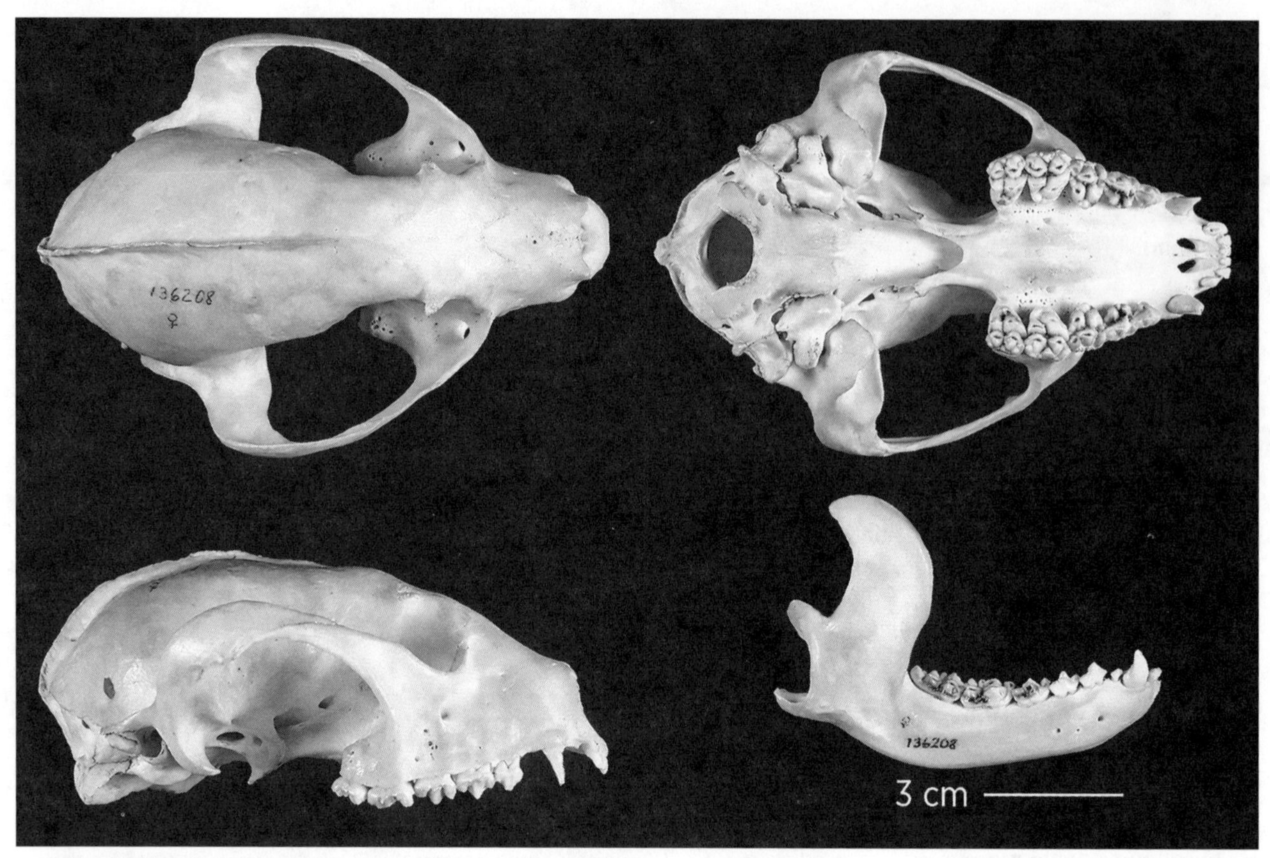

FIG. 452. Ailuridae; skull and mandible of the red panda, *Ailurus fulgens*.

Family MEPHITIDAE (skunks and stink badgers)

Small (body length 200–400 mm; mass 200 g–1 kg [spotted skunks, *Spilogale*]) to medium-sized (body length 300–500 mm; mass up to 4.5 kg [hog-nosed skunks, *Conepatus*]); body broad and squat; limbs short; tail short in stink badgers (*Mydaus*) but long (about equal to body length) and densely haired in all skunks; feet plantigrade, digits 5/5, claws robust, well suited for digging; body color strongly aposematic, black with broad to narrow white stripes or spots, with white extending onto tail in most or tail with broad white tip. All members have very well-developed anal glands that produce noxious odors for threat deterrence. Opportunistic omnivores, dominant portion of diet is animal matter. Nocturnal; terrestrial; non-territorial, generally forage solitarily but may den collectively. Employ induced ovulation and delayed implantation. Habitat ranges from open forests to grasslands, meadow systems, and rocky montane areas; elevation generally sea level to about 1,800 m, but range to over 4,000 m; some species highly commensal.

The Mephitidae were long considered a subfamily of the Mustelidae, but have now been elevated to family rank and placed as the basal clade within Musteloidea (see Gatesy et al. 2017 and fig. 421).

RECOGNITION CHARACTERS:

1. **usually distinctively marked with black and white aposematic patterns**
2. foot posture plantigrade
3. digits 5/5; both pollex and hallux in contact with ground
4. claws well developed, non-retractile
5. ears with skin of outside rim folded to form a small purse (bursa) in some species
6. perineal glands absent
7. anal glands in enlarged anal sacs; anal pouches absent
8. **skull somewhat delicate, rounded; rostrum usually moderately long**
9. postorbital processes of frontal small (<125% of postorbital width), sometimes nearly absent (*Conepatus*)

10. postorbital constriction considerably posterior to postorbital processes
11. sagittal crest small, single
12. palate usually extends to about posterior edge of tooth row (cf. Mustelidae, in which the palate extends further posteriorly)
13. auditory bulla smaller than in mustelids, and not oriented anteroposteriorly
14. mastoid process reduced, smaller than paroccipital process
15. paroccipital process very short, not cupped around posterior edge of bulla
16. enlarged ectotympanic; very small, bipartite entotympanic; no true septum dividing components
17. no alisphenoid canal
18. bladelike carnassials usually well developed
19. **last upper molar notably square, not strongly dumbbell-shaped**
20. baculum present, small

FIG. 453. Mephitidae; western spotted skunk, *Spilogale gracilis* (*top*), and striped skunk, *Mephitis mephitis* (*bottom*).

DENTAL FORMULA:

Conepatus: $\frac{3}{3}\ \frac{1}{1}\ \frac{2}{3}\ \frac{1}{2} = 32$

Mephitis, Mydaus, Spilogale: $\frac{3}{3}\ \frac{1}{1}\ \frac{3}{3}\ \frac{1}{2} = 34$

TAXONOMIC DIVERSITY: About 12 species in four genera allocated to two subfamilies:

Mephitinae—*Conepatus* (hog-nosed skunks), *Mephitis* (striped and hooded skunks), and *Spilogale* (spotted skunks)

Myadinae—*Mydaus* (stink badgers)

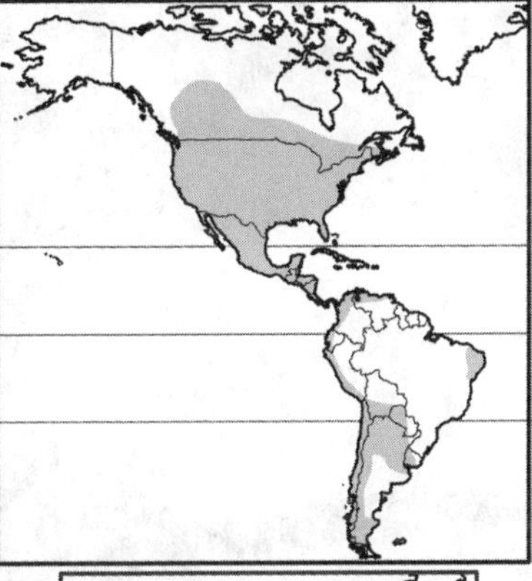

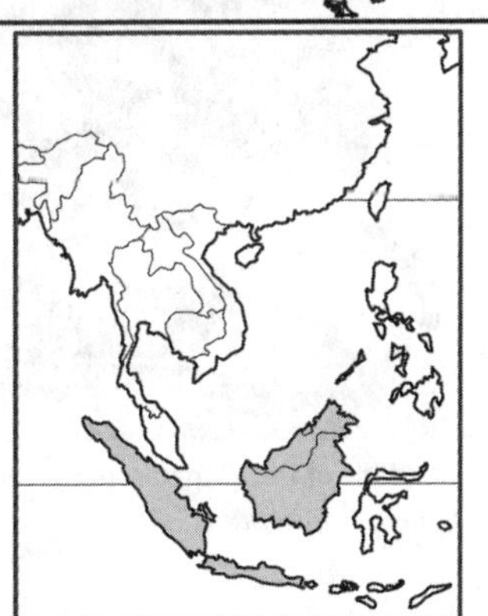

FIG. 454. Geographic range of family Mephitidae: North and South America (*Conepatus, Mephitis, Spilogale*); Indonesian islands of Sumatra, Java, and Borneo and Palawan islands of Philippines (*Mydaus*).

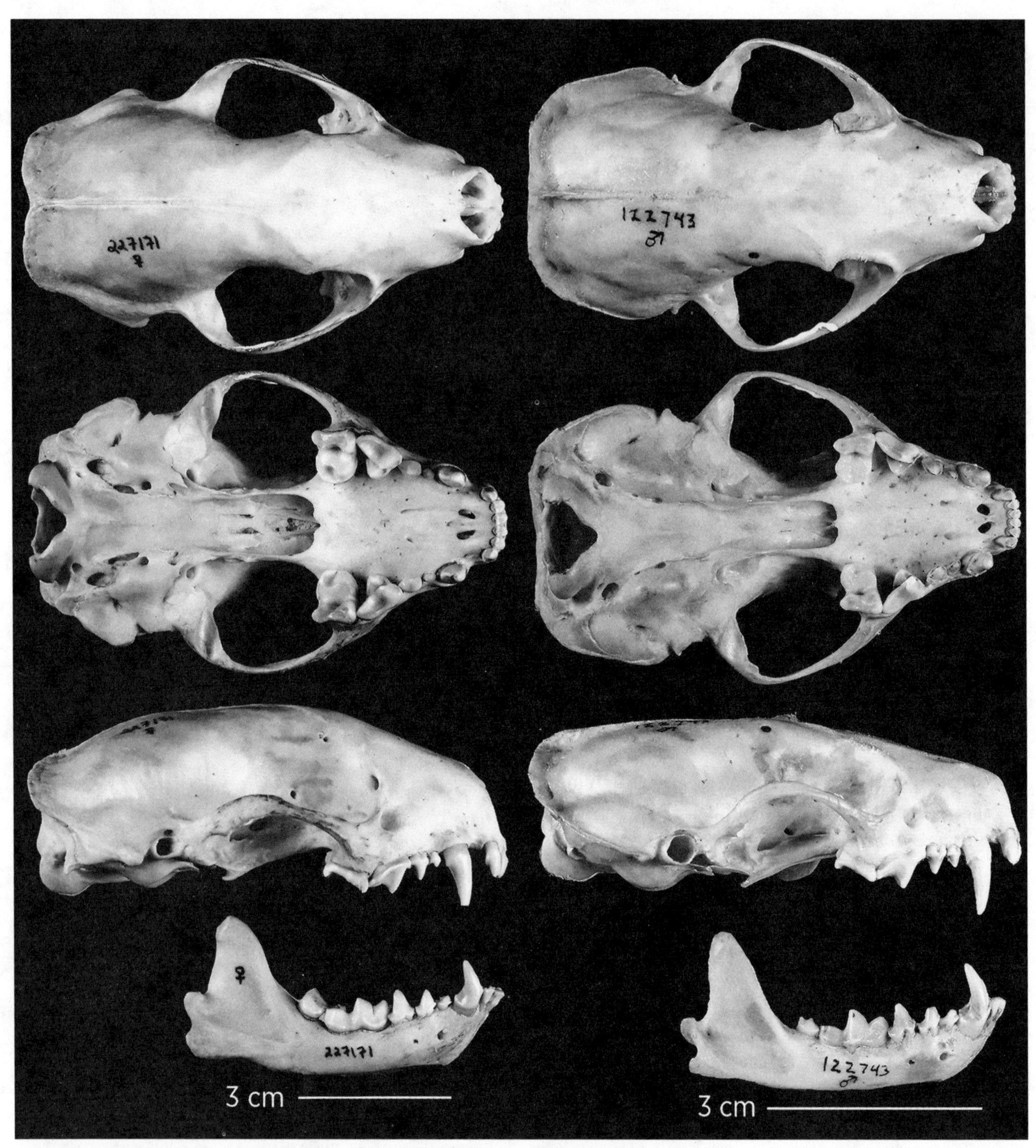

FIG. 455. Mephitidae; skull and mandible of the striped skunk, *Mephitis mephitis* (*left*), and the Eastern spotted skunk, *Spilogale putorius* (*right*).

Family MUSTELIDAE (weasels, stoats, ferrets, otters, badgers, martens, fisher, wolverine)

Size from smallest carnivorans (*Mustela*, 35–70 g) to medium-sized (*Enhydra*, 37 kg); total length 15–220 cm; males larger than females; tail variable in length, but usually long (15–550 mm); ears small and rounded, **with bursae in some species**; legs typically short; body stocky (badgers) or elongate and gracile (weasels, otters); anal sacs (and associated anal glands) usually well developed; color mixed white, brown, or black, sometimes with contrasting spots or stripes (*Ictonyx*, *Meles*, *Taxidea*); some species of *Mustela* employ seasonal molt, changing coat color from winter (white) to summer (brown). Primarily meat eaters, very active and agile hunters; otters consume fish or freshwater or marine invertebrates; some eat fruit and nuts (*Eira*, *Meles*), insects, or honey (*Mellivora*). Most are solitary and terrestrial, moving with scampering or bounding locomotion; some are agile climbers (*Eira*, *Martes*). Females of some species employ induced ovulation and delayed implantation. Habits range from terrestrial to semiarboreal to aquatic in both freshwater and salt water; occupy most terrestrial biomes, from Arctic to tropical forests.

DIAGNOSTIC CHARACTER:

- **postmandibular process often prominent and curved around mandibular fossa, often locking lower jaw into place**

RECOGNITION CHARACTERS:

1. foot posture plantigrade (badgers), semi-digitigrade (martens), or digitigrade (weasels); partially webbed in otters
2. digits 5/5; both pollex and hallux in contact with ground
3. claws well developed, semi-retractile
4. perineal glands absent
5. anal glands in enlarged anal sacs; anal pouches absent
6. **skull blocky, robust, often flattened; rostrum short**
7. postorbital processes of frontal long (>125% of postorbital width) or short (e.g., *Arctonyx*, *Enhydra*, *Martes*)
8. postorbital constriction considerably posterior to postorbital processes
9. sagittal crest variable; small in most genera but especially large in *Gulo*; may be either single or double (e.g., *Eira*, *Martes*)

FIG. 456. Mustelidae; North American river otter, *Lontra canadensis* (Lutrinae; *upper left*), American mink, *Neovison vison* (Mustelinae; *upper right*), American badger, *Taxidea taxus* (Taxidiinae; *middle right*), and wolverine, *Gulo gulo* (Martinae; *bottom*).

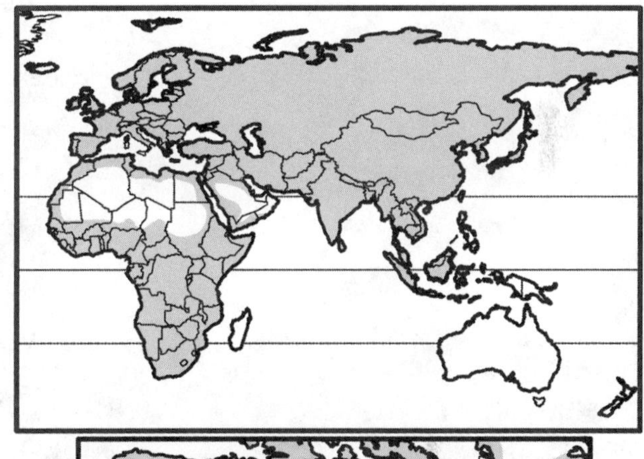

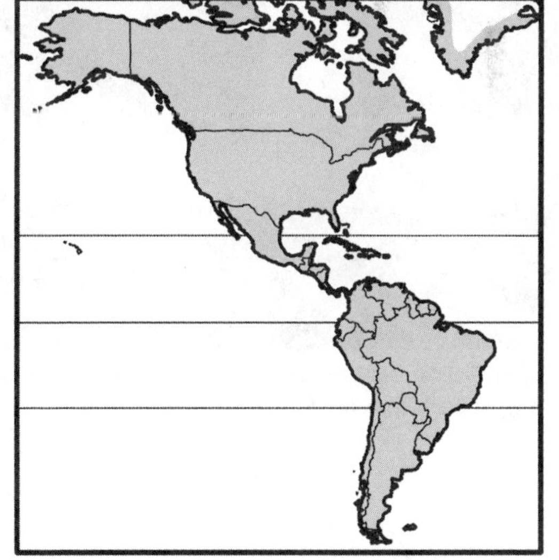

FIG. 457. Geographic range of family Mustelidae: all continents except Australia and Antarctica; absent from islands of New Guinea, New Zealand, and Madagascar.

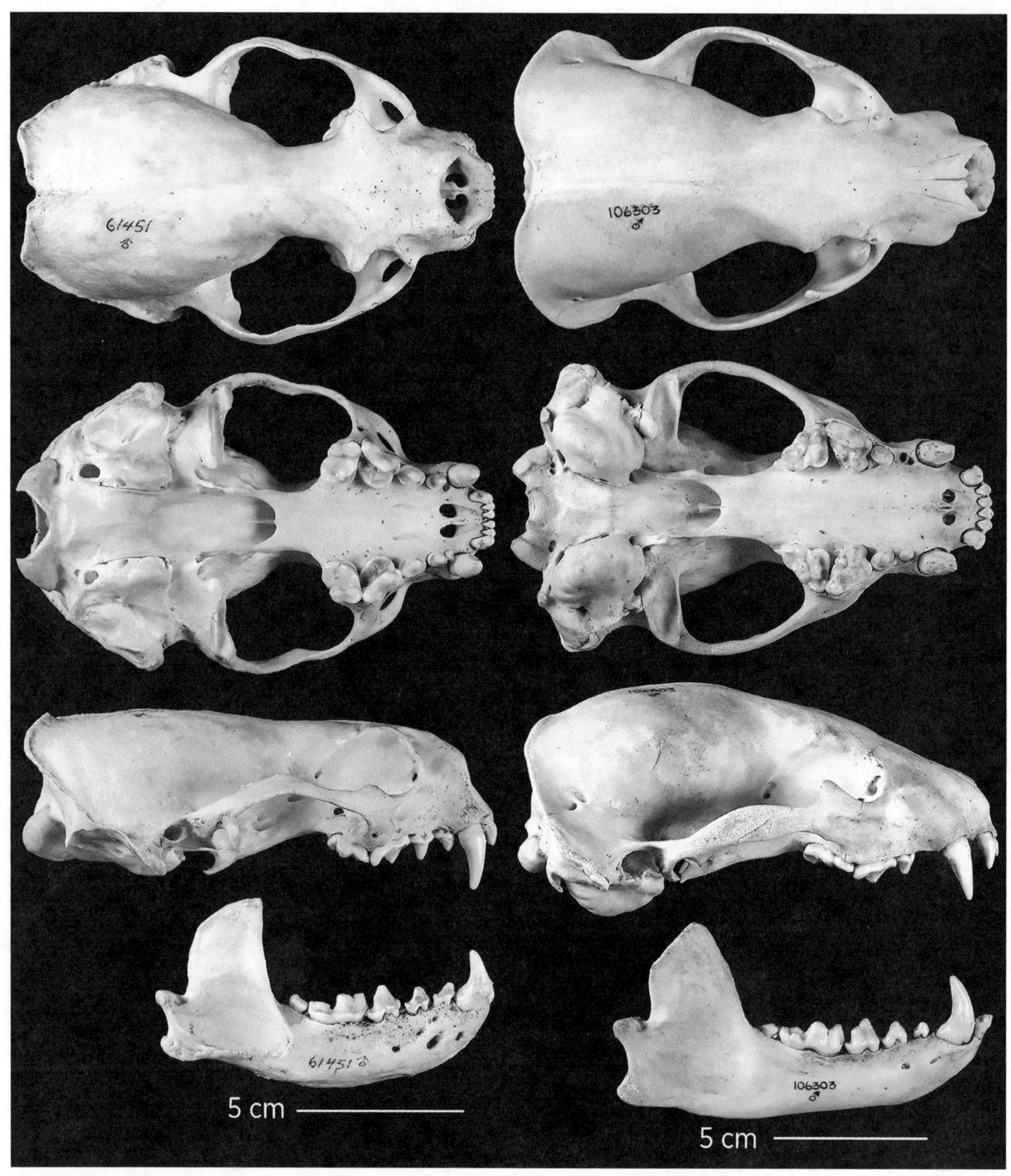

FIG. 458. Mustelidae; skull and mandible of the American river otter, *Lontra canadensis* (Lutrinae; *left*), and the American badger, *Taxidea taxus* (Taxidiinae; *right*).

10. **palate usually extends well past posterior edge of tooth row** (cf. Mephitidae)
11. paroccipital process long (*Arctonyx*, *Eira*, *Melivora*) or short (most genera), generally free from bulla
12. enlarged ectotympanic; very small, bipartite entotympanic; no true septum dividing components
13. no alisphenoid canal
14. bladelike carnassials usually well developed (not in *Enhydra*)
15. **upper molar relatively large, usually dumbbell-shaped or squarish**
16. baculum present, well developed

DENTAL FORMULA: Usually $\frac{3}{3}\ \frac{1}{1}\ \frac{3}{3}\ \frac{1}{2} = 34$; sea otters (*Enhydra*) usually have only two lower incisors, and their dentition may be as variable as $\frac{3}{2\text{–}3}\ \frac{1}{1}\ \frac{2\text{–}4}{2\text{–}4}\ \frac{1}{1\text{–}2} = 26\text{–}38$

TAXONOMIC DIVERSITY: 22 genera and about 57 species traditionally divided into five (Simpson 1945) or two (Wozencraft 2005) subfamilies, but molecular analyses support eight distinct clades, each now given subfamily rank (synopsis from Larivière and Jennings 2009; see also Koepfli et al. 2008; Sato et al. 2012):

Galictidinae—*Galictis* (grisons), *Ictonyx* (striped polecat, zorilla), *Poecilogale* (African striped weasel), and *Vormela* (marbled polecat)

Helictidinae—*Melogale* (ferret badgers)

Lutrinae—*Aonyx* (African clawless and Asian small-clawed otters), *Enhydra* (sea otter), *Hydrictis* (speckled-necked otter), *Lontra* (American river otters), *Lutra* (Old World otters), *Lutrogale* (smooth-coated otter), and *Pteronura* (giant otter)

Martinae—*Eira* (tayra), *Gulo* (wolverine), *Martes* (martens), and *Pekania* (fisher)

Melinae—*Arctonyx* (hog badger) and *Meles* (Eurasian badgers)

Mellivorinae—*Mellivora* (honey badger)

Mustelinae—*Lyncodon* (Patagonian weasel), *Mustela* (weasels), and *Neovison* (American mink)

Taxidiinae—*Taxidea* (American badger)

Family PROCYONIDAE (coati, kinkajou, olingo, raccoon, ringtail)

Medium-sized (total length 60–135 cm; mass 0.8–22 kg); muzzle long and narrow or short and broad, snout somewhat flexible, especially in *Nasua*; ears small to medium in length; tail long, usually 20–70 cm, usually ringed, prehensile in *Potos*; color typically contrasting combinations of white with gray, brown, or reddish brown. All are omnivorous, feeding on roots, shoots, nuts, fruit, crustaceans, and small vertebrates. All are excellent climbers of vertical rocky outcrops or trees. Most are solitary, some travel in family groups or larger bands (*Nasua*). Occupy most biomes in New World temperate and tropical regions, especially near water.

RECOGNITION CHARACTERS:

1. **tail long, usually ringed with alternating black and light-colored bands**
2. **foot posture plantigrade or semi-plantigrade**
3. digits 5/5; both pollex and hallux in contact with ground
4. claws prominent, most nonretractile (retractile in *Bassariscus*)
5. perineal glands absent
6. anal sacs present; anal pouches absent

FIG. 459. Procyonidae; ringtail, *Bassariscus astutus* (*top*), kinkajou, *Potos flavus* (*middle*), and raccoon, *Procyon lotor* (*bottom*).

FIG. 460.
Geographic range of family Procyonidae: restricted to the temperate and tropical regions of North and South America.

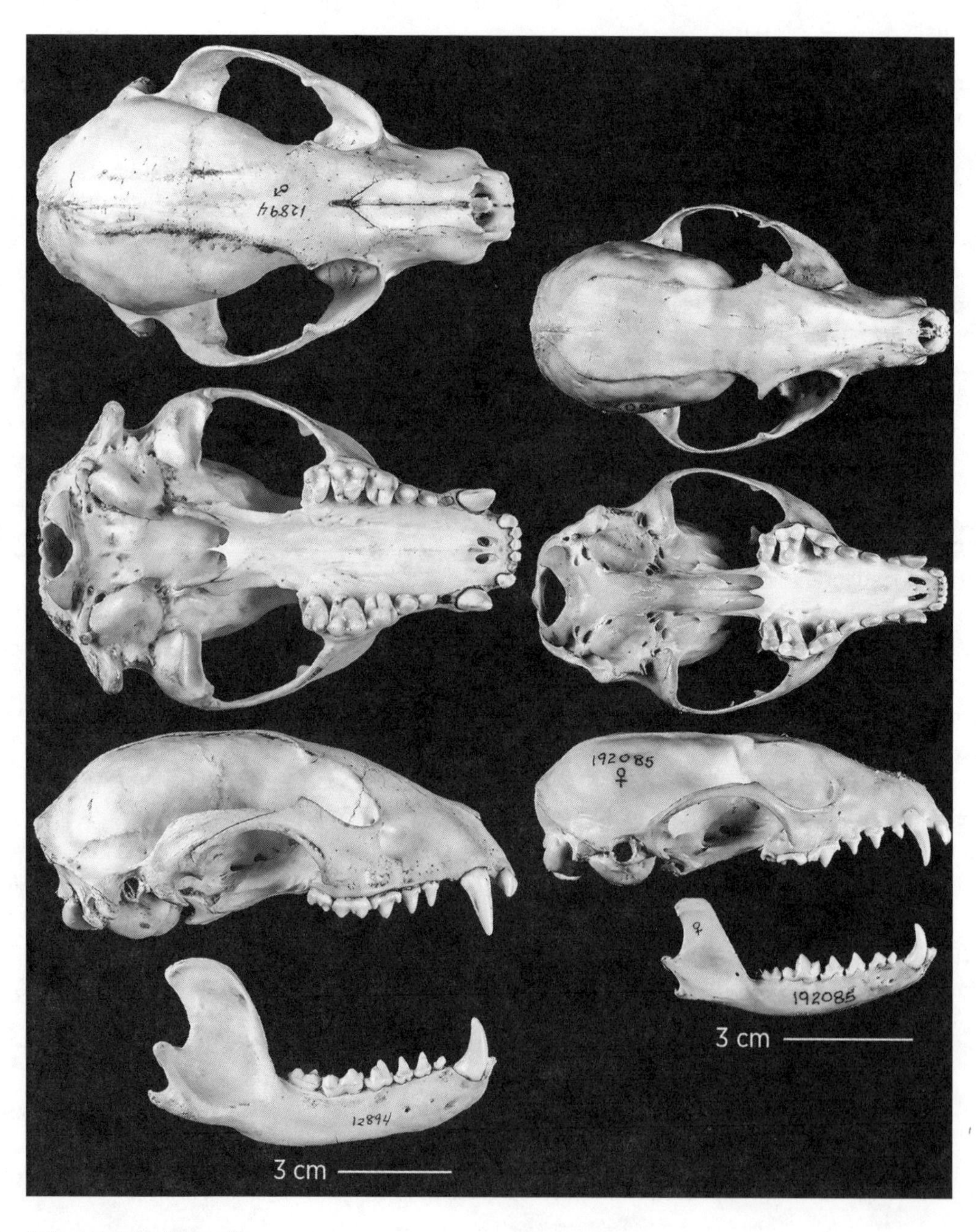

FIG. 461. Procyonidae; skull and mandible of the raccoon, *Procyon lotor* (*left*), and the ringtail, *Bassariscus astutus* (*right*).

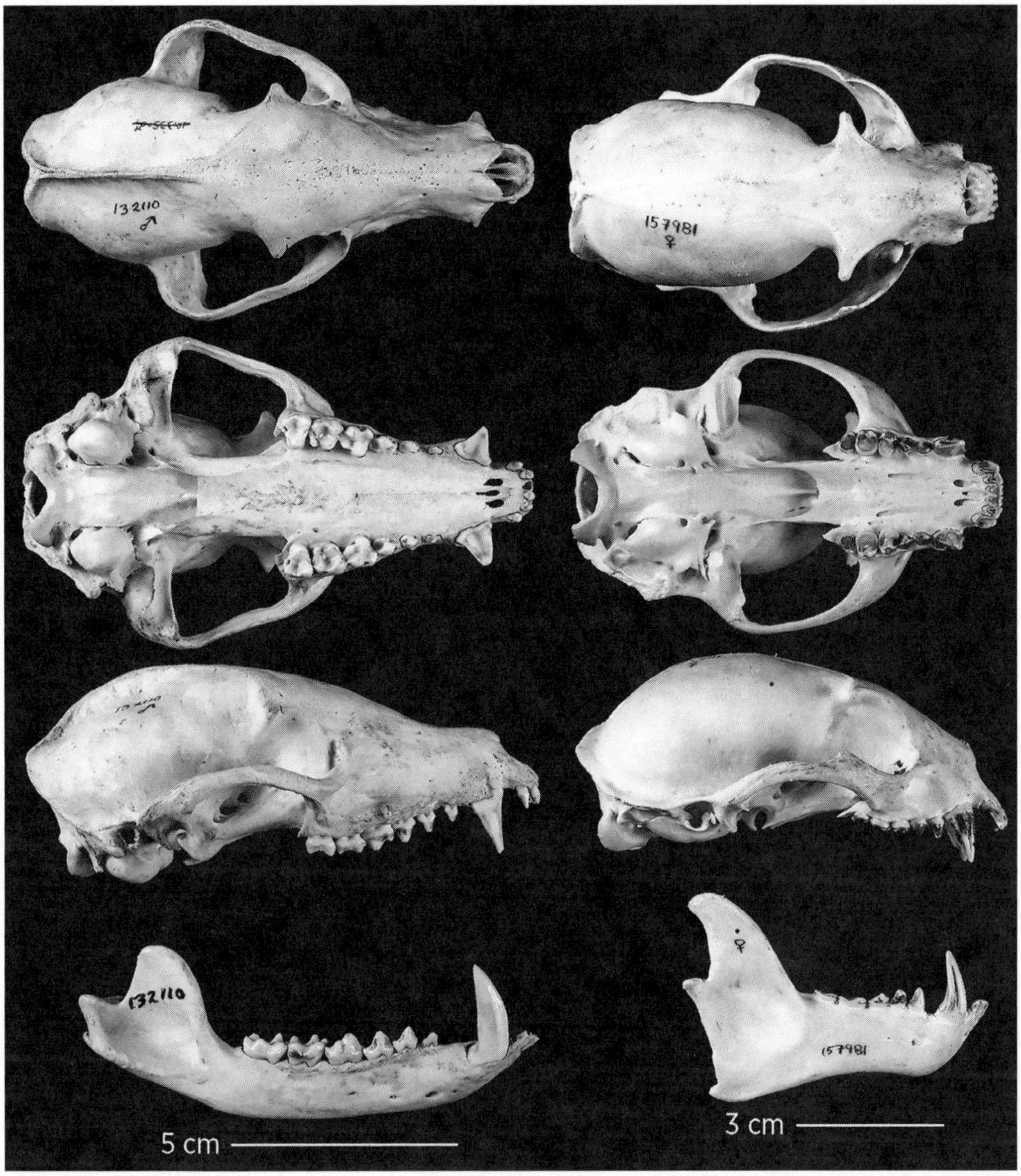

FIG. 462. Procyonidae; skull and mandible of the white-nosed coati, *Nasua narica* (*left*), and the kinkajou, *Potos flavus* (*right*).

7. **skull robust, usually elongate, rostrum relatively short** (except in *Nasua*)
8. postorbital processes of frontal long (>125% of postorbital width), short in *Procyon*
9. postorbital constriction immediately posterior to postorbital processes
10. sagittal crest large and single (*Nasua*), small and single (*Procyon*), or small and double (*Bassaricyon*, *Bassariscus*, *Potos*)
11. paroccipital process very short, not covering posterior surface of bulla, or very short, cupped around posterior surface (*Potos*)
12. enlarged ectotympanic; very small, bipartite entotympanic; no true septum dividing components
13. no alisphenoid canal
14. cheek teeth bunodont
15. **carnassials poorly developed** (except *Bassariscus*), upper molars usually with protocone and hypocone
16. **last upper molar relatively large**
17. baculum present, well developed

DENTAL FORMULA: Generally $\frac{3}{3}\ \frac{1}{1}\ \frac{4}{4}\ \frac{2}{2} = 40$, but premolars 3/3 in *Potos*

TAXONOMIC DIVERSITY: 12 species in six genera: *Bassariscus* (ringtail, cacomistle), *Bassaricyon* (olingos, olinguito), *Nasua* (coatis, coatimundi), *Nasuella* (mountain coati), *Potos* (kinkajou), and *Procyon* (raccoons).

Clade PINNIPEDIA

The final clade of carnivorans, Pinnipedia, was historically treated as a suborder (e.g., Simpson 1945) or as an order (e.g., Stains 1967, Hall 1981). It includes three families that are highly adapted to a marine existence.

Family ODOBENIDAE (walrus)

Very large (total length up to 4 m); strong male-biased sexual dimorphism (mass of females 860 kg, males 1,270 kg); body thick and swollen in appearance; no external pinnae; no free tail; hind limbs can be placed under body; swim by means of alternate strokes of rear flippers, with forelimbs used as rudders or as paddles during low-speed swimming; both forelimbs and hind limbs used awkwardly on land or ice; adult coloration grayish brown, virtually hairless. Feed chiefly on mollusks and other bottom-living invertebrates, which they extract from shells by suction. More or less gregarious, assembling in herds of up to 2,000 individuals; seldom seen in the open ocean, but stay close to floating ice or shoreline; move seasonally with ice floes. Occur in shallow marine waters and ice floes around the rim of the Arctic Ocean.

FIG. 463. Odobenidae; walrus, *Odobenus rosmarus*.

DIAGNOSTIC CHARACTERS:

- **upper canine enormous, tusklike**
- **no lower incisors**

RECOGNITION CHARACTERS:

1. body uniform in color, never spotted
2. hind limbs capable of being turned forward
3. nails present on all digits of forefoot and hind foot; skin of flipper not extending distally beyond nails
4. testes abdominal (testicond)
5. postorbital processes of frontal absent
6. sagittal crest absent
7. **mastoid process enormous, bound to bulla**
8. occipital condyles flare widely, placed below foramen magnum
9. upper incisors small, conical
10. alisphenoid canal present
11. **cheek teeth simple, peg-like**
12. baculum present, well developed

DENTAL FORMULA: $\frac{1\text{–}2}{0}\ \frac{1}{1}\ \frac{3\text{–}4}{3\text{–}4}\ \frac{0}{0} = 18\text{–}24$

TAXONOMIC DIVERSITY: The family is monotypic, with a single living species, *Odobenus rosmarus*.

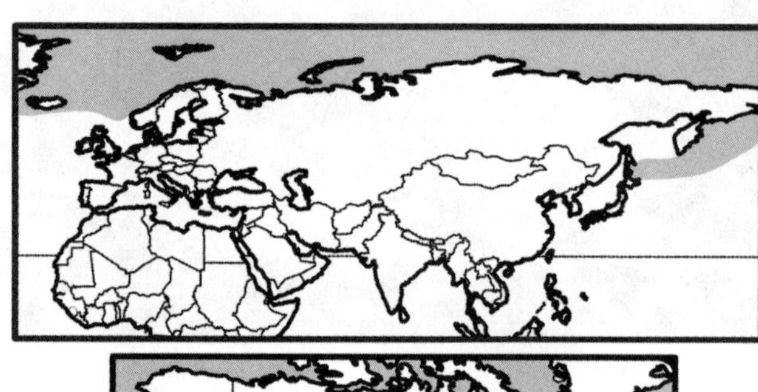

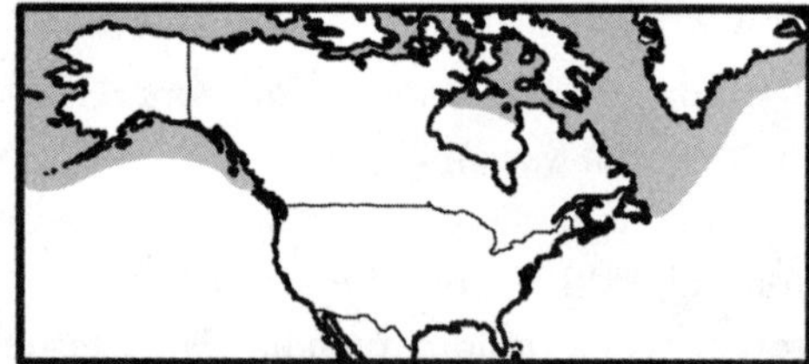

FIG. 464. Geographic range of family Odobenidae: coastal regions of the Arctic Ocean and adjacent seas.

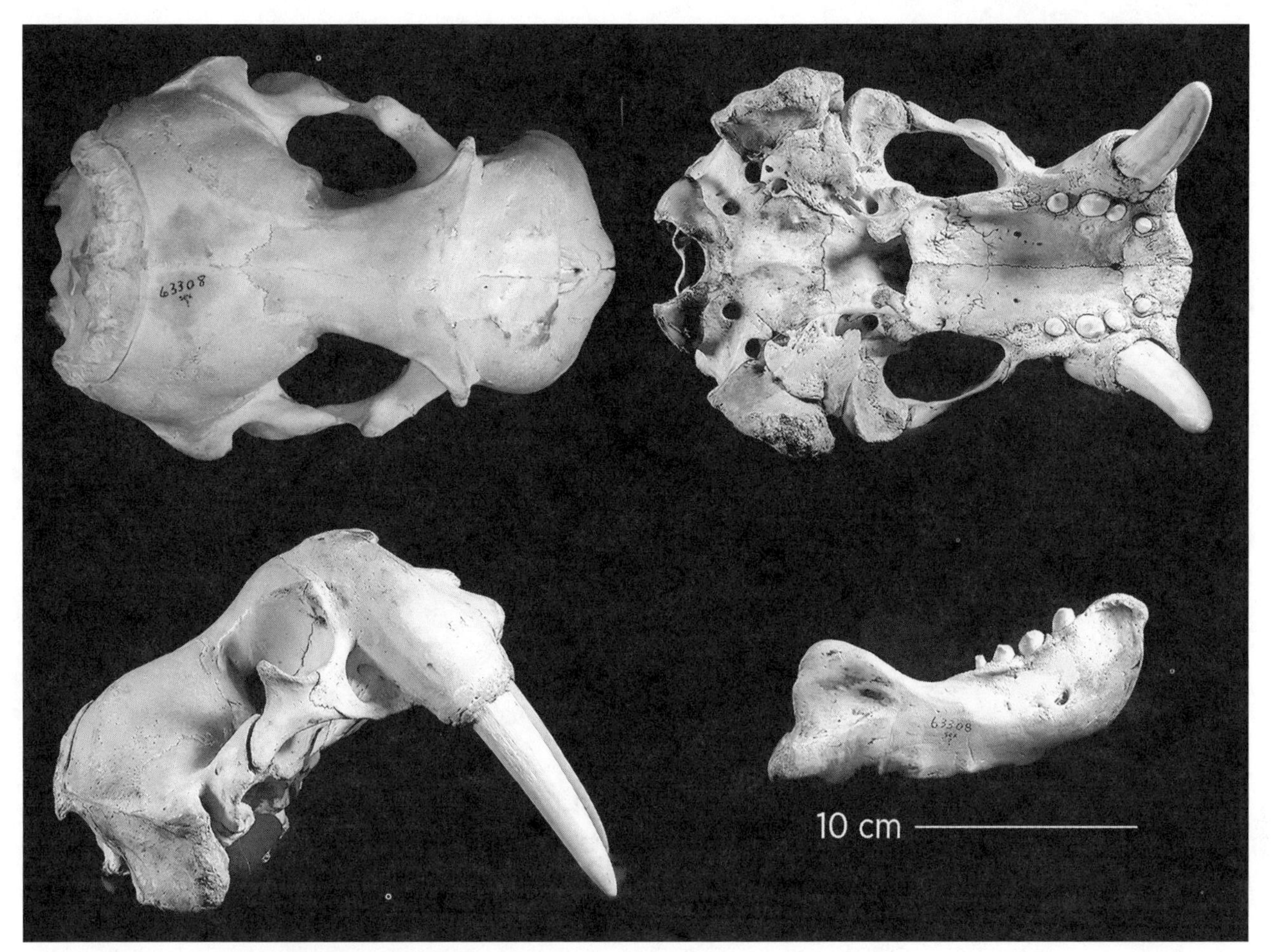

FIG. 465. Odobenidae; skull and mandible of the walrus, *Odobenus rosmarus*.

Family OTARIIDAE (eared seals: fur seals, sea lions, Steller's sea lion)

Large (total length 1.5–3.5 m), with strong male-biased sexual dimorphism (mass of Arctocephalinae is 60–120 kg in females, 270–300 kg in males; that of Otariinae is 90–270 kg in females, 280–1,000 kg in males); body slender and elongate; external ears (pinnae) present (absent in other pinnipeds); tail small but distinct, hind limbs (flippers) used in terrestrial locomotion, can be rotated forward (under) to support body (unlike Phocidae), but gait slow and awkward; coloration generally uniform, pale brown to black (some exhibit countershading, with a lighter chest and throat). Diet consists primarily of small fish and squid. Highly gregarious, form huge colonies during breeding season when males guard harems; some migrate very long distances (up to 9,000 km in *Callorhinus*). Molt annually over a period of weeks to a month or more, but not confined to land or ice during this time. Occur in marine waters, on islands, and in coastal areas in subpolar, temperate, and subtropical regions; absent from Antarctica and the Arctic Ocean, but are likely to expand north as the sea ice disappears due to climate change.

DIAGNOSTIC CHARACTERS:

- **pinnae present** (absent in other pinnipeds)
- **supraorbital process well developed** (absent or rudimentary in other pinnipeds)

RECOGNITION CHARACTERS:

1. body relatively uniform in color, never spotted or banded
2. **hind limbs capable of being turned forward**
3. nails generally absent from forefeet, but prominent on middle three digits of hind feet; skin of flipper extends distally beyond nails
4. testes scrotal
5. postorbital processes of frontal long (>125% of postorbital width)

FIG. 466. Otariidae: Otariinae; Steller sea lion, *Eumetopias jubatus* (*male and female; bottom left*), and California sea lion, *Zalophus californianus* (*male and female; top right*).

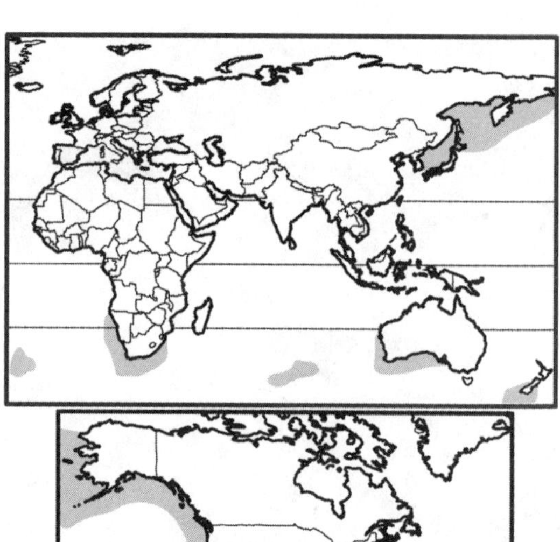

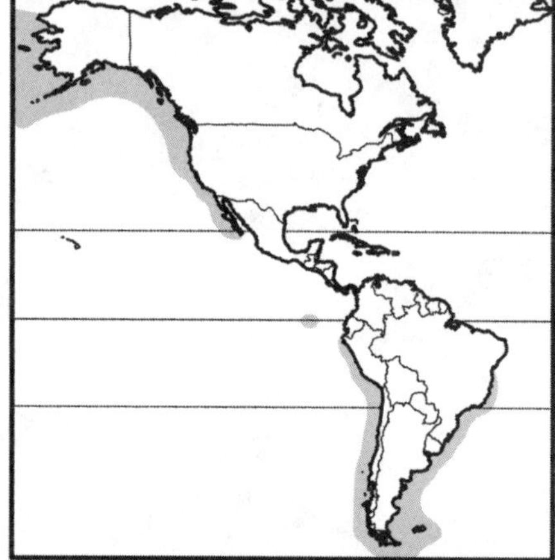

FIG. 467. Geographic range of family Otariidae: coasts of northeastern Asia, western North America, western and southeastern South America, southern Africa, southern Australia, New Zealand, and most predominantly southern oceanic islands.

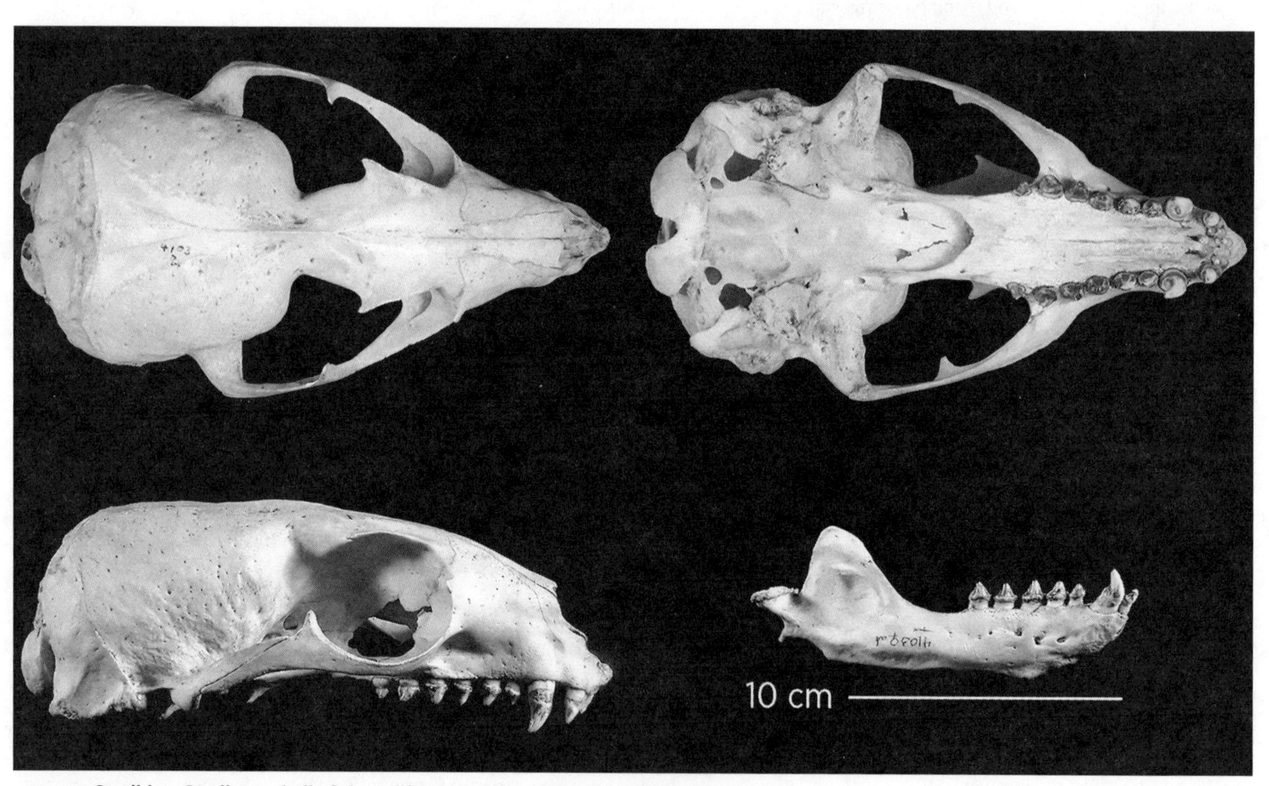

FIG. 468. Otariidae: Otariinae; skull of the California sea lion, *Zalophus californianus*.

6. postorbital constriction considerably posterior to postorbital processes
7. sagittal crest large, single
8. alisphenoid canal present
9. mastoid process separate from auditory bulla
10. medial two upper incisors with transverse groove; 3rd upper incisor caniniform
11. cheek teeth homodont, usually with only one large cusp
12. baculum present, well developed

DENTAL FORMULA: $\frac{3}{2}\ \frac{1}{1}\ \frac{4}{4}\ \frac{1\text{–}3}{1} = 34\text{–}38$

TAXONOMIC DIVERSITY: About 15 species in seven genera, usually divided into two subfamilies:

Arctocephalinae—*Arctocephalus* (southern fur seals) and *Callorhinus* (northern fur seal)

Otariinae—*Eumetopias* (Steller sea lion), *Neophoca* (Australian sea lion), *Otaria* (South American sea lion), *Phocarctos* (New Zealand sea lion), and *Zalophus* (California and Galapagos sea lions)

Family PHOCIDAE (true, earless, or hair seals)

Large to very large, with males much larger (*Cystophora*, *Mirounga*), slightly larger (*Halichoerus*, *Histriophoca*), equal to (*Pagophilus*, *Phoca*, *Pusa*), or smaller than females (*Hydrurga*, *Leptonychotes*, *Lobodon*); mass of adults 80–450 kg (male *Mirounga* up to 3,600 kg); tail stubby; front flipper short, much less than one-fourth length of body; claws well developed; hind limbs used mainly for swimming, cannot be placed under body and thus are not used for terrestrial movement, which is accomplished by muscular contractions along the entire body; color varies from silver through brown to black, with most distinctly spotted and two banded (*Pagophilus*, *Histriophoca*). Diet is mainly fish, squid, octopus, and shellfish, but specialized feeders take macroplankton (*Lobodon*) and penguins (*Hydrurga*). Social organization ranges from solitary (*Hydrurga*) to highly gregarious (*Mirounga*, *Pagophilus*). Shed hair during spring or summer, generally remaining on land or ice while new pelage grows; elephant seals and monk seals also shed external skin such that large patches of skin and hair slough off ("catastrophic molt"). Occur mainly in coastal and pelagic regions of polar, temperate, and subtropical seas; some move up rivers, and some occur only in landlocked freshwater lakes (*Pusa*).

DIAGNOSTIC CHARACTER:

- **no alisphenoid canal** (in contrast to Otariidae and Odobenidae)

RECOGNITION CHARACTERS:

1. **body usually spotted or banded** (uniformly colored in some—Baikal seal, monk seals, elephant seals)
2. **hind limbs extended posteriorly, incapable of forward rotation**
3. claws present on digits of forefoot and usually on hind foot; skin on flipper not extending beyond nails
4. testes abdominal (testicond)
5. postorbital processes of frontal absent or poorly developed
6. sagittal crest absent, small and double (some old *Phoca*), or small and single (*Hydruga*, *Leptonychotes*, *Mirounga*)
7. mastoid process bound to inflated auditory bulla
8. upper incisors not grooved
9. **cheek teeth usually with three longitudinal cusps**
10. baculum present, well developed

DENTAL FORMULA: $\frac{2\text{–}3}{1\text{–}2}\ \frac{1}{1}\ \frac{4}{4}\ \frac{0\text{–}2}{0\text{–}2} = 26\text{–}36$

FIG. 469. Phocidae; ringed seal, *Pusa hispida* (*top*), hooded seal, *Cystophora cristata* (*adult male and female; middle*), and northern elephant seal, *Mirounga angustirostris* (*adult male, adult female, pup; bottom*).

TAXONOMIC DIVERSITY: 18 species in 13 genera (traditionally divided into two or three subfamilies, but current taxonomy subject to ongoing revision): *Cystophora* (hooded seal), *Erignathus* (bearded seal), *Halichoerus* (gray seal), *Histriophoca* (ribbon seal), *Hydruga* (leopard seal), *Leptonychotes* (Weddell seal), *Lobodon* (crab-eater seal), *Mirounga* (elephant seals), *Monachus* (monk seals), *Ommatophoca* (Ross seal), *Pagophilus* (harp seal), *Phoca* (harbor and spotted seals), and *Pusa* (Baikal, Caspian, and ringed seals).

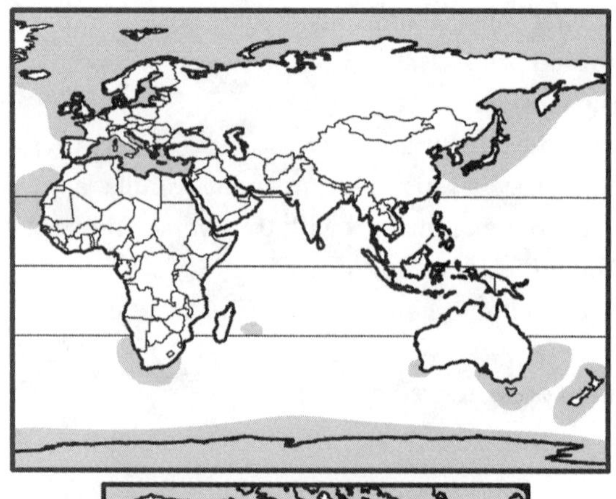

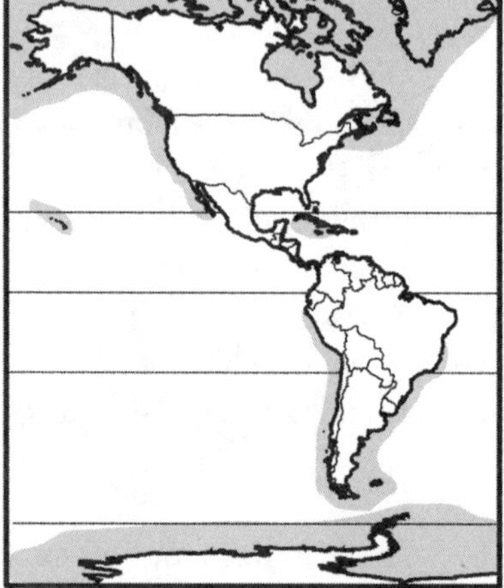

FIG. 470. Geographic range of family Phocidae: ice fronts and coastlines of polar and temperate parts of the oceans and adjoining seas of the world, but extending to some tropical island areas.

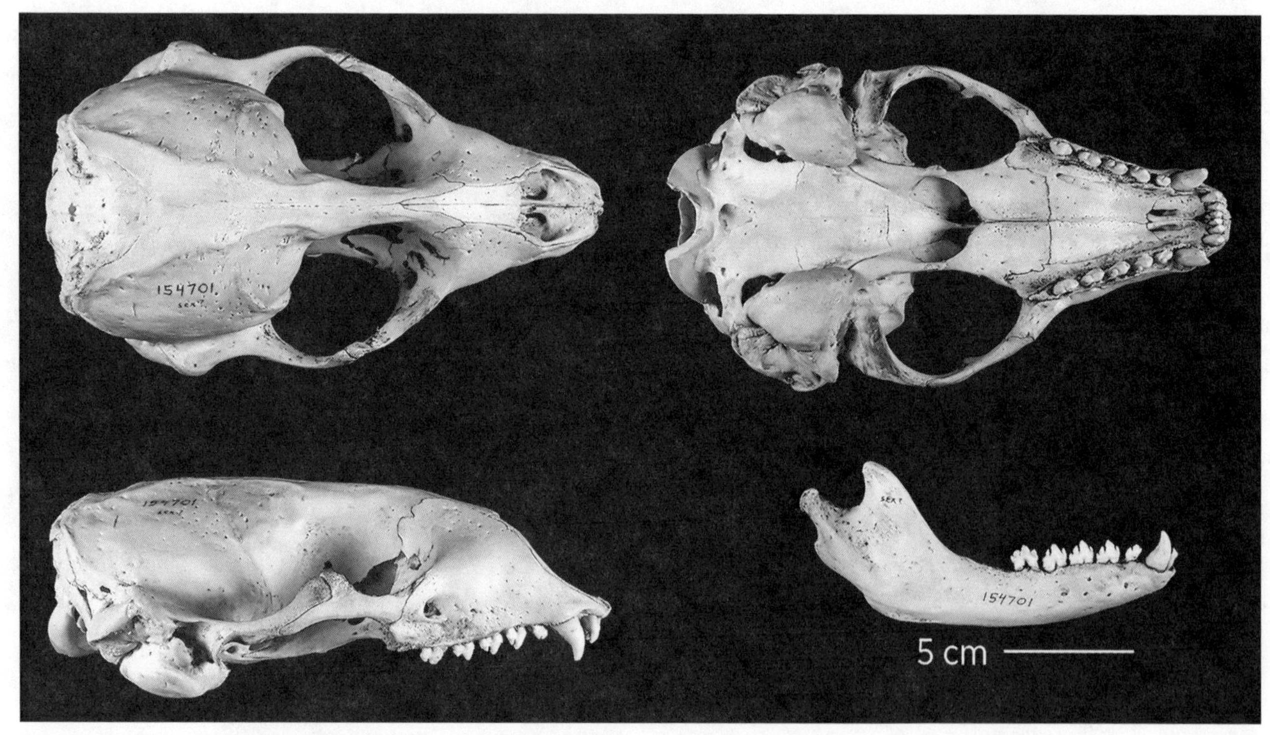

FIG. 471. Phocidae; skull and mandible of a harbor seal, *Phoca vitulina*.

Clade EUUNGULATA

The clade Euungulata, or "true" ungulates, contains the perissodactyls and the cetartiodactyls; the latter clade unites the traditional Artiodactyla with Cetacea (see figs. 94 and 95). This grouping contrasts sharply with earlier classifications of the "ungulates." Simpson (1945), for example, placed the perissodactyls and artiodactyls (exclusive of cetaceans) in separate superorders largely reflecting their differences in foot structure (mesaxonic versus paraxonic; see below). Alternatively, McKenna and Bell (1997) placed both perissodactyls and artiodactyls in the grandorder Ungulata, a paraphyletic group that also included the tubulidentates, hyraxes, sirenians, elephants, and cetaceans. Molecular sequence data firmly support the association of perissodactyls and artiodactyls + cetaceans in a single clade, and the latter relationship is now supported by well-preserved transitional fossils (Gingerich et al. 2001). In a major departure from the traditional linkage of Perissodactyla and Artiodactyla (or Cetartiodactyla), the expanded molecular analysis presented by Gatesy et al. (2017) suggests that Perissodactyla is basal to Chiroptera.

Order PERISSODACTYLA

The perissodactyls have been recognized unambiguously for more than two centuries as a unified phylogenetic assemblage diagnosed by the following traits:

DIAGNOSTIC CHARACTERS:

- **foot posture unguligrade**
- **foot structure mesaxonic—axis of symmetry passes through middle (3rd) digit, which is larger than other digits** (fig. 472)

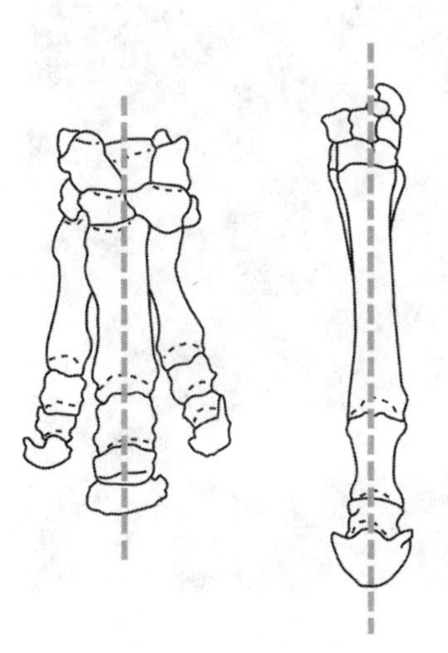

FIG. 472. Forefeet of a rhinoceros (*left*) and a horse (*right*), with axis of symmetry passing down leg through 3rd digit (dashed line). Redrawn from Berta et al. (2015).

- digits usually 1/1 or 3/3 (4/3 in tapirs, but 4th digit on forefoot is smaller than other three—see family account below)
- calcaneus does not articulate with fibula
- **alisphenoid canal present**
- nasals wide posteriorly
- cheek teeth lophodont, π-shaped with longitudinal ectoloph and transverse protoloph and metaloph (see fig. 18 and accompanying description)
- **femur with a 3rd trochanter** (fig. 473)

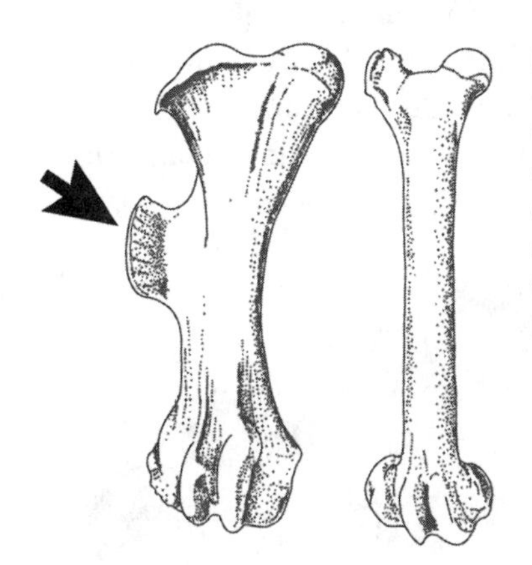

FIG. 473. Femur of a horse (Perissodactyla), with 3rd trochanter (*left*; arrow), compared with the femur of a deer (Artiodactyla; *right*), which lacks the 3rd trochanter. Redrawn from Lawlor (1979).

- **astragalus with single pulley-like proximal facet** (tibial pulley, which has two distinctively oblique trochlear ridges) articulating with distal end of tibia and a flattened distal surface articulating with the navicular and cuboid (fig. 474)

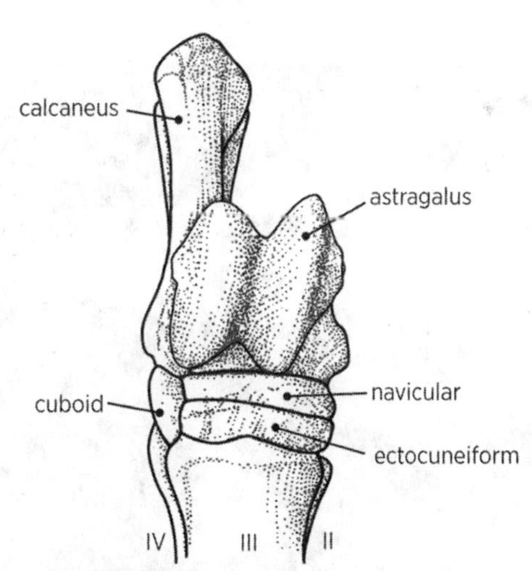

FIG. 474. Single-pulley astragalus of a horse (see fig. 486 for illustration of and comparison to double-pulley condition of Artiodactyla). Redrawn from Vaughan et al. (2015).

Family TAPIRIDAE (tapirs)

Large (height at shoulder 75–120 cm; mass to more than 300 kg); heavyset with short, stout legs and short tail; large calloused pads and oval hooves support weight; skin extremely thick, hair straight and sparse; eyes small; no head ornamentation; upper lip and nostrils elongate into short muscular proboscis; color uniform brown to gray in New World *Tapirus*; head, neck, and limbs black and body white in Asiatic *Acrocodia*; young of all species with pale longitudinal stripes and spots on back and sides. Mostly nocturnal; chiefly solitary, shy; inhabitants of wet primary forest from lowlands to highlands in tropics; excellent swimmers and divers, readily entering water to feed and rest; aquatic vegetation and tree foliage primary foods.

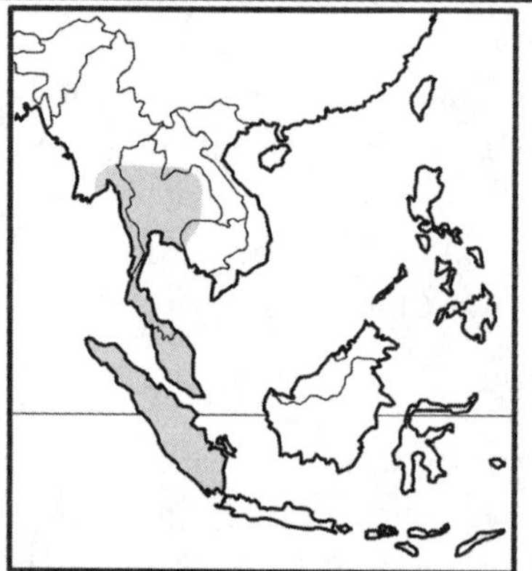

FIG. 476. Geographic range of family Tapiridae: New World lowland and montane tropics (southern Mexico to Argentina) and Southeast Asian tropics.

FIG. 475. Tapiridae; Baird's tapir, *Tapirus bairdii*.

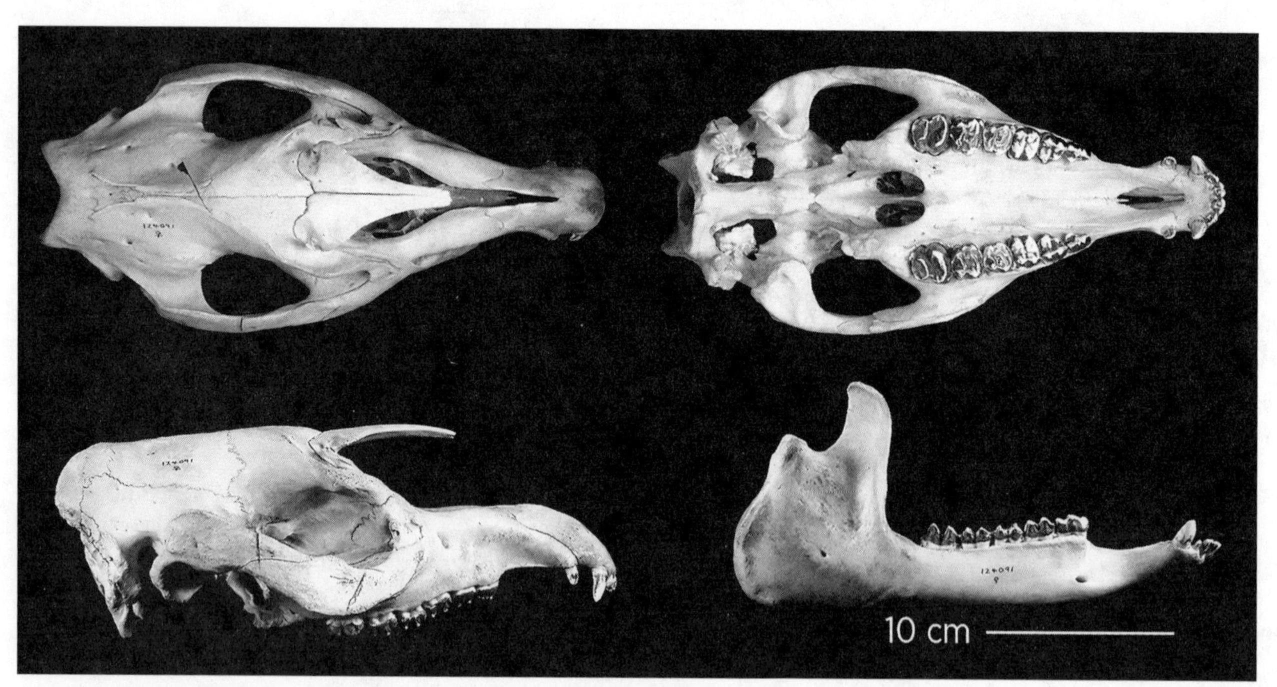

FIG. 477. Tapiridae; skull and mandible of the mountain tapir, *Tapirus pinchaque*.

DIAGNOSTIC CHARACTERS:

- **forefoot with four digits** (but 4th reduced), **hind foot with three**
- **snout modified into movable proboscis**
- **nasal opening of skull very large and recessed**

RECOGNITION CHARACTERS:

1. occipital crest of skull small or absent
2. **nasals short, triangular, projecting freely**
3. incisors chisel-shaped; canines well developed and conical
4. cheek teeth not homodont (1st premolars do not closely resemble other premolars and molars)
5. cheek teeth brachydont with simple π-shaped lophodonty—uppers with ectoloph, protoloph, and metaloph; lowers with transverse protolophid and metalophid; sides of lophs steep, but primarily involved in vertical crushing

DENTAL FORMULA: $\frac{3}{3}\ \frac{1}{1}\ \frac{4}{3\text{–}4}\ \frac{3}{3} = 42\text{–}44$

TAXONOMIC DIVERSITY: Two genera, one in the New World (*Tapirus*, with four species) and one in the Old World (*Acrocodia*, with a single species).

Family RHINOCEROTIDAE (rhinoceroses)

Size very large and ponderous (height at shoulder 110–200 cm; mass 900–3,600 kg); legs short and stout, graviportal limb posture; **tail thin, short, terminating in tuft of stiff bristles**; neck short and thick; head large, elongate, and concave in lateral profile; ears erect, oval, and slightly tufted, placed at back of head; dermal horns composed of agglutinated keratinous fibers without bony core, attached medially on nasals or, if two horns, on nasals and frontals; upper lip prehensile, extending beyond lower lip; **skin thick, coarse, often in platelike folds; hair sparse** to generally absent over body; color drab brown, black, or gray. Generally nocturnal, solitary (except *Diceros*, which is moderately social); all species frequent wallows and water holes for bathing, where even less social species will tolerate conspecifics; herbivorous grazers and browsers. Habitat ranges from open grasslands to shrublands and savannas; Asian species are forest dwelling and relatively poorly studied.

DIAGNOSTIC CHARACTERS:

- **three digits on all feet**
- **head concave dorsally** (more or less flat in other perissodactyls)
- **one or two simple horns located near snout** (all other perissodactyls lack cranial ornamentation)
- **temporal fossa exceptionally large**

RECOGNITION CHARACTERS:

1. skull low, elongate, **occipital crest well developed**; orbital rim incomplete
2. nasals large, thickened, and rugose for support of horn, and free of contact with premaxillae
3. cheek teeth usually homodont, with relatively simple π-shaped lophodont pattern; uppers with protoloph and metaloph oblique, gently curved ridges connecting with thick ectoloph

FIG. 478. Rhinocerotidae: Dicerotinae; white rhinoceros, *Ceratotherium simum*.

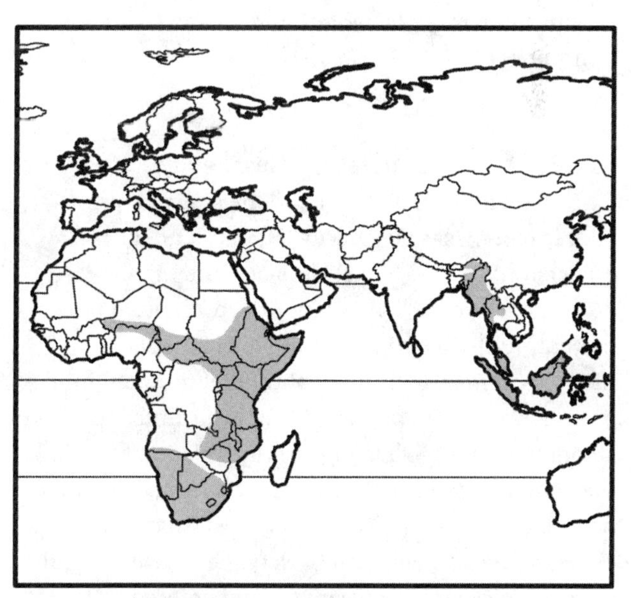

FIG. 479. Geographic range of family Rhinocerotidae: tropical grasslands of Africa; tropical forests of Southeast Asia, from eastern India to Borneo.

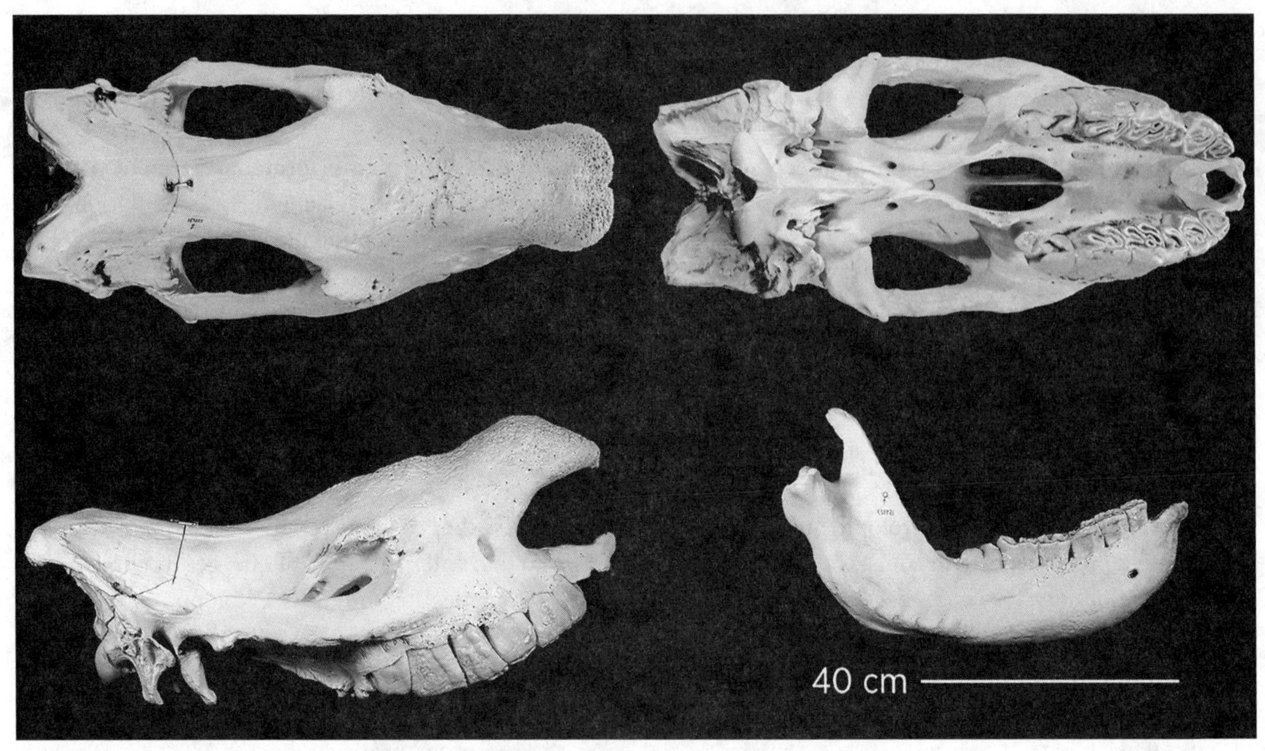

FIG. 480. Rhinocerotidae: Dicerotinae; skull and mandible of the white rhinoceros, *Ceratotherium simum*.

DENTAL FORMULA: $\frac{0\text{–}2}{0\text{–}1}\ \frac{0}{0\text{–}1}\ \frac{3\text{–}4}{3\text{–}4}\ \frac{3}{3} = 24\text{–}34$

TAXONOMIC DIVERSITY: Five species in four living genera placed by some authors in two subfamilies:

Dicerotinae—*Ceratotherium* (square-lipped or white rhinoceros) and *Diceros* (black rhinoceros)

Rhinocerotinae—*Dicerorhinus* (Sumatran [two-horned] rhinoceros) and *Rhinoceros* (Indian and Javan [one-horned] rhinoceroses)

Family EQUIDAE (horses, asses, zebras)

Large (height at shoulder in undomesticated taxa 80–145 cm; mass 200–500 kg); **limbs slender**, especially distal elements, elongate, and highly developed for cursorial locomotion; distal limb bones partially (ulna to radius) or completely (fibula to tibia) fused in most species; both forelimbs and hind limbs have a "stay apparatus" that locks limbs in a standing position when muscles are relaxed, allowing the animal to sleep while standing; skin smooth, well haired, tail and mane bushy; head, especially muzzle, elongate; eyes and ears set back on head; eyes partially stereoscopic, ears large, conical, and pointed; color of wild species pale drab to reddish tan or striped brown

FIG. 481. Equidae; domesticated horse, *Equus caballus* (*Equus* group).

to black on white. Nocturnal or diurnal; herbivorous grazers. Generally highly social, gregarious, somewhat wandering, some migratory, gather seasonally into large herds; stable family group is basic social unit. Inhabit open to moderately wooded grasslands and desert scrub where short grasses predominate and water is available.

DIAGNOSTIC CHARACTERS:

- **feet with only one distinct digit**
- **mane present, bushy**
- **orbit enclosed by postorbital bar** (continuous with temporal fossa in other perissodactyls)

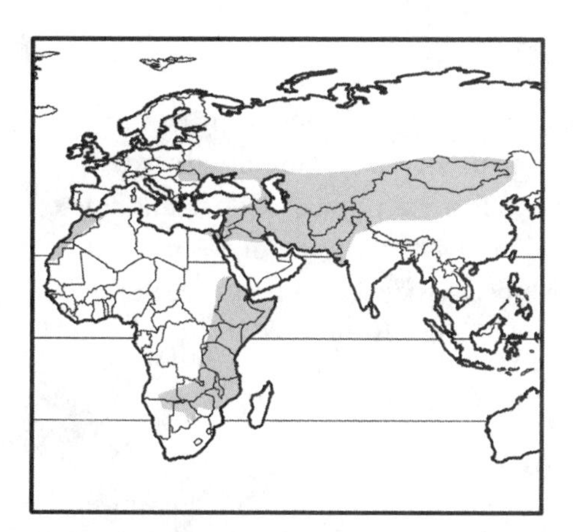

FIG. 482. Geographic range of family Equidae: native range Africa and southwestern and central Asia; cosmopolitan via introduction by humans.

RECOGNITION CHARACTERS:

1. skull with short cranium and elongate rostrum; occipital crest small or absent
2. orbits small, enclosed in bony ring
3. **nasal bones long and narrow**
4. incisors broad; canines small, if present
5. premolars and molars extremely hypsodont, ever-growing until advanced age
6. **cheek teeth homodont, forming uniform grinding platform with complex π-shaped lophodont pattern of exposed enamel, dentine, and cement**

DENTAL FORMULA: $\frac{3}{3}\ \frac{0\text{–}1}{0\text{–}1}\ \frac{3\text{–}4}{3}\ \frac{3}{3} = 36\text{–}42$

TAXONOMIC DIVERSITY: Eight living species in the single genus *Equus*, currently allocated among six species groups:

Asinus group—*E. asinus* (ass)

Dolichohippus group—*E. grevyi* (Grevy's zebra)

Equus group—*E. caballus* (horse, including Przewalski's horse)

Hemionus group—*E. hemionus* (onager) and *E. kiang* (kiang)

Hippotigris group—*E. zebra* (mountain zebra)

Quagga group—*E. quagga* (quagga) and *E. burchellii* (Burchell's zebra)

FIG. 483. Equidae; skull and mandible of the domestic horse, *Equus caballus* (*Equus* group).

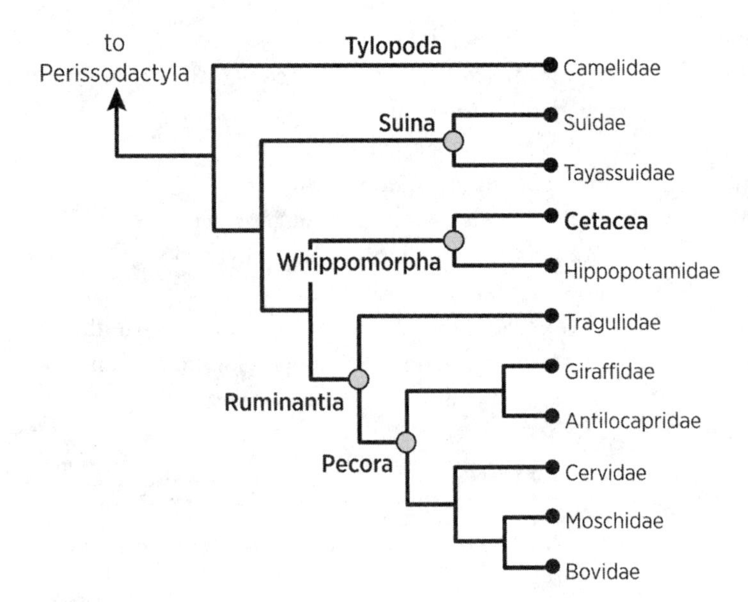

FIG. 484. Phylogenetic hypothesis of relationships among families of cetartiodactylan ungulates. The sister relationship between the Cetacea and the Hippopotamidae (which constitute the suborder Whippomorpha; see fig. 94) is identified, as are the four major subgroups (either suborders or infraorders) of the traditional order Artiodactyla. The tree is derived from supermatrix analyses of both morphological and DNA sequences (Hassanin and Douzery 2003; Fernández and Vrba 2005; Agnarrson and May-Collado 2008; Meredith et al. 2011; Hassanin et al. 2012; Gatesy et al. 2013, 2017). The position of the Antilocapridae within the Ruminantia is uncertain: it has been placed as sister to the Giraffidae (as here), as sister to the Cervidae, or as basal to the Giraffidae and other Pecora families. Similarly, the relationships between the Cervidae, Moschidae, and Bovidae are uncertain, with the Moschidae alternatively linked either to the Cervidae or to the Bovidae (as here).

Superorder CETARTIODACTYLA

As indicated in the introduction to infraclass Eutheria or Placentalia, recent efforts to resolve higher-level relationships among eutherian lineages have resulted in the recognition that whales (Cetacea) likely share ancestors with the Hippopotamidae, a family in the order Artiodactyla (fig. 484). This conclusion was originally based on molecular data, but it was the discovery of Eocene whale fossils with paraxonic feet and a double-pulley astragalus—features that largely define the Artiodactyla—that led to its general acceptance (Gingerich et al. 2001). To reflect this association, taxonomists coined Cetartiodactyla as a clade containing both orders, in recognition that the traditional Artiodactyla is a paraphyletic grouping, although Asher and Helgen (2010) have argued that the name change is not necessary and is inconsistent with other cases in which novel radiations have retained independent names (e.g., Dinosauria and Aves, Synapsida and Mammalia).

Because living cetaceans and artiodactyls are each easily diagnosed by a set of separate morphological attributes, we follow current usage here (Grubb 2005; Mead and Brownell 2005) and organize the accounts that follow by dividing the Cetartiodactyla into the traditional orders Artiodactyla and Cetacea.

The Cetartiodactyla includes a remarkable diversity and richness of species inhabiting all the world's oceans and all landmasses save Antarctica. Perhaps because of this diversity, taxonomists are still trying to understand the patterns and timing of radiations in this group, and even resolution of associations within each order remains a work in progress. In the following accounts, we present a consensus view based on the current understanding; with few exceptions, we do not address the numerous competing hypotheses for ordinal and familial relationships, as they do not influence our ability to describe and identify lineages.

Order ARTIODACTYLA

The order Artiodactyla comprises the most successful living ungulates. The Perissodactyla originated in the Paleocene and reached their greatest diversity in the Eocene. The Artiodactyla first appeared in the Eocene and radiated greatly in the Oligocene and especially the Miocene. Since then, however, the perissodactyls have steadily declined, while the artiodactyls have remained diverse. Artiodactyls currently constitute the most diverse and abundant group of large mammals, far overshadowing perissodactyls in taxonomic diversity, numerical abundance, and ecological influence. Of the 36 families (and about 591 genera) of artiodactyls present in the Cenozoic, 10 (and about 85 genera) survive to the present. In contrast, the perissodactyls, represented by 14 Cenozoic families (and about 238 genera), are reduced today to only 3 families and 7 genera (numbers tabulated from McKenna and Bell 1997). This greater diversity of artiodactyls may reflect their more efficient digestive strategy (foregut, or ruminant, digestion, vs. hindgut, or cecant, digestion in perissodactyls), which allows them to be more selective foragers and possibly allows for greater niche segregation (and hence increased species diversity).

Like perissodactyls, artiodactyls have limbs that are highly modified for a cursorial existence; these modifications include reduction or loss of the clavicle, extension of distal limb elements, reduction in the number of digits, and partial or complete fusion of distal limb elements and of metapodial bones. A hallmark feature defining artiodactyls is the double-pulley astragalus (fig. 486; contrast this with the single pulley of the perissodactyls, shown in fig. 474), and whereas perissodactyls have mesaxonic foot structure, that of artiodactyls is paraxonic.

DIAGNOSTIC CHARACTERS:

- **foot posture digitigrade (camels) or unguligrade**
- **foot structure paraxonic—two principal (weight-bearing) digits present, nearly equal in size and symmetrical in shape, with axis of symmetry passing between 3rd and 4th digits** (fig.485)

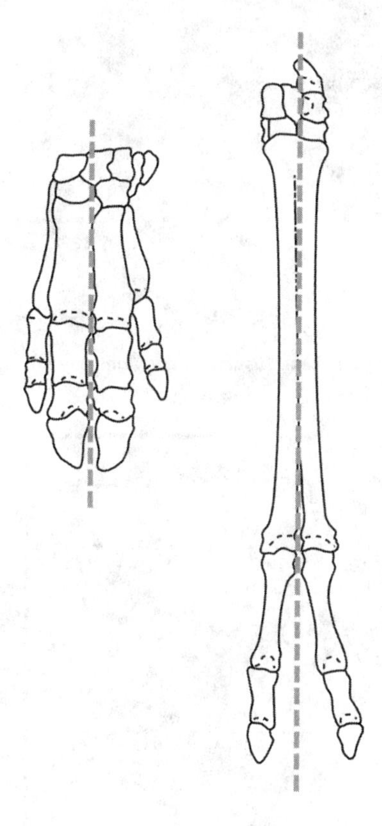

FIG. 485. Forefeet of a pig (*left*) and a guanaco (*right*), with axis of symmetry passing down leg between 3rd and 4th digits (dashed line). Redrawn from Berta et al. (2015).

- two or four digits on all feet (three toes on hind foot of peccaries, Tayassuidae); no pollex or hallux
- femur without a 3rd trochanter (see fig. 473 for illustration and comparison to an equid femur with a 3rd trochanter)
- **astragalus with pulley-like surface both above** (with vertical trochlear ridges articulating with tibia) **and below** (articulating with cuboid and navicular, which become fused into the cubonavicular in Ruminantia, fig. 486)

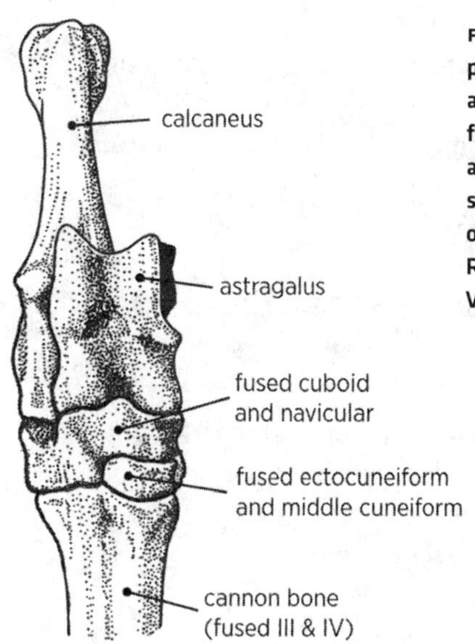

FIG. 486. Double-pulley astragalus of a deer (see fig. 474 for illustration of and comparison to single-pulley condition of Perissodactyla). Redrawn from Vaughan et al. (2015).

- calcaneus articulating with fibula
- **no alisphenoid canal**
- nasals usually not wider posteriorly
- cheek teeth either bunodont (pigs, peccaries, hippopotamuses) or selenodont

PHYLOGENY AND CLASSIFICATION OF ARTIODACTYLA

The classification of the Artiodactyla, even excluding Cetacea, has received substantial revision in recent years, and additional changes are likely in the near future as problematic nodes in the tree become fully resolved. For consistency here, we provide the classification espoused by the various consensus supermatrix phylogenetic analyses (see figs. 94 and 484) and used by the account authors in Wilson and Mittermeier (2011). This view retains the three subordinal groups of the traditional Artiodactyla (Suina [or Suiformes], Tylopoda, and Ruminantia), but removes the Hippopotamidae from its traditional placement within the Suina and places this family in the infraorder Ancodonta (which groups hippopotamuses with their extinct cousins) of the suborder Whippomorpha (the clade that includes both hippopotamuses and cetaceans). We also retain the traditional division of the ruminant infraorder Pecora into three superfamilies, although, as we noted above, several of the family-level phylogenetic linkages remain unresolved.

Suborder Suina
Family Suidae (pigs)
Family Tayassuidae (peccaries)
Suborder Whippomorpha
Infraorder Ancodonta
Family Hippopotamidae (hippopotamuses)
Suborder Tylopoda
Family Camelidae (camels, vicuña, guanaco, llama, alpaca)
Suborder Ruminantia
Infraorder Tragulina
Family Tragulidae (chevrotain, mouse deer)
Infraorder Pecora
Superfamily Cervoidea
Family Antilocapridae (pronghorn)
Family Cervidae (deer, elk, caribou, moose, reindeer)
Family Moschidae (musk deer)
Superfamily Giraffoidea
Family Giraffidae (giraffe, okapi)
Superfamily Bovoidea
Family Bovidae (bison, muskox, goats, sheep, antelope, cattle)

Suborder SUINA

The suborder Suina includes taxa with bunodont molars, tusklike canines, and feet usually retaining all four toes with complete and separate digits. The skull contrasts with those of other artiodactyls in having a **posterior extension of the squamosal bone that meets the exoccipital bone and conceals the mastoid bone.** Pigs have a nonruminant type of stomach with two chambers, but employ cecant fermentation (babirusa may have primitive foregut fermentation); peccaries also have a multi-chambered stomach but are foregut fermenters.

RECOGNITION CHARACTERS:

1. foot posture unguligrade
2. digits 4/3 (Tayassuidae) or 4/4 (Suidae)
3. hooves present
4. **cannon bone lacking in all feet** (partial fusion of metatarsals in Tayassuidae)
5. **no horns or antlers**
6. **postorbital bar never complete**
7. **mastoid not exposed, obscured by broad contact of squamosal and occipital bones**
8. **one to three pairs of upper incisors present**
9. canines tusklike, sharp-edged
10. **cheek teeth bunodont** and generally brachydont
11. stomach with two (Suidae) or two to three (Tayassuidae) chambers

Family SUIDAE (pigs, hogs)

Medium-sized (body length 110–200 cm; mass 60–220 kg); body barrel-shaped, build sturdy; limbs short; **sparsely haired or with coarse bristles**, dorsal mane sometimes present; **muzzle elongate and broad** (especially so in warthogs [*Phacochoerus*]); snout highly mobile, with disclike cartilage for support, and terminal nostrils; eyes small, positioned high on head; ears variable in size, but typically small and pointed; tail short, typically with distal tuft of bristles, especially well developed in warthogs. Dentition highly variable, even

FIG. 487. Suidae: Suinae: Suini; wild boar, *Sus scrofa* (*adult and young*).

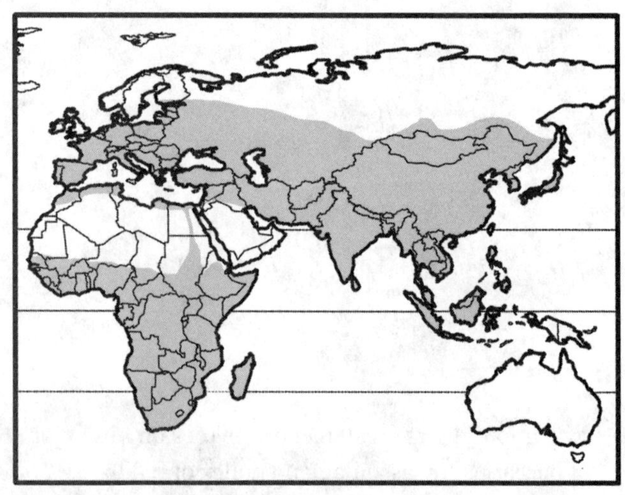

FIG. 488. Geographic range of family Suidae: Old World—central and southern Europe, Africa (excluding Sahara), Madagascar, Middle East through India to southern China and Southeast Asia, including Philippine and Indonesian islands [free-ranging by introduction in United States, South America, New Guinea, New Zealand, and possibly Madagascar].

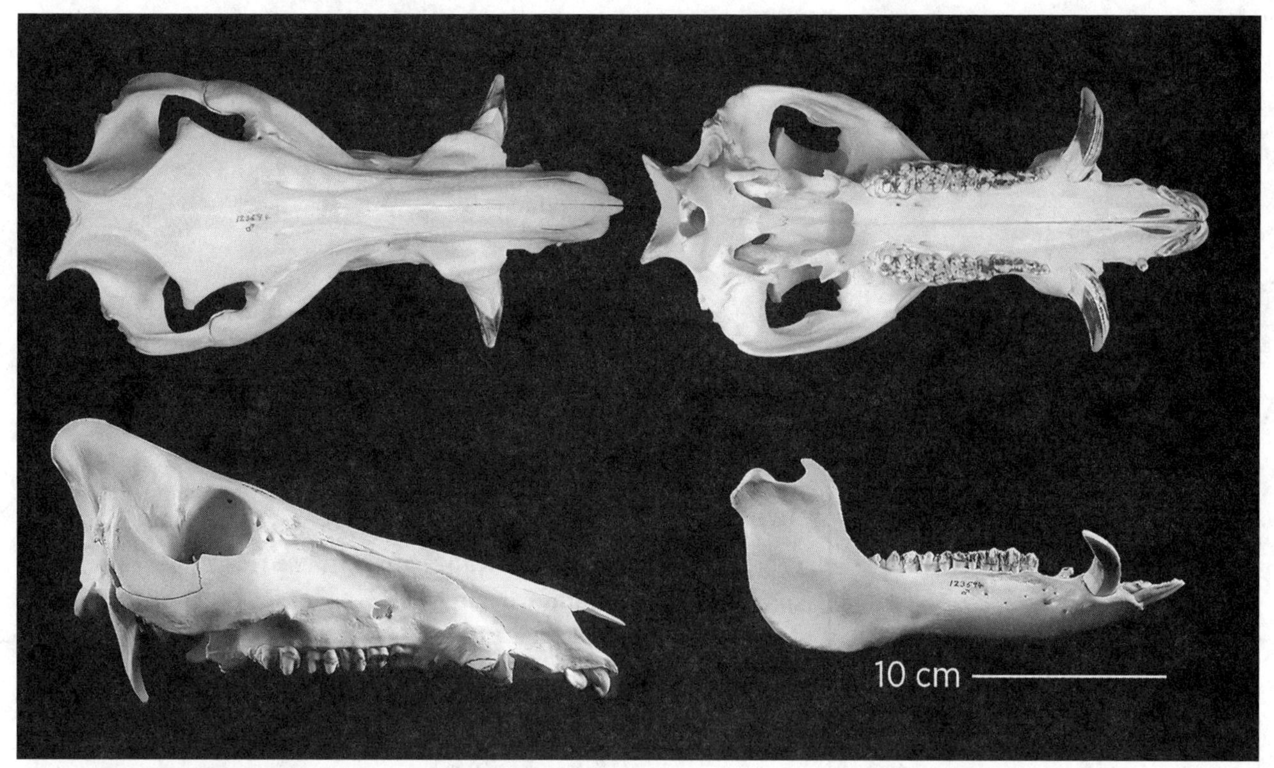

FIG. 489. Suidae: Suinae: Suini; skull and mandible of the wild boar, *Sus scrofa*.

within species, with reduction or loss of both upper incisors and premolars; well-developed canines present in both sexes; upper canines flare laterally as formidable tusks (those of male *Babyrousa* grow dorsally through skin, curving back toward the face). Highly gregarious, usually in family groups but some in large herds; may be nocturnal or diurnal; diet omnivorous and highly variable, including vertebrates and invertebrates, leaves, fruit, rhizomes, tubers, and fungi; root for food using snout. Habitat is generally thickets and closed temperate or tropical forests, frequently along riparian corridors, except for warthogs, which occupy open woodlands or savannas.

RECOGNITION CHARACTERS:

1. digits 4/4, but usually only two functional in locomotion (both lateral toes small)
2. **snout elongate, mobile, flattened at end**
3. nostrils opening anteriorly
4. **no fusion of metacarpals or metatarsals** (no cannon bone)
5. paroccipital process elongate
6. no ventral flange on angular process of lower jaw
7. **canines usually with sharp edges; upper canines larger than lower canines, directed outward and upward**

DENTAL FORMULA: $\frac{0\text{–}3}{0\text{–}3}\ \frac{1}{1}\ \frac{2\text{–}4}{1\text{–}4}\ \frac{3}{3} = 26\text{–}44$

TAXONOMIC DIVERSITY: Six genera placed in four tribes in the single subfamily, Suinae:

Babyrousini—*Babyrousa* (babirusas)
Phacochoerini—*Phacochoerus* (warthogs)
Potamochoerini—*Hylochoerus* (giant forest pig) and *Potamochoerus* (bush pig and red river hog)
Suini—*Porcula* (pygmy hog) and *Sus* (warty pigs, wild boar, domesticated pig, and bearded pigs)

Family TAYASSUIDAE (peccaries)

Small to medium-sized (body length 75–105 cm; mass 15–30 kg); piglike, body sturdy; snout highly mobile, with rounded, disclike cartilage; nostrils open anteriorly; tail greatly reduced; **coat with coarse bristles, grizzled grayish to blackish**, off-white to yellowish on cheeks and on collar that extends over shoulders to throat (in collared peccary), with white lips and lower jaw but no collar (white-lipped peccary), or brownish gray with faint collar of lighter hairs over shoulders (Chacoan peccary); well-developed middorsal scent gland in lumbar region. Highly gregarious, with group size varying from a few individuals to more than 50, and with group home range (i.e., no individual home ranges); principally diurnal and omnivorous, including more plant materials in the diet than do suids but also consuming a wide variety of invertebrates and vertebrates. Habitat preference is thick brush in biomes ranging from desert scrub and thorn forest to tropical lowland forest.

FIG. 490. Tayassuidae; white-lipped peccary, *Tayassu pecari*.

FIG. 491. Geographic range of family Tayassuidae: tropical rainforest, deciduous forest, and arid shrublands of the New World, from southwestern United States to Argentina.

RECOGNITION CHARACTERS:

1. digits 4/3, but only two functional in locomotion
2. snout elongate, mobile, flattened at end
3. nostrils opening anteriorly
4. **middle two metatarsals fused proximally** (e.g., partial cannon bone); metacarpals free
5. paroccipital process small
6. no ventral flange on angular process of lower jaw
7. **canines with sharp cutting edges; upper canines approximately same size as lower canines, directed downward**

DENTAL FORMULA: $\frac{2}{3}\ \frac{1}{1}\ \frac{3}{3}\ \frac{3}{3} = 38$

TAXONOMIC DIVERSITY: Three monotypic genera: *Catagonus* (Chacoan peccary), *Pecari* (collared peccary or javelina), and *Tayassu* (white-lipped peccary).

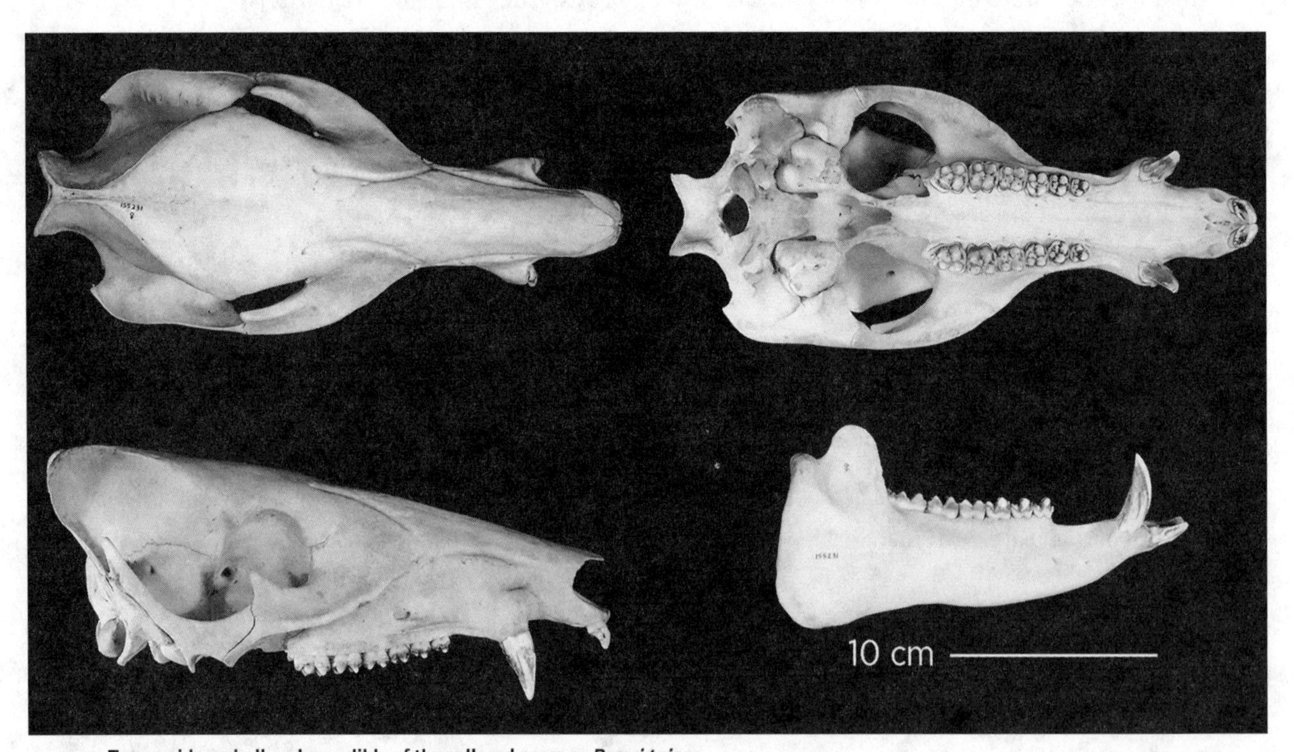

FIG. 492. Tayassuidae; skull and mandible of the collared peccary, *Pecari tajacu*.

Suborder WHIPPOMORPHA

The suborder Whippomorpha unites the whales (Cetacea) with the hippopotamuses and their extinct relatives (Ancodonta). The name was coined as a Latinization of a colloquial term for the "whippo" hypothesis that whales and hippopotamuses were related (see Waddell et al. 1999). We present the Cetacea independently, reflecting the tremendous ecomorphological evolution associated with their adaptation to a fully aquatic lifestyle and the associated character states that unite them with, yet complicate their simple presentation as part of, the Cetartiodactyla.

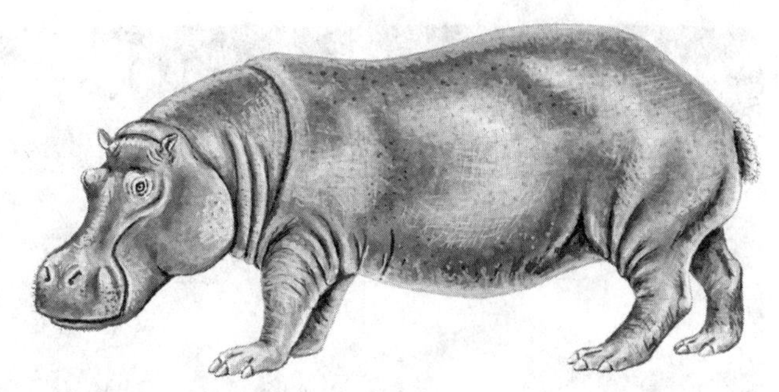

FIG. 493. Hippopotamidae; common hippopotamus, *Hippopotamus amphibius*.

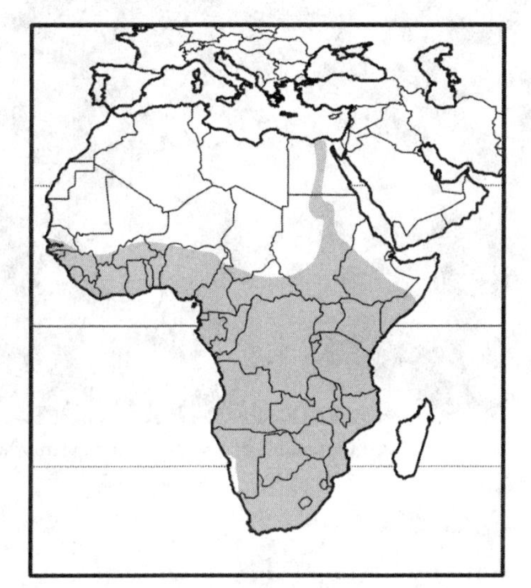

FIG. 494. Geographic range of family Hippopotamidae: sub-Saharan tropical Africa and Nile River basin.

Infraorder ANCODONTA

Recently divided from its traditional placement within the suborder Suina (Simpson 1945), the infraorder Ancodonta includes the extant hippopotamuses and their close but now extinct relatives. Like the Suina, this group is characterized by bunodont molars, tusklike canines, and feet retaining all four toes with complete digits and separate metapodials (no cannon bones). The ancodont skull, like that of the Suina and in contrast with those of other artiodactyls, presents a posterior extension of the squamosal bone that meets the exoccipital bone and conceals the mastoid bone. Hippopotamuses have a four-chambered stomach and are foregut fermenters.

Family HIPPOPOTAMIDAE (hippopotamus, pygmy hippopotamus)

Medium-sized to very large (body length up to 4 m, height at shoulder up to 1.5 m, and mass up to 4,500 kg in *Hippopotamus*; length 2 m, shoulder height 1 m, mass 200–270 kg in *Choeropsis*); head and **muzzle very broad; nostrils dorsal on snout**, closable; eyes dorsolateral, protruding; ears small; legs short and thick; tail short and laterally flattened, with coarse hairs forming terminal tuft; **skin thick, almost hairless**, dark brown dorsally and pinkish ventrally. Hippopotamuses lack sweat glands and possess a glandular epidermis that exudes an oily pink liquid that serves both as sunscreen and as an antibiotic to keep wounds from becoming infected. Gregarious, occurring in herds of 10–40 individuals of both sexes; amphibious, principally in freshwater systems but also live in brackish estuaries; swim and dive readily, foraging on bottom for aquatic plants, with submergence times up to 6 minutes; dorsal position of eyes and nostrils allows individuals to stay below water surface with only top of head exposed; travel on land at night to forage on grasses and other plants. Inhabit major waterways, estuaries, lakes, and stagnant swamps in deserts, savannas, woodlands, and forests.

DIAGNOSTIC CHARACTERS:

- **mouth very large, extremely wide gape**
- **incisors large, directed anteriorly** (especially inner pair of mandible), **tusklike**

RECOGNITION CHARACTERS:

1. digits 4/4, all functional in locomotion
2. no fusion of metacarpals or metatarsals
3. orbits elevated, postorbital bar incomplete
4. paroccipital process small
5. **angular process of lower jaw with large ventral flange**
6. **canines very large, tusklike; upper canines much smaller than lower canines, directed downward**

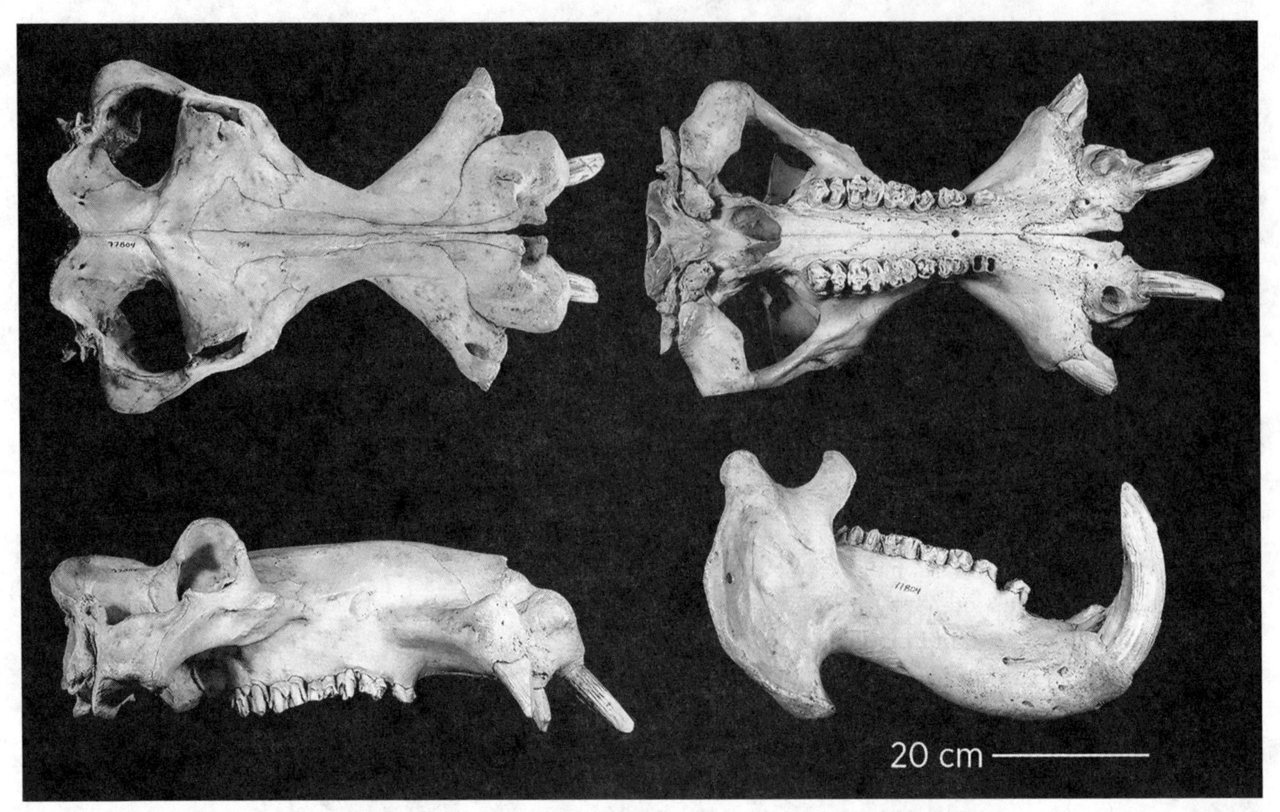

FIG. 495. Hippopotamidae; skull and mandible of the common hippopotamus, *Hippopotamus amphibius*.

DENTAL FORMULA: $\frac{2}{1\text{–}2}\ \frac{1}{1}\ \frac{3\text{–}4}{3\text{–}4}\ \frac{3}{3} = 34\text{–}40$

TAXONOMIC DIVERSITY: Two monotypic genera: *Choeropsis* (pygmy hippopotamus) and *Hippopotamus* (common hippopotamus).

Suborder TYLOPODA

The suborder Tylopoda includes the camels and their relatives—primitive ruminant, selenodont ungulates that are apparently intermediate between suines and ruminants. They share with ruminants the reduction or loss of upper incisors, fusion of the ecto- and mesocuneiform bones in the ankle, and a multichambered, ruminating stomach. Tylopods also share with ruminants the loss of the trapezium from the carpus. Tylopods, however, do not exhibit fusion of the magnum and trapezoid bones or of the cuboid and navicular, as in ruminants. The only extant tylopods are the camels and their relatives, a group that originated in North America and spread to Eurasia, Africa, and South America during the Pleistocene over land bridges; they are now restricted to arid and semiarid regions of the Old World (*Camelus*) and to South America (*Lama*, *Vicugna*), and are extinct in North America, but have been introduced to Australia.

Family CAMELIDAE (camels, llamas, vicuñas, guanacos, alpacas)

Medium-sized to large (height at shoulder 70–130 cm, mass 60–75 kg in llama and relatives; height 190–230 cm [at hump], mass 450–650 kg in camels); head relatively small, muzzle slender; nostrils nearly horizontal slits; lip cleft; ears small and rounded (*Camelus*) or pointed (*Lama* and *Vicugna*); neck long and relatively thin; one (dromedary) or two (Bactrian camel) humps in *Camelus*; hindquarters contracted, thighs appearing somewhat separated from torso. Unlike those of all other ungulates, camelid feet are digitigrade, not unguligrade, and present a broad foot pad (hence they may also be considered subunguligrade) that provides effective support on soft, sandy soils (note that Tragulidae may also be digitigrade at times); facilitating this adaptation, the 3rd and 4th metapodials are fused proximally (forming a partial cannon bone). but are separate and flare laterally at their distal ends, and the subsequent phalanges are laterally expanded to

FIG. 496. Camelidae: Aucheniini; guanaco, *Lama guanicoe*.

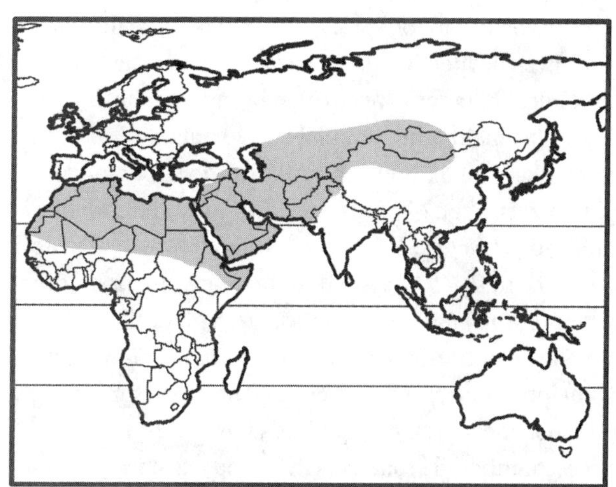

FIG. 497. Geographic range of family Camelidae: Old World (Camelini)—wild populations limited to Gobi Desert of central Asia; New World (Aucheniini)—restricted to the Andes, from Peru to Tierra del Fuego; domestic stock of both tribes introduced worldwide.

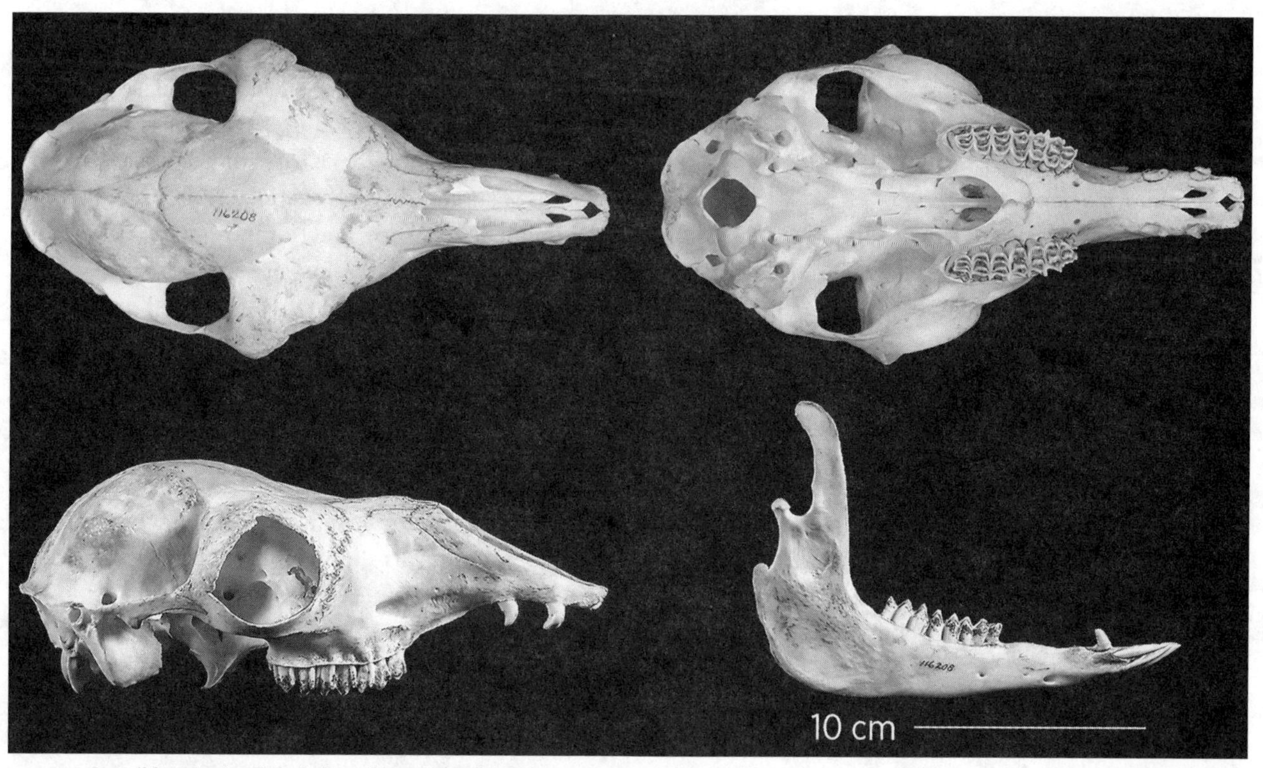

FIG. 498. Camelidae: Aucheniini; skull and mandible of the vicuña, *Vicugna vicugna*.

support the broad foot pads; foot pad is more flexible and with a deeper cleft in *Camelus* than in other camelids. Phalanges are covered by small nails, not hooves (as in all other ungulates). Dorsal hair long and fleece-like (*Lama*), short and silky (*Vicugna*), or long and tufted on top of head, neck, hump, and elbow of forelimb (*Camelus*); color variable, from brown to yellow (*Camelus*) to black, gray, yellow, tan, or white (including combinations) in *Lama* and *Vicugna*. The skull of camelids can be distinguished from that of ruminants by the presence of upper incisors (three pairs in young animals, one in adults). Camelids have a ruminant, three-chambered stomach. Gregarious, live in harem herds of about 25 females and a single male; non-herd males may be solitary or in bachelor groups; diurnal and herbivorous. Inhabit arid deserts and grasslands. *Camelus* includes two Old World species, with wild populations persisting only in the Gobi Desert of Asia. In South America, *Lama* and *Vicugna* occur in the Andes and Patagonian steppe from Peru to Tierra del Fuego.

RECOGNITION CHARACTERS:

1. **foot posture digitigrade** (all other living artiodactyls are unguligrade, except tragulids on occasion)
2. digits 2/2
3. nails present (no hooves)
4. **cannon bone in all limbs, but fusion not complete at distal end**
5. **no horns or antlers**
6. postorbital bar complete
7. mastoid exposed
8. one pair of upper incisors present in adults, caniniform
9. canines present, not sharp-edged
10. molars selenodont
11. stomach with three chambers

DENTAL FORMULA: $\frac{1}{1\text{–}3}\ \frac{1}{1}\ \frac{1\text{–}3}{1\text{–}2}\ \frac{3}{3} = 28\text{–}34$

TAXONOMIC DIVERSITY: Five or six species in three genera allocated to two tribes:

Camelini—*Camelus* (dromedary and Bactrian camel)

Aucheniini (= Lamini)—*Lama* (wild guanaco, domesticated llama, possibly domesticated alpaca) and *Vicugna* (wild vicuña, probably domesticated alpaca)

Although Lamini is widely given as the tribal name for New World camelids, Grubb (2005) cites ICZN Article 40.2 in support of his argument that Aucheniini has priority over Lamini. The alpha taxonomy of New World camelids is unclear, probably reflecting a complex history of domestication, hybridization, and lack of written history in the Andes (Wheeler 1995). Most authorities accept that there are four species, but their phylogenetic relationships are obscured. The domesticated llama and alpaca, until recently believed to have derived from the guanaco, now are thought to have distinct origins—llama from guanaco and alpaca from vicuña (Kadwell et al. 2001) Whether these four species represent two genera (wild *Lama guanaco* and domesticated *L. glama* versus wild *Vicugna vicugna* and domesticated *V. pacos*; Franklin 2011) or one (subsumed under *Lama*; Groves and Grubb 2011) remains under discussion.

Suborder RUMINANTIA

Ruminants are the most evolutionarily derived group of artiodactyls as well as the most taxonomically diverse and abundant. All members are strict herbivores, and all are modified for highly cursorial locomotion. Ruminants regurgitate food for additional mastication (rumination, or "chewing their cud"); the "stomach" is complex, with four chambers that support microbial cellulose digestion (i.e., they are all foregut fermenters). The first three chambers (rumen, reticulum, omasum) are esophageal in origin and provide a neutral pH where microbial fermentation occurs (the last of these is poorly developed in Tragulina), whereas the final chamber (abomasum) is the true stomach and the location of standard acid digestion. All ruminants have selenodont cheek teeth, and the anterior dentition is variously specialized by loss or reduction of the upper incisors, by the development of incisiform lower canines, and commonly by the loss of upper canines. The skull differs from those of members of the Suiformes in the **exposure of the mastoid bone between the squamosal and exoccipital bones**. Antlers or horns, often large and complex structures, are present in most species; these structures tend to be smaller or absent in females. The limbs show a pronounced trend toward elongation of the distal segments; reduction, fusion, or loss of the fibula; fusion of the carpals and tarsals; and perfection of the two-toed foot. **All ruminants display fusion of the navicular and cuboid bones**, forming the cubonavicular, which articulates with the distal pulley of the astragalus (see fig. 486); the cubonavicular, in turn, is fused with other ankle bones in some families. Similarly, in the forelimb, the **magnum and trapezoid are fused**, and the trapezium has been lost.

The suborder Ruminantia is traditionally divided into the hornless ruminants (infraorder Tragulina) and those possessing some form of cranial ornamentation (infraorder Pecora). Ruminant families generally are readily distinguished externally. Cranial features that separate most groups include presence or absence of upper canines, of a conspicuous gap between the nasal and lacrimal bones (nasolacrimal gap), of a depression in the lacrimal bone anterior to the orbit (lacrimal depression), and of one versus two lacrimal foramina as well as their location inside, on the edge of, or outside the orbit.

RECOGNITION CHARACTERS:

1. foot posture unguligrade (sometimes digitigrade in tragulids)
2. **functional digits usually 2/2, side toes reduced or absent**
3. hooves present
4. **cannon bone usually present in all feet** (not in forelimbs of tragulids, where the metacarpals remain either unfused or only partially fused)
5. **navicular and cuboid bones fused (cubonavicular); magnum and trapezoid fused**
6. **horns or antlers usually present, at least in males** (not in tragulids or moschids)
7. postorbital bar complete
8. **mastoid exposed**
9. **no upper incisors**
10. if canines present, not sharp-edged
11. molars selenodont
12. stomach complex, four-chambered (functionally three-chambered in Tragulidae)

Infraorder TRAGULINA

Tragulina includes the hornless ruminants, represented today by only one family. Tragulines have a functionally three-chambered stomach, as the final chamber (the omasum) is poorly developed, in contrast to the condition found in pecorans.

Family TRAGULIDAE (chevrotains, mouse deer)

Small (body length 45–100 cm; height at shoulder 20–40 cm; mass 2.5–15 kg); head small; no horns or antlers; snout narrow with hairless muzzle; eyes large; ears generally small and pointed (*Moschiola*, *Tragulus*) or large and round (*Hyemoschus*); neck short; back arched, with humped, compact rump; legs long and thin; tail short; no facial or pedal glands. Coat fine, longer along back, sometimes with dorsal mane; colored dark to chestnut brown above, white below with pattern of white spots and stripes along body in some species. Solitary, pairing only in breeding season; crepuscular to nocturnal, secretive and wary; primarily herbivorous but do eat insects and small vertebrates; ruminating stomach. Inhabit tropical forests, mangrove thickets, and swamps.

FIG. 499. Tragulidae; Sri Lankan spotted chevrotain, *Moschiola meminna*.

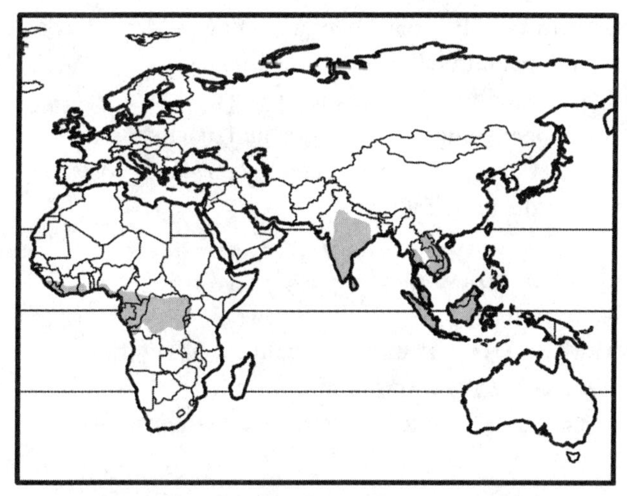

FIG. 500. Geographic range of family Tragulidae: central and western Africa (*Hyemoschus*, water chevrotain); India and Southeast Asia (*Moschiola*, *Tragulus*, mouse deer).

FIG. 501. **Tragulidae; skull and mandible of the lesser mouse deer, *Tragulus javanicus*.**

RECOGNITION CHARACTERS:

1. **no horns or antlers**
2. **digits 4/4**; side toes slender, but well developed; generally unguligrade, but digitigrade at times
3. hind leg with 3rd and 4th metatarsals fused (cannon bone), but foreleg metacarpals either unfused (African species) or only partially fused (Asian species)
4. navicular, cuboid, and ectocuneiform bones fused
5. no lacrimal depression
6. nasal and lacrimal bones adjoining or with little or no separation (i.e., no nasolacrimal gap)
7. one lacrimal foramen present, inside orbit
8. **upper canines present, long and tusklike in males**
9. lateral (vertical) surfaces of molars smooth in texture (cf. Giraffidae)

DENTAL FORMULA: $\frac{0}{3}\ \frac{1}{1}\ \frac{3}{3}\ \frac{3}{3} = 34$

TAXONOMIC DIVERSITY: Eight species in three genera: *Hyemoschus* (water chevrotain), *Moschiola* (spotted chevrotains), and *Tragulus* (mouse deer or chevrotains).

Infraorder PECORA

The Pecora include the so-called horned ruminants—those with some form of cranial ornamentation. They also **lack upper incisors** and have **incisiform lower canines**. All pecorans have a fully functional four-chambered stomach.

Superfamily CERVOIDEA

Family ANTILOCAPRIDAE (pronghorn)

Medium-sized (shoulder height 80–105 cm; mass 49–70 kg); body compact; face elongate; eye set deep in orbit; ears long, pointed; bifurcated horns present in both sexes (smaller and may lack prong in females); tail short; legs long and slender; facial, body, and pedal glands present. Gregarious in mixed herds in winter, solitary or in smaller bands at other times; adult males territorial most of year, younger males form bachelor herds; doe-fawn groups in summer. Diurnal, migrate opportunistically based on local conditions; feed almost exclusively on forbs and browse. Very agile, fastest land mammal in North America, with sustained speeds of 50 km/hr and maximum speed of 90 km/hr. Characteristic white rump patch raised when alarmed. Habitat includes open grasslands and semidesert shrublands.

DIAGNOSTIC CHARACTER:

- **pronghorns, not antlers, present**; bone core covered by keratinized sheath that is forked in mature individuals; sheath shed annually (unique characteristic of family; fig. 503); both sexes have horns, but those in females are smaller, may lack prong

RECOGNITION CHARACTERS:

1. **digits 2/2**; no lateral digits present
2. **lacrimal depression absent**
3. nasal and lacrimal bones separated by large opening (**nasolacrimal gap**)
4. **two lacrimal foramina present, on side of orbit**
5. upper canines absent
6. lateral (vertical) surfaces of molars smooth in texture (cf. Giraffidae)
7. cheek teeth hypsodont

DENTAL FORMULA: $\frac{0}{3}\ \frac{0}{1}\ \frac{3}{3}\ \frac{3}{3} = 32$

TAXONOMIC DIVERSITY: The family is monotypic, with a single species, *Antilocapra americana*.

FIG. 502. Antilocapridae; pronghorn, *Antilocapra americana*.

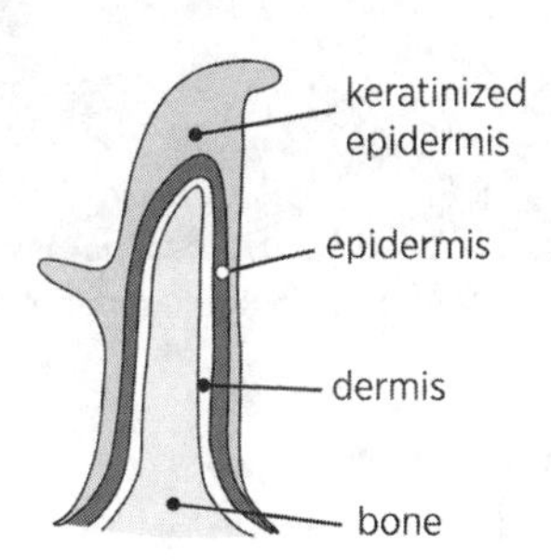

FIG. 503. Cross-section through a pronghorn. Redrawn from Gunderson (1976).

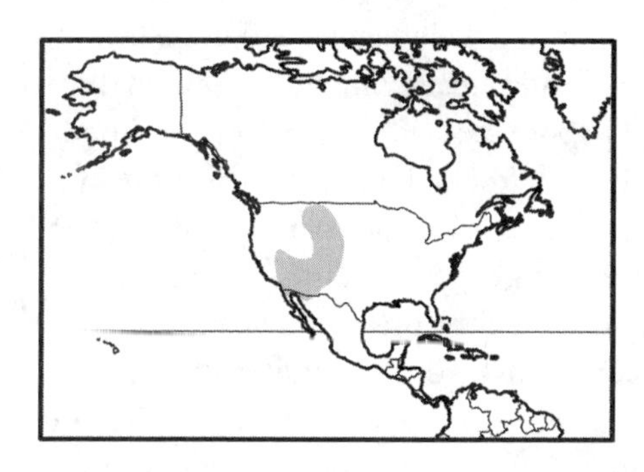

FIG. 504. Geographic range of family Antilocapridae: temperate arid and semiarid grasslands of western North America.

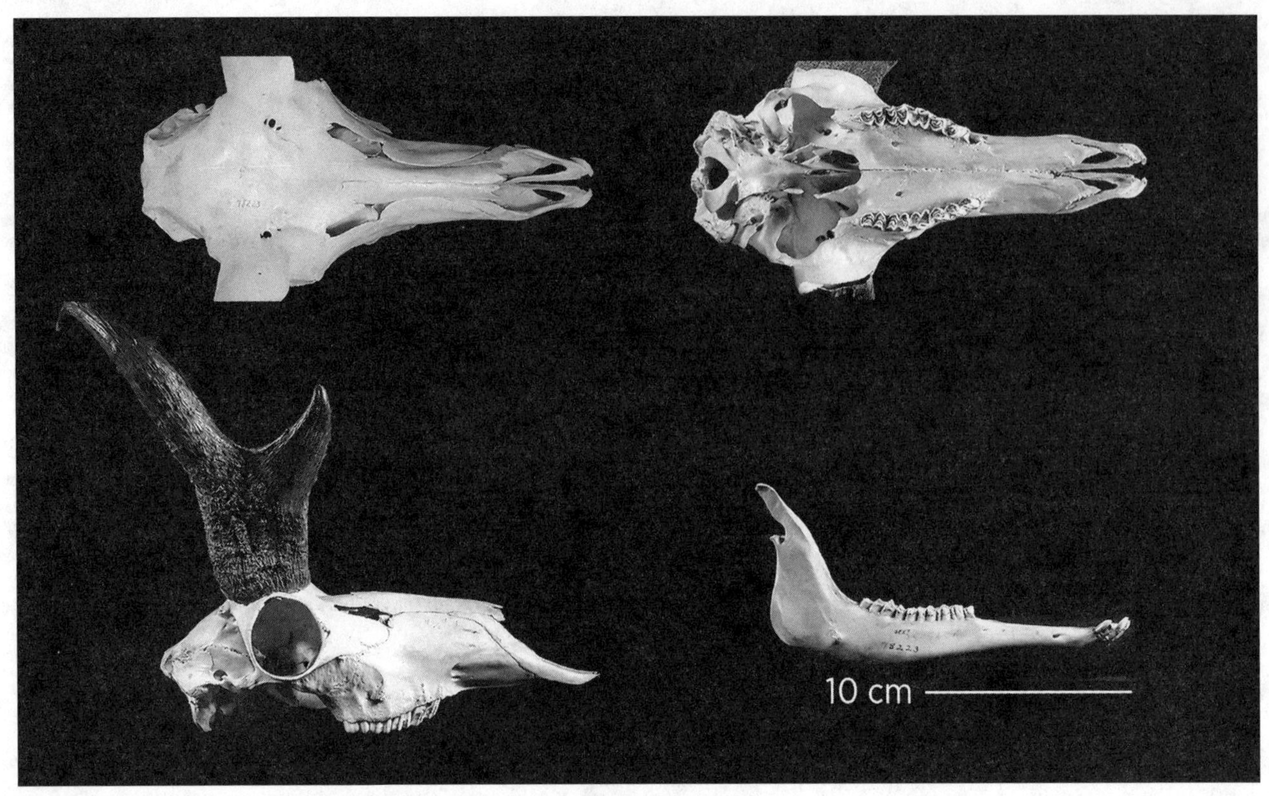

FIG. 505. Antilocapridae; skull and mandible of the pronghorn, *Antilocapra americana*.

Family CERVIDAE (deer, elk, muntjac, caribou, moose, brockets, pudu)

Small to large (shoulder height 0.3 m [*Pudu*] to 2.4 m [*Alces*]; mass 10–800 kg); body slender to sturdy; face elongate; ears large and erect; bony antlers present in males only (but in both sexes in *Rangifer*), grown from permanent pedicel on frontal bone and shed annually, vary from short to massive, simple to multibranched or palmate; legs short to long; facial, tarsal, metatarsal, or interdigital glands may be present. Generally gregarious, occurring in herds from a few to many individuals; some species form male-dominated harems in breeding season; some migrate seasonally, covering very long distances (*Rangifer*); antlers are used for defense and male-male combat during breeding season; herbivorous, feeding on all classes of vegetation; mainly crepuscular but can be both diurnal and nocturnal. Occupy most terrestrial biomes, from tundra and deserts to both temperate and tropical forests; summer ranges may be distinct from winter ranges.

DIAGNOSTIC CHARACTERS:

- **antlers in males only** (except *Rangifer*), **usually complexly branched** (spikelike in most small genera), **bony core sheathed in skin and fur ("velvet") while growing; velvet and antler shed separately, usually annually** (fig. 507)
- antlers grow from a pedicel, a non-deciduous bony projection of the frontal; shedding occurs at the contact between the pedicel and the antler, termed the "burr"
- **digits 4/4; side toes small and nonfunctional**
- **lacrimal depression present**
- **two lacrimal foramina present at front edge of, or outside, orbit**

RECOGNITION CHARACTERS:

1. nasal and lacrimal separated by large oblong opening or fenestration (**nasolacrimal gap**)
2. upper canines present in some taxa, either small and rounded or tusklike
3. lateral (vertical) surfaces of molars smooth in texture (cf. Giraffidae)

DENTAL FORMULA: $\frac{0}{3}\ \frac{0\text{–}1}{1}\ \frac{3}{3}\ \frac{3}{3} = 32\text{–}34$

TAXONOMIC DIVERSITY: About 53 species in 18 genera divided into two subfamilies based on the structure of the lateral (2nd and 5th) metacarpal bones, which are reduced either to only the proximal portion (a condition termed plesiometacarpal) or only the distal portion (telemetacarpal); each subfamily has multiple tribes (from Mattioli 2011; see also Gilbert et al. 2006):

Cervinae (= Plesiometacarpalia):
- Cervini—*Axis* (hog, Bawean, and Calamian deer, chital), *Cervus* (elk, red deer, sika deer, wapiti), *Dama* (fallow deer), *Elaphurus* (Père David's deer), *Rucervus* (barasingha, brow-antlered deer), and *Rusa* (spotted deer, sambar, Javan deer)
- Muntiacini—*Elaphodes* (tufted deer) and *Muntiacus* (muntjacs)

Capreolinae (= Telemetacarpalia or Odocoileinae):
- Alceini—*Alces* (moose)
- Capreolini—*Capreolus* (roe deer) and *Hydropotes* (Chinese water deer)
- Odocoileini—*Blastocerus* (marsh deer), *Hippocamelus* (North Andean and South Andean huemules), *Mazama* (brockets), *Odocoileus* (mule and white-tailed deer), *Ozotoceros* (pampas deer), *Pudu* (pudus), and *Rangifer* (reindeer or caribou)

FIG. 506. Cervidae; moose, *Alces alces* (Capreolinae: Alceini, *adult male and calf*; *top*), and wapiti, or elk, *Cervus elaphus* (Cervinae: Cervini, *adult male and calf*; *bottom*).

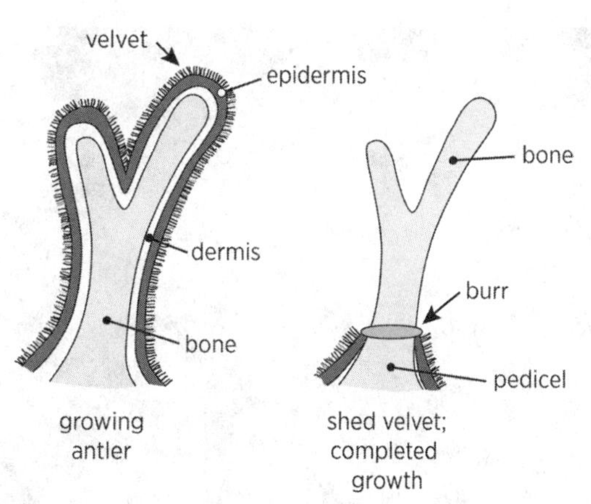

FIG. 507. Cross-section of a cervid antler in velvet (*left*) and fully grown (*right*). Redrawn from Gunderson (1976).

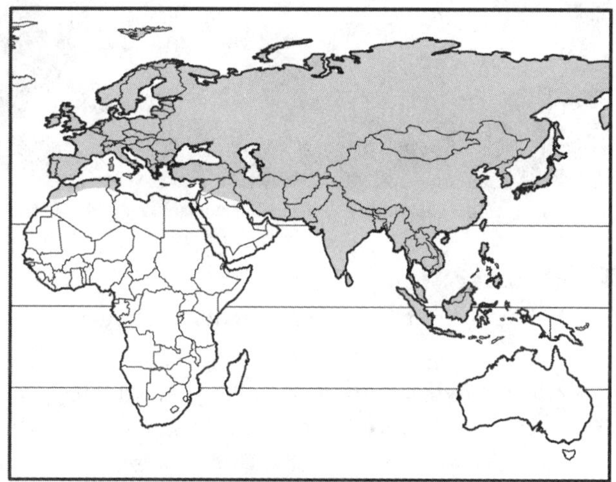

FIG. 508. Geographic range of family Cervidae: worldwide, excluding all but extreme northern Africa and all of Australasia (introduced into Australia and New Zealand).

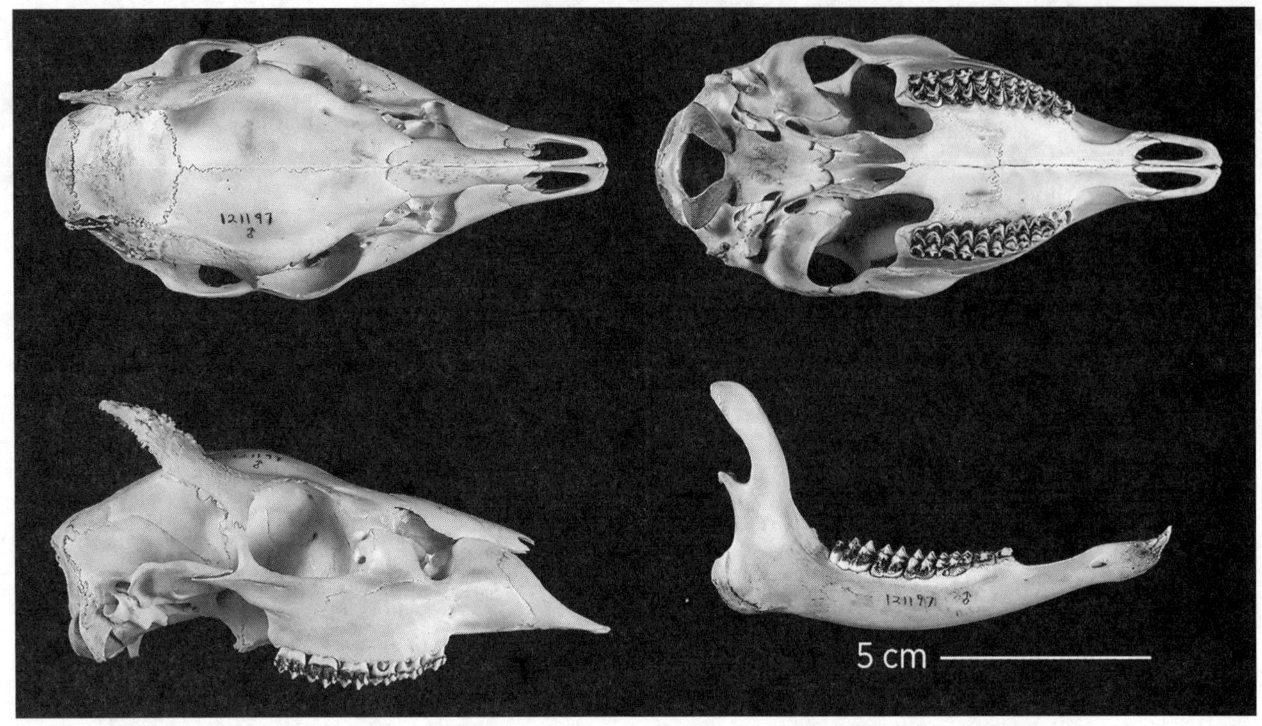

FIG. 509. Cervidae: Capreolinae: Odocoileini; skull and mandible of the South American red brocket, *Mazama americana*.

Family MOSCHIDAE (musk deer)

Small (body length 70–100 cm; height at shoulder 50–70 cm; mass 9–17 kg); no horns or antlers; long, rabbitlike upright ears; tail short; hind legs longer than forelegs; feet adapted for climbing in rough terrain with long, slender, and pointed hooves, including lateral toes that almost touch ground; **males with greatly enlarged upper canines**, forming saber-like tusks, movable in alveoli by muscle band; canines of females smaller; musk gland on abdomen between genitals and umbilicus in males; pelage very coarse, thick, and slightly waved; hairs weakly attached to skin so come away easily. Solitary, territorial, and secretive, nocturnal to crepuscular; primarily browsers, but diet also includes grazed grasses, mosses, and lichens. Inhabit lower montane forests and brushlands.

Earlier classifications placed the Moschidae as a subfamily of the Cervidae, but musk deer share few of the distinctive cervid characters, and molecular studies link them to the Bovidae.

RECOGNITION CHARACTERS:

1. **no horns or antlers**
2. digits 4/4, lateral toes functional
3. **lacrimal depression present**
4. nasal and lacrimal separated by large oblong opening or fenestration (**nasolacrimal gap**)
5. **one lacrimal foramen present at front edge of orbit** (resembling condition in Tragulidae and Bovidae)
6. **upper canines well developed, saber-like**
7. lateral (vertical) surfaces of molars smooth in texture (cf. Giraffidae)

DENTAL FORMULA: $\frac{0}{3}\ \frac{1}{1}\ \frac{3}{3}\ \frac{3}{3} = 34$

TAXONOMIC DIVERSITY: The family is monotypic, with seven species in the single genus *Moschus*.

FIG. 510. Moschidae; Siberian musk deer, *Moschus moschiferus*.

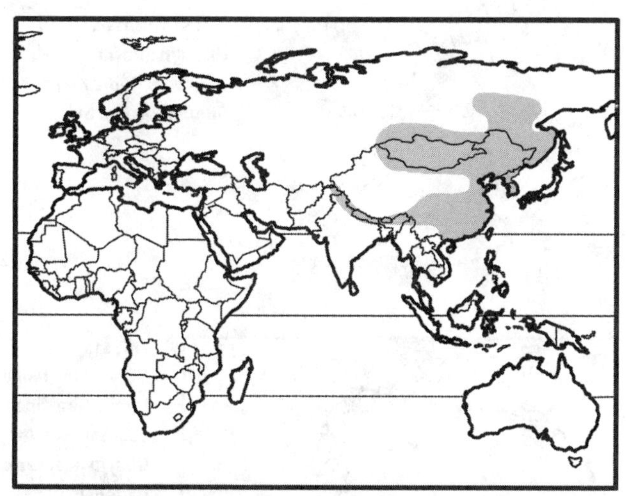

FIG. 511. Geographic range of family Moschidae: central Asia, from Afghanistan east to China and southern Siberia.

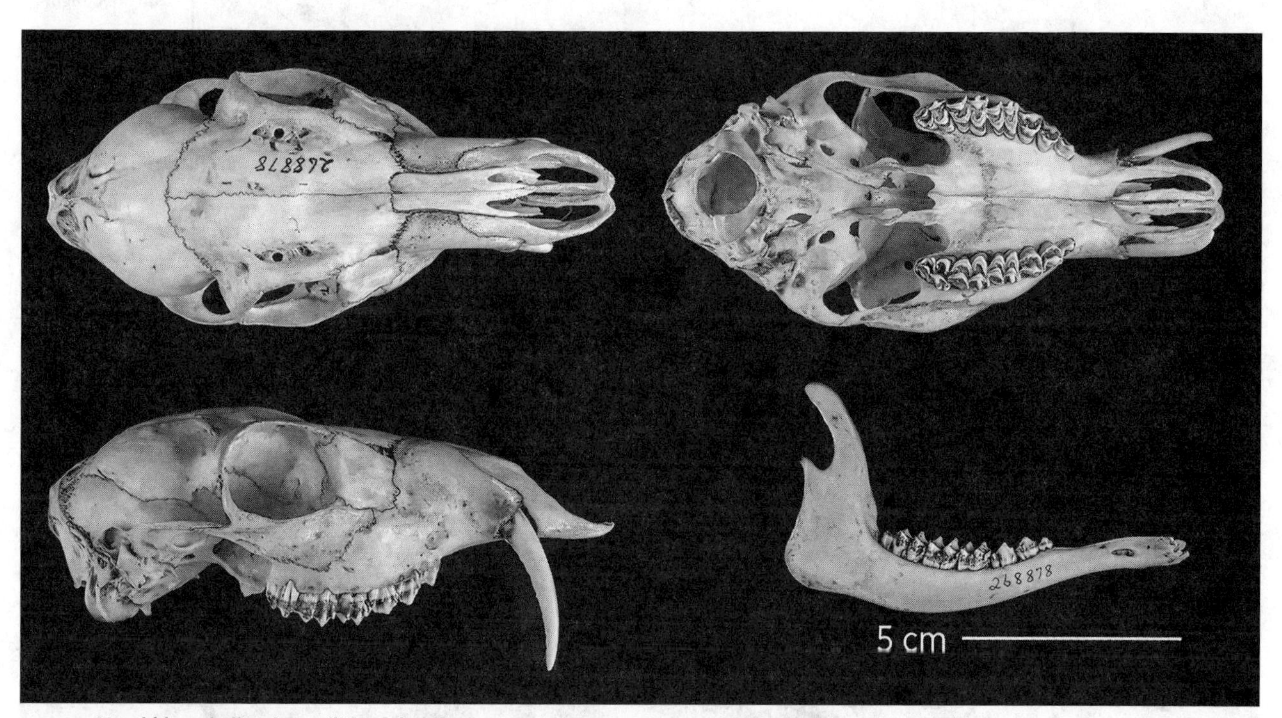

FIG. 512. Moschidae; skull and mandible of the Siberian musk deer, *Moschus moschiferus*.

Superfamily GIRAFFOIDEA

Family GIRAFFIDAE (giraffes, okapi)

Medium-sized to large (standing height from 1.5–1.7 m and mass 200 kg [*Okapia*] to 4.5–5.8 m and 500–750 kg [*Giraffa*]); head small in relation to body, long; upper lip highly mobile, tongue long and mobile; nostrils can be closed; eyes large; ears large and pointed; two permanent bony "horns," or ossicones, develop from distinct centers of ossification in the dermis, secondarily fusing to the skull at the frontal-parietal sutures (i.e., not frontal projections as in other artiodactyls), covered by skin in both sexes but balding in males (fig. 514); up to three additional prominences on frontals and posterior part of nasals; **neck long**; back sloping posteriorly; **tail long, tufted at tip**; hide thick, colored through shades of brown into various patterns of blotches or stripes, underparts white; mane short; glands absent, except for interdigital glands in *Okapia*. Giraffes gregarious, occurring in herds of up to 25 individuals; long-distance runners, moving up to 50 km/hr; herbivorous, feed primarily on leguminous shrub and tree species; occupy open

FIG. 513. Giraffidae; giraffe, *Giraffa camelopardalis*.

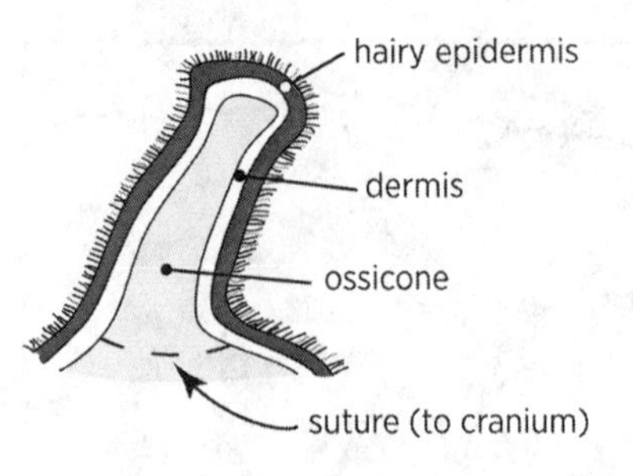

FIG. 514. Cross section through giraffe ossicone. Redrawn from Gunderson (1976).

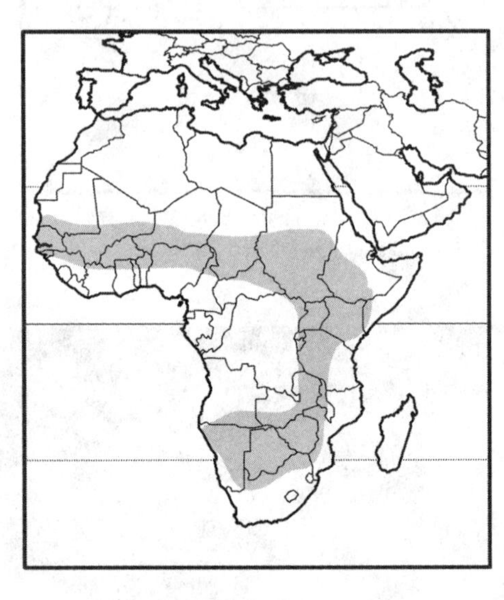

FIG. 515. Geographic range of family Giraffidae: sub-Saharan Africa; *Giraffa* in savannas, *Okapia* in dense rainforest of the Democratic Republic of the Congo.

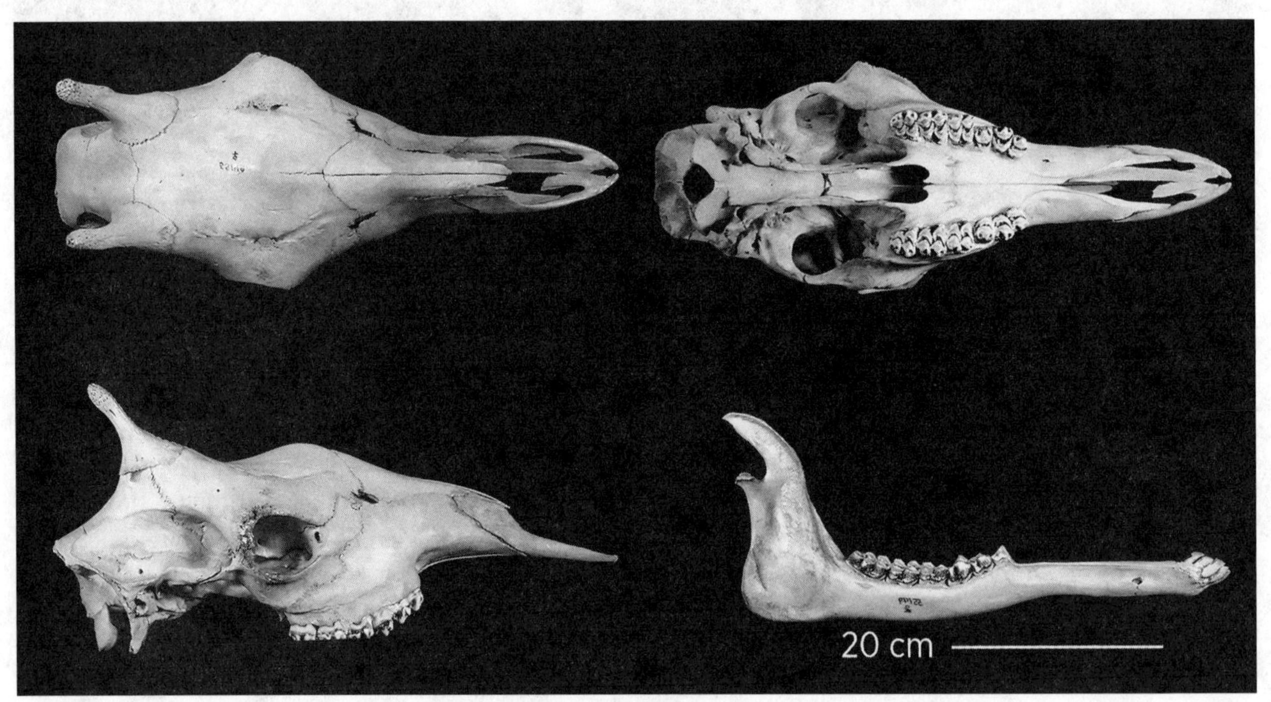

FIG. 516. Giraffidae; skull and mandible of the giraffe, *Giraffa camelopardalis*.

woodland savannas and grasslands. Okapi secretive and solitary, rarely in pairs; keen hearing and sense of smell, but poor eyesight; herbivorous, consuming primarily browse but with some grasses; inhabit dense tropical rainforest.

DIAGNOSTIC CHARACTERS:

- **two "horns" (ossicones) composed of dermally derived bone located over frontal-parietal sutures** (head ornaments in other ruminants are outgrowths of frontal bone) (fig. 514)
- **additional ossicones often present medial and anterior to others**

RECOGNITION CHARACTERS:

1. **digits 2/2**; no side toes
2. **no lacrimal depression**
3. nasal and lacrimal bones distinctly separated (**nasolacrimal gap**)
4. **one lacrimal foramen present, small, inside orbit**
5. upper canines absent
6. **lateral surfaces of molars rough in texture**

DENTAL FORMULA: $\frac{0}{3}\ \frac{0}{1}\ \frac{3}{3}\ \frac{3}{3} = 32$

TAXONOMIC DIVERSITY: Two genera: *Giraffa* (giraffes), with four species, and the monotypic *Okapia* (okapi).

Superfamily BOVOIDEA

Family BOVIDAE (antelope, bison, buffalo, cattle, goats, muskox, sheep)

Small to large (shoulder height from 25 cm and mass 3 kg [*Neotragus*] to 2 m and 1,350 kg [*Bos*]); body form varies considerably; head, face, and ears vary in relative proportions; two frontal horns (four in *Tetracerus*) always present; horns straight, recurved, or spiraled, often exhibiting growth phases by nodules or annuli; neck long and gracile to very short and thickened; tail long or short, usually not tufted; legs long and thin or short and thick; pelage diverse, short and smooth to rough and shaggy, variously colored and marked, usually paler ventrally; facial and pedal glands common, interdigitals in gregarious species. Solitary to highly gregarious, territorial to nomadic, some with long seasonal migrations. Predominantly crepuscular, but some either diurnal or nocturnal; herbivorous, most exhibit either grazing or browsing preferences. Range widely across tundra to desert and from rocky mountainous regions to dense lowland tropical forests; most occupy grasslands, scrublands, or open savanna biomes.

DIAGNOSTIC CHARACTERS:

- **horns, projecting from frontal bones, always present in males, variable in females** (when present, usually smaller than in males), **permanent, unbranched, covered with hardened keratin sheath that grows from base; sheath and bony core non-deciduous (cf. Antilocapridae)** (figs. 518, 519)
- **single (rarely two) lacrimal foramen present, usually on inside of orbital rim**

FIG. 517. Bovidae; bison, *Bos bison* (Bovinae: Bovini; *adult male and calf*; *top*); bighorn sheep, *Ovis canadensis* (Antilopinae: Caprini; *middle*), and muskox, *Ovibos moschatus* (Antilopinae: Caprini; *bottom*).

FIG. 518. Heads, horn shapes, and relative sizes of selected bovids: *A*, common duiker (*Sylvicapra grimmia*); *B*, springbok (*Antidorcas marsupialis*); *C*, waterbuck (*Kobus ellipsiprymnus*); *D*, greater kudu (*Tragelaphus strepsiceros*); *E*, gemsbok (*Oryx gazella*); *F*, blue wildebeest (*Connochaetes taurinus*).

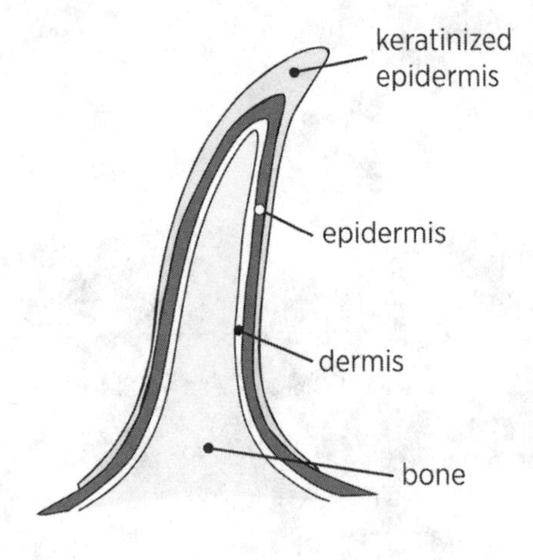

FIG. 519. Cross-section through a bovid horn. Redrawn from Gunderson (1976).

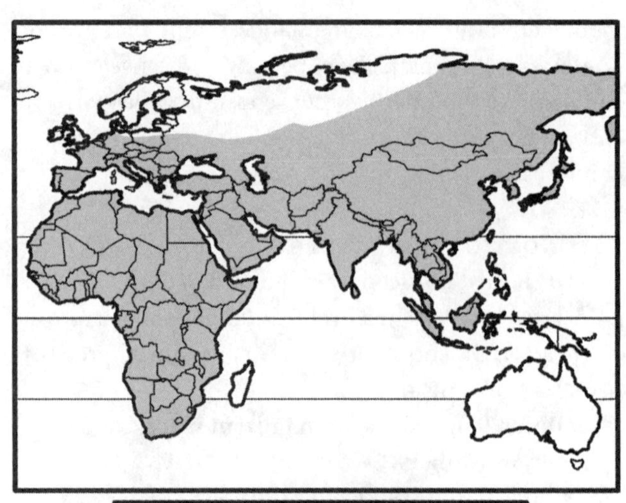

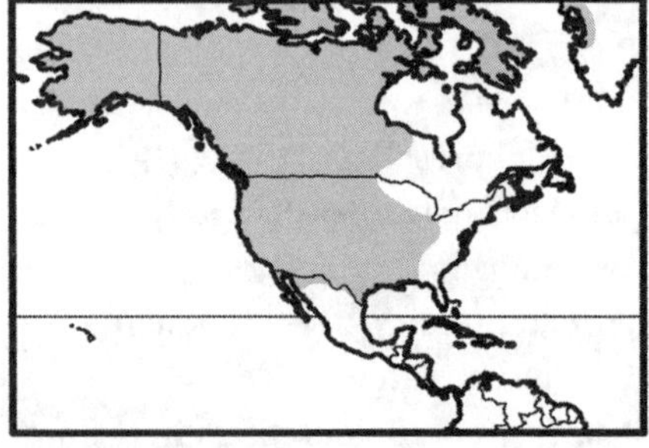

FIG. 520. Geographic range of family Bovidae: North America, Eurasia, Africa (absent from South America and Australasia); by far the greatest diversity occurs in Africa, secondarily (primarily sheep and goats) in Asia.

RECOGNITION CHARACTERS:

1. **digits usually 4/4, but side toes small and nonfunctional** (occasionally absent)
2. lacrimal depression present or absent
3. nasal and lacrimal bones separated (nasolacrimal gap) or united (no gap)
4. upper canines absent
5. vertical (lateral) surfaces of molars smooth in texture (cf. Giraffidae)
6. stomach with four chambers

DENTAL FORMULA: $\frac{0}{3}\ \frac{0}{1}\ \frac{3}{2\text{–}3}\ \frac{3}{3} = 30\text{–}32$

TAXONOMIC DIVERSITY: 54 genera and about 280 species divided into two subfamilies, one with nine tribes and the other with three tribes (from Groves and Leslie 2011; see Hassanin et al. 2012 and Bibi 2013 for phylogenetic analyses) (table 8):

Table 8. Taxonomy and distribution of subfamilies, tribes, and genera of Bovidae.

Subfamily/Tribe	Genus	Number of species	Common name	Distribution
Antilopinae				
Aepycerotini	*Aepyceros*	2	impalas	Africa
Alcelaphini	*Alcelaphus*	7	hartebeest	Africa
	Beatragus	1	herola	Africa
	Connochaetes	5	wildebeest	Africa
	Damaliscus	10	bontebok, blesbok, topi, tsessebe	Africa
Antilopini	*Ammodorcas*	1	dibatag	Africa
	Antidorcas	3	springbok	Africa
	Antilope	1	black buck	Africa
	Dorcatragus	1	beira	Africa
	Eudorcas	5	gazelles	Africa
	Gazella	20	gazelles	Africa, Asia
	Litocranius	2	gerenuk	Africa
	Madoqua	12	dik-dik	Africa
	Nanger	5	gazelles	Africa
	Ourebia	4	oribi	Africa
	Procapra	2	gazelle	central Asia
	Raphicerus	3	grysbok, steenbok	Africa
	Saiga	2	saiga	Asia
Caprini	*Ammotragus*	1	aoudad	northern Africa
	Arabitragus	1	Arabian tahr	Arabian Peninsula
	Budorcas	4	takin	Asia
	Capra	9	goat, ibex, markhor, tur	Europe, central Asia, Arabia
	Capricornis	7	serow	Asia
	Hemitragus	1	Himalayan tahr	Himalayan region
	Nemorhaedus	6	goral	Asia
	Nilgiritragus	1	nilgiri tahr	India
	Oreamnos	1	mountain goat	North America
	Ovibos	1	muskox	North America, Eurasia
	Ovis	20	argali, sheep	Northern Hemisphere
	Pantholops	1	chiru	central Asia
	Pseudois	2	blue sheep	central Asia
	Rupicapra	6	chamois	Europe, western Asia
Cephalophini	*Cephalophus*	26	duiker	Africa
	Philantomba	12	duiker	Africa
	Sylvicapra	3	duiker	Africa
Hippotragini	*Addax*	1	addax	Africa
	Hippotragus	3	roan and sable antelopes	Africa
	Oryx	6	oryx	Africa, Arabian Peninsula
Neotragini	*Neotragus*	5	antelope, suni	Africa
Oreotragini	*Oreotragus*	11	klipspringer	Africa
Reduncini	*Kobus*	12	kob, lechwe, puku, waterbuck	Africa
	Pelea	1	rhebok	Africa
	Redunca	9	reedbuck	Africa

Table 8. Continued

Subfamily/Tribe	*Genus*	*Number of species*	*Common name*	*Distribution*
Bovinae				
Boselaphini	*Boselaphus*	1	nilgai	India
	Tetracerus	1	four-horned antelope	India
Bovini	*Bos*	6	cattle, bison, gaur, yak	worldwide
	Bubalus	2	Asiatic buffalo	Asia
	Pseudoryx	1	saola	SE Asia
	Syncerus	4	African buffalo	Africa
Tragelaphini	*Ammelaphus*	2	lesser kudus	Africa
	Nyala	1	nyala	Africa
	Strepsiceros	4	kudus	Africa
	Taurotragus	2	elands	Africa
	Tragelaphus	15	bongo, bushbuck, gedemsa, sitatunga	Africa

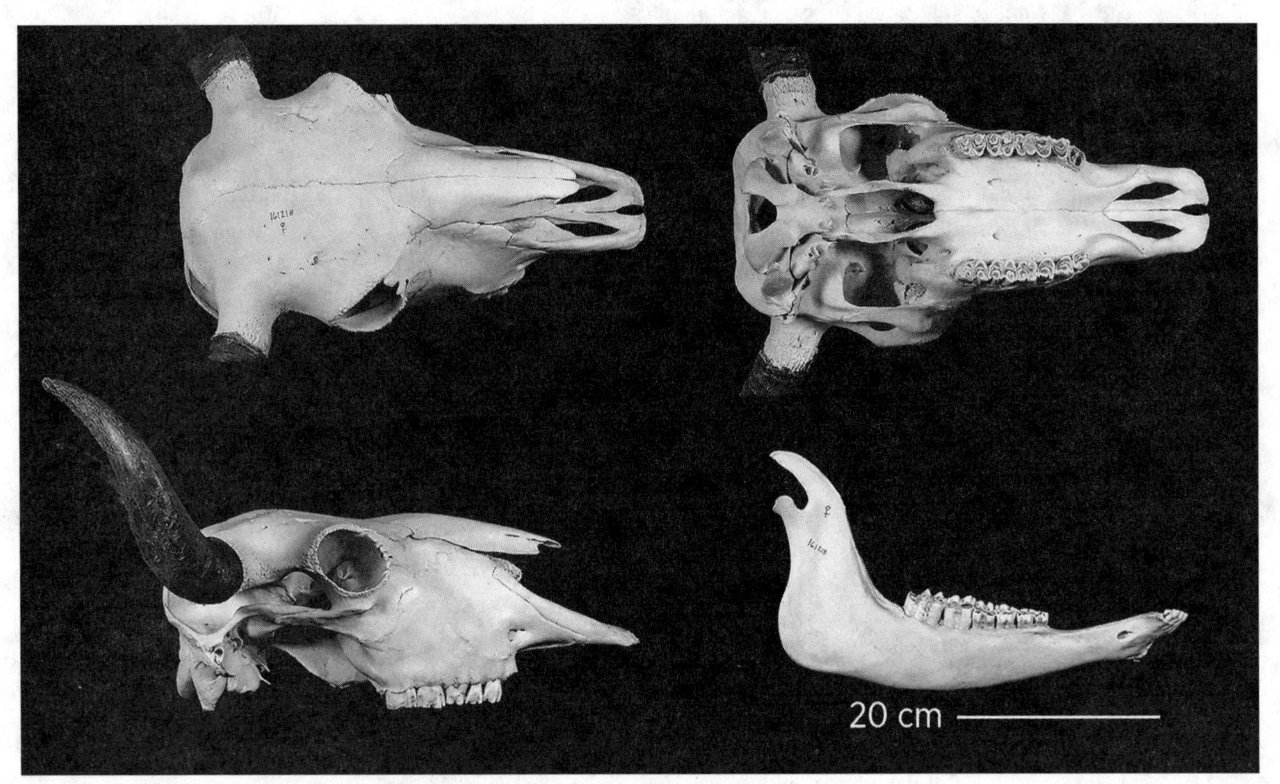

FIG. 521. Bovidae: Bovinae: Bovini; skull and mandible of the American bison, *Bos bison*.

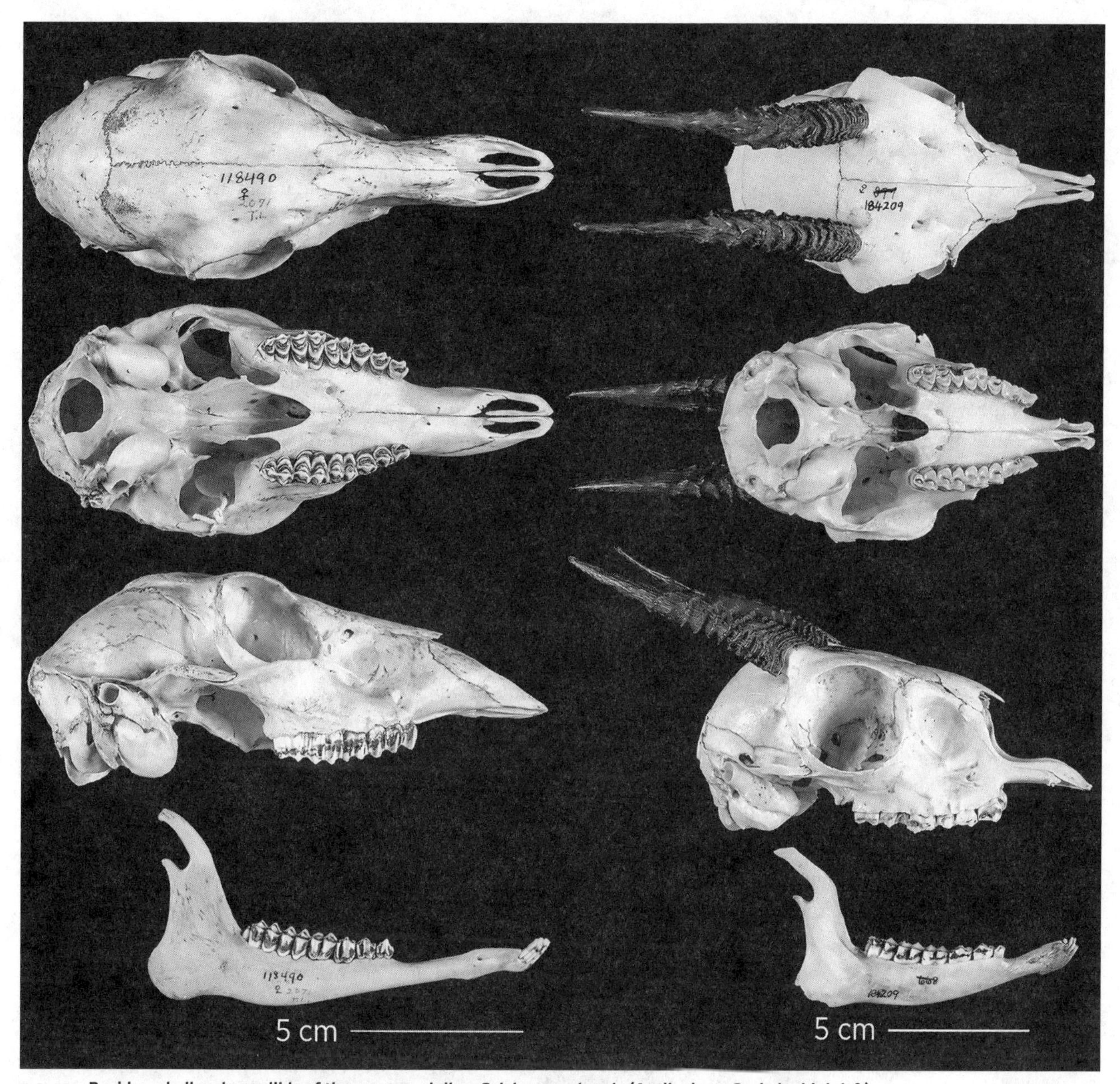

FIG. 522. Bovidae; skull and mandible of the common duiker, *Sylvicapra grimmia* (Antilopinae: Cephalophini; *left*), and Kirk's dik-dik, *Madoqua kirkii* (Antilopinae: Antilopini; *right*).

Clade CETACEA (= Cete)

Although it is traditionally treated as an order of mammals, we conservatively call the Cetacea a clade in recognition that it is almost certainly housed within the order Artiodactyla. Whales, along with sirenians, are notable for being the mammals most fully adapted to aquatic life. They are completely aquatic; the body is fusiform (cigar-shaped), nearly hairless, insulated by thick blubber, and lacks sebaceous glands. Most cetacean vertebrae have high neural spines, and the cervical vertebrae are highly compressed, often fused. The clavicle is absent, the forelimbs (flippers) are paddle-shaped, and no external digits or claws are present. Little movement is possible between the joints distal to the shoulder. The proximal segments of the forelimb are short, whereas the digits frequently are unusually long because of the development of more phalanges per digit than the basic eutherian number (2–3–3–3–3), a condition termed hyperphalangy. The hind limbs are vestigial, do not attach to the axial skeleton, and are not visible externally. The flukes (tail fins) are horizontally oriented and, like the dorsal fins (when present), have a fibrous cartilaginous core but lack skeletal support; propulsion involves vertical undulations of the body. To aid in streamlining the body, the single pair of mammae lies flat along the abdomen, and the teats are enclosed within slits adjacent to the urogenital opening. In males, the testes remain abdominal and the penis is fully retractable into the body.

Whales include the largest vertebrate, living or extinct (the blue whale, *Balaenoptera musculus*, at approximately 31 m in length and 160 metric tons in mass), but the range in body size across all members of the clade is extensive. The smallest cetacean is the critically endangered vaquita (*Phocoena sinus*, 120–150 cm, 30–48 kg), found in the Gulf of California and threatened by fisheries.

The cetacean skull is highly modified as a result of the posterior migration of the external nares and by the associated expansion of rostral elements (premaxillary, maxillary, and vomer) both anteriorly and dorsoposteriorly and of posterior elements (most notably the supraoccipital bones) anteroposteriorly (fig. 523). Termed "telescoping," this process provides for a longer rostrum, but also results in the maxillary and premaxillary forming the roof of the skull, and in both overlapping and compressing bones such as the nasal, frontal, and parietal posteriorly; moreover, these latter bones are superimposed and compressed caudally by the occipital. This process results in great overlapping of these migrated elements, and as such resembles the sliding elements of old-fashioned collapsible telescopes, hence the name.

Telescoping differs fundamentally between the two major groups of whales, with the maxillary bones extending mostly ventrocaudally in baleen whales (Mysticeti), but dorsocaudally in toothed whales (Odontoceti), with substantial ramifications for facial musculature (see Rommel et al. 2008 for a particularly clear explanation). The tympanoperiotic bone (housing the middle and inner ear) is not braced against adjacent bones of the skull in most cetaceans and is partly insulated from the rest of the cranium by surrounding air sinuses and soft tissue.

ORIGIN AND RELATIONSHIPS OF WHALES

Whales have been linked phylogenetically to a group of primitive, extinct carnivorous ungulates, the mesonychians, with which early whales share a number of morphological traits, including tooth structure, digitigrade stance, and paraxonic foot structure. Because of their unique form and adaptations to an aquatic environment, Simpson (1945) placed all whales in their own cohort, Mutica. However, as noted in the introduction to the superorder Cetartiodactyla, due to advances in molecular systematics and recent fossil discoveries, whales are now recognized either as a sister group to the Artiodactyla or, more likely, as derived from stem hippopotamids. Gingerich et al. (2001) reported on an Eocene whale that included sufficient limb bones to confirm that it had a "double-pulley" astragalus, a feature until then unique to the artiodactyls. This discovery effectively solidified the relationship between these two orders. The node linking Cetacea and Artiodactyla is called the Cetartiodactyla; deeper within this clade, the Whippomorpha includes the hippopotamuses and whales and is a sister group to the Ruminantia (see figs. 94 and 484). Molecular studies increasingly support the hippopotamid association, and if whales are most closely associated with hippopotamuses, then the Artiodactyla *sensu stricto* is paraphyletic unless cetaceans are included; hence, some authors refer to the order Cetartiodactyla as including both groups. Rose (2006), Uhen (2010), and Berta et al. (2015), provide useful summaries of this subject.

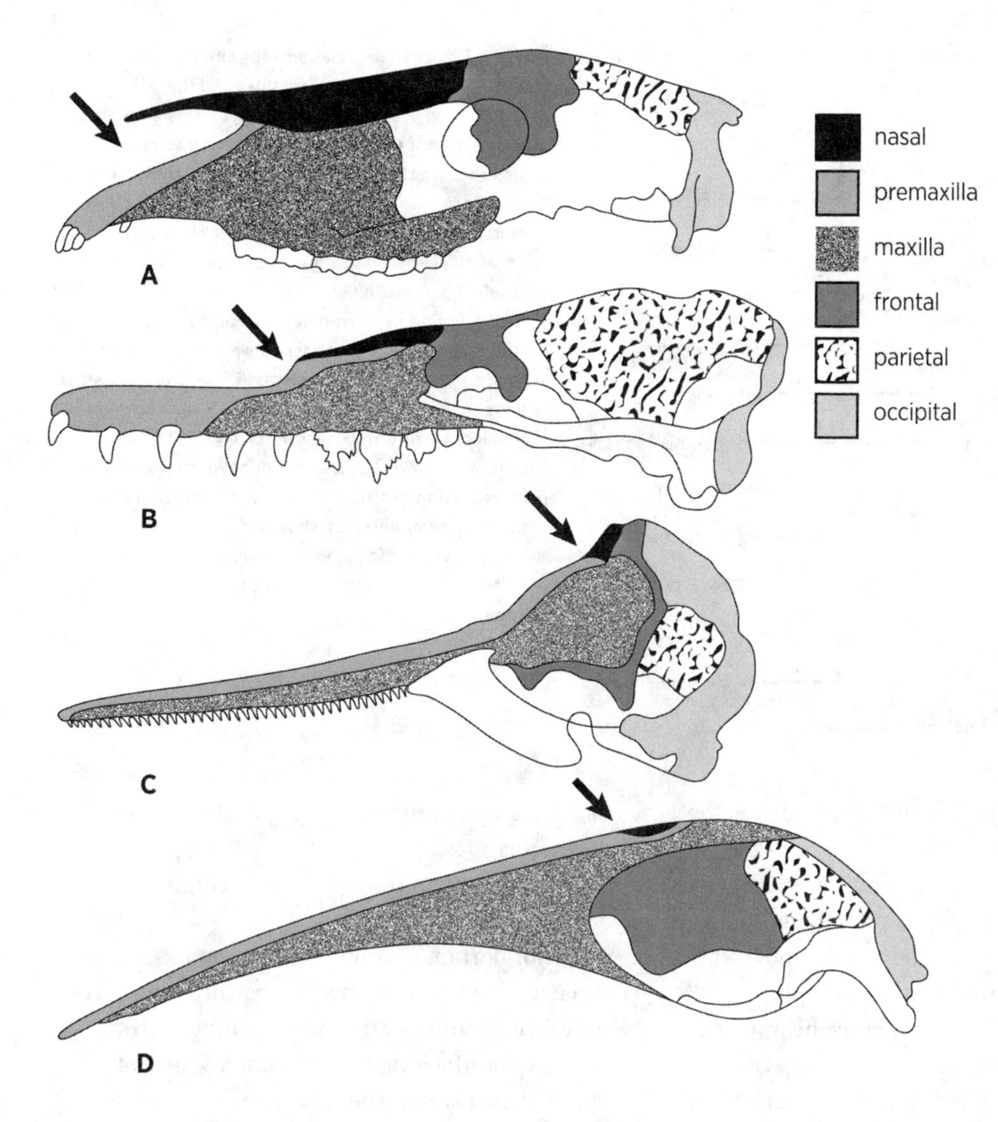

FIG. 523. Telescoping of the skull in cetaceans (position of external nares is indicated by arrows). *A*, a terrestrial, non-cetacean mammal (*Equus*, a horse); *B*, a fossil archaeocete whale (*Basilosaurus*), note the heterodont dentition; *C*, a modern odontocete whale (*Delphinus*, a dolphin); and *D*, a modern mysticete whale (*Balaenoptera*, a rorqual whale). Redrawn from Feldhamer et al. (2015).

CLASSIFICATION OF CETACEA

The Cetacea as an order is traditionally divided into two suborders, one for the baleen, or mysticete, whales (suborder Mysticeti) and a second for the toothed, or odontocete, whales (suborder Odontoceti). We use this division of whales, which is supported by most recent data, with modification of the total number of families recognized (see below). Because we recognize the subordinate position of Cetacea within Artiodactyla, we refer to both Cetacea and its subordinate groupings as clades to reflect their monophyletic nature:

Subclade Mysticeti (baleen whales)
- Family Balaenidae (right whales)
- Family Balaenopteridae (rorquals)
- Family Eschrichtiidae (gray whale)
- Family Neobalaenidae (pygmy right whale)

Subclade Odontoceti (toothed whales)
- Family Delphinidae (dolphins)
- Family Iniidae (Amazon and Orinoco river dolphins)
- Family Lipotidae (baiji)
- Family Monodontidae (narwhal, beluga)
- Family Phocoenidae (porpoises)
- Family Physeteridae (sperm whales [including Kogiidae])
- Family Platanistidae (Ganges and Indus river dolphins)
- Family Pontoporiidae (franciscana)
- Family Ziphiidae (beaked whales)

Familial relationships within both subclades remain unclear, in part because selection for an aerodynamic swimming lifestyle has led to some convergences in morphology. This is particularly true in the more diverse Odontoceti, traditionally divided into three to four clades or superfamilies, including (e.g., McKenna and Bell 1997, Rice 1998) Physeteroidea (sperm whales), Ziphoidea (beaked whales, often included in Physeteroidea), Delphinoidea (Delphinidae,

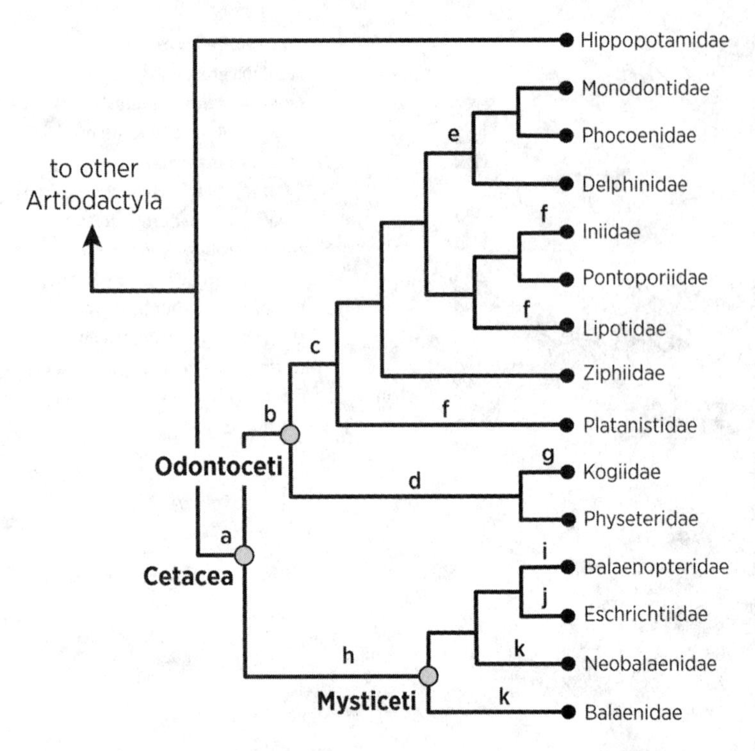

FIG. 524. Phylogenetic relationships among cetacean families and between those families and the Hippopotamidae of the Artiodactyla. Letters at internal branches identify a few hypothesized character appearances in the evolutionary history of these lineages. Cete clade: a = no tooth replacement, major reduction of hind limbs, hind feet lost and hind limbs entirely internal, thickened wall around middle ear. Odontocete subclade: b = single blowhole but nasal passages separate, melon hypertrophied, evolution of echolocation; c = single blowhole, nasal passages merged; d = extreme asymmetry in size of nasal passages (right passage small relative to left), only one nasal bone; e = major brain expansion; f = return to freshwater; g = both nasal bones absent. Mysticete subclade: h = origin of baleen to filter food, reduction in size of mandibular foramen; i = tongue reduced to primarily connective tissue, numerous ventral throat grooves or pleats, ventral throat pouch, feeding by engulfment; j = feeding by benthic suction; k = feeding by skimming.

Phocoenidae, Monodontidae), and Platanistoidea (which historically included all river dolphins, but in recognition of the paraphyly of this group, now includes only the Platanistidae; remaining families are now treated within Delphinoidea). Contemporary molecular approaches are gradually providing clarity to this complex radiation, but resolution remains incomplete. Hypothesized relationships among these groups are depicted in figure 524, which is based on combined morphological and molecular character analyses (from Gatesy et al. 2013; see also Hassanin et al. 2012). While this tree is fully resolved, several of the nodes remain uncertain.

Subclade MYSTICETI (baleen whales)

RECOGNITION CHARACTERS:

1. **no teeth** (teeth present in fetus, lost before birth)
2. **baleen present**, either continuous anteriorly (Balaenopteridae) or with an anterior gap (all other families) (fig. 525)
3. two external nasal openings (blowholes) present, slit-like, located anterior to eye
4. facial profile of skull convex, with no fatty "melon" present
5. **skull more or less symmetrical**
6. nasals roofing part of the nasal passage
7. nasal passage simple (lacking complex system of diverticula)
8. nasals higher than frontals (frontals higher in odontocetes)
9. posterior portion of vomer exposed on basicranium and covering the basisphenoid/basioccipital suture
10. **maxilla extending posteriorly as a long, narrow process, interlocking over frontal, not spread outward over supraorbital process**
11. auditory bulla (tympanoperiotic bones) attached to skull
12. lower jaw loosely joined by ligaments at symphysis
13. mandibular condyle directed upward
14. sternum consists of single bone
15. do not echolocate

Family BALAENIDAE (right whales, bowhead whales)

Right and bowhead whales feed largely on planktonic crustaceans (krill) and mollusks, often skimming these foods at relatively slow speed from the surface of the water. They are most common near coastlines or near pack ice; the southern right whale (*Eubalaena australis*) makes long annual migrations from temperate or tropical waters to spend the austral summer in Antarctic waters; the northern right whales (*E. glacialis* and *E. japonica*) remain in the North Pacific or North Atlantic,

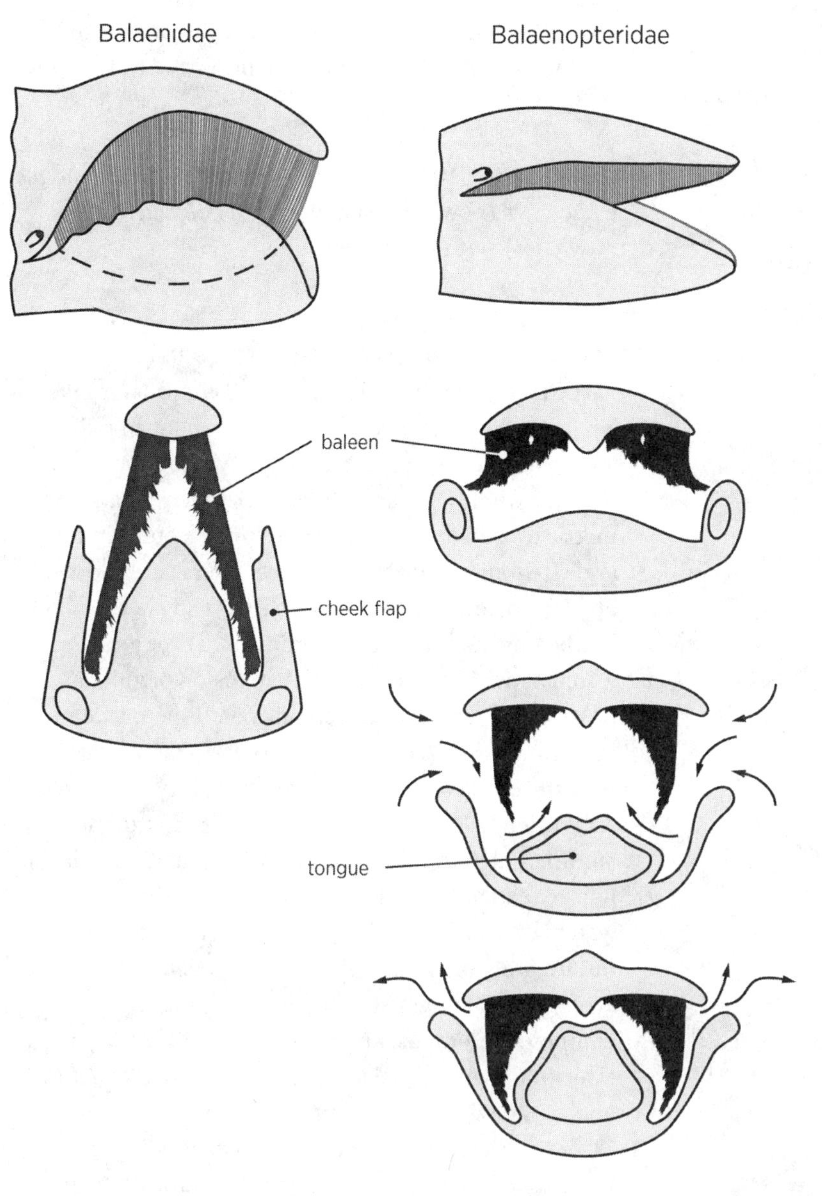

FIG. 525. Baleen types: *left*, long baleen of a right whale (Balaenidae); *right*, short, stout baleen of a rorqual (Balaenopteridae). Arrows at lower right indicate pattern of water flow into and out of mouth during feeding. Modified from Pivorunas (1979) and Feldhamer et al. (2015).

respectively, year-round. The bowhead whale (*Balaena mysticetus*) never ventures far south of the Arctic Circle.

DIAGNOSTIC CHARACTERS:

- **head huge, making up nearly one-third of body length**
- **skull strongly arched (bowed) in lateral view**

FIG. 526. Balaenidae; bowhead, *Balaena mysticetus*.

RECOGNITION CHARACTERS:

1. **color dark gray to black** (white patches often on chin)
2. **body chunky, robust** (up to 18.5 m long and 67 metric tons)
3. flippers short and rounded, with five digits (four in other cetaceans)
4. dorsal fin absent
5. **rostrum arched to accommodate long baleen plates**
6. **no longitudinal grooves, or furrows, in skin of throat**
7. **baleen plates long and narrow** (up to 350 separate plates up to 4 m in length); fold on floor of mouth when jaws are closed

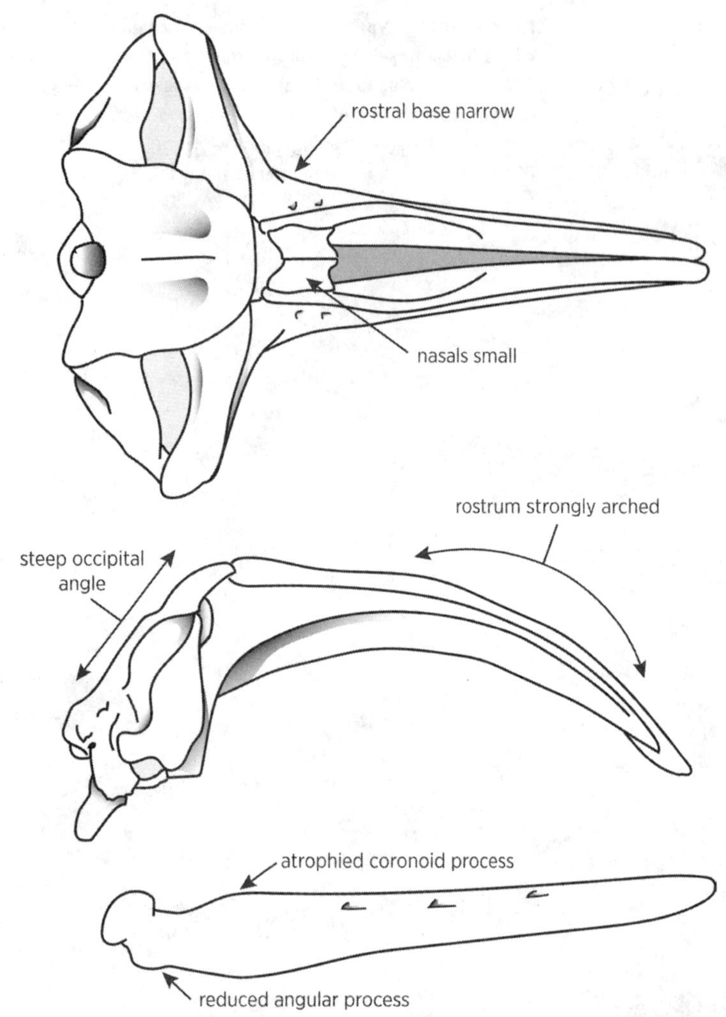

FIG. 527. Balaenidae; skull and mandible of a bowhead whale, *Balaena mysticetus*; salient characters distinguishing a balaenid whale from other Mysticeti are indicated. Redrawn from Marx et al. (2016).

8. base of rostrum relatively narrow
9. nasals small
10. **posterior margins of nasal and premaxilla not extending beyond level of anterior margin of supraorbital process of frontal**
11. frontal scarcely visible at crest of skull
12. **maxilla without elongate process extending posteriorly**
13. anterior margin of parietal behind posterior margins of premaxilla, maxilla, and nasal
14. supraoccipital extending anteriorly beyond zygomatic process of squamosal
15. supraoccipital angle very steep
16. coronoid process of mandible atrophied
17. angular process of mandible reduced
18. cervical vertebrae fused

RANGE: Arctic, Antarctic, and temperate waters of both hemispheres, as far south as Florida and as far north as southern Brazil.

TAXONOMIC DIVERSITY: Two genera: the monotypic *Balaena* (bowhead whale), and *Eubalaena* (right whales), with three species.

Family BALAENOPTERIDAE (rorquals)

Vary in size from the relatively small common minke whale (*Balaenoptera acutorostrata*, at about 10 m in length) to the largest vertebrate known to have existed, the blue whale (*B. musculus*, at 31 m in length and 160 metric tons in mass). Body form varies from slender and streamlined to chunky. Baleen plates are short and broad, and the skin of the throat and chest is marked by numerous longitudinal pleats or grooves, which allow for the enormous expansion of gular and post-throat region during engulfment foraging. Some species typically feed in cold, high-latitude waters near the edges of the ice, where upwelling, nutrient-rich waters allow for great growths of plankton in summer; planktonic crustaceans and small schooling fish are eaten. They then migrate toward equatorial waters in winter; wintering adults usually don't feed, but live off stored blubber. The humpback whale, *Megaptera*, is notable for its spectacular leaps and complex, melodious, and varied songs. In spite of their size, unknown rorqual species may yet exist; genetic analyses distinguished the Antarctic minke whale (*B. bonaerensis*) from the common minke whale only in the 1990s; even more surprising, Omurai's whale (*B. omurai*) was discovered in 2003 when genetic analysis confirmed that a whale stranded in the Sea of Japan was a distinct species.

DIAGNOSTIC CHARACTERS:

- **numerous longitudinal grooves in skin of throat** (Eschrichtiidae may have two to four such grooves)
- **rows of baleen continuous anteriorly** (separated at anterior end of mouth in other mysticetes)
- **lower jaw conspicuously bowed outward** (more or less straight in other mysticetes)

RECOGNITION CHARACTERS:

1. body slender (length 10–31 m)
2. **color gray or black above, with varying amounts of white below**

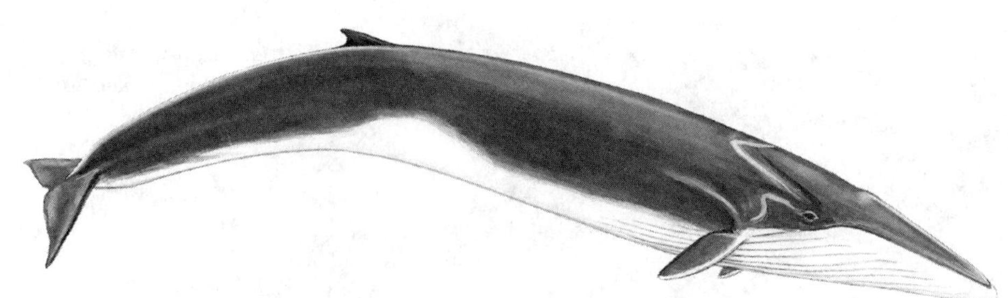

FIG. 528. Balaenopteridae; fin whale, *Balaenoptera physalus*.

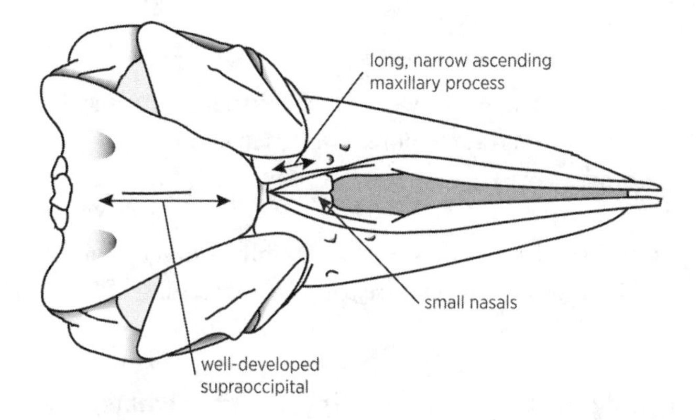

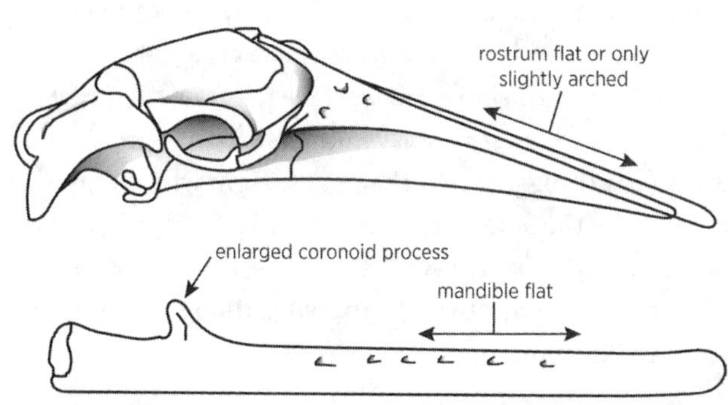

FIG. 529. Balaenopteridae; skull and mandible of the common minke whale, *Balaenoptera acutorostrata*; salient characters distinguishing a balaenopterid whale from other Mysticeti are indicated. Redrawn from Marx et al. (2016).

3. dorsal fin present, sickle-shaped (*Balaenoptera*) or small (*Megaptera*)
4. **baleen plates short, broad**
5. **rostrum relatively flat and broad at base**
6. nasals small; either not exposed or only barely exposed on skull roof
7. nasal and premaxilla extending posteriorly beyond level of anterior margin of supraorbital process of frontal
8. frontals scarcely or not at all visible at crest of skull
9. maxilla with elongate posterior process
10. **parietal extending anteriorly beyond posterior margins of premaxilla, maxilla, and nasal**
11. supraoccipital extending anteriorly beyond zygomatic process of squamosal
12. cervical vertebrae not fused

RANGE: Oceans of the world, summers in Arctic and Antarctic waters, and winters in more tropical waters.

TAXONOMIC DIVERSITY: Two genera: *Balaenoptera*, with seven species, and the monotypic *Megaptera*.

Family ESCHRICHTIIDAE (gray whales)

Fairly large (up to 15 m in length; mass up to 32 metric tons) with a slender body sporting dorsal protuberances, but no dorsal fin per se, relatively small head, narrow and arched rostrum; baleen plates are short, and telescoping of the skull is not extreme; nasal bones are large, and the frontals are broadly visible on the roof of the skull. These whales are restricted to the northern Pacific (an Atlantic population became extinct in the seventeenth or early eighteenth century), migrating between Arctic waters in summer and the coast of Baja California or Korea and Japan in winter. They live on stored lipids through winter, but in summer they will forage continuously throughout the 24-hour day, feeding in shallow (<50 m) waters, where they employ suction feeding on bottom-dwelling amphipods, polychaete worms, and bivalves. Their foraging creates characteristic "feeding pits" up to 1 × 3 m and 0.5 m deep. Molecular analyses are equivocal as to whether eschrichtiids are a sister taxon to the balaenopterids or nested within the latter.

RECOGNITION CHARACTERS:

1. body slender (length 11–15 m)
2. **color gray to black with white mottling**
3. **dorsal fin absent** (only small bumps present)
4. **throat with two (occasionally three or four) longitudinal grooves in skin**
5. **baleen plates short, narrow**
6. rostrum relatively narrow, slightly arched
7. **nasals large**
8. nasal and premaxilla extending posteriorly beyond

FIG. 530. Eschrichtiidae; gray whale, *Eschrichtius robustus*.

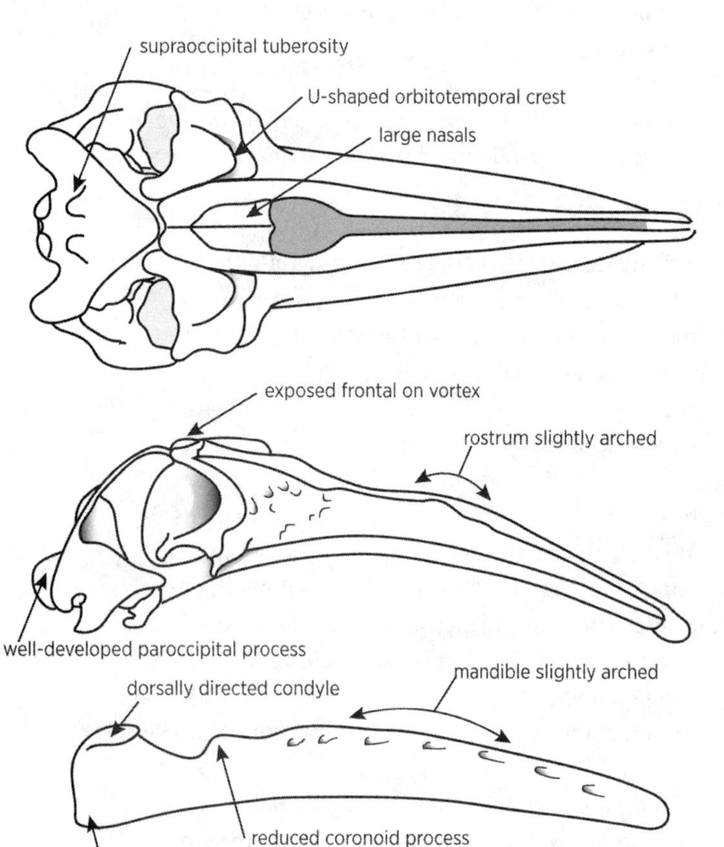

FIG. 531. Eschrichtiidae; skull and mandible of a California gray whale, *Eschrichtius robustus*; salient characters distinguishing an eschrichtiid whale from other Mysticeti are indicated. Redrawn from Marx et al. (2016).

level of anterior margin of supraorbital process of frontal

9. **frontal broadly exposed at crest of skull**
10. maxilla with elongate posterior process
11. anterior margin of parietal behind posterior margins of premaxilla, maxilla, and nasal
12. **supraoccipital not extending anteriorly beyond zygomatic process of squamosal**
13. paired tuberosities on the occipital bone
14. cervical vertebrate not fused

RANGE: North Pacific Ocean; give birth in subtropical lagoons, migrate to boreal waters in summer to forage.

TAXONOMIC DIVERSITY: The family is monotypic, with a single species, *Eschrichtius robustus*.

Family NEOBALAENIDAE (pygmy right whale)

Feed primarily on copepods; apparently do not make long-distance migrations; and, unlike true right whales, do not breach or slap water surface with the tail fluke. General biology very poorly known. Fordyce and Marx (2012) suggest that the pygmy right whale may actually be the sole extant member of the Cetotheriidae (subfamily Neobalaeninae), the earliest toothless, baleen-bearing mysticetes otherwise thought extinct by the late Pliocene.

RECOGNITION CHARACTERS: Similar to those of the true right whales (Balaenidae), the family in which this whale has usually been placed. Pygmy right whales differ from balaenids in the following features:

1. small size (length 5–6 m)
2. larger number of ribs (34, more than any other cetacean)
3. slender body form (like that of rorquals)
4. narrow rather than broad and more rounded flippers
5. four digits (as in all mysticetes except Balaenidae, which have five)
6. small dorsal fin present
7. skull only slightly bowed in lateral view
8. base of rostrum very wide relative to tip
9. supraorbital shield thrust anteriorly
10. supraorbital process not ventrally developed and without lateral expansion
11. U- rather than M-shaped anterior margins of palatine bones
12. postmandibular process not bulbous

FIG. 532. Neobalaenidae; pygmy right whale, *Caperea marginata*.

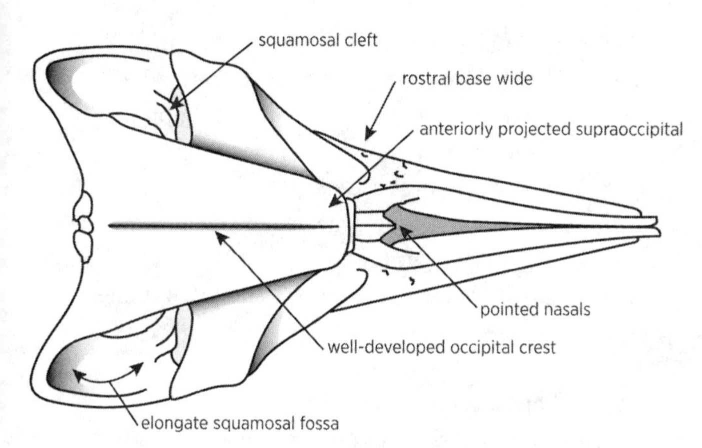

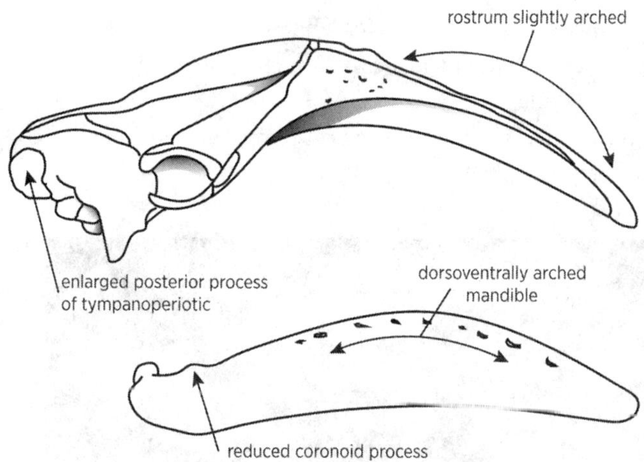

FIG. 533. Neobalaenidae; skull and mandible of the pygmy right whale, *Caperea marginata*; salient characters distinguishing a neobalaenid whale from other Mysticeti are indicated. Redrawn from Marx et al. (2016).

RANGE: Marine waters of Southern Hemisphere.

TAXONOMIC DIVERSITY: The family is monotypic, with a single species, *Caperea marginata*.

Subclade ODONTOCETI (toothed whales)

RECOGNITION CHARACTERS:

1. **teeth present; simple, homodont, monophyodont**
2. **no baleen**
3. one nasal opening (blowhole) present, crescent-shaped, located posterior to eye (except in sperm whales)
4. facial profile of skull concave, with depression occupied by a fatty organ ("melon") used in echolocation
5. **skull usually asymmetrical**
6. nasals greatly reduced, not roofing any part of nasal passage
7. nasal passage with a complex system of diverticula
8. frontals higher than nasals (lower in mysticetes)
9. **maxilla extending posteriorly as a large broad process, not interlocking with frontal, instead spreading outward over portion of supraorbital process**
10. auditory bulla (tympanoperiotic bones) not attached to skull
11. lower jaws firmly fused at symphysis
12. mandibular condyle directed posteriorly
13. sternum consists of three or more bones
14. echolocate

Family DELPHINIDAE (dolphins, orca)

The largest and most diverse group of cetaceans. Inhabit all oceans and some large rivers and estuaries in southern Asia, Africa, and South America. The facial depression of the skull is large, and the frontal and maxillary bones roof over the reduced temporal fossa. The "melon" is well developed and gives many delphinids a forehead that bulges prominently behind the beaklike snout. Some lack a beak and have a rounded profile (orca, pilot whale). Feed characteristically by making shallow dives and surfacing several times a minute, but diet (and associated morphology) variable; most consume fish or cephalopods, while some consume large amounts of crustaceans, and many orca populations specialize on other marine mammals (some argue they evolved as predators of great whales). Rapid swimmers; some species regularly leap from the water during feeding and traveling. Typically highly gregarious, and assemblages of some 100,000 individuals have been observed. Schooling behavior enhances the effectiveness of food searching, prey capture, and predator avoidance. Exceptional echolocators; some evidence that they use intense sound pulses to stun prey.

FIG. 534. **Delphinidae; bottlenose dolphin, *Tursiops truncatus* (Delphininae; *top*), orca, *Orcinus orca* (Orcininae; *middle*), and short-finned pilot whale, *Globicephala macrorhynchus* (Globicephalinae; *bottom*).**

RECOGNITION CHARACTERS:

1. body slender (length 1.5–9.5 m)
2. dorsal fin usually present (absent in *Lissodelphis*), pointed or falcate in shape
3. no longitudinal grooves in skin of throat
4. snout variable; with a distinct beak sharply differentiated from forehead or with a bulging forehead and no beak; or with a long snout merging continuously with forehead
5. two to six of cervical vertebrae usually fused
6. skull only slightly asymmetrical
7. left premaxilla reduced posteriorly such that it does not contact the nasal
8. maxilla expanded posteriorly
9. occipital crest not particularly prominent
10. rostrum variable (short or long, narrow or broad)
11. no boss on premaxilla (cf. Phocoenidae)
12. pterygoid and palatine forming parallel shelves on each side of nasal passage
13. symphysis of lower jaw short to moderately long
14. teeth simple, conical

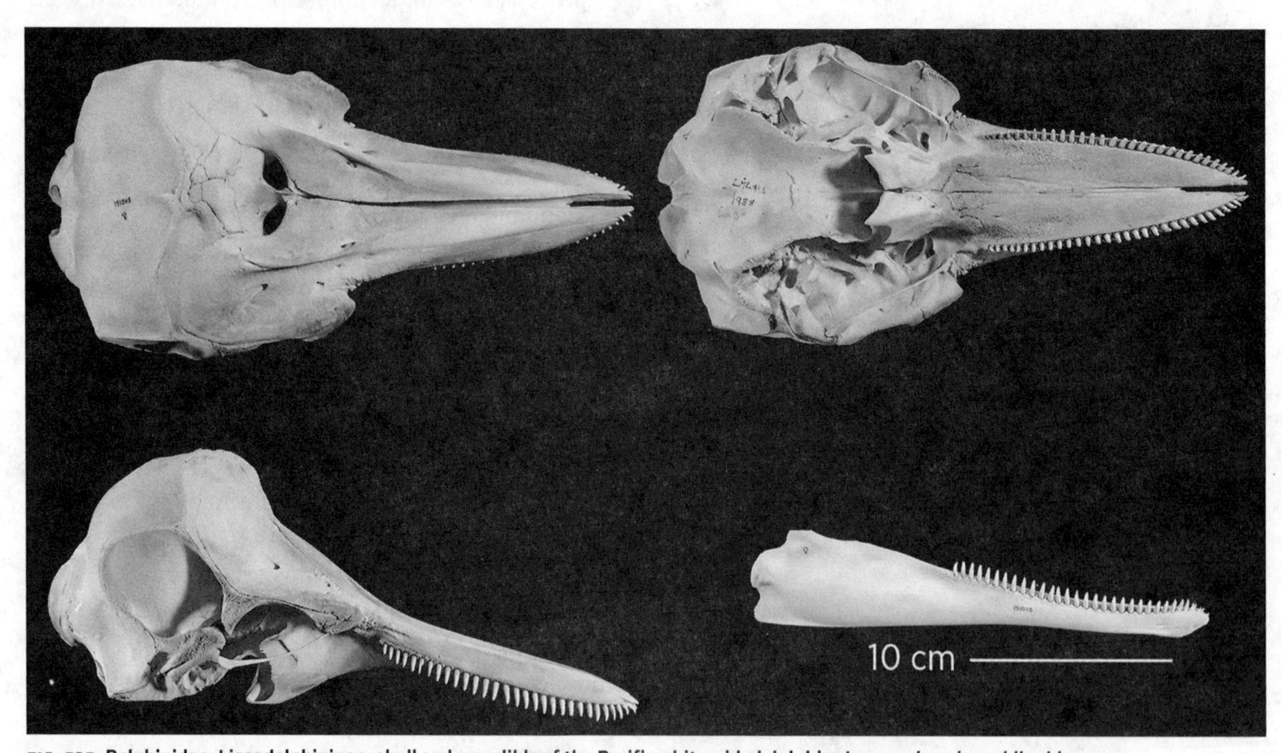

FIG. 535. **Delphinidae: Lissodelphininae; skull and mandible of the Pacific white-sided dolphin, *Lagenorhynchus obliquidens*.**

RANGE: All oceans of the world, as well as some large, mostly tropical rivers.

DENTAL FORMULA: $\frac{0\text{–}65}{2\text{–}58}$
(identity of individual teeth uncertain)

TAXONOMIC DIVERSITY: 36 species in 17 genera; explosive radiation of this lineage in the late Miocene resulted in a number of species with unclear relationships, in turn confusing subfamilial organization; allocated to four to seven or more subfamilies; we follow one recent synthesis below (LeDuc 2009).

Delphininae—*Delphinus* (common dolphins), *Lagenodelphis* (Fraser's dolphin), *Sousa* (humpbacked dolphin), *Stenella* (striped, spotted, and spinner dolphins), and *Tursiops* (bottlenose dolphins)

Globicephalinae—*Feresa* (pygmy killer whale), *Globicephala* (pilot whales), *Grampus* (Risso's dolphin), *Peponocephala* (melon-headed whale), and *Pseudorca* (false killer whale)

Lissodelphininae—*Cephalorhynchus* (Southern Hemisphere dolphins), *Lissodelphis* (right whale dolphins), and *Sagmatias* (= *Lagenorhynchus* in part; white-sided and dusky dolphins)

Orcininae—*Orcaella* (Irrawaddy dolphin, Australian snubfin dolphin) and *Orcinus* (orca)

Stenoninae—*Sotalia* (tucuxi, Guiana dolphin) and *Steno* (rough-toothed dolphin)

Families INIIDAE, LIPOTIDAE, PLATANISTIDAE, and PONTOPORIIDAE (river dolphins)

As noted above, river dolphins as a group are not monophyletic. *Inia* (Amazon and Orinoco river dolphins) and *Pontoporia* (franciscana) are sister taxa, and belong to a larger clade of dolphins and porpoises (the Delphinoidea) that also includes *Lipotes* (baiji). On the other hand, *Platanista* (the Indus and Ganges river dolphins) appears to be either the sister taxon to the beaked whales (Ziphiidae) or basal to a clade that includes all odontocetes except sperm whales. Recent work has thus placed the four river dolphin genera in separate families (see fig. 524, table 9).

Except for *Pontoporia*, which occurs in the coastal waters from southern Brazil through Uruguay to northern Argentina, these species are adapted to living in freshwater river systems, typically in highly turbid waters where visibility is minimal. Thus, the eyes of all species are reduced, and both navigation and food detection are primarily by echolocation. *Platanista* lacks eye lenses and usually swims on its side, often with its foreflipper in contact with the river bottom. All river dolphins eat fish and crustaceans.

The baiji, or Chinese river dolphin, is probably now extinct, and with it the family Lipotidae. Ultimate causal factors are believed to be illegal bottom-set longline fisheries with multiple hooks in the 1980s, and possibly electrocution from illegal electrofishing in the 1990s; these factors led to a decline from as many as 6,000 animals in the 1950s to only a few hundred by 1970. The coup de grace, however, was dealt by the development of extensive hydropower, flood control, and irrigation projects, which led to catastrophic habitat loss.

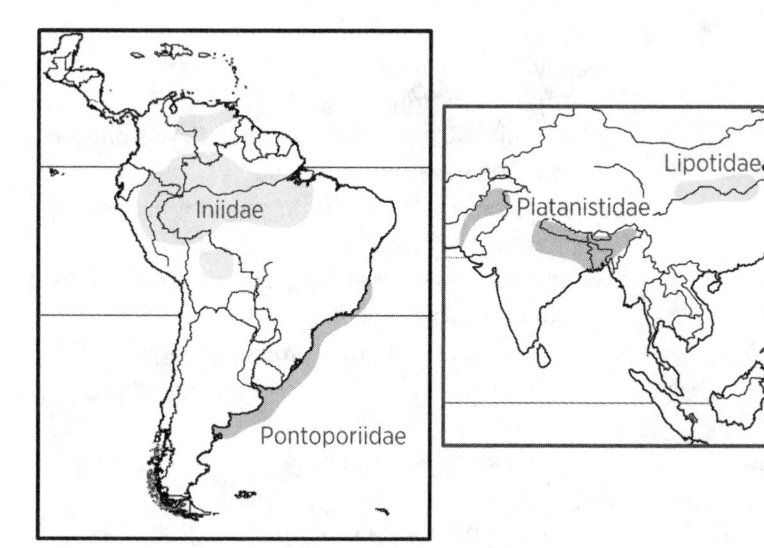

FIG. 536. Geographic ranges of the four river dolphin families: South America, Orinoco and Amazon basins (Iniidae); southern Atlantic coast from central Brazil south to northern Argentina (Pontoporiidae); Yangtze basin, China (Lipotidae—probably extinct); Indus and Ganges basins, south-central Asia (Platanistidae).

FIG. 537. Ganges river dolphin, *Platanista gangetica* (Platanistidae; *top right*), and Amazon river dolphin, *Inia geoffrensis* (Iniidae; *bottom left*).

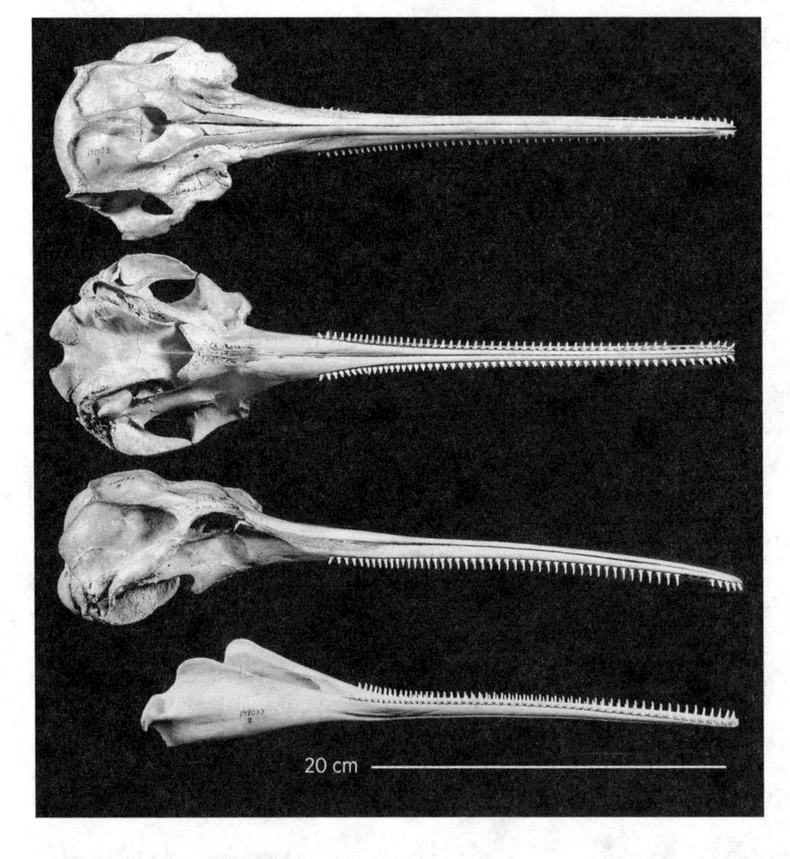

FIG. 538. Pontoporiidae; skull and mandible of the franciscana (or La Plata dolphin), *Pontoporia blainvillei*.

DIAGNOSTIC CHARACTER:

- **lateral margin of maxilla with prominent longitudinal crest**

RECOGNITION CHARACTERS:

1. body slender (length 1.5–3 m)
2. dorsal fin present, low, obtuse
3. no longitudinal grooves in skin of throat
4. **snout long, slender, sharply differentiated from bulging forehead**
5. cervical vertebrae separate
6. skull only slightly asymmetrical (except in *Platanista*)
7. **maxilla narrow, not greatly expanded posteriorly**
8. occipital crest poorly developed
9. **rostrum very narrow, long**
10. **premaxilla with prominent swelling (boss) anterior to nasal opening**
11. pterygoid and palatine forming parallel shelves adjacent to nasal passage
12. **symphysis of lower jaws long (constituting half the length of the paired dentaries)**
13. teeth simple, conical (with prominent ridge around teeth in *Inia*)

DENTAL FORMULA: See table 9.

TAXONOMIC DIVERSITY: See table 9.

Table 9. Some features useful in distinguishing among families of river dolphins.

	Iniidae	*Lipotidae*	*Platanistidae*	*Pontoporiidae*
diversity	*Inia geoffrensis* (Amazon river dolphin), *I. boliviensis* (Bolivian boto), *I. araguaiaensis* (Araguian boto)	*Lipotes vexillifer* (baiji, Chinese river dolphin)	*Platanista gangetica* (Indus and Ganges river dolphins)	*Pontoporia blainvillei* (franciscana or La Plata dolphin)
distribution	Amazon and Orinoco basins; freshwater	middle and lower part of Yangtze basin; freshwater; probably extinct	Indus and Ganges basins; freshwater	coastal waters, southern Brazil to central Argentina (Golfo San Matias)
appearance	largest of river dolphins; young dark gray, adults pink or pinkish	bluish gray, whitish beneath	light grayish brown, paler beneath	light brown, paler beneath
dorsal fin	low, obtuse	low, obtuse	low, obtuse	triangular
blowhole	transverse crescent (similar to delphinids)	longitudinal but somewhat rectangular	longitudinal slit	transverse crescent (similar to delphinids)
eyes	small	small	small, lacking lenses (hence blind)	small
rostrum, beak	rostrum and mandible prominent, long, robust; stiff vibrissae on top of rostrum	long beak, upturned snout	long, narrow beak, sharply differentiated forehead	long beak
teeth	25/24 to 34/33; heterodont; front teeth conical, rear teeth with lingual ridge	32/31 to 35/36; homodont	26/26 to 37/35; small, shorter posteriorly	50/48 to 61/61, tiny; homodont
skull	premaxillae displaced laterally, do not contact nasals; pterygoids not enlarged; palatines separated by vomer at midline of palate	pterygoids not enlarged; palatines in contact along midline of palate	facial crest on external border of maxillae; pterygoids greatly enlarged, completely covering palatines in ventral aspect	skull nearly symmetrical; pterygoids not enlarged, not covering palatines
digestive system	forestomach present; main stomach simple	no forestomach; main stomach divided into 3 chambers	forestomach present; main stomach divided into 2 chambers; cecum present (absent in all other odontocetes)	no forestomach; main stomach simple

Family MONODONTIDAE (beluga, narwhal)

Small to medium-sized whales, with mass up to 2 metric tons. The narwhal (*Monodon*) is remarkable for its long (up to 2.7 m), straight, forward-directed tusk (upper incisor), which is larger in males and presumably subject to sexual selection; normally only the right tooth emerges, but rarely the left does as well, in which case it is invariably shorter. The beluga (*Delphinapterus*) is called the "white whale" because of its pale gray to white coloration. Both species occur in the Arctic Ocean, the Bering and Okhotsk seas, Hudson Bay, the St. Lawrence River, and in some large rivers in Siberia and Alaska. The facial depression of the skull is large, and the maxillary and frontal bones roof over the reduced temporal fossa; the zygomatic process of the squamosal bone is strongly reduced. The beluga has up to 11/11 teeth whose affinities remain unclear; the narwhal has 1/0 teeth. Both are highly gregarious species that feed largely on fish, both bottom-dwelling species and those that live at intermediate depths. Male narwhals "fence" with their tusks. Both species are highly vocal; the beluga is referred to as the "sea canary."

RECOGNITION CHARACTERS:

1. body slender (length 3–6 m)
2. **no dorsal fin** (although a low dorsal ridge is present)
3. no longitudinal grooves in skin of throat
4. **snout short, broad** (*Delphinapterus*) **or indistinct** (*Monodon*)

FIG. 539. Monodontidae; beluga, *Delphinapterus leucas* (*top right*), and narwhal, *Monodon monoceros* (*bottom left*).

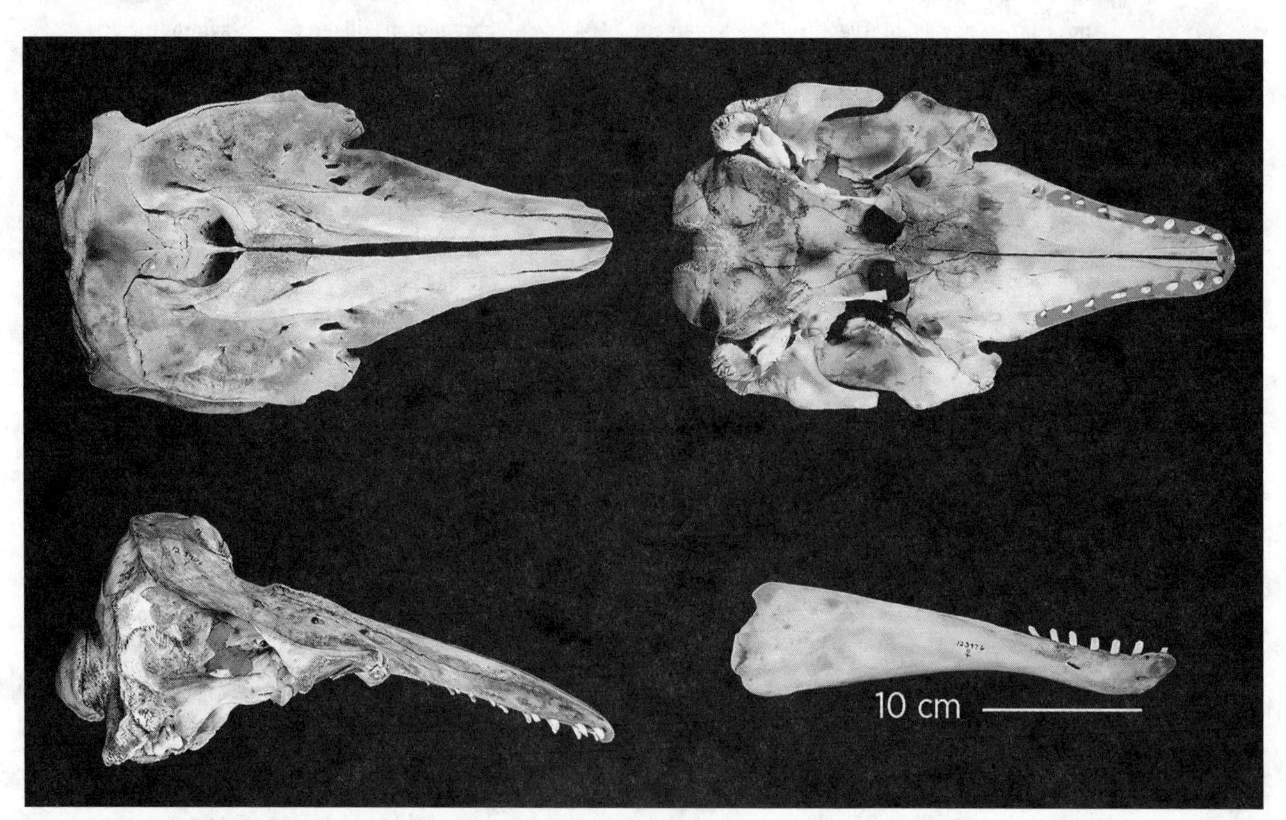

FIG. 540. Monodontidae; skull and mandible of the beluga, *Delphinapterus leucas*.

5. cervical vertebrae separate
6. skull only slightly asymmetrical
7. maxilla expanded posteriorly
8. occipital crest not prominent
9. rostrum broad
10. no boss on premaxilla
11. pterygoid and palatine form parallel shelves adjacent to nasal passage
12. symphysis of lower jaw short
13. teeth simple, conical (except tusk of *Monodon*)

RANGE: Arctic Ocean and adjacent seas, estuaries, and coastal rivers

DENTAL FORMULA: $\frac{1}{0}$ (*Monodon*) or $\frac{5}{2}$ to $\frac{11}{11}$ (*Delphinapterus*) (identity of individual teeth uncertain).

TAXONOMIC DIVERSITY: Two monospecific genera: *Delphinapterus* (beluga) and *Monodon* (narwhal).

Family PHOCOENIDAE (porpoises)

Porpoises occur widely in the coastal waters of all oceans and connected seas of the Northern Hemisphere, as well as in some coastal waters of South America and some rivers in southeastern Asia. Some, such as the common porpoises (*Phocoena*) inhabit coastal waters; others, such as Dall's porpoise (*Phocoenoides dalli*), generally inhabit deeper pelagic areas. They are small cetaceans, weighing 90–120 kg. The jaws are short and there is no beak; the dorsal fin is low or absent. Porpoises may be highly gregarious, with schools of 100 or more individuals common; they are shier and more reclusive than delphinids, however, performing fewer showy displays such as breaching and spy-hopping and, with the exception of Dall's porpoise, do not tend to bow-ride. A variety of foods are taken, including cuttlefish and squid, crustaceans, and fish.

RECOGNITION CHARACTERS:
1. body slender to relatively robust (length 1.2–1.4 m)
2. dorsal fin usually present (absent in *Neophocaena*)
3. no longitudinal grooves in skin of throat
4. **snout without distinct beak**
5. three to seven of cervical vertebrae fused
6. skull only slightly asymmetrical (note similar-sized nares in fig. 542)
7. maxilla expanded posteriorly
8. occipital crest not particularly prominent
9. rostrum relatively short and broad
10. **premaxilla with prominent swelling (boss) anterior to nasal opening**
11. pterygoid and palatine form parallel shelves adjacent to nasal passage
12. symphysis of lower jaw short
13. **teeth usually spade-like, with two to three poorly defined cusps arranged longitudinally**

RANGE: Coastal waters of all oceans and connecting seas; two species pelagic, one in the North Pacific (Dall's porpoise) and another in subantarctic waters (spectacled porpoise, *Phocoena dioptrica*).

DENTAL FORMULA: $\frac{15}{30}$ (identity of individual teeth uncertain)

TAXONOMIC DIVERSITY: About seven species in three genera: *Neophocaena* (finless porpoises), *Phocoena* (common porpoises), and *Phocoenoides* (Dall's porpoise).

FIG. 541. Phocoenidae; Dall's porpoise, *Phocoenoides dalli*.

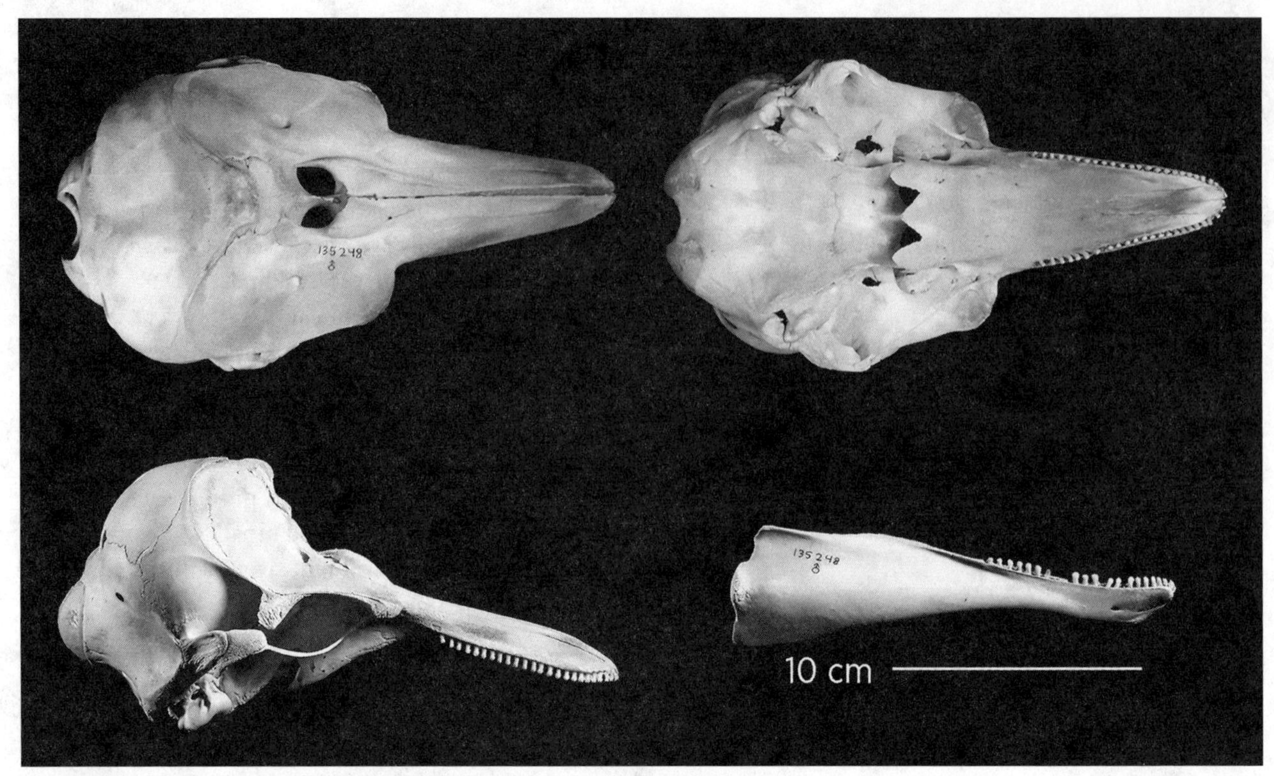

FIG. 542. Phocoenidae; skull and mandible of the harbor porpoise, *Phocoena phocoena*.

Family PHYSETERIDAE (including KOGIIDAE) (sperm and pygmy sperm whales)

The pygmy and dwarf sperm whales (*Kogia*) are placed in a separate family, Kogiidae, by some authors to accommodate a number of differences between them and the sperm whale, *Physeter*; other authors treat these two groups as separate subfamilies of Physeteridae. Regardless, these two groups are well recognized as constituting a monophyletic clade (Physeteroidea), and for convenience, we treat them together.

Physeterids may become very large (20 m in length, 53 metric tons in mass). The head is huge, the rostrum truncated, broad, and flat, and the facial region contains a massive spermaceti organ, which contains great quantities of oil and is thought to play an important role in thermoregulation. *Physeter* are highly social, with groups as large as 1,000 individuals recorded; females group with young, males form bachelor schools. Generally forage on squid in open ocean, often at great depths where little or no light penetrates and where navigation and prey detection is by echolocation (depths of 1,000 m are common, depths over 1,300 m have been recorded, and depths over 3 km are suspected based on stomach contents). Males migrate to polar seas in summer; females and young stay in temperate and tropical waters year-round.

Kogiids are much smaller (maximum 4 m in length, 320 kg in mass) and are porpoise- or sharklike in appearance. Kogiids are united in lacking nasal bones, having teeth without enamel, and having a distinct sagittal septum in the supracranial basin of the skull.

DIAGNOSTIC CHARACTER:

- **snout very large, broad, blunt, undifferentiated from rest of head**

RECOGNITION CHARACTERS:

1. **body robust** (length 2–20 m)
2. dorsal fin present; sickle-shaped (*Kogia*) or reduced (*Physeter*)
3. **throat with longitudinal grooves indistinct to absent** (*Kogia*) **or numerous and short** (*Physeter*)
4. six to seven cervical vertebrae, all fused in *Kogia*, all but atlas fused in *Physeter*
5. **skull strongly asymmetrical; left nasal passage much larger than right one; right premaxilla enlarged**
6. strongly developed supracranial basin (houses spermaceti organ)
7. **maxilla expanded posteriorly**
8. **occipital crest prominent**
9. rostrum short (Kogiidae) or long (Physeteridae), broad

FIG. 543. **Sperm whale, *Physeter macrocephalus* (Physeteridae; *top*) and pygmy sperm whale, *Kogia breviceps* (Kogiidae; *bottom*).**

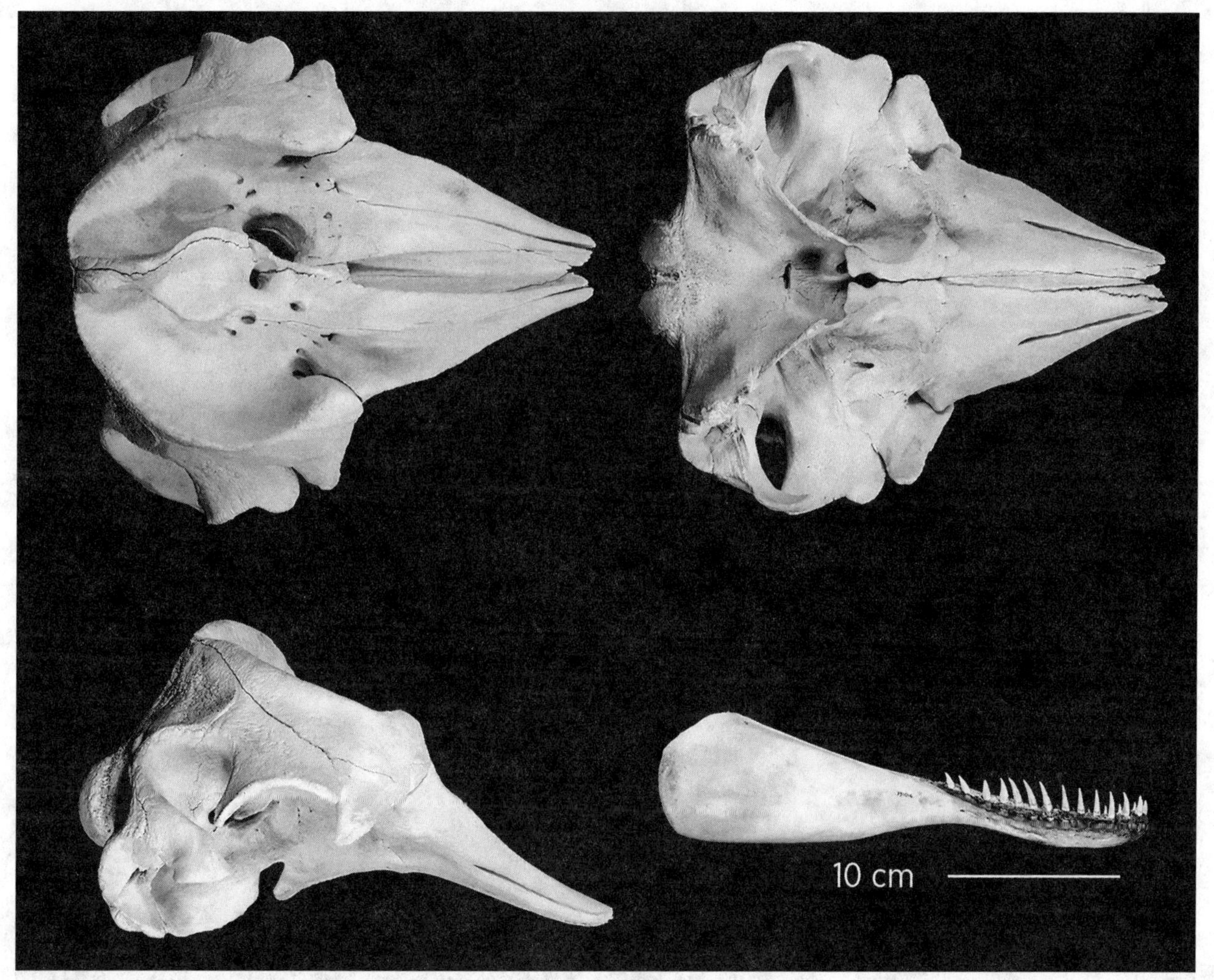

FIG. 544. **Kogiidae; skull and mandible of the pygmy sperm whale, *Kogia breviceps*.**

10. no boss on premaxilla
11. pterygoid and palatine not forming parallel shelves
12. symphysis of lower jaws relatively short (*Kogia*) or long (constituting one-third of length of the paired dentaries in *Physeter*)
13. teeth simple, conical

RANGE: Oceans of world.

DENTAL FORMULA: $\frac{0}{9\text{–}30}$
(identity of individual teeth uncertain)

TAXONOMIC DIVERSITY: Three species in two genera: *Kogia* (pygmy and dwarf sperm whales) and *Physeter* (sperm whale).

Family ZIPHIIDAE (beaked whales)

A very poorly known group of whales, with some species never seen alive. Medium-sized whales (4–12.5 m, up to 11.5 metric tons); snout usually long and narrow, with prominently bulging forehead in many. Only *Tasmacetus* has teeth on the upper jaw. These whales are deep divers of open oceans (some to nearly 3 km) that can stay submerged for prolonged periods (>2 hours). They appear to be highly social and travel in schools in which all members surface and dive in synchrony. Primary food is deep-water squid and fish. Annual migrations are known for some species.

DIAGNOSTIC CHARACTER:

- **posterior margin of fluke without deep notch** (notch present in other families of odontocetes)

RECOGNITION CHARACTERS:

1. body slender to moderately robust (length 4–12.5 m)
2. dorsal fin present, small
3. **one or two pairs of longitudinal grooves in skin of throat**
4. snout long and narrow, usually sharply differentiated from bulging forehead (*Hyperoodon* and *Berardius*) or forming a continuous profile with cranium
5. two to seven of cervical vertebrae fused
6. skull slightly to strongly asymmetrical
7. maxilla expanded posteriorly
8. occipital crest prominent
9. **rostrum very narrow, deep, with open groove between premaxillae in dorsal view** (closed in older animals by dorsal intrusion of vomer)

FIG. 545. Ziphiidae; Cuvier's beaked whale, *Ziphius cavirostris* (Ziphiinae; *top*), and Stejneger's beaked whale, *Mesoplodon stejnegeri* (Hyperoodontinae; *bottom*).

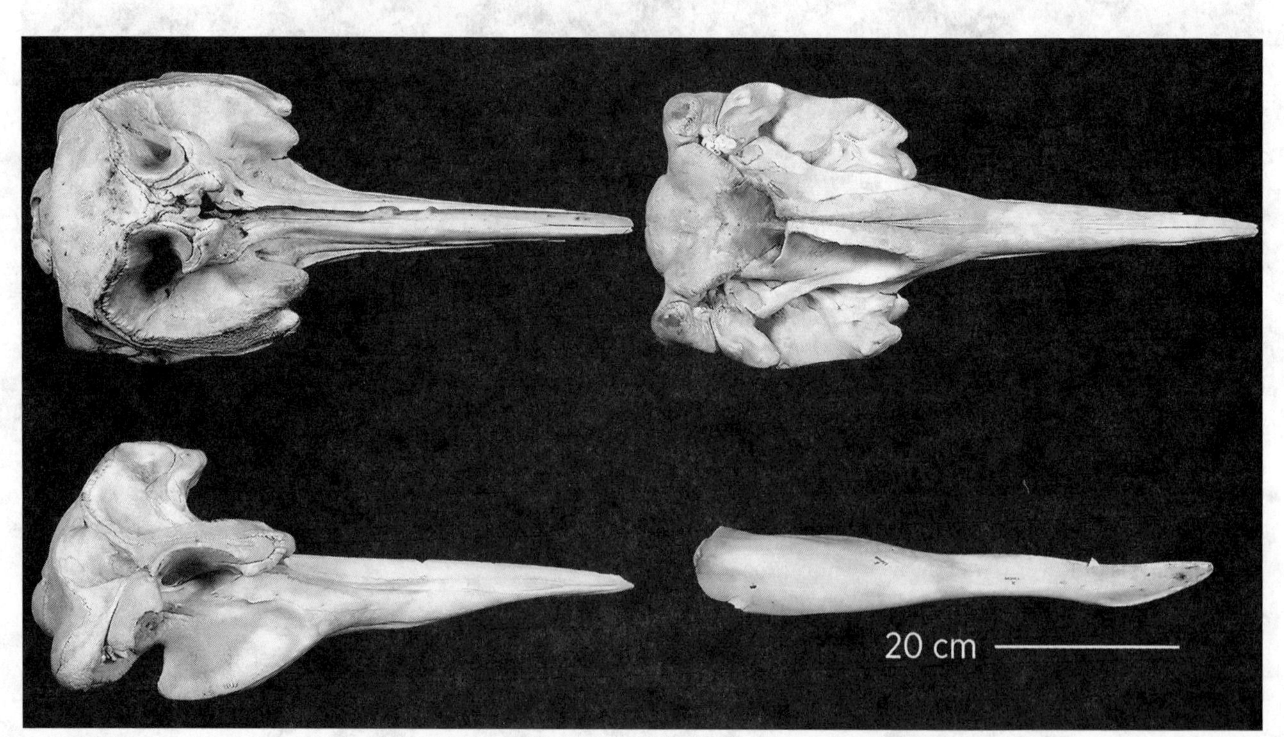

FIG. 546. Ziphiidae: Hyperoodontinae; skull and mandible of Hubbs's beaked whale, *Mesoplodon carlhubbsi*.

10. no boss on premaxilla
11. pterygoid and palatine not forming parallel shelves
12. symphysis of lower jaw short
13. teeth simple, conical or compressed

RANGE: Oceans of the world; deep seas beyond the continental shelf.

DENTAL FORMULA: Usually $\frac{0}{1}$, but $\frac{0}{2}$ in *Berardius* and $\frac{19}{27}$ in *Tasmacetus* (identity of individual teeth uncertain)

TAXONOMIC DIVERSITY: Six genera with about 22 species, grouped by some into two subfamilies:

Hyperoodontinae—*Hyperoodon* (bottlenose whales; two species, cooler waters of North Atlantic [northern bottlenose whale] and circumpolar southern oceans [southern bottlenose whale]), *Indopacetus* (Longman's beaked whale; Indian Ocean and tropical/subtropical Pacific Ocean), and *Mesoplodon* (beaked whales; 15 species; global temperate and tropical oceans)

Ziphiinae—*Berardius* (four-toothed whales; two species, circumpolar southern ocean [Arnoux's beaked whale] and cooler waters of North Pacific and Bering Sea [Baird's beaked whale]), *Tasmacetus* (Shepherd's beaked whale; circumpolar southern Pacific, Atlantic, and Indian oceans), and *Ziphius* (Cuvier's beaked whale; all oceans except high latitudes)

Glossary

Note: Many terms are defined in the text; for the most part, those terms are not included here.

Aerial—pertaining to flying; bats are the only mammals with powered flight, but many other mammals glide to escape predators or travel within forested habitat.

Antitragus—a (usually) small projection of skin that arises at the ventral margin of the pinna, typically pointing dorsally or anteriorly and toward the tragus. Found in many species, but particularly notable in bats.

Arboreal—living in trees (e.g., many primates, squirrels, sloths).

Bone—durable supportive tissue consisting of cells distributed in a matrix of fibrous protein (collagen) and salts (primarily calcium and phosphate).

Calcar—a cartilaginous rod extending from the ankle in many bats that serves to support the uropatagium (tail membrane). Several rodents (e.g., flying squirrels) have a similar structure that extends from either the wrist or the elbow (or both) to support a gliding membrane.

Canal—a perforation, or foramen, that tends to be elongated as a tube (e.g., alisphenoid canal, infraorbital canal).

Cartilage—relatively soft supportive tissue consisting of rounded cells in a matrix of polysaccharides and fibrous protein (collagen). Many bones develop from a cartilaginous precursor.

Claw—a keratinized dermal structure at the tips of the digits of many mammals; usually long, curved, and sharply pointed; serves to protect the tip of the digit and functions in both defense and movement (climbing, digging) (cf. nail, hoof).

Cloaca—a common chamber into which reproductive, urinary, and digestive products enter before leaving the body; often the reproductive and urinary tracts combine first in a urogenital canal, which joins the digestive tract to form the cloaca.

Crepuscular—active at twilight (dusk or dawn, or both).

Cursorial—pertaining to a running habit (e.g., artiodactyls, perissodactyls).

Cusp—a projection, or bump, on the occlusal surface of a tooth.

Cuspidate—pertaining to a tooth with prominent cusps.

Deflected—bent outward or laterally (e.g., away from the midline) (cf. inflected).

Dewclaws (or dew hooves)—reduced, clawed or hoofed lateral digits found in many carnivores and artiodactyls; located just above the main functional digits; presumably reflect evolutionary reduction in the face of increased dominance of a subset (usually 1 or 2) of digits.

Digastric muscle—a muscle that serves both to depress the lower jaw (and thus open the mouth) and aid in anterior-posterior chewing motion; comprised of two parts: an anterior belly that originates on the inner surface of the mandible and a posterior belly that originates on the mastoid or paroccipital process, both of which insert on the hyoid.

Digits—fingers or toes.

Fenestrated—having an irregular network of holes or perforations.

Foramen (pl. foramina)—a hole, opening, or perforation through bone (e.g., foramen magnum, incisive foramen).

Fossa (pl. fossae)—a depression that generally forms a site for muscle attachment or bone articulation (e.g., mandibular fossa, masseteric fossa, temporal fossa).

Fossorial—often used interchangeably with subterranean to denote animals that spend the majority of their lives in self-dug tunnel systems below ground. Derived from the Latin *fossor*, which means "digger"; some authorities thus use *fossorial* only to identify animals that exhibit any anatomical specializations for digging, and thus which would include those that range across the fossorial-subterranean to semifossorial spectrum (cf. semifossorial).

Guard hair—the prominent, coarse hair characterizing the outer fur of mammals; may be modified for protection or communication; spines, bristles, and mane hairs all are examples.

Heterodont—having dentition with varying tooth forms (e.g., incisors, canines, molars) (cf. homodont).

Heterothermic—having a variable body temperature; the variation may be regional (e.g., in limbs of many animals living in cold environments) or involve the whole body (e.g., torpor, hibernation).

Holarctic region—biogeographic region comprising both the Palearctic and Nearctic regions.

Homodont—having dentition in which all teeth are similar in structure (e.g., many porpoises) (cf. heterodont).

Hoof (pl. hooves)—keratinized dermal structure at the tip of a digit that completely encloses the tip of the phalanx (e.g., in horses, deer) (cf. nail, claw).

Inflected—bent inward or medially (e.g., toward the midline) (cf. deflected).

Insertion—the site of attachment of a muscle (usually on a bone) on the more movable of the two bones or other elements that are joined by the muscle (e.g., the temporal muscle inserts on the coronoid process of the dentary, and has its origin in the temporal fossa of the skull) (cf. origin).

Intermembral index—ratio of the forelimbs to the hind limbs (hence, [humerus + radius]/[femur + tibia] × 100); useful in predicting locomotor patterns in primates. Values near 100 typify quadrupedal species, whereas values below and above 100 generally characterize species favoring their hind limbs (leaping, bipedal) and forelimbs (brachiating), respectively.

Ischial callosities—a pair of naked, highly keratinized patches of skin on the rump of some primates.

Masseter muscle—one of the principal muscles employed to close the jaw, often consisting of several bands (lateral, medial, etc.); originates on and adjacent to the zygomatic arch and inserts mostly on the masseteric fossa of the lower jaw. Particularly important in herbivores.

Nail—keratinized dermal structure at the tip of a digit; unlike claws or hooves, nails are usually short, flat, blunt, and do not enclose (partially or completely) the tip of the phalanx (e.g., primates) (cf. claw, hoof).

Nearctic region—biogeographic region that includes the temperate and Arctic regions of the New World (North America and temperate Central America).

Neotropical region—biogeographic region that includes the tropics (and southern temperate regions) of the New World (South America, tropical Central America, and the Greater and Lesser Antilles).

Occlusal—pertaining to the biting surface of a tooth where contact is made when the jaws close.

Origin—the site of attachment of a muscle (usually on a bone) on the less movable of the two bones or other elements that are joined by the muscle (e.g., the masseter muscles originate on (and adjacent to) the zygomatic arch and insert on the dentary) (cf. insertion).

Palearctic region—biogeographic region containing the temperate and arctic regions of the Old World (Europe, Africa north of the Sahara, Asia excluding southern tropical regions).

Pelage—fur, hair, as a unit.

Pentadactyl—five-toed.

Perforate—pierced by an opening or hole.

Pinna (pl. pinnae)—the flap of skin and cartilage that surrounds the external ear.

Plagiaulacoid—general term for bladelike teeth, usually restricted to the grooved, bladelike teeth found among living mammals in some marsupials (e.g., order Diprotodontia).

Prismatic—prism-like; as in the cheek teeth of arvicoline rodents, which have sharply angular ridges on their occlusal surfaces.

Procumbent—Protruding anteriorly, as do the incisors of some mammals, notably many marsupials, insectivores, rodents, and primates.

Rhinarium—the naked pad of the nose of many mammals (e.g., dogs, lower primates).

Saltatorial—pertaining to hopping (e.g., kangaroos).

Scansorial—pertaining to climbing by use of claws (e.g., squirrels, cats).

Scutes—flat bony plates of dermal tissue covered by epidermis and forming the outer shell of armadillos.

Semifossorial—having a partially fossorial (or subterranean) habit (e.g., ground squirrels); refers to species that spend more time above ground (foraging, socializing, etc.) than fossorial species do and lack many of the extensive morphological adaptations associated with burrowing (cf. fossorial).

Sesamoid—a bone formed in a tendon (e.g., patella, baculum).

Subterranean—living in self-dug burrows and rarely venturing onto the ground surface (e.g., most moles, gophers); often used synonymously with fossorial.

Syndactylous—having two or more digits that are bound together in a common tube of skin, while the underlying bones remain distinct.

Talon (-id)—a posterior "heel" or expansion on an upper (lower) cheek tooth that gives the tooth a square outline and expands the crushing surface.

Temporal muscle—one of the principal muscles employed to close the jaw; originates on the posterodorsal and lateral portions of the braincase (the temporal fossa) and inserts chiefly on the coronoid process of the mandible. Particularly large in carnivores.

Tragus—a projection of skin that arises at the anteroventral margin of the pinna; found in most bats and in springhares (Rodentia, Pedetidae).

Trigon (-id)—the triangular, three-cusped portion of an

upper (lower) tribosphenic molar; includes paracone (-id), metacone (-id), and protocone (-id).

Tuberculo-sectorial—a tritubercular tooth with sharp cutting edges (cusps).

Underfur—soft, often wooly insulative hairs in mammalian pelage that provide the bulk of hair and serve to insulate.

Uropatagium—a membrane between the tail and hind limb of a bat (= interfemoral membrane).

Vibrissae—long whiskers specialized for tactile reception; commonly located in the facial region.

Literature Cited

Agnarrson, I., and L. J. May-Collado. 2008. The phylogeny of Cetartiodactyla: the importance of dense taxon sampling, missing data, and the remarkable promise of cytochrome *b* to provide reliable species-level phylogenies. Molecular Phylogenetics and Evolution 48:964–985.

Altringham, J. D. 2011. Bats: From Evolution to Conservation. Oxford University Press, Oxford, UK, xi + 324 + 34 plates.

Archer, M. 1984. The Australian marsupial radiation. Pp. 633–808 in M. Archer and G. Clayton, Vertebrate Zoogeography and Evolution in Australasia: Animals in Space and Time. Hesperian Press, Carlisle, Western Australia. xxiv + 1203 pp.

Asher, R. J., and K. M. Helgen. 2010. Nomenclature and placental mammal phylogeny. BMC Evolutionary Biology 10:102.

Asher, R. J., I. Horovitz, and M. R. Sánchez-Villagra. 2004. First combined cladistic analysis of marsupial mammal interrelationships. Molecular Phylogenetics and Evolution 33:240–250.

Berta, A., J. L. Sumich, and K. M. Kovacs. 2015. Marine Mammals: Evolutionary Biology. 3rd ed. Academic Press, New York, 726 pp.

Bezuidenhout, A. J., and H. E. Evans. 2005. The Anatomy of the Woodchuck (*Marmota monax*). Special Publications of the American Society of Mammalogists 13, 180 pp.

Bi, S., Y. Wang, J. Guan, X. Sheng, and J. Meng. 2014. Three new Jurassic euharamiyidan species reinforce early divergence of mammals. Nature 514:579–584.

Bibi, F. 2013. A multi-calibrated mitochondrial phylogeny of extant Bovidae (Artiodactyla, Ruminantia) and the importance of the fossil record in systematics. BMC Evolutionary Biology 13:166.

Bininda-Emonds, O. R. P., M. Cardillo, K. E. Jones, R. D. E. MacPhee, R. M. D. Beck, R. Grenyter, S. A. Price et al. 2007. The delayed rise of present-day mammals. Nature 446:507–512.

Brown, J. C. 1971. The description of mammals. 1. The external characters of the head. Mammal Review 1:151–168.

Brown, J. C., and D. W. Yalden. 1973. The description of mammals. 2. Limbs and locomotion of terrestrial mammals. Mammal Review 3:107–134.

Brusatte, S., and Z.-X. Luo. 2016. Ascent of the mammals. Scientific American, June, 28–35.

Burgin, C. J., J. P. Colella, P. L. Kahn, and N. S. Upham. 2018. How many species of mammals are there? Journal of Mammalogy 99:1–14.

Case, J. A., F. J. Goin, and M. O. Woodburne. 2005. "South American" marsupials from the late Cretaceous of North America and the origin of marsupial cohorts. Journal of Mammalian Evolution 12:461–494.

Chaisson, R. 1989. Laboratory Anatomy of the Cat. W. C. Brown, Dubuque, IA, ix + 148 pp.

Corbet, G. B. 1978. The Mammals of the Palaearctic Region: A Taxonomic Review. British Museum (Natural History) and Cornell University Press, Ithaca, New York, 314 pp.

Corbet, G. B., and J. E. Hill. 1986. A World List of Mammalian Species. 2nd ed. Facts on File Publications, New York, 254 pp.

Courcelle, M., M.-K. Tilak, Y. L. R. Leite, E. J. P. Douzery, and P.-H. Fabre. 2019. Digging for the spiny rat and hutia phylogeny using a gene capture approach, with the description of a new mammal subfamily. Molecular Phylogenetics and Evolution 136:241–253.

Dawson, M. R., L. Marivoux, C. K. Li, C. Beard, and G. Métais. 2006. *Laonastes* and the "Lazarus effect" in Recent mammals. Science 311:1456–1458.

Delsuc, F., G. C. Gibb, M. Kuch, G. Billet, L. Hautier, J. Southon, J.-M. Rouillard, J. C. Fernicola, S. F. Vizcaíno, R. D. E. MacPhee, and H. N. Poinar. 2016. The phylogenetic affinities of the extinct glyptodonts. Current Biology 26:R155–R156.

Delsuc, F., M. Kuch, G. C. Gibb, E. Karpinski, D. Hackenberger, P. Szpak, J. G. Martínez, J. I. Mead, H. G. McDonald, R. D. E. MacPhee, G. Billet, L. Hautier, and H. N. Poinar. 2019. Ancient mitogenomes reveal the evolutionary history and biogeography of sloths. Current Biology 29:2031–2042.e2036.

Denys, C., J. Michaux, F. Catzeflis, S. Ducrocq, and P. Chevret. 1995. Morphological and molecular data against the monophyly of Dendromurinae (Muridae: Rodentia). Bonner Zoologische Beitraege 45:173–190.

Denys, C., P. J. Taylor, and K. P. Alpin. 2017. Family Muridae (true mice and rats, gerbils and relatives). Pp. 536–884 in D. E. Wilson, T. E. Lacher, Jr., and R. A. Mittermeier (eds.), Handbook of Mammals of the World, Vol. 7, Rodents II. Lynx Edicions, Barcelona, 1008 pp.

Díaz, M. M., R. M. Barquez, and D. H. Verzi. 2015. Genus *Tympanoctomys* Yepes, 1942. Pp. 1043–1048 in J. L. Patton, U. F. J. Pardiñas, and G. D'Elía (eds.), South American Mammals, Vol. 2, Rodents. University of Chicago Press, Chicago, IL, xxvi + 1336 pp.

dos Reis, M., J. Inque, M. Hasegawa, R. J. Asher, P. C. J. Donoghue, and Z. Yang. 2012. Phylogenomic datasets provide both precision and accuracy in estimating the timescale of placental mammal phylogeny. Proceedings of the Royal Society of London B: Biological Sciences 279:3491–3500.

Erbajeva. M. A. 1988. Cenozoic Pikas (Taxonomy, Systematics, Phylogeny). Nauka, Moscow. 224 pp (in Russian).

Erbajeva. M. A. 1994. Phylogeny and evolution of Ochotonidae with emphasis on Asian ochotonids. National Science Museum Monographs (Tokyo) 8:1–13.

Esselstyn, J. A., A. S. Achmadi, and K. C. Rowe. 2012. Evolutionary novelty in a rat with no molars. Biology Letters. 8:990–993.

Esselstyn, J. A., C. H. Oliveros, M. T. Swanson, and B. C. Faircloth. 2017. Investigating difficult nodes in the placental mammal tree with expanded taxon sampling and thousands of ultraconserved elements. Genome Biology and Evolution 9:2308–2321.

Evans, A. R., and G. D. Sanson. 2003. The tooth of perfection: functional and spatial constraints on mammalian tooth shape. Biological Journal of the Linnean Society 78:173–191.

Everson, K. M., V. Soarimalala, S. M. Goodman, and L. E. Olson. 2016. Multiple loci and complete taxonomic sampling resolve the phylogeny and biogeographic history of tenrecs (Mammalia: Tenrecidae) and reveal higher species rates in Madagascar's humid forests. Systematic Biology 65:890–909.

Fabre, P.-H., L. Hautier, D. Dimitrov, and E. J. P. Douzery. 2012. A glimpse on the pattern of rodent diversification: a phylogenetic approach. BMC Evolutionary Biology 12: 88.

Fabre, P.-H., L. Hautier, and E. J. P. Douzery. 2015. A synopsis of rodent molecular phylogenetics, systematics and biogeography. Pp. 19–69 in P. G. Cox and L. Hautier (eds.), Evolution of the Rodents: Advances in Phylogeny, Functional Morphology and Development. Cambridge University Press, Cambridge, UK, xiv + 611 pp.

Fabre, P.-H., J. L. Patton, and Y. L. R. Leite. 2016. Family Echimyidae. Pp. 552–641 in D. E. Wilson, T. E. Lacher Jr., and R. A. Mittermeier (eds.), Handbook of Mammals of the World, Vol. 6, Lagomorphs and Rodents 1. Lynx Editions, Barcelona, 987 pp.

Feldhamer, G. A., L. C. Drickamer, S. H. Vessey, J. F. Merritt, and C. Krajewski. 2015. Mammalogy: Adaptation, Diversity, and Ecology. 4th ed. Johns Hopkins University Press, Baltimore, xiii + 747 pp.

Fernández, M. H., and E. S. Vrba. 2005. A complete estimate of the phylogenetic relationships in Ruminantia: a dated species-level supertree of the extant ruminants. Biological Review 80:269–302.

Fleagle, J. G. 2013. Primate Adaptation and Evolution. Academic Press, San Diego, x + 441 pp.

Flynn, J. J., J. A. Finarelli, and M. Spaulding. 2010. Phylogeny of the Carnivora and Carnivoramorpha, and the use of the fossil record to enhance understanding of evolutionary transformations. Pp. 25–52 in A. Goswami and A. Friscia (eds.), Carnivoran Evolution: New Views on Phylogeny, Form, and Function. Cambridge University Press, Cambridge, UK, xiii + 492 pp.

Foley, N. M., V. D. Thong, P. Soisook, S. M. Goodman, K. N. Armstrong, D. S. Jacobs, S. J. Puechmaille, and E. C. Teeling. 2014. How and why overcome the impediments to resolution: lessons from rhinolophid and hipposiderid bats. Molecular Biology and Evolution 32:313–333.

Fordyce, R. E., and F. G. Marx. 2012. The pygmy right whale *Caperea marginata*: the last of the cetotheres. Proceedings of the Royal Society of London B: Biological Sciences 280. doi:10.1098/rspb.2012.2645.

Franklin, W. L. 2011. Family Camelidae (camels). Pp. 206–247 in D. E. Wilson and R. A. Mittermeier (eds.), Handbook of the Mammals of the World, Vol. 2, Hoofed Mammals. Lynx Edicions, Barcelona, 885 pp.

Garbino, G. S. T., and C. C. de Aquino. 2018. Evolutionary significance of the entepicondylar foramen of the humerus in New World monkeys (Platyrrhini). Journal of Mammalian Evolution 25:141–151.

Gardner, A. L. 2005. Order Pilosa, Suborder Vermilingua. Pp. 100–103 in D. E. Wilson and D. A. Reeder (eds.), Mammal Species of the World: A Taxonomic and Geographic Reference, 3rd ed. Johns Hopkins University Press, Baltimore, 1:xxxv + 1–743.

Garshelis, D. L. 2009. Family Ursidae. Pp. 448–497 in D. E. Wilson and R. A. Mittermeier (eds.), Handbook of the Mammals of the World, Vol. 1, Carnivores. Lynx Edicions, Barcelona, 727 pp.

Gatesy, J., J. H. Geisler, J. Chang, C. Buell, A. Berta, R. W. Meredith, M. S. Springer, and M. R. McGowen. 2013. A phylogenetic blueprint for a modern whale. Molecular Phylogenetics and Evolution 66:479–506.

Gatesy, J., R. W. Meredith, J. E. Janecka, M. P. Simmons, W. J. Murphy, and M. S. Springer. 2017. Resolution of a concatenation/coalescence kerfuffle: partitioned coalescence support and a robust family-level tree for Mammalia. Cladistics 33:295–332.

Gaubert, P. 2009. Family Prionodontidae (linsangs). Pp. 170–173 in D. E. Wilson and R. A. Mittermeier (eds.), Handbook of the Mammals of the World, Vol. 1, Carnivores. Lynx Edicions, Barcelona, 727 pp.

Gaubert, P., W. C. Wozencraft, P. Cordeiro-Estrela, and G. Veron. 2005. Mosaics of convergences and noise in morphological phylogenies: what's in a viverrid-like carnivoran? Systematic Biology 54:865–894.

Gaudin, T. J., R. J. Emry, and J. R. Wible. 2009. The phylogeny of living and extinct pangolins (Mammalia, Pholidota) and associated taxa: a morphology-based analysis. Journal of Mammalian Evolution 16:235–305.

Gilbert, C., A. Ropiquet, and A. Hassanin. 2006. Mitochondrial and nuclear phylogenies of Cervidae (Mammalia, Ruminantia): systematics, morphology, and biogeography. Molecular Phylogenetics and Evolution 40:101–117.

Gilchrist, J. S., A. P. Jennings, G. Veron, and P. Cavallini. 2009. Family Herpestidae. Pp. 262–329 in D. E. Wilson and R. A. Mittermeier (eds.), Handbook of the Mammals of the World, Vol. 1, Carnivores. Lynx Edicions, Barcelona, 727 pp.

Gingerich, P. D., M. U. Haq, I. L. Zalmout, I. H. Khan, and M. S. Malakani. 2001. Origin of whales from early artiodactyls: hands and feet of Eocene Protocetidae from Pakistan. Science 293:2239–2242.

Grinnell, J., J. S. Dixon, and J. M. Linsdale. 1937. Fur-Bearing Mammals of California: Their Natural History, Systematic Status, and Relations to Man. Vols. 1 & 2. University of California Press, Berkeley, 777 pp.

Groves, C. P. 2005a. Order Peramelemorphia. Pp. 38–42 in D. E. Wilson and D. A. Reeder (eds.), Mammal Species of the World: A Taxonomic and Geographic Reference, 3rd ed. Johns Hopkins University Press, Baltimore, 1:xxxv + 1–743.

Groves, C. P. 2005b. Order Primates. Pp. 111–284 in D. E. Wilson and D. A. Reeder (eds.), Mammal Species of the World: A Taxonomic and Geographic Reference, 3rd ed. Johns Hopkins University Press, Baltimore, 1:xxxv + 1–743.

Groves, C. P., and P. Grubb. 2011. Ungulate Taxonomy. Johns Hopkins University Press, Baltimore, ix + 316 pp.

Groves, C. P., and D. M. Leslie, Jr. 2011. Family Bovidae. Pp. 444–779 in D. E. Wilson and R. A. Mittermeier (eds.), Handbook of the Mammals of the World, Vol. 2, Hoofed Mammals. Lynx Edicions, Barcelona, 885 pp.

Grubb, P. 2005. Order Artiodactyla. Pp. 637–722 in D. E. Wilson and D. A. Reeder (eds.), Mammal Species of the World: A Taxonomic and Geographic Reference, 3rd ed. Johns Hopkins University Press, Baltimore, 1:xxxv + 1–743.

Gunderson, H. L. 1976. Mammalogy. McGraw-Hill, New York, vii + 483 pp.

Gunnell, G. F., N. B. Simmons, and E. R. Seiffert. 2014. New Myzopodidae (Chiroptera) from the late Paleogene of Egypt: emended family diagnosis and biogeographic origins of Noctilionoidea. PLoS ONE 9(2).e86712.

Hall, E. R. 1981. The mammals of North America. 2nd ed. Vols. 1 & 2. John Wiley and Sons, New York, 1:xv + 1–600 + 90; 2:vi + 601–1181 + 90.

Hartenberger, J. L. 1985. The order Rodentia: major questions on their evolutionary origin, relationships and suprafamilial systematics. Pp. 1–33 in W. P. Luckett and J. L. Hartenberger (eds.), Evolutionary Relationships among Rodents: A Multidisciplinary Analysis. Plenum Press, New York, xiii + 1–721.

Hassanin, A., and E. J. P. Douzery. 2003. Molecular and morphological phylogenies of Ruminantia and the alternative position of the Moschidae. Systematic Biology 52:206–228.

Hassanin, A., F. Delsuc, A. Ropiquet, C. Hammer, B. J. van Vuuren, C. Matthee, M. Ruiz-Garcia et al. 2012. Pattern and timing of diversification of Cetartiodactyla (Mammalia, Laurasiatheria), as revealed by a comprehensive analysis of mitochondrial genomes. Comptes Rendus Biologies 335:32–50.

Hassanin, A., J.-P. Hugot, and B. J. van Vuuren. 2015. Comparison of mitochondrial genome sequences of pangolins (Mammalia, Pholidota). Comptes Rendus Biologies 338:260–265.

Helgen, K. M. 2005a. Family Aplodontiidae. P. 753 in D. E. Wilson and D. A. Reeder (eds.), Mammal Species of the World: A Taxonomic and Geographic Reference, 3rd ed. Johns Hopkins University Press, Baltimore, 2:xx + 745–2,142.

Helgen, K. M. 2005b. Order Scandentia. Pp. 104–109 in D. E. Wilson and D. A. Reeder (eds.), Mammal Species of the World: A Taxonomic and Geographic Reference, 3rd ed. Johns Hopkins University Press, Baltimore, 1:xxxv + 1–743.

Helgen, K. M., F. C. Cole, L. E. Helgen, and D. E. Wilson.

2009. Generic revision in the Holarctic ground squirrel genus *Spermophilus*. Journal of Mammalogy 90:270–305.

Heritage, S., D. Fernández, H. M. Sallam, D. T. Cronin, J. M. Esara Eschube, and E. R. Seiffert. 2016. Ancient phylogenetic divergence of the enigmatic African rodent *Zenkerella* and the origin of anomalurid gliding. PeerJ 4:e2320. doi:10.7717/peerj.2320.

Hershkovitz, P. 1962. Evolution of Neotropical cricetine rodents (Muridae) with special reference to the phyllotine group. Fieldiana: Zoology 46:1–524.

Hershkovitz, P. 1977. Living New World Monkeys (Platyrrhini). Vol. 1. University of Chicago Press, Chicago, IL, xiv + 1117 pp.

Hill, J. E. 1974. A new family, genus and species of bat (Mammalia: Chiroptera) from Thailand. Bulletin, British Museum of Natural History, Zoological Series 27:301–336.

Holden, M. E., T. Cserkész, and G. G. Musser. 2017. Family Sminthidae (birch mice). Pp. 22–48 in D. E. Wilson, T. E. Lacher, Jr., and R. A. Mittermeier (eds.), Handbook of the Mammals of the World, Vol. 7, Rodents II. Lynx Edicions, Barcelona, 1008 pp.

Honacki, J. H., Kinman, K. E., and J. W. Koeppl (eds.). 1982. Mammal Species of the World: A Taxonomic and Geographic Reference. Allen Press and Association of Systematics Collections, Lawrence, KS, ix + 1–694.

Huchon, D., and E. P. Douzery. 2001. From the Old World to the New World: a molecular chronicle of the phylogeny and biogeography of hystricognath rodents. Molecular Phylogenetics and Evolution 20:238–251.

Hunt, R. M., Jr., and R. H. Tedford. 1993. Phylogenetic relationships within aeluroid Carnivora. Pp. 53–73 in F. S. Szalay, M. J. Novacek, and M. C. McKenna (eds.), Mammal Phylogeny: Placentals. Springer-Verlag, New York, xi + 321 pp.

Hunter, J. P., and J. Jernvall. 1995. The hypocone as a key innovation in mammals. Proceedings of the National Academy of Sciences USA 92:10718–10722.

Hutterer, R. 2005a. Order Erinaceomorpha. Pp. 212–219 in D. E. Wilson and D. A. Reeder (eds.), Mammal Species of the World: A Taxonomic and Geographic Reference, 3rd ed. Johns Hopkins University Press, Baltimore, 1:xxxv + 1–743.

Hutterer, R. 2005b. Order Soricomorpha. Pp. 220–311 in D. E. Wilson and D. A. Reeder (eds.), Mammal Species of the World: A Taxonomic and Geographic Reference, 3rd ed. Johns Hopkins University Press, Baltimore, 1:xxxv + 1–743.

Jackson, S., and C. Groves. 2015. Taxonomy of Australian Marsupials. CSIRO Publishing, Clayton South, Australia, 529 pp.

Janecka, J. E., W. Miller, T. H. Pringle, F. Wiens, A. Zitzmann, K. M. Helgen, M. S. Springer, and W. J. Murphy. 2007. Molecular and genomic data identify the closest living relative of Primates. Science 318:792–794.

Jenkins, P. D., C. W. Kilpatrick, M. F. Robinson, and R. J. Timmins. 2005. Morphological and molecular investigations of a new family, genus and species of rodent (Mammalia: Rodentia: Hystricognatha) from Lao PDR. Systematics and Biodiversity 2:419–454.

Jones, F. W. 1923. The Mammals of South Australia, Part 1. Government Printer, Adelaide.

Jones, K. E., A. Purvis, A. MacLarnon, O. R. P. Bininda-Emonds, and N. B. Simmons. 2002. A phylogenetic supertree of the bats (Mammalia: Chiroptera). Biological Review 77:223–259.

Kadwell, M., M. Fernández, H. F. Stanley, R. Baldi, J. C. Wheeler, R. Rosadio, and M. W. Bruford. 2001. Genetic analysis reveals the wild ancestors of the llama and alpaca. Proceedings of the Royal Society of London B: Biological Sciences 268:2575–2584.

Keynes, R. D. 2001. Charles Darwin's *Beagle* Diary. Cambridge University Press, Cambridge, UK, xi + 466.

Kielan-Jaworowska, Z., R. L. Cifelli, and Z.-X. Luo. 2004. Mammals from the Age of Dinosaurs: Origins, Evolution, and Structure. Columbia University Press, New York, xv + 630 pp.

Kingdon, J. 1979. East African Mammals: An Atlas of Evolution in Africa. Vol. 3, Part B. Large Mammals. Academic Press, London, 436 pp.

Kingdon, J., B. Agwanda, M. Kinnaird, T. O'Brien, C. Holland, T. Gheysens, M. Boulet-Audet, and F. Vollrath. 2012. A poisonous surprise under the coat of the African crested rat. Proceedings of the Royal Society of London B: Biological Sciences 279:675–680.

Kirsch, J. A., J. M. Hutcheon, D. C. Byrnes, and B. D. Lloyd. 1998. Affinities and historical zoogeography of the New Zealand Short-tailed bat, *Mystacina tuberculata* Gray 1843, inferred from DNA-hybridization comparisons. Journal of Mammalian Evolution 5:33–64.

Kitchener, A. C., C. Breitenmoser-Würsten, E. Eizirik, A. Gentry, L. Werdelin, A. Wilting, N. Yamaguchi, et al. 2017. A revised taxonomy of the Felidae. The final report of the Cat Classification Task Force of the IUCN/SSC Cat Specialist Group. Cat News Special Issue 11, 80 pp.

Klingener, D. 1964. The comparative myology of four dipodoid rodents (genera *Zapus*, *Napeozapus*, *Sicista*, and *Jaculus*). Miscellaneous Publications, Museum of Zoology, University of Michigan 124:1–100.

Koepfli, K. P., K. A. Deere, G. J. Slater, C. Begg, K. Begg, L. Grassman, M. Lucherini, G. Veron, and R. K. Wayne. 2008. Multigene phylogeny of the Mustelidae: resolving relationships, tempo and biogeographic history of a mammalian adaptive radiation. BMC Biology 6:10. doi:10.1 186/1741-7007-6-10.

Korth, W. W. 1994. The Tertiary Record of Rodents in North America. Plenum, New York, 319 pp.

Lack, J. B., Z. P. Roehrs, C. E. Stanley, Jr., M. Ruedi, and R. A. Van Den Bussche. 2010. Molecular phylogenetics of *Myotis* indicate familial-level divergence for the genus *Cistugo* (Chiroptera). Journal of Mammalogy 91:976–992.

Landry, S. O., Jr. 1999. A proposal for a new classification and nomenclature for the Glires (Lagomorpha and Rodentia). Mitteilungen des Museums für Naturkunde, Berlin, Zoologische Reihe 75:283–316.

Larivière, S., and A. P. Jennings. 2009. Family Mustelidae. Pp. 564–658 in D. E. Wilson and R. A. Mittermeier (eds.), Handbook of the Mammals of the World, Vol. 1, Carnivores. Lynx Edicions, Barcelona, 727 pp.

Lawlor, T. E. 1979. Handbook to the Orders and Families of Living Mammals. Mad River Press, Eureka, CA, 327 pp.

Lebedev, V. S., A. A. Bannikova, M. Pagès, J. Pisano, J. R. Michaux, and G. I. Shenbrot. 2013. Molecular phylogeny and systematics of Dipodoidea: a test of morphology-based hypotheses. Zoologica Scripta 42:231–249.

LeDuc, R. 2009. Delphinids, Overview. Pp. 298–302 in W. F. Perrin, B. Würsig, and J. G. M. Thewissen (eds.), Encyclopedia of Marine Mammals, 2nd ed. Academic Press, New York, 1352 pp.

Lin, J., G. Chen, L. Gu, Y. Shen, M. Zheng, W. Zheng, X. Hu, X. Zhang, Y. Qiou, X. Liu, and C. Jiang. 2014. Phylogenetic affinity of tree shrews to Glires is attributed to fast evolution rate. Molecular Phylogenetics and Evolution 71:193–200.

Luo, Z.-X. 2007. Transformation and diversification in early mammal evolution. Nature 450:1011–1019.

Luo, Z.-X. 2011. Developmental patterns in Mesozoic evolution of mammal ears. Annual Reviews of Ecology, Evolution, and Systematics 42:355–380.

Luo, Z.-X., R. L. Cifelli, and Z. Kielan-Jaworowska. 2001. Dual origin of tribosphenic mammals. Nature 409: 53–57.

Luo, Z.-X., C.-X. Yuan, Q.-J. Meng, and Q. Ji. 2011. A Jurassic eutherian mammal and divergence of marsupials and placentals. Nature 476:442–445.

Macdonald, D. W. 1984. The Encyclopedia of Mammals. Facts on File, New York. 934 pp.

Macdonald, D. W. 1985. The carnivores: order Carnivora. Pp. 619–722 in R. E. Brown and D. W. Macdonald (eds.), Social Odours in Mammals, Vol. 2. Clarendon Press, Oxford, xi + 882 pp.

MacPhee, R. D. E. 2011. Basicranial morphology and relationships of Antillean Heptaxodontidae (Rodentia, Ctenohystrica, Caviomorpha). Bulletin of the American Museum of Natural History 363:1–70.

Marshall, L. G., J. A. Case, and M. O. Woodburne. 1990. Phylogenetic relationships of the families of marsupials. Pp. 433–505 in H. H. Genoways (ed.), Current Mammalogy, Vol. 2. Plenum Press, New York.

Martin, R. D. 1990. Primate Origins and Evolution: A Phylogenetic Reconstruction. Princeton University Press, Princeton, NJ, xiv + 804 pp.

Marx, F. G., O. Lambert, and M. D. Uhen. 2016. Cetacean Paleobiology. John Wiley & Sons, Chichester, Great Britain, 336 pp.

Mattioli, S. 2011. Family Cervidae. Pp. 350–443 in D. E. Wilson and R. A. Mittermeier (eds.), Handbook of the Mammals of the World, Vol. 2, Hoofed Mammals. Barcelona: Lynx Edicions, 885 pp.

McDowell. S. B. 1958. The Greater Antillean insectivores. Bulletin of the American Museum of Natural History 115:115–124.

McKenna, M. C., and S. K. Bell. 1997. Classification of Mammals above the Species Level. Columbia University Press, New York, xii + 631 pp.

McLaughlin, C. A., and R. B. Chaisson. 1990. Laboratory Manual of the Rabbit. W. C. Brown, Dubuque, IA, xii + 112 pp.

Mead, J. G., and R. L. Brownell, Jr. 2005. Order Cetacea. Pp. 723–743 in D. E. Wilson and D. A. Reeder (eds.), Mammal Species of the World: A Taxonomic and Geographic Reference, 3rd ed. Johns Hopkins University Press, Baltimore, 1:xxxv + 1–743.

Mead, J. G., and R. E. Fordyce. 2009. The therian skull: a lexicon with emphasis on the Odontocetes. Smithsonian Contributions to Zoology, 627:xi + 261 pp.

Meredith, R. W., M. Westerman, and M. S. Springer. 2009. A phylogeny of Diprotodontia (Marsupialia) based on sequences for five nuclear genes. Molecular Phylogenetics and Evolution 51:554–571.

Meredith, R. W., J. E. Janecka, J. Gatesy, O. A. Ryder, C. A. Fisher, E. C. Teeling, A. Goodbla et al. 2011. Impacts of the Cretaceous Terrestrial Revolution and KPg extinction on mammal diversification. Science 334:521–524.

Michaux, J. R., and G. I. Shenbrot. 2017. Family Dipodidae (jerboas). Pp. 62–100 in D. E. Wilson, T. E. Lacher, Jr., and R. A. Mittermeier (eds.), Handbook of the Mammals of the World, Vol. 7, Rodents II. Lynx Edicions, Barcelona, 1008 pp.

Miller, G. S. 1907. The families and genera of bats. Bulletin of the United States National Museum 57:1–283.

Miller-Butterworth, C. M., W. J. Murphy, S. J. O'Brien, D. S. Jacobs, M. S. Springer, and E. C. Teeling. 2007. A family matter: conclusive resolution of the taxonomic position of the long-fingered bats, *Miniopterus*. Molecular Biology and Evolution 24:1553–1561.

Mittermeier, R. A., A. B. Rylands, and D. E. Wilson. 2013. Handbook of the Mammals of the World. Vol. 3. Primates. Lynx Edicions, Barcelona, 952 pp.

Moojen, J. 1948. Speciation in the Brazilian spiny rats (genus *Proechimys*, family Echimyidae). University of Kansas Publications, Museum of Natural History 1:301–406.

Musser, G. M. 2014. A systematic review of Sulawesi *Bunomys* (Muridae, Murinae) with the description of two new species. Bulletin of the American Museum of Natural History 332:1–313.

Musser, G. M., and M. D. Carleton. 2005. Superfamily Muroidea. Pp. 894–1531 in D. E. Wilson and D. M. Reeder (eds.), Mammal Species of the World: A Taxonomic and Geographic Reference, 3rd ed. Johns Hopkins University Press, Baltimore, 2:xx + 745–2,142.

Nater, A., M. P. Mattle-Greminger, A. Nurcahyo, M. G. Nowak, M. de Manuel, T. Desai, C. Groves et al. 2017. Morphometric, behavioral, and genomic evidence for a new orangutan species. Current Biology 27:3487–3498.

Nyakatura, K., and O. R. P. Bininda-Emonds. 2012. Updating the evolutionary history of Carnivora (Mammalia): a new species-level supertree complete with divergence time estimates. BMC Biology 10:12.

O'Leary, M. A., J. I. Block, J. J. Flynn, T. J. Gaudin, A. Giallombardo, N. P. Giannini, S. L. Goldberg et al. 2013. The placental mammal ancestor and the post-K-Pg radiation of placentals. Science 339:662–667.

Patterson, B. D., and R. W. Norris. 2016. Towards a unified nomenclature for ground squirrels: the status of the Holarctic chipmunks. Mammalia 80:241–251.

Patterson, B. D., and N. S. Upham. 2014. A newly recognized family from the Horn of Africa, the Heterocephalidae (Rodentia: Ctenohystrica). Zoological Journal of the Linnean Society 172:942–963.

Patton, J. L., M. N. F. da Silva, and J. R. Malcolm. 2000. Mammals of the Rio Juruá and the evolutionary and ecological diversification of Amazonia. Bulletin of the American Museum of Natural History 244:3–306.

Patton, J. L., U. F. J. Pardiñas, and G. D'Elía (eds.) 2015. South American Mammals. Vol. 2. Rodents. University of Chicago Press, Chicago, IL, xxvi + 1336 pp.

Pauli, J. N., J. E. Mendoza, S. A. Steffan, C. C. Carey, P. J. Weimer, and M. Z. Peery. 2014. A syndrome of mutualism reinforces the lifestyle of a sloth. Proceedings of the Royal Society of London B: Biological Sciences 281:20133006.

Perelman, P., W. E. Johnson, C. Roos, H. N. Seuánez, J. E. Horvath, M. A. M. Moreira, B. Kessing et al. 2011. A molecular phylogeny of living primates. PLoS Genetics 7(3):e1001342.

Pierson, E. D., V. M. Sarich, J. M. Lowenstein, M. J. Daniel, and W. E. Rainey. 1986. A molecular link between bats of New Zealand and South America. Nature 323:60–63.

Pisano, J., F. L. Condamine, V. Lebedev, A. Bannikova, J.-P. Quére, G. I. Shenbrot, M. Pagès, and J. R. Michaux. 2015. Out of Himalaya: the impact of past Asian environmental changes on the evolutionary and biogeographical history of Dipodoidea (Rodentia). Journal of Biogeography 42:856–870.

Pivorunas, A. 1979. The feeding mechanisms of baleen whales. American Scientist 67:432–440.

Presslee, S., G. J. Slater, F. Pujos, A. M. Forasiepi, R. Fischer, K. Molloy, M. Mackie, J. V. Olsen, A. Kramarz, M. Taglioretti, F. Scaglia, M. Lezcano, J. L. Lanata, J. Southon, R. Feranec, J. Bloch, A. Hajduk, F. M. Martin, R. Salas Gismondi, M. Reguero, C. de Muizon, A. Greenwood, B. T. Chait, K. Penkman, M. Collins, and R. D. E. MacPhee. 2019. Palaeoproteomics resolves sloth relationships. Nature Ecology & Evolution 3:1121–1130.

Radke, W. J., and R. B. Chaisson. 1998. Laboratory Anatomy of the Mink. WCB/McGraw Hill, Boston, MA.

Reilly, S. M., and T. D. White. 2003. Hypaxial motor patterns and the function of epipubic bones in primitive mammals. Science 299:400–402.

Rice, D. W. 1998. Marine mammals of the world: systematics and distribution. Society for Marine Mammalogy Special Publications 4:1–231.

Roberts, T. E., H. C. Lanier, E. J. Sargis, and L. E. Olson. 2011. Molecular phylogeny of treeshrews (Mammalia: Scandentia) and the timescale of diversification in Southeast Asia. Molecular Phylogenetics and Evolution 60:358–372.

Roehrs, Z. P., J. B. Lack, and R. Van de Bussche. 2010. Tribal phylogenetic relationships within Vespertilioninae

(Chiroptera: Vespertilionidae) based on mitochondrial and nuclear sequence data. Journal of Mammalogy 91: 1073–1092.

Romer, A. S. 1966. Vertebrate Paleontology. 3rd ed. University of Chicago Press, Chicago, IL, viii + 468 pp.

Rommel, S. A., D. A. Pabst, and W. A. McLellan. 2008. Skull anatomy. Pp. 1033–1046 in W. F. Perrin, B. Würsig, and J. G. M. Thewissen (eds.), Encyclopedia of Marine Mammals, 2nd ed. Academic Press, New York, 1352 pp.

Rose, K. D. 2006. The Beginning of the Age of Mammals. Johns Hopkins University Press, Baltimore, xiv + 428.

Rossi, R. V., R. S. Voss, and D. P. Lunde. 2010. A revision of the didelphid marsupial genus *Marmosa*. Part 1. The species in Tate's "*mexicana*" and "*mitis*" sections and other closely related forms. Bulletin of the American Museum of Natural History 334:1–83.

Rowe, T. 1988. Definition, diagnosis, and origin of Mammalia. Journal of Vertebrate Paleontology 8:241–264.

Rowe, T. 1996. Coevolution of the mammalian middle ear and neocortex. Science 273:651–654.

Sato, J. J., M. Wolsan, F. J. Prevosti, G. D'Elía, C. Begg, K. Begg, T. Hosoda, K. L. Campbell, and H. Suzuki. 2012. Evolutionary and biogeographic history of weasel-like carnivorans (Musteloidea). Molecular Phylogenetics and Evolution 63:745–757.

Schenk, J. J., K. C. Rowe, and S. J. Steppan. 2013. Ecological opportunity and incumbency in the diversification of repeated continental colonizations by muroid rodents. Systematic Biology 62:837–864.

Schultz, N. G., M. Lough-Stevens, E. Abreu, T. Orr, and M. D. Dean. 2016. The baculum was gained and lost multiple times during mammalian evolution. Integrative and Comparative Biology 56:644–656. doi:10.1093/icb/icw034.

Simmons, N. B. 1998. A reappraisal of interfamilial relationships of bats. Pp. 1–26 in T. H. Kunz and P. A. Racey (eds.), Bat Biology and Conservation. Smithsonian Institution Press, Washington, DC, 365 pp.

Simmons, N. B. 2005. Order Chiroptera. Pp. 312–529 in D. E. Wilson and D. M. Reeder (eds.), Mammal Species of the World: A Taxonomic and Geographic Reference, 3rd ed. Johns Hopkins University Press, Baltimore, 1:xxxv + 1–743.

Simmons, N. B., and J. H. Geisler. 1998. Phylogenetic relationships of *Icaronycteris*, *Archeonycteris*, *Hassianycteris*, and *Palaeochiropteryx* to extant bat lineages, with comments on the evolution of echolocation and foraging strategies in microchiroptera. Bulletin of the American Museum of Natural History 235:1–182.

Simpson, G. G. 1945. The principles of classification and a classification of mammals. Bulletin of the American Museum of Natural History 85:i-xvi, 1–350.

Solari, S., R. A. Medellín, B. Rodríguez-Herrera, E. R. Dumont, and S. F. Burneo. 2019. Family Phyllostomidae (New World leaf-nosed bats). Pp. 444–583 in D. E. Wilson and R. A. Mittermeier (eds.), Handbook of Mammals of the World, Vol. 9, Bats. Lynx Edicions, Barcelona, 1009 pp.

Stains, H. J. 1967. Carnivores and pinnipeds. Pp. 325–354 in S. Anderson and J. K. Jones, Jr. (eds.), Recent Mammals of the World: A Synopsis of Families. Ronald Press, New York, vii + 453 pp.

Stains, H. J. 1984. Carnivores. Pp. 491–521 in S. Anderson and J. K. Jones, Jr. (eds.), Orders and Families of Recent Mammals of the World. John Wiley & Sons, New York. Xii + 686 pp.

Stanhope, M. J., W. G. Waddell, O. Madsen, W. de Jong, S. B. Hedges, G. C. Cleven, D. Kao, and M. S. Springer. 1998. Molecular evidence for multiple origins of Insectivora and for a new order of endemic African insectivore mammals. Proceedings of the National Academy of Sciences USA 95:9967–9972.

Steppan, S. J., and J. J. Schenk. 2017. Muroid rodent phylogenetics: 900-species tree reveals increasing diversification rates. PLoS ONE 12(8):e0183070.

Sunquist, M. E., and F. C. Sunquist. 2009. Family Felidae. Pp. 54–169 in D. E. Wilson and R. A. Mittermeier (eds.), Handbook of the Mammals of the World, Vol. 1, Carnivores. Lynx Edicions, Barcelona, 727 pp.

Teeling, E. C., S. Dool, and M. S. Springer. 2012. Phylogenies, fossils and functional genes: the evolution of echolocation in bats. Pp. 1–22 in G. F. Gunnell and N. B. Simmons (eds.), Evolutionary History of Bats: Fossils, Molecules and Morphology. Cambridge University Press, Cambridge, UK, xii + 560 pp.

Tejedor, A. 2011. Systematics of funnel-eared bats (Chiroptera: Natalidae). Bulletin of the American Museum of Natural History 353:1–140.

Thorington, R. W., Jr., J. L. Koprowski, M. A. Steele, and J. F. Whatton. 2012. Squirrels of the World. Johns Hopkins University Press, Baltimore, 459 pp.

Uhen, M. D. 2010. The origin(s) of whales. Annual Review of Earth and Planetary Science 38:189–219.

Ungar, P. S. 2010. Mammal Teeth: Origin, Evolution, and Diversity. Johns Hopkins University Press, Baltimore, xii + 304 pp.

Upham, N. S., and B. D. Patterson. 2015. Evolution of caviomorph rodents: a complete phylogeny and timescale for living genera. Pp. 63–120 in A. I. Vassallo and D. Antenucci (eds.), Biology of Caviomorph Rodents: Diversity and Evolution. SAREM Series A, Investigaciones Mastozoológicas, Buenos Aires, Argentina, ix + 329 pp.

Vaughan, T. A., J. M. Ryan, and N. J. Czaplewski. 2015. Mammalogy. 6th ed. Jones and Bartlett Learning, Burlington, MA, xii + 755 pp.

Veron, G., C. Bonillo, A. Hassanin, and A. P. Jennings. 2017. Molecular systematics and biogeography of the Hemigalinae civets (Mammalia, Carnivora). European Journal of Taxonomy 285:1–20.

von Koenigswald, W. 2011. Diversity of hypsodont teeth in mammalian dentitions—construction and classification. Palaeontographica Abteilung A. Palaeozoologie-Stratigraphie 294:63–94.

Voss, R. S., and S. A. Jansa. 2009. Phylogenetic relationships and classification of didelphid marsupials, an extant radiation of New World metatherian mammals. Bulletin of the American Museum of Natural History 322:1–177.

Voss, R. S., E. E. Gutiérrez, S. Solari, R. V. Rossi, and S. A. Jansa. 2014. Phylogenetic relationships of mouse opossums (Didelphidae, *Marmosa*) with a revised subgeneric classification and notes on sympatric diversity. American Museum Novitates 3817:1–27.

Waddell, P. J., N. Okada, and M. Hasegawa. 1999. Towards resolving the interordinal relationships of placental mammals. Systematic Biology 48:1–5.

Weil, A. 2003. Teeth as tools. Nature 422:128.

Weimann, B., M. A. Edwards, and C. N. Jass. 2014. Identification of the baculum in American pika (*Ochotona princeps*: Lagomorpha) from southwestern Alberta, Canada. Journal of Mammalogy 95:284–289.

Weksler, M. 2006. Phylogenetic relationships of oryzomyine rodents (Muroidea: Sigmodontinae): separate and combined analyses of morphological and molecular data. Bulletin of the American Museum of Natural History 296:1–149.

Wheeler, J. C. 1995. Evolution and present situation of the South American Camelidae. Biological Journal of the Linnean Society 54:271–295.

Wheeler, J. C. 2012. South American camelids—past, present and future. Journal of Camelid Science 5:1–24.

Whitaker, J. O., Jr. 2017. Family Zapodidae (jumping mice). Pp. 50–61 in D. E. Wilson, T. E. Lacher, Jr., and R. A. Mittermeier (eds.), Handbook of the Mammals of the World, Vol. 7, Rodents II. Lynx Edicions, Barcelona, 1008 pp.

Wildman, D. E., N. M. Jameson, J. C. Opazo, and S. V. Yi. 2009. A fully resolved genus level phylogeny of Neotropical primates (Platyrrhini). Molecular Phylogenetics and Evolution 53:694–702.

Williamson, E. A., F. G. Maisels, and C. P. Groves. 2013. Family Hominidae (Great Apes). Pp. 792–854 in R. A. Mittermeier, A. B. Rylands, and D. E. Wilson (eds.), Handbook of the Mammals of the World, Vol. 3, Primates. Barcelona: Lynx Edicions, Barcelona, 951 pp.

Williamson, T. E., S. L. Brusatte, and G. P. Wilson. 2014. The origin and early evolution of metatherian mammals: the Cretaceous record. ZooKeys 465:1–76.

Wilson, D. E., and R. A. Mittermeier (eds.). 2009. Handbook of the Mammals of the World. Vol. 1. Carnivores. Lynx Edicions, Barcelona, 727 pp.

Wilson, D. E., and R. A. Mittermeier (eds.). 2011. Handbook of the Mammals of the World. Vol. 2. Hoofed Mammals. Lynx Edicions, Barcelona, 885 pp.

Wilson, D. E., and R. A. Mittermeier (eds.). 2015. Handbook of the Mammals of the World. Vol. 5. Monotremes and Marsupials. Lynx Edicions, Barcelona, 799 pp.

Wilson, D. E., and D. M. Reeder. 2005. Mammal Species of the World: A Taxonomic and Geographic Reference. 3rd ed. Johns Hopkins University Press, Baltimore, 1:xxxv + 1–743, 2:xx + 745–2142.

Wilson, D. E., T. E. Lacher, Jr., and R. A. Mittermeier. 2017. Handbook of Mammals of the World. Vol. 7. Rodents II. Lynx Edicions, Barcelona, 1008 pp.

Wozencraft, C. W. 2005. Order Carnivora. Pp. 532–628 in D. E. Wilson and D. M. Reeder (eds.), Mammal Species of the World: A Taxonomic and Geographic Reference, 3rd ed. Johns Hopkins University Press, Baltimore, 1:xxxv + 1–743.

Index to Taxonomic Names above the Genus Level

The following index includes all taxonomic entities addressed in this Manual above the level of genera; that is, Class through Tribe. Pages followed by "t" or "f" refer to tables or figures, respectively, while bold font highlights major discussions or treatments for selected taxa.